W9-BRS-013

EARTH SCIENCE

The Physical Setting

Second Edition

Thomas McGuire
Earth Science Educator

AMSCO

Amsco School Publications, Inc.
315 Hudson Street, New York, N.Y. 10013

In addition to this book, Amsco School Publications, Inc. publishes the following books written by Thomas McGuire: *Reviewing Earth Science: The Physical Setting, Third Edition; Laboratory and Skills Manual—Earth Science: The Physical Setting; Earth Science: Reviewing the Essentials.*

This book is dedicated to the thousands of students I have known in 32 years of teaching. I hope they have learned nearly as much from me as I have from them.

The publisher would like to thank the following educators who reviewed the manuscript.

Diana K. Harding
Retired
New York State
Education Department

Lee Kowalsky
Earth Science Teacher
Proctor Senior High School
Utica, New York

Janet Iadanza
Earth Science Teacher
Rockwood Park School
Howard Beach, New York

Daniel B. Kujawinski, Ph.D.
Educational Consultant
Brant, New York

Cover Design: Meghan J. Shupe
Text Design: AGT/Howard Petlack
Composition: Northeastern Graphic, Inc.
Art: Hadel Studio and Northeastern Graphic, Inc.
Cover Photo: Eruption of the Eyjafjallajokull volcano in Iceland;
© Associated Press/Arnar Thorisson/Helicopter.is
Back Cover Photos: Left to right: Portage Glacier, Anchorage, Alaska; M 81 galaxy in Ursa Major, courtesy of NASA; Raplee Anticline, Mexican Hat, Utah; North Clear Creek Falls in the San Juan Range of the Rocky Mountains, Colorado (Portage Glacier, Raplee Anticline, North Clear Creek Falls, courtesy of Thomas McGuire.)

Please visit our Web site at: **www.amscopub.com**

When ordering this book, please specify
R 797 H *or* EARTH SCIENCE: THE PHYSICAL SETTING, SECOND EDITION, HARDBOUND
ISBN 978-1-56765-946-7
New York City Number 56765-946-6

Copyright © 2011 by Amsco School Publications, Inc.
No part of this book may be reproduced in any form without written permission from the publisher.

Printed in the United States of America

1 2 3 4 5 6 7 8 9 10   10 11 12 13 14

To the Student

Earth Science: The Physical Setting, Second Edition, which follows the New York State Core Curriculum, which is based on National Standards, is an introduction to the study of Earth Science. The specific standards covered in each chapter are listed in the table of contents and next to the text to which they apply and next to each Student Activity. With this book, you can gain a firm understanding of the fundamental concepts of Earth Science—a base from which you may confidently proceed to further studies in science and enjoy a deeper appreciation of the world around you. You also will need to become familiar with the 2010 *Earth Science Reference Tables,* a document prepared by the New York State Education Department. You will find the individual tables within the appropriate chapters of this text. You can obtain a copy of the entire document from your teacher or it can be downloaded from the State Education Web site: *http://www.emsc.nysed.gov/osa/reftable/earth science-rt/esrt2010-engw.pdf*

This book is designed to make learning easier for you. Many special features that stimulate interest, enrich understanding, encourage you to evaluate your progress, and enable you to review the concepts are provided. These features include:

1. **Carefully selected, logically organized content.** This book offers an introductory Earth Science course stripped of unnecessary details that lead to confusion. It covers the New York State Core Curriculum for the Physical Setting—Earth Science.

2. **Clear understandable presentation.** Although you will meet many new scientific terms in this book, you will find that the language is generally clear and easy to read. Each new term is carefully defined and will soon become part of your Earth Science vocabulary. The illustrations and photographs also aid in your understanding, since they, like the rest of the content, have been carefully designed to clarify concepts. Words in **boldface** are defined in place and in the Glossary. Words in *italics* are important science words you already should know.

3. **Introduction.** An introductory section at the beginning of each chapter sets the stage for the rest of the chapter. Here you will find a list of Words to Know and the learning objectives for the chapter.

4. **Step-by-step solutions to problems followed by practice.** Problem solving is presented logically, one step at a time. Sample solutions to all types of Earth Science problems are provided. These sample problems will help you approach mathematical problems logically. To enhance your newly acquired skill, you will find practice problems following most sample problems.

5. **Internet sites.** Within the chapters are the URLs, or Web addresses, of various internet sites that provide additional information or activities.

6. **End-of-chapter review questions.** The Regents-style, Part A, multiple-choice questions at the end of each chapter help you to review and assess your grasp of the content. The open-ended questions provide practice in answering questions found in Part B and Part C of the Regents exam. To answer some of these questions you may need to refer to the *Earth Science Reference Tables* or the tables found in the chapters.

7. **Appendices.** Appendix A introduces you to laboratory safety. In Appendix B, you will be presented with a format to follow when preparing laboratory reports. Appendix C reviews the International System of Units. Appendix D lists the physical constants important to Earth Science. Appendix E explores the use of graphs in science.

8. **Glossary.** This section contains all the **boldfaced** words found in the text along with their definitions.

The study of Earth Science can be both stimulating and challenging. The author sincerely hopes that this book will increase your enjoyment of this science.

Contents

Unit 1 Earth Measures and Models   **1**

1 • SCIENCE AND PLANET EARTH   **2**

1: MA 1, 2, 3; 1: SI 1, 2, 3; 1: ED 1; 2: IS 1, 2, 3; 6: ST 1; 6: M 2;
6: M & S 3; 6: P OF C 5; 6: O 6; 7: C 1; 7: S 2

What Is Science? / What Is Earth Science? / How Is Earth Science
Related to Other Sciences? / Why Study Earth Science? / How Do
Scientists Gather and Analyze Information? / How Is Density
Determined? / How Do Scientists Make and Use Graphs? / How Is
Technology Changing the Way Scientists Work?

Student Activities: Good Science and Bad Science; Exponential
Notation in the Real World; Making Estimations; Density of Solids;
The Thickness of Aluminum Foil; Making a Graph of the Revolution
of the Planets; An Internet Scavenger Hunt

2 • EARTH'S DIMENSIONS AND NAVIGATION   **31**

1: MA 1, 2, 3; 1: SI 3; 2: IS:1; 4: 1.1C, D, F, I; 4: 1.2C, 4: 2.1J;
6: M 2; 6: M & S 3;

What Is Earth's Shape? / What Are Earth's Parts? / How Is Location
Determined?

Student Activities: How Round Is Earth? Pie Graphs of Earth's
Spheres; Interpreting Reference Tables; Determining Your Latitude;
Finding Solar Noon; Determining Your Longitude; Reading Latitude
and Longitude on Maps

3 • MODELS AND MAPS   **56**

1: MA 1, 3; 1: SI 1; 1: ED 1; 2: IS 1 4: 2.1Q; 6: ST 1; 6: M 2;
6: M & S 3; 7: S 2

What Is a Model? / What Are Fields? / What Is a Topographic Map?

Student Activities: Models in Daily Life; A Map to Your Home;
Making a Water Compass; Characteristics of Isolines; A Temperature
Field; Making a Topographic Model; Reading Your Local Topographic
Map; A Profile on a Local Topographic Map; Interpreting Isoline Maps;
Rescue and Evacuation Planning

Unit 2 Minerals, Rocks, and Resources   **84**

4 • INVESTIGATING MINERALS   **85**

1: MA 1; 1: SI 1, 3; 4: 3.1A, B; 6: M 2

What Are Minerals? / How Can We Identify Minerals by Their Observable Properties? / What Are the Most Common Minerals?

Student Activities: Solids, Liquids, and Gases; Luster of Common Objects; Cleavage and Fracture of Household Substances; Mineral Identification

5 ● THE FORMATION OF ROCKS   114
1: SI 1, 2, 3; 4: 2.1M, W; 4: 3.1B, C; 6: ST 1, 6: M 2;
6: P OF C 5

What Are Rocks? / How Are Igneous Rocks Classified? / How do Sedimentary Rocks Form? / How Do Metamorphic Rocks Form? / What Is the Rock Cycle?

Student Activities: Making a Rock Collection; Classification; Identification of Igneous Rocks; Identification of Sedimentary Rocks; Identification of Metamorphic Rocks

6 ● MANAGING NATURAL RESOURCES   144
2: IS 1, 2, 3; 4: 3.1C; 6: ST 1; 6: E & S 4; 6: O 6; 7: C 1

How Does the Global Marketplace Work? / What Are Nonrenewable Resources? / What Are Our Most Important Renewable Resources? / How Can We Conserve Resources? / What Are the Effects of Environmental Pollution?

Student Activities: Adopt a Resource; Water Use in the Home

Unit 3 Weathering and Erosion   164
7 ● WEATHERING AND SOILS   165
1: MA 1, 2, 3; 1: SI 1, 2, 3; 1: ED 1; 4: 2.1S; 6: P OF C 5
Why Does Weathering Occur? / How Does Soil Form?

Student Activities: Rock Abrasion; Calculating Surface Area; Chemical Weathering and Temperature

8 ● EROSION AND DEPOSITION   184
1: SI 1, 2, 3; 4: 2.1T, U, V; 6: ST 1; 6: E & S 4; 6: P OF C 5
What Is Erosion? / What Is Deposition? / When Is There Equilibrium Between Erosion and Deposition?

Student Activities: What Is Graded Bedding? What's in Sediment?

Unit 4 Water Shapes Earth's Surface   204
9 ● STREAM DYNAMICS   205
1: MA 1, 2, 3; 1: SI 1, 3; 1: ED 1; 4: 1.2G; 4: 2.1P, T, U, V;
6: ST 1; 6: M 2; 6: M & S 3; 7: C 1

What Does the Hudson River Mean to New York State? / What Is a River System? / How Do We Measure Streams? / What Is a Drainage Pattern?

Student Activities: Drainage of the School Grounds; Modeling a Stream System; Water Velocity; Measuring Stream Discharge

10 • GROUNDWATER 227

1: MA 1, 2, 3; 1: SI 1, 2, 3; 1: ED 1 4: 1.2G; 6: M 2; 7: C 1; 7: S 2

Where Does the Water in Streams Come From? / Where Is Earth's Water? / Does Groundwater Occur in Specific Zones? / How Does Groundwater Move? / Where Is Groundwater Available? / What Are Some Groundwater Problems?

Student Activities: Observing Condensation; Groundwater Model; Comparing the Porosity of Different Materials; Capillarity of Sediments

11 • OCEANS AND COASTAL PROCESSES 247

1: MA 1; 1: SI 1, 2, 3; 2: IS 1; 4: 1.1A, I; 4: 1.2 C, F, G; 4: 2.1A, B, L, O, U; 4: 2.2B, C, D; 6: ST 1; 6: P OF C 5; 6: E & S 6; 7: C 1; 7: S 2

Why Is Earth Called the Blue Planet? / What Makes Ocean Water Different? / How Can We Investigate the Oceans? / Why Does the Water in the Ocean Circulate? / What Causes the Tides? / How Do Coastlines Change? / How Should We Manage Active Shorelines?

Student Activities: The Density of Seawater; Observing Gyres; Extremes of Tidal Ranges; Graphing the Tides; Zoning for Coastal Preservation

12 • GLACIERS 270

4: 2.1T, U, V; 6: M 2; 6: P OF C 5; 7: S 2

How Did Scientists Discover That Glaciers Once Covered New York State? / What Is a Glacier? / How Do Glaciers Cause Erosion? / How Can We Recognize Deposition by Glaciers? / How Can We Recognize Deposition by Meltwater? / What Are Ice Ages?

Student Activities: Snow to Ice; A Model of a Glacier; Inventory of Glacial Features

13 • LANDSCAPES 289

2: IS 1; 4: 2.1R; 6: ST 1

What Are New York's Natural Wonders? / What Are Landscapes? / What Factors Influence Landscape Development? / What Are the Landscapes of New York State?

Student Activities: Investigating Landforms; Landscape Boundaries; Landforms of New York State

Unit 5 Earth's Internal Heat Engine — 305

14 • EARTHQUAKES AND EARTH'S INTERIOR — 306
1: MA 1; 4: 2.1A, B, J, K, L; 4: 2.2B; 6: ST 1; 6: M 2; 6: M & S 3

What Causes Earthquakes? / How Are Earthquakes Measured? / How Do Earthquakes Radiate Energy? / How Are Earthquakes Located? / What Is Inside Earth?

Student Activities: Adopt an Earthquake; Modeling Seismic Waves; Earth's Internal Structure to Scale

15 • PLATE TECTONICS — 334
1: SI 1, 2, 3; 2: IS 1; 4: 2.1J, K, L, M, N; 6: M 2; 6: P OF C 5

Do Continents Move? / What Is Earth's Internal Structure? / What Happens at Plate Boundaries? / Does Earth's Geography Change?

Student Activities: Matching Shorelines; Experimenting With Oobleck; Zones of Crustal Activity; Graphing Hawaiian Volcanoes

16 • GEOLOGIC HAZARDS — 361
1: ED 1; 2: IS 1; 4: 2.1L, T; 7: S 2

What Is A Geologic Hazard?

Student Activities: Make Your School Safer; Devising an Earthquake Preparedness Plan; Adopt a Volcano; Evaluating School Property

Unit 6 Revealing Earth's History — 380

17 • SEQUENCING GEOLOGIC EVENTS — 381
1: MA 1,2; 4: 1.2J; 4: 2.1M,P,W; 6: ST 1; 6: M 2; 6: M & S 3

How Can We Determine the Sequence of Events? / How Can We Interpret Geologic Profiles? / How Do Geologists Establish Absolute Time?

Student Activities: Relative and Absolute Time; Local Rock Features; Symbols and Rocks; A Classroom Model of Radioactive Decay

18 • FOSSILS AND GEOLOGIC TIME — 407
1: MA 1, 2; 2: IS 1; 4: 1.2H, I, J; 6: M 2; 6: M & S 3

What Are Fossils? / How Did Life Begin on Earth? / What Is Organic Evolution? / How Has Geologic Time Been Divided? / What Is the Geologic History of New York State? / How Do Geologists Correlate Rock Layers?

Student Activities: Nearby Fossil Beds; Interpreting Fossil Footprints; Variations Within a Species; An Extinct Species; Geologic Time Line

Unit 7 The Dynamic Atmosphere   437

19 • WEATHER VARIABLES AND HEATING OF THE ATMOSPHERE   438

1: MA 1, 2; 1: SI 1; 2 IS 1; 4: 2.1A, B, C, D, E; 4: 2.2A, B;
6: M 2; 6: M & S 3; 6: P OF C 5

What Are the Elements of Weather? / How Does the Sun Warm Earth? / How Does Solar Energy Circulate Over Earth?

Student Activities: Making a Thermometer; Making a Barometer; Making a Wind Gauge and Wind Vane; Recording Weather Variables; Observing Refraction

20 • HUMIDITY, CLOUDS, AND ATMOSPHERIC ENERGY   467

1: MA 1, 3; 1: SI 1, 2; 1: ED 1; 4: 2.1A, B, C, E, F;
4: 2.2A, B, C; 6: M 2; 6: O 6

Let It Snow / How Does the Atmosphere Store Energy? / How Does Water Vapor Enter the Atmosphere? / How Do We Measure Water in the Atmosphere? / How Do Clouds Form?

Student Activities: Rate of Evaporation; Removing Moisture From Air; A Stationary Hygrometer; Modeling Temperature Changes; Homemade Clouds

21 • AIR PRESSURE AND WINDS   494

1: MA 1, 2, 3; 1: SI 1, 2, 3; 4: 2.1B, C, D, E, F, G, H, I;
4: 2.2A, B, C; 6: M 2; 6: P OF C 5

Fast as the Wind / What Causes Winds? / Why Do Local Winds Occur? / What Causes Regional Winds?

Student Activities: The Force of Air Pressure; Pressure and Depth; Observing Convection; Movement of Pressure Systems; The Coriolis Effect

22 • WEATHER MAPS   521

2: IS 1; 4: 2.1C, E, F, G, H; 4: 2.2B, C; 6: M 2; 6: P OF C 5

Weather Forecasting / What Are Air Masses? / What Are Mid-Latitude Cyclones and Anticylones? / How Are Weather Fronts Identified? / How Do Mid-Latitude Cyclones Evolve? / How Are Weather Data Shown on Weather Maps? / How Are Weather Maps Drawn and Used?

Student Activities: Identifying Air Masses; Stages of Cyclonic Development; Current Station Models; Drawing Weather Fronts; Reliability of Weather Forecasts

23 • WEATHER HAZARDS AND THE CHANGING ATMOSPHERE 550

1: MA 1, 2, 3; 1: SI 1, 2, 3; 1: ED 1; 2: IS 1; 4: 2.1C, D, E, F, H; 4: 2.2D; 6: M 2; 6: O 6; 7: C 1; 7: S 2

The Cost of Natural Disasters / What Weather Events Pose Hazards? / How Can We Protect Ourselves From Weather Hazards? / How Is Earth's Atmosphere Changing?

Activities: Lightning Distance; Storm Survival; Hurricane Tracking; Comprehensive Emergency Planning; Community Planning Map; Carbon Capture

24 • PATTERNS OF CLIMATE 578

1: MA 1, 2, 3; 1: SI 1, 2, 3; 2: IS 1; 4: 2.1G, H, I; 4: 2.2C, D; 6: ST 1; 6: M 2; 6: M & S 3

Are Climates Changing? / What Is Climate? / How Does Latitude Affect Climate? / What Other Geographic Factors Affect Climate? / What Geographic Features of New York State Affect the Local Climate? / How Is Climate Shown on Graphs?

Student Activities: Effects of Rising Sea Levels; Graphing Average Monthly Temperature; Modeling Temperature Change; Climates and Ocean Currents

Unit 8 Earth and Beyond 604

25 • EARTH, SUN, AND SEASONS 605

1: SI 1, 2, 3; 1: ED 1; 2: IS 1, 2, 3; 4: 1.1A, C, D, E, F, G, H; 6: M 2; 6: P OF C 5

What Is Astronomy? / How Can We Describe the Position of Celestial Objects? / Our Internal Clock / What Is the Sun's Apparent Path Across the Sky? / How Does the Sun's Path Change With the Seasons? / Does the Sun's Path Depend on the Observer's Location? / What Is Really Moving, Sun or Earth? / How Do Earth's Motions Affect the Appearance of Other Celestial Objects? / How Do Earth's Motions Seem to Affect the Stars?

Student Activities: Astrology on Trial; The Length of a Shadow; Constructing a Sundial; Tracing the Daily Path of the Sun; Observing the Sun; Modeling Earth Motions; The Big Dipper and Polaris; Locating Major Constellations; Photographing Star Trails, Celestial Observations

26 • EARTH, ITS MOON, AND ORBITS 639

1: MA 1, 2, 3; 4: 1.1A, B, F; 6: M 2; 6: E & S 4; 6: P OF C 5; 7: C 1; 7: S 2

Apollo Program / What Is the History of Earth's Moon? / How Can We Describe Orbits? / What Determines a Satellite's Orbit? / Why Does the Moon Show Phases? / What Is an Eclipse?

Student Activities: Orbit of the Moon; The Next Lunar Eclipses

27 • THE SOLAR SYSTEM 661
1: ED 1; 2: IS 1; 4: 1.1A, B; 4: 1.2A, B, C, D; 6: M 2;
6: M & S 3; 6: O 6; 7: C 1; 7: S 2

Colonizing Space / What Is the Origin of the Solar System? / What Properties Do the Planets Share? / How Are the Planets Grouped? / What Other Objects Orbit the Sun?

Student Activities: Graphing Solar System Data; Design a Landing Module; The Solar System to Scale

28 • STARS AND THE UNIVERSE 681
1: SI 1, 2, 3; 1: ED 1; 4: 1.2A, B; 6: M 2; 6: P OF C 5;
7: C 1; 7: S 2

How Do Astronomers Investigate Stars? / The Search for Extraterrestrial Life / What Is a Star? / How Are Stars Classified? / How Do Stars Evolve? / What Is the Structure of the Universe? / What Is the History of the Universe? / What Is the Future of the Universe?

Student Activities: Making a Spectrum; Making a Telescope; Light Intensity and Distance; A Model of the Big Bang

APPENDICES 708
Appendix A: Laboratory Safety 708
Appendix B: A Format for Laboratory Reports 709
Appendix C: The International System of Units 710
Appendix D: Physical Constants 711
Appendix E: Graphs in Science 711
Appendix F: Earth Science Reference Tables 712

GLOSSARY 713

INDEX 727

PHOTO CREDITS 740

UNIT 1
Earth Measures and Models

The spherical shape of Earth was proposed by Egyptian and Greek mathematicians and philosophers, as well as by learned people from other cultures, more than 3000 years ago. However, before the Age of European Exploration (the fifteenth and sixteenth centuries), most people still thought that Earth was flat. Sailors stayed close to the coast so they would not "sail over the edge." For most people living at this time, their families and the things they needed to live were close to home. Long-distance travel involved hardships and dangers, so few people traveled far from home.

The Italian explorer and navigator Christopher Columbus understood that Earth is round. However, Columbus also believed that Earth was much smaller than it really is. This led him to guess that he could reach Asia by sailing a relatively short distance west from Europe. Fortunately for him, the Americas are located where Columbus expected to find Asia. After four very hazardous voyages, Columbus thought he had found a new route to Asia. He did not realize that he had sailed only part of the way to Asia, and was very lucky to have survived his voyages. He had unknowingly reached a "new world" on his way to Asia, but died unaware of his actual discovery.

1

Science and Planet Earth

WORDS TO KNOW

astronomy	geology	oceanography
coordinate system	inference	percent deviation
density	meteorology	science
Earth science	observation	scientific (exponential) notation
ecology		

This chapter will help you answer the following questions:

❶ What do we mean by Earth science?

❷ Why is Earth science important?

❸ How do scientists make, use, analyze, and communicate their observations and measurements?

WHAT IS SCIENCE?

1: SCIENTIFIC INQUIRY 1, 2, 3

Science has played a central role in the advancement of civilization. The Latin origin of the word *science* (*scire*) can be translated as "to know." While some people might think of scientific conclusions as unchanging facts, our understanding of nature is never complete. As the understanding of nature grows, old ideas that no longer fit our observations are discarded. The so-called facts of science are often temporary, whereas the methods of science (observation and

FIGURE 1-1. Students and professional geoscientists investigate Earth through field work and laboratory analysis. These students are learning about rock that millions of years ago intruded into older rock. Field observations reveal that intruding material was a molten liquid that has since cooled and solidified.

analysis) are permanent. Therefore, science is a way of making and using observations. (See Figure 1-1.)

Science often attempts to answer such questions as: Why do we see the moon on some nights, but not on others? What causes clouds to form? Why are there violent storms, earthquakes, and volcanoes? How can people wisely use Earth's resources and still preserve the most valued qualities of our natural environment? What is the history of Earth and the universe? If we could see our planet from far away in space, we might be more aware of what a tiny part of the universe we occupy.

Rational thought and clear logic support the best scientific ideas. Scientists often use numbers and mathematics (data) to present their results because mathematics is considered to be straightforward, logical, and consistent. These qualities are valued in scientific work. Visit the following Web site to see "From the Sky" by Yann Arthus-Bertrand: *http://www.scribd.com/doc/3914640/Yann-ArthusBertrand-From-The-Sky*

Scientific discoveries need to be *verifiable*. This means that other scientists who investigate the same issues and make their own observations should arrive at similar conclusions. For example, when a climate prediction is supported by the work of many scientists and/or by computer models, the prediction is considered to be more reliable.

Science at Work

Alfred Wegener proposed his theory of continental drift in the early 1900s; it was based on indirect evidence. During his lifetime, he could not find enough evidence to convince most other Earth scientists that continents move over Earth's surface. However, new evidence gathered by other scientists working 50 years later gave renewed support to his ideas. Today, plate tectonics, as the theory is now known, is supported by precise measurements. This is a good example of how the efforts of many scientists and new observations resulted in a new way of thinking about how our planet works.

Science can therefore be defined as a universal and ongoing method of gathering, organizing, analyzing, testing, and using information about our world. Science provides a structure to investigate questions and to discover relationships and patterns and form conclusions. The reasoning behind the conclusions should be clear. Scientists will continue to evaluate and modify conclusions. The body of knowledge of science, even as presented in this book, is simply the best current understanding of how the world works.

STUDENT ACTIVITY 1-1 —GOOD SCIENCE AND BAD SCIENCE

2: INFORMATION
SYSTEMS 1, 2, 3

Sometimes it is easier to understand science if you look for what is *not* science. Tabloids are newspapers that emphasize entertainment. They publish questionable stories that other media do not report. Bring your teacher a science article from a questionable news source. Your teacher will display the stories for the class to discuss. What are the qualities of these stories that make them a poor source of scientific information?

WHAT IS EARTH SCIENCE?

The science you study in school is often divided into four branches: life science (biology), physics, chemistry, and Earth science. (See Figure 1-2.) **Earth science** applies the tools of the other sciences to study our planet, including its rocky portion, its oceans, its atmosphere, its interior, and its surroundings in space.

Earth science can be divided into several branches. **Geology** is the study of the rocky portion of Earth, its interior and surface processes. Geologists investigate the processes that shape the land, and they study Earth materials, such as minerals and rocks. (See Figure 1-3 on page 6.) They also actively search for natural resources, including valuable minerals, ores, and fossil fuels.

Meteorology is the study of the atmosphere and how it changes. Meteorologists predict weather and help us to deal with weather-related disasters and other atmospheric events that affect our lives. They also investigate climatic (long-term weather) changes.

Oceanography is the study of the oceans that cover most of Earth's surface. Oceanographers investigate ocean currents, how the oceans affect weather and coastlines, the composition and properties of seawater, and the best ways to manage marine resources.

Astronomy is the study of Earth's motions and motions of objects beyond Earth, such as planets and stars. Astronomers consider such questions as: Is Earth the only planet of its kind? How

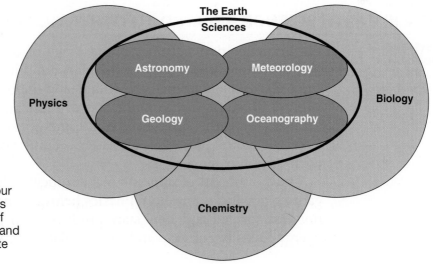

FIGURE 1-2. The four major Earth sciences draw on the fields of physics, chemistry, and biology to investigate Earth systems.

FIGURE 1-3.
Dr. Frederick Vollmer, a geologist from New York, educates students onsite about the Sierra Nevada mountains.

big is the universe? When did the universe begin, and how will it end?

Many Earth scientists are involved in **ecology**, or environmental science, which seeks to understand how living things interact with their natural setting. They observe how the natural environment changes, how those changes are likely to affect living things, and how people can preserve the most important characteristics of the natural environment.

HOW IS EARTH SCIENCE RELATED TO OTHER SCIENCES?

6: SYSTEMS
THINKING 1
7: CONNECTIONS 1
7: STRATEGIES 2

Earth science draws from a broad range of other sciences. This helps present a broad view of the planet and its place in the universe. Earth scientists need to understand the principles of chemistry to investigate the composition of rocks and how they form. Changes in weather are caused by the energy exchanges. By knowing the chemical properties of matter, scientists can investigate the composition of stars.

The movements of stars and planets obey the laws of physics regarding gravity and motion. Physics helps us understand how the universe formed and how stars produce such great amounts of energy. Density currents and the circulation of fluids control the atmosphere, the oceans, and even changes deep within our planet. Nuclear physics has allowed scientists to measure the age of Earth with remarkable accuracy.

The Earth sciences also make use of the principles of biology and, in turn, support the life sciences. Organic evolution helps us understand the history of Earth. At the same time, fossils found in rock are the primary evidence for evolutionary biology. The relationships between the physical (nonliving) planet and life-forms are the basis for environmental biology. Only recently, have some people understood how changes in Earth and changes in life-forms have influenced each other throughout long periods of geologic time.

WHY STUDY EARTH SCIENCE?

1: SCIENTIFIC INQUIRY 1, 2, 3
6: SYSTEMS THINKING 1
6: OPTIMIZATION 6
7: STRATEGIES 2

Although some students may become professional Earth scientists, it is more likely that you will find work in another area. No matter what career you choose, Earth science will affect your life. Everyone needs to know how to prepare for changes in weather, climate, seasons, and earth movements. We also need to become educated citizens who make important decisions in a democratic world.

Natural disasters are rare events, but when they occur they can cause great loss of life and property. To limit loss, people can prepare for hurricanes, tornadoes, floods, volcanic eruptions, earthquakes, and climate shifts. Humans can survive the effects of cold and drought if they plan ahead. However, people need to know how likely these events are and how best to avoid their devastating consequences. We need to know the answers to many questions, such as: How will humans be affected by general changes in climate? Can these changes be prevented? Will a large asteroid or comet strike Earth, and how will it affect Earth's inhabitants?

Our civilization depends on the wise use of natural resources. Freshwater, metals, and fossil fuels are among the great variety of materials that have supported a growing world economy. These resources have brought us wealth and comfort. We need to know: How much of these materials are available for use? What will happen if these materials run out? What is the environmental effect of extracting, refining, and using these resources?

These issues affect all of us regardless of our profession. As citizens and consumers, we make decisions, and as citizens, we elect governments that need to consider these issues.

Environmental Awareness

As you will read in future chapters, the natural environments of this planet have changed dramatically over the past 4.6 billion years. They continue to change, and we must expect change in the future. Humans do not control the environment; we are a part of environmental change. With or without human influences, change will occur. However, we can certainly affect how those changes occur.

For example, there has been a campaign to use paper cups rather than polystyrene foam cups. Most people think of paper as a less harmful material. However, some studies have found that the environmental effect of paper cups can be greater than that of polystyrene foam cups. Sometimes the full story is not as simple as we are led to believe. For more information visit the Web site: *http://www.newton.dep.anl.gov/askasci/gen99/gen99498.htm*

On the other hand, it is clear that human-caused emissions of carbon dioxide are having a global effect on the environment. But, unlike the choice between paper or foam cups, the solution here is very difficult. People enjoy the convenience of using cars and the comfort of indoor heating and cooling. Alternative energy choices such as nuclear and renewable energy can be unpopular or too expensive. What can we do when people prefer to use cheaper or less controversial alternatives now rather than following long-term economic and scientific choices? Complex or potentially costly issues such as these are our most challenging problems.

There is an environmental saying, "Think globally and act locally." This is good idea. It stresses personal responsibility for environmental consequences. We currently see that rapidly developing economies throughout the world are placing new demands on Earth's limited resources and creating more waste. More than ever, nations must learn to work cooperatively to solve global environmental problems.

Working with Science

FIGURE 1-4.
Cynthia Chandley

CYNTHIA CHANDLEY: Water Rights Lawyer

Cynthia Chandley is not an Earth scientist, but she knows how important it is to understand our Earth. (See Figure 1-4.) She earned a degree in geology. After several years of working in the mining industry, she attended law school and became an environmental lawyer. Ms. Chandley now works as a water rights lawyer. She says, "I constantly use my geoscience background to influence the use and preservation of an essential resource. However, these issues go well beyond my profession. Everyone needs to understand our planet's interconnected systems to determine how our resources can be most effectively managed for ourselves and for future generations."

HOW DO SCIENTISTS GATHER AND ANALYZE INFORMATION?

1: MATHEMATICAL
ANALYSIS 1
1: SCIENTIFIC
INQUIRY 2, 3
2: INFORMATION
SYSTEMS 1, 2
6: MAGNITUDE AND
SCALE 3

You receive information about your surroundings through your five senses: sight, touch, smell, taste, and hearing. To do their work, scientists must make use of information gathered using their senses. These bits of information are **observations**. Some observations are qualitative. They involve relative terms, such as long or short, bright or dim, hot or cold, loud or soft, red or blue, and they compare observations without using numbers or measurements. Other observations are quantitative. When you say that the time is 26 seconds past 10 o'clock in the morning you are being very specific. Quantitative comes from the word *quantity,* meaning "how many." Therefore, quantitative observations include numbers and units of measure.

Scientists use measurements to determine precise values that have the same meaning to everyone. Measurements often are made with instruments that extend our senses. Microscopes and telescopes allow the observation of things too small, too far away, or too dim to be visible without these instruments. (See Figure 1-5 on page 10.) Balance scales, metersticks, clocks, and thermometers allow scientists and students to make more accurate observations than they could make without the use of instruments. To use an interactive online stopwatch, visit the Web site: *http://www.shodor.org/interactivate/activities/stopwatch/*

FIGURE 1-5.
Instruments help people make better observations.

People accept many things as fact, even if they have not observed them directly. An **inference** is a conclusion based on observations. When many rocks at the bottom of a cliff are similar in composition to the rock that makes up the cliff, it is reasonable to infer the rocks probably broke away from the cliff.

Scientists often make inferences. When scientists observe geological events producing rocks in one location and find similar rocks in other locations, they make inferences about past events. Although they did not witness these events, they infer that the same processes occurred in both places. No person can see the future. Therefore, all predictions are inferences. In general, scientists prefer direct observations to inferences.

Exponential Notation

Scientists deal with data that range from the sizes of subatomic particles to the size of the universe. If you measure the universe in subatomic units you end up with a number that has about 40 zeros. How can this range of values be expressed without using numbers that are difficult to write and even more difficult to work with? Scientists use **exponential notation**, sometimes called **scientific notation**. Exponential notation uses powers of 10 to express numbers that would be more difficult to write or read using standard decimal numbers.

Numbers written in exponential notation take the form of $c \times 10^e$, where c (the coefficient) is always a number equal to or greater than

1 but less than 10 and *e* (the exponent) is the power of 10. Being able to understand and use exponential notation is very important. Any number can be expressed in exponential notation by following these two steps.

Step 1: Change the original number to a number equal to or greater than 1 but less than 10 by moving the decimal point to the right or left.

Step 2: Assign a power of 10 (exponent) equal to the number of places that the decimal point was moved.

A good way to remember whether the power of 10 will be positive or negative is to keep in mind that positive exponents mean numbers greater than 1, usually large numbers. Negative exponents mean numbers less than 1, which are sometimes called decimal numbers. Once you get used to it, it becomes easy. The following Web site may help you understand very large and small distances and the need for exponential notation: *http://www.powersof10. com/index.php?mod=register_film*

STUDENT ACTIVITY 1-2 —EXPONENTIAL NOTATION IN THE REAL WORLD

1: MATHEMATICAL ANALYSIS 1
6: MAGNITUDE AND SCALE 3

Make a list of 5 to 10 very large or very small values, document their use, and translate them into scientific (exponential) notation. Your examples must come from printed or Internet sources, such as *http://www.infoplease.com/ipa/A0854973.html*, outside your Earth science course materials.

For each example you present, include the following:

1. The value expressed as a standard number. (If units of measure are given, be sure to include them.)
2. What is being expressed. (For example, it might be the size of a particular kind of atom.)
3. The same value expressed in exponential notation.
4. Where you found the value. Please give enough information so that another person could find it easily.

Visit the following Web site to see the formal mathematical rules of calculation and expressing numbers: *http://www.vendian.org/ envelope/dir0/exponential_notation.html*

SAMPLE PROBLEMS

Problem 1 The age of the universe is estimated to be about 13,700,000,000 years; express this number in exponential notation.

Solution   *Step 1:* Change the original number to a number equal to or greater than 1, but less than 10 by moving the decimal point to the right or left. (Zeros that appear outside nonzero digits can be left out.) In this case, you get 1.37.

             Step 2: Assign a power of 10 (exponent) equal to the number of places that the decimal point was moved. This decimal point was moved to the left 10 places, making the power of 10 a positive number. So the age of our planet is 1.37×10^{10} years.

Problem 2 Light with a wavelength of 0.00004503 centimeter (cm) appears blue. Express this value in scientific notation.

Solution   *Step 1:* After moving the decimal point five places to the right, the coefficient becomes 4.503. The zero before the 3 is kept because it appears between nonzero digits. This zero is needed to establish the number's value.

             Step 2: When the decimal point is moved right, the exponent is a negative number. The power of 10 is -5. The number is 4.503×10^{-5} cm.

The International System of Measurement

Over the course of time, different countries developed their own systems of measurements. The inch and the pound originated in England. There were no international standards until the European nations established a system now known as the "International System of Units." This system is called "SI," based on its name in French, *System Internationale*. SI units are now used nearly everywhere in the world except the United States. SI is similar to the metric system.

In a temperature-controlled vault in France, a metal bar has been marked with two scratches that are exactly 1 meter (m) apart. In the past, it was the definition of meter, and all devices used to measure length were based on that standard. Everyone knew the length of a meter and everyone's meter was the same. Today the meter is defined as a certain number of wavelengths of light emitted by krypton-86 under specific laboratory conditions. The advantage of this change is the standard length can be created anywhere.

In everyday life, people often use a system of measures called "United States Customary Measures." Units such as the mile, the pound, and the degree Fahrenheit have been in use in this country for many years. Most Americans are familiar with them and resist change. As this country becomes part of a world economy, SI units will gradually replace the United States Customary units, as suggested in Figure 1-6. Many beverages are now sold in liters. A variety of manufactured goods created for world markets are also measured in SI units.

Scientists everywhere use SI units for several reasons:

- They are universal. Scientists do not need to translate units when they communicate with their colleagues in other countries.
- SI units are related by factors of 10. For example, there are 10 millimeters (mm) in a centimeter (cm), 10 centimeters in a decimeter (dm), and 10 decimeters in a meter (m).
- Scientific instruments sold around the world are generally marked in SI units.

You can find an on-line, Earth science related, metric conversion calculators at the following Web site: *http://www.worldwidemetric.com/metcal.htm*

FIGURE 1-6. In some places in the United States, road signs with SI (metric) units are replacing signs with U.S. Customary Measures.

USING SI UNITS As you perform laboratory activities, you will often need to make measurements. You will be using metric rulers that measure in centimeters. In addition, you will use balances that measure mass in kilograms. You may be asked to use your measurements to find another quantity such as volume or density. To find volume, multiply the object's length by its width by its height.

SAMPLE PROBLEM

Problem   A storage tank is 10 m long by 8 m wide by 3 m high. How much water can it hold?

Solution   You can calculate the volume by multiplying the length by the width by the height:

$$\text{Volume} = \text{length} \times \text{width} \times \text{height}$$
$$= 10 \text{ m} \times 8 \text{ m} \times 3 \text{ m}$$
$$= 240 \text{ m}^3$$

Practice Problem 1
A rectangular bar of soap measures 6.1 cm by 4.2 cm by 2.1 cm. Find the volume of the bar of soap to the nearest tenth of a cubic centimeter.

Estimation

Estimation is a valuable skill for anyone, but especially for scientists. If you want to know whether a measurement or calculation is correct, it is very helpful to estimate the value. If your estimate and the determined value are not close, you may need to give some more thought to your procedure. Suppose you wanted to guess how many people were in a particular crowd, such as the sports event shown in Figure 1-7. You could count the number of people in one section and multiply by the number of sections.

If you heard that a backpacker on the Appalachian Trail walked from Maine to Georgia in five days, it is easy to figure out that this could not be true. The distance is more than 1600 km. That would

FIGURE 1-7. If you did not have access to ticket sales information, how could you estimate the number of people at a Major League sporting event?

mean walking more than 300 km a day. This is clearly not possible, especially on a mountainous trail. Most likely the walk took five months. Walking about 16 km a day is reasonable.

There are many convenient ways to make estimates. For example, you could estimate the height of a ceiling by observing that a person 1.7 m tall extends half way to the ceiling. Therefore your estimate of the ceiling height would be about 3.5 m.

STUDENT ACTIVITY 1-3 —MAKING ESTIMATIONS

1: ENGINEERING
DESIGN 1
1: MATHEMATICAL
ANALYSIS 3
7: STRATEGIES 2

Working in groups, estimate the volume of your classroom or your school building. Do not use any measuring instruments. Number the steps of your procedure 1, 2, 3, . . . etc. For each step, briefly explain exactly how you got your data and made your calculations. Please use only SI (metric) units.

HOW IS DENSITY DETERMINED?

1: MATHEMATICAL
ANALYSIS 1, 2, 3
1: ENGINEERING
DESIGN 1
6: MODELS 2
6: MAGNITUDE AND
SCALE 3
6: PATTERNS OF
CHANGE 5

Density is the concentration of matter, or the ratio of mass to volume. Substances such as lead or gold that are very dense are heavy for their size. Materials that we consider light, such as air or polystyrene foam, are relatively low in density. Objects made of the same solid material usually have about the same density. (Density can change with temperature or pressure as a substance expands or contracts.)

Density is an important property of matter. For example, differences in density are responsible for winds and ocean currents. Density is defined as mass per unit volume. For example, if the mass of an object is 30 grams (g) and its volume is 10 cubic centimeters (cm³), then its density is 30 g divided by 10 cm³, or 3 g/cm³. The formula for calculating density is given on page 1 of the *Earth Science Reference Tables.*

Within the chapters of this book you will find the components of the *Earth Science Reference Tables:* tables, graphs, maps, physical values, and mathematical equations that you will need throughout this course. You do not need to memorize any of the information in the *Reference Tables* because this document will always be available to you for classroom work, labs, and tests. However, you should become familiar with the *Reference Tables* so you know when to use them.

Density is generally expressed in units of mass divided by units of volume:

$$\text{Density} = \frac{\text{mass}}{\text{volume}}$$

Note that the units are carried through the calculation, yielding the proper unit of density such as grams per cubic centimeter (g/cm³). The following Sample Problem will show you how to use this equation.

SAMPLE PROBLEM

Problem   What is the density of an object that has a volume of 20 cm³ and a mass of 8 g?

Solution

$$\text{Density} = \frac{\text{mass}}{\text{volume}}$$

$$= \frac{8 \text{ g}}{20 \text{ cm}^3}$$

$$= 0.4 \, g/cm^3$$

Practice Problem 2

If a 135-g crystal sphere has a volume of 50 cm³, what is its density?

Water, with a density of 1 g/cm³, is used as a standard of density. Therefore, whether an object sinks or floats in water can be used to estimate density. If an object is less dense than water, the object will float in water. If the object is more dense than water, the object will sink. Most wood floats in water because it is less dense than water. Iron, glass, and most rocks sink because they are more dense than water. The idea of density will come up many times in Earth science and it will be discussed as it is applied in later chapters.

The instrument shown in Figure 1-8 is a Galileo thermometer. It is named for the Italian scientist who invented it. This thermometer is based on the principle that the density of water changes slightly with changes in temperature. As the water in the column becomes warmer and less dense, more of the glass spheres inside the tube sink to the bottom. Therefore, the number of weighted spheres that float depends on the temperature of the water. Reading the number attached to the lowest sphere that floats gives the temperature.

You can demonstrate the relative density of liquids by first pouring corn syrup, then water, followed by cooking oil, and finally alcohol into a glass cylinder. You must be careful not to mix the liquids, which will remain layered in order of density as shown in Figure 1-9.

FIGURE 1-8. This instrument is a Galileo thermometer. As the water inside becomes warmer or cooler, it changes in density. So the number of glass spheres that float changes.

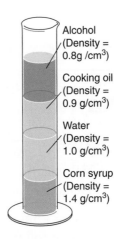

Alcohol
(Density =
0.8g /cm³)

Cooking oil
(Density =
0.9 g/cm³)

Water
(Density =
1.0 g/cm³)

Corn syrup
(Density =
1.4 g/cm³)

FIGURE 1-9. Liquids in graduated cylinder

If a rubber stopper with a density of 1.2 g/cm³ were added, it would sink through the alcohol, cooking oil, and water layers. The stopper would remain suspended between the water and the corn syrup. Rubber is more dense than water, so it sinks in water. Corn syrup is more dense than rubber. Therefore, the rubber stopper would float on top of the corn syrup layer.

STUDENT ACTIVITY 1-4 —DENSITY OF SOLIDS

1: MATHEMATICAL
ANALYSIS 1, 2, 3
6: MODELS 2

Density can be used to identify different substances. In general, no matter how much you have of a certain substance, its density is the same. Rather than measuring density directly, usually the mass is measured, and the volume is determined so that density (mass/volume) can be calculated. The equation volume = length × width × height is used to determine the volume of rectangular solids. There are also equations that can be used to determine the volume of other regular solids, such as spheres.

Your teacher will supply you with a variety of objects to measure. Construct a data table that lists the name, mass, volume, and density of each object. Measure the mass and determine the volume of each object, then calculate the density of each. Be sure to use SI (metric) measurements.

After you have calculated the density of each sample, place a star next to the name of those that will float in water. How can you tell that they will float?

STUDENT ACTIVITY 1-5 —THE THICKNESS OF ALUMINUM FOIL

1: MATHEMATICAL
ANALYSIS 1, 2, 3
1: ENGINEERING
DESIGN 1

Given the following:

1. $\text{Density} = \dfrac{\text{mass}}{\text{volume}}$
2. Volume = length × width × thickness
3. Density of aluminum = 2.7 g/cm³

Materials: a metric ruler, a kilogram scale, a small square piece of aluminum foil (about 30 cm on each side).

Problem: Determine the thickness of the aluminum foil to the nearest ten-thousandth of a centimeter (two significant figures).

Hint: Combine the two equations above into a single equation with one unknown. Then substitute measurements, to solve for thickness. (In this problem, thickness takes the place of height as a third dimension.)

Errors in Measurement

No matter how carefully a measurement is made, it is likely that there will be some error. Using measuring instruments more carefully or using more precise instruments can reduce error, but error can never be eliminated. In general, errors are reduced to the point at which they are not important or at which it is not worth the effort to make them smaller. Sometimes measurements are used in calculations, such as the determination of density. In these cases, any errors in measurement will result in errors in the calculated value.

Percent deviation (sometimes called percent error) is a useful way to compare the size of an error with the size of what is being measured. For example, an error of 1 cm in the width of this book is a large error. But an error of 1 cm in the distance to the moon would be a very small error. They are both errors of 1 cm. However, because the book is so much smaller, this 1-cm error is far more significant. The equation for percent deviation is

$$\text{Deviation (\%)} = \frac{\text{difference from accepted value}}{\text{accepted value}} \times 100$$

The Sample Problem on the next page will show you how to use this equation.

SAMPLE PROBLEM

Problem A student estimated the height of a tree to be 15 m. However, careful measurement showed the true height was 20 m. What was the percent deviation?

Solution

$$\text{Deviation}\,(\%) = \frac{\text{difference from accepted value}}{\text{accepted value}} \times 100$$

$$= \frac{20\,\text{m} - 15\,\text{m}}{20\,\text{m}} \times 100$$

$$= \frac{5\,\text{m}}{20\,\text{m}} \times 100$$

$$= 25\%$$

Notice the following features in this calculation:

1. The calculation starts with the complete algebraic equation. The only numbers that show in this first step are constants used in every application of the formula.
2. Values are substituted into the formula, including numbers and their associated units of measure.
3. The steps to the solution are organized so that they are easy to follow, leading to the answer at the end. (Be sure to include the proper units of measure.)

Practice Problem 3

A student determined the density of a stone to be 3.6 g/cm³. The accepted value is 3.0 g/cm³. What was the student's percent deviation?

HOW DO SCIENTISTS MAKE AND USE GRAPHS?

1: MATHEMATICAL
ANALYSIS 1, 2, 3
6: MODELS 2
6: PATTERNS OF
CHANGE 5
7: STRATEGIES 2

A graph is a visual way to organize and present data. Instead of reading paragraphs of information or studying columns of figures, a graph makes comparisons between variables easier because it is more visual. Unlike a data table, a graph allows the reader to see changes in data, to understand relationships between variables within the data, and to see trends or patterns.

Line Graphs

A line graph, such as the one in Figure 1-10, shows how a measured quantity changes with time, distance, or some other variable. Line graphs are constructed by plotting data on a **coordinate system**, a grid on which each location has a unique designation defined by the intersection of two lines. A coordinate system is set up on vertical and horizontal axes. The horizontal (*x*) axis is usually used for the independent variable. It usually indicates a uniform change, such as hours, years, or centimeters. Normally, the regular change expected in the independent variable is well understood. The vertical (*y*) axis is used for the dependent variable. It usually indicates the amount of the measured quantity being studied, such as temperature, height, or population. The values of the dependent variable are what you are trying to find. The graph shows how the dependent variable changes with respect to the independent variable.

The rise or fall of the line in Figure 1-10 shows the increase or decrease in the price of copper over a 5-year period. When the line on the graph moves upward and to the right, it represents a continuous increase. When the line on the graph moves downward and to the right, it indicates a continuous decrease. A horizontal line on the graph represents no change. The steeper the line segment rises

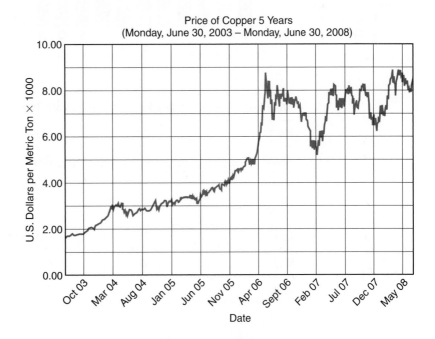

FIGURE 1-10. This line graph shows the price of copper has increased dramatically since China and India have joined the world economic markets. What was the price of copper in May 2008?

to the right, the greater the slope of the segment, and the greater the increase in price. Likewise, the steeper the line segment falls to the right, the greater the decrease in price. Some line graphs are straight lines; others are curved lines.

Pie and Bar Graphs

Sometimes, a line graph is not the best graph to use when organizing and presenting data. In Earth science, bar and pie graphs are often used. The bar graph is useful when comparing similar measurements taken at different times or in different places. For example, the bar graph in Figure 1-11, which is based on the data in Table 1-2, compares monthly rainfall, or precipitation (PPT), in millimeters (mm) over the period of 1 year.

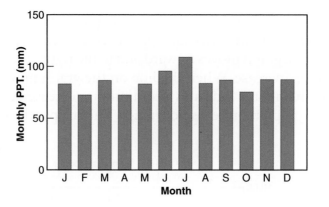

FIGURE 1-11.
This bar graph represents the average monthly precipitation for Lake Placid, New York.

Table 1-2. Average Monthly Precipitation for Lake Placid, New York

Month	PPT (mm)	Month	PPT (mm)
January	81	July	107
February	71	August	84
March	86	September	86
April	71	October	74
May	81	November	86
June	94	December	86

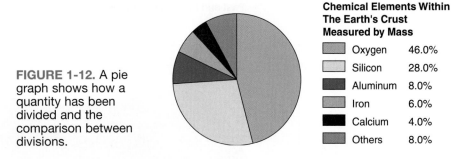

Chemical Elements Within The Earth's Crust Measured by Mass

	Oxygen	46.0%
	Silicon	28.0%
	Aluminum	8.0%
	Iron	6.0%
	Calcium	4.0%
	Others	8.0%

FIGURE 1-12. A pie graph shows how a quantity has been divided and the comparison between divisions.

The pie graph is used to show how a certain quantity has been divided into several parts, as well as to show the comparisons between these parts. The pie graph in Figure 1-12 shows the most abundant chemical elements in the rocks of Earth's crust by percent by mass.

Guidelines for Making Graphs

Graphs are all around us. They are especially common in news and in advertising where it is important to communicate information quickly. Sometimes, in the effort to keep the graph simple, it may contain errors. When you construct graphs in science, you should follow these guidelines:

- Keep in mind that the purpose of a graph is to provide information. The graph should have a title to explain the relationships represented. All important information should be presented as clearly and simply as possible. The axes should be labeled with quantity and units. One axis might be time in years whereas the other is human population. The graph on the next page from the United Nations gives its prediction of world population for the next several centuries. (See Figure 1-13 on page 24.)
- The independent variable should be plotted on the horizontal axis. Usually, data shows how one factor responds to changes in the other. For example, in Figure 1-13, it is clear that the human population does not determine the passage of time, but the world population depends on when (time) it is determined. In this case, time is the independent variable and population is the dependent variable. Time (the year, month, etc.) belongs on the horizontal axis.

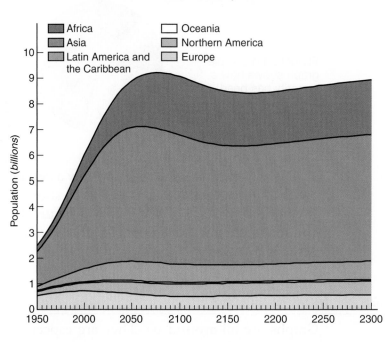

Past and Predicted World Population 1950–2300

Legend:
- Africa
- Asia
- Latin America and the Caribbean
- Oceania
- Northern America
- Europe

FIGURE 1-13.
The United Nations predicts that the world population will stabilize as third-world countries become more economically developed.

• Fit the graph to the data. Design the vertical and horizontal axes so that the data reasonably fills the graph but does not go beyond the number scales on the two axes.

(Appendix E at the back of this book also contains a guide for constructing graphs.)

STUDENT ACTIVITY 1-6 —MAKING A GRAPH OF THE REVOLUTION OF THE PLANETS

1: MATHEMATICAL ANALYSIS 2

Does the distance of a planet from the sun affect how long it takes to make one orbit of the sun? You can investigate this question by drawing a graph.

On page 665, you will find a table "Solar System Data." Use the data in this table to graph the relationship between the distance of a planet from the sun and its period of revolution. Label each data point with the name of the object from Mercury through Neptune. (Do not include the sun or Earth's moon.)

As a follow up, you might try graphing planetary distance and other factors in this table.

HOW IS TECHNOLOGY CHANGING THE WAY SCIENTISTS WORK?

1: SCIENTIFIC
INQUIRY 1, 2, 3
2: INFORMATION
SYSTEMS 1

How science is "done" has always depended on the tools available. Some tools have revolutionized Earth science. Computers provide a good example. When computers are used with other devices, they can perform many functions. Computers help us analyze data from satellites, produce and edit images taken with cameras, and quickly access information on the World Wide Web. The first electronic computers filled whole rooms, and were so expensive that only a few research centers could afford them. Today, laptop computers have computing power equal to that of a supercomputer of the 1970s.

Connecting computers in networks has reached the point at which you can almost instantly access information stored in millions of computers all over the world. This is accomplished by the World Wide Web connected by the Internet. It allows all of us to communicate faster than ever before.

STUDENT ACTIVITY 1-7 —AN INTERNET SCAVENGER HUNT

2: INFORMATION
SYSTEMS 1, 2

In a scavenger hunt, the goal is to collect a variety of unrelated objects. In this case, the "objects" will be bits of information. Each example will require two responses: (1) give the answer to the question, and (2) record where on the Internet you found it; that is, provide the Internet address (URL) and/or the name of the Internet site. Find answers to as many as you can.

1. What is the weather like today in Phoenix, Arizona?
2. Where and when has a very large earthquake occurred in the past 6 months?
3. Other than the sun, what is the nearest star to Earth?
4. What is the human population of New York City?
5. What is the current value of gold per ounce?
6. How many sunspots were recorded in 2007?
7. What name was given to the third tropical storm in the Atlantic Ocean last year?
8. What is the chemical composition of emeralds?

GIS and GPS are two of the most useful, recent technological advances for the Earth sciences. The Geographic Information System

FIGURE 1-14. A global positioning system receiver processes signals from satellites to determine your precise location on Earth.

(GIS) is a visual resource that allows you to plot the spatial relationships of data. Because GIS is based on information in computers all over the world, a wide variety of information can be retrieved and mapped. It also can be updated regularly. For example, GIS data can be used to make maps that show soil types, vegetation, or land ownership.

A Global Positioning System (GPS) unit is a device that receives and analyzes information sent by satellites. The information allows you to determine your location with remarkable accuracy, as shown in Figure 1-14. Installed in your car, a GPS unit can direct you to an unfamiliar location in real time. The GPS is so accurate that it has been used to measure the slow movement of continents over Earth's surface and even changes in the heights of some mountains.

Inquiry in Science

Many people would say that an inquiring mind is the most important asset humans have. Using observations, information resources, and a variety of analytical tools, people can often make important discoveries by asking the right questions and following productive leads. As long as there is the curiosity to ask questions and the will to find the answers to them, science will help find those answers.

CHAPTER REVIEW QUESTIONS

Part A

1. Some scientists estimate that age of the sun and solar system is about 4.65×10^9 years. Which choice correctly expresses this value?

   (1) 465 years   (3) 465,000,000 years
   (2) 465,000 years   (4) 4,650,000,000 years

2. The average diameter of the Milky Way galaxy is about 115,000 light-years. Which choice below correctly expresses this in scientific notation?

   (1) 1×10^5 light-years
   (2) 115×10^5 light-years
   (3) 1.15×10^5 light-years
   (4) 2.3×10^{10} light-years

3. A student recorded information about a rock sample. The four statements are true, which one is an observation?

   (1) If placed in water, the rock will sink.
   (2) The rock is billions of years old.
   (3) The rock has a mass of 93.5 g.
   (4) The rock formed by natural events.

4. A student made the following notes about the current weather conditions. Which statement is most likely an inference?

   (1) The temperature 3 hours ago was 20°C.
   (2) The current air pressure is 1000.4 millibars.
   (3) The sky is completely overcast with clouds.
   (4) Rain will occur in the next 6 hours.

5. An object has a mass of 46.5 g and a volume of 15.5 cm³. What is the density of the object?

   (1) 0.3 g/cm³
   (2) 46.5 g/cm³
   (3) 3.0 g/cm³
   (4) $720.8.$ g/cm³

6. A student measured the mass of a rock as 20 g. But the actual mass of the rock is 30 g. What was the student's percent deviation?

   (1) 25   (3) 33
   (2) 44   (4) 50

7. Which is usually considered a division of Earth science?

   (1) chemistry   (3) physics
   (2) geology   (4) biology

Part B

8. Why do scientists use graphs to present data?

   (1) Graphs do not contain errors.
   (2) Graphs make data easier to understand.
   (3) Graphs take less room than data tables.
   (4) Graphs make papers easier to get published.

9. The data table below shows the mass and volume of three samples of the same mineral. (You may use a separate paper for your calculations.)

DATA TABLE

Sample	Mass (g)	Volume (cm³)	Density (g/cm³)
A	50	25	
B	100	50	
C	150	75	

Which graph best represents the relationship between the density and the volume of these mineral samples?

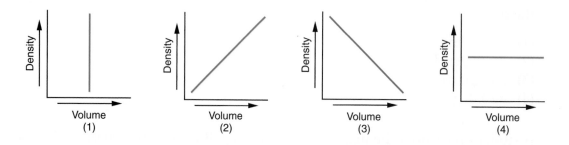

10. Why do we use percent deviation rather than simply expressing the size of the error itself?

    (1) Percent deviation gives more information than the value of the error itself.
    (2) If there is no error, percent deviation makes this clearer.
    (3) Percent deviation emphasizes the importance of errors.
    (4) Sometimes the value of the error itself is not known.

11. The density of quartz is 2.7 g/cm³. If a sample of quartz has a mass of 81 g, what is its volume?

    (1) 0.03 cm³   (3) 11.1 g
    (2) 8.1 g   (4) 30 cm³

12. Gold has a density of 19.3 g/cm³. A prospector found a gold nugget with a volume of 10 cm³. What was the mass of the nugget?

    (1) 1.93 g   (3) 19.3 g
    (2) 193 g   (4) 1930 g

13. Pumice is an unusual rock because some samples float on water. What does this tell you about pumice?

    (1) Pumice can be less dense than water.
    (2) Pumice is most common in high mountain locations.
    (3) Pumice is usually found in very small pieces.
    (4) Pumice absorbs water.

14. The density of granite is 2.7 g/cm³. If a large sample of granite is cut in half, what will be the density of each of the pieces?

    (1) 1.35 g/cm³   (3) 5.4 g/cm³
    (2) 2.7 g/cm³   (4) 27 g/cm³

15. If two leading scientists are investigating the same question and they reach similar conclusions, what does this show?

    (1) They probably changed their results to agree.
    (2) The conclusion they both made is probably in error.
    (3) Their scientific work showed a lack of originality.
    (4) Their conclusions have a good chance of being correct.

Part C

Base your answers to questions 16 through 19 on the following information and the data table.

The snowline is the lowest elevation at which snow remains on the ground all year. The data table on page 30 shows the elevation of the snowline at different latitudes in the Northern Hemisphere.

Latitude (°N)	Elevation of Snowline (m)	Latitude (°N)	Elevation of Snowline (m)
0	5400	50	1600
10	4900	65	500
25	3800	80	100
35	3100	90	0

16. On a properly labeled grid, plot the latitude and elevation of the snowline for the locations in the data table. Give the graph a title. Use a dot for each point and connect the dots with a line.

17. Mt. Mitchell, in North Carolina, is located at 36°N and has a peak elevation of 2037 m. Plot the latitude and elevation of Mt. Mitchell on your graph. Use a plus sign (+) to mark this point.

18. Using your graph, determine (to the *nearest whole degree*) the lowest latitude at which a peak with the same elevation as Mt. Mitchell would have permanent snow.

19. State the relationship between latitude and elevation of the snowline.

20. The diagram below shows three liquids of different densities in a 100-mL cylinder. A sphere of oak wood about half the diameter of the cylinder and a silver sphere of the same size are dropped in the cylinder without mixing the liquids. The oak wood has a density of 0.9 g/cm³, and the density of silver is 10.5 g/cm³. Where will the spheres come to rest?

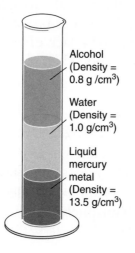

Alcohol
(Density =
0.8 g /cm³)

Water
(Density =
1.0 g/cm³)

Liquid
mercury
metal
(Density =
13.5 g/cm³)

2 Earth's Dimensions and Navigation

WORDS TO KNOW

atmosphere	latitude	prime meridian
axis	lithosphere	stratosphere
equator	longitude	terrestrial coordinate
geosphere	mesosphere	thermosphere
Greenwich Mean Time	oblate	troposphere
hydrosphere		

This chapter will help you answer the following questions:

1. How do we know that Earth is a sphere?
2. What layers of Earth do we know best?
3. How can we tell where we are on Earth when there are no familiar landmarks?

WHAT IS EARTH'S SHAPE?

Evidence of Earth's Shape

1 MATHEMATICAL
ANALYSIS 1
4: 1.1c
4: 1.2c

Although Earth looks flat and endless, there were some ancient scholars who understood that Earth is a gigantic sphere. The scholars came to this conclusion because they noticed that as a ship sails away to sea, it seems to disappear from the bottom first. Ships appear to sail over and below the horizon as shown in Figure 2-1 on page 32.

FIGURE 2-1. Over the ocean, the horizon looks flat. However, as a ship sails away, it seems to disappear hull first, as it gradually dips below the horizon.

Another indication of Earth's shape comes from observing the moon. During an eclipse of the moon, Earth's shadow moves over the surface of the moon. The edge of that shadow is always a uniformly curved line. Ancient Greeks knew that the only shape that always casts a uniformly curved shadow is a sphere. The reason for their thinking is illustrated in Figure 2-2.

If you know someone who lives hundreds or thousands of miles away to the east or west, you know that person's local time is different from yours. If it is noon in New York, it is only 9 A.M. for a person in California. At the same time, people in Europe are having their evening meal. For a person in central Asia or Australia, it

FIGURE 2-2. During an eclipse of the moon, Earth's shadow gradually covers the full moon. Notice that this shadow, as always, appears as a dark circle whose diameter is about three times the diameter of the moon.

might be midnight. When time differences over the whole planet are considered, it is clear that Earth is a gigantic sphere.

There is also evidence of Earth's shape in the observation of distant objects in the night sky. A person at the North Pole sees the North Star, Polaris, directly overhead. To a person located farther south, Polaris appears lower in the sky. In fact, at the equator, Polaris is on the horizon. (See Figure 2-3.)

The **equator** is an imaginary line that circles Earth half way between the North and South poles. South of the equator, Polaris

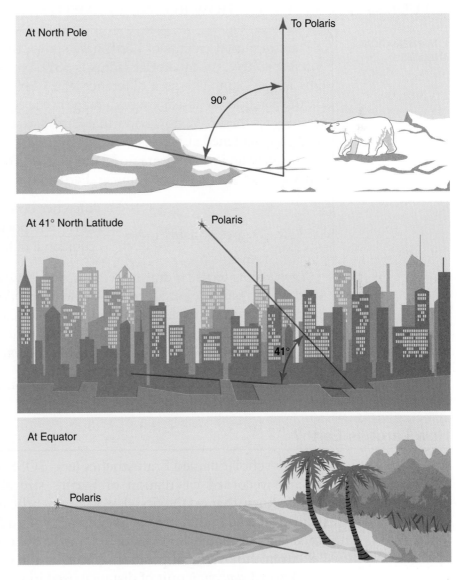

FIGURE 2-3. The altitude of Polaris in the northern sky depends on your latitude.

is not visible at all. Observers south of the equator can see the stars of the Southern Cross, which is never visible in New York. These observations support the idea of a spherical planet.

The Apollo program explored the moon in the late 1960s. During these missions, astronauts flew far enough from Earth to be able to see the entire planet. The astronauts took photographs that show that our planet is a nearly spherical object orbiting in the vastness of space.

STUDENT ACTIVITY 2-1 —HOW ROUND IS EARTH?

1: MATHEMATICAL ANALYSIS 1

Careful measurements of Earth have shown that it is not a perfect sphere; its equatorial radius is 6378 km, and its polar radius is 6357 km. This is a difference of 21 km (about 13 miles). Earth's rotation on its axis causes a bulge at the equator. How much of a bulge is there? We now know that Earth's exact shape is **oblate**, or slightly flattened at the poles.

To calculate Earth's degree of flattening, use the following formula. If the result is a large number, Earth is not very round.

$$\text{Degree of flattening} = \frac{\text{difference between equatorial and polar radii}}{\text{equatorial radius}}$$

Your task is to use a drawing compass to draw two circles centered on the same point. Draw one circle proportional to the polar radius and a second circle proportional to the equatorial radius. You will need to establish and use a scale so that your drawings will fit on a sheet of paper. Compare these two circles and state how far from round Earth would appear from space.

How Large Is Earth?

A Greek scholar named Eratosthenes (era-TOSS-then-ease) made the first recorded calculation of Earth's size about 2000 years ago. He knew that on the first day of summer the noon sun was directly overhead at the town of Syene in Egypt. In Alexandria, 5000 *stadia* (approximately 800 km, or 500 mi) to the north, the sun was 7.2° from the overhead position. (*Stadia* is the plural form of the Greek word *stadion,* a unit of distance used in Eratosthenes' time.)

Since 7.2° is $\frac{1}{50}$ of a circle, Eratosthenes reasoned that the distance around the Earth must be 50 × 5000 *stadia,* or 250,000 *stadia.* Although the exact length of a *stadion* is not known, Eratosthenes' figure appears to be remarkably close to the more accurate measurements made today. Visit this Web site to see astronomer Carl Sagan explain how Eratosthenes calculated Earth's circumference: *http://www.youtube.com/watch?v=0JHEqBLG650*

WHAT ARE EARTH'S PARTS?

2: INFORMATION
SYSTEMS 1
4: 1.1i
4: 2.1j

Based on differences in composition, Earth can be divided into three parts. These parts form spheres, one inside the other, separated by differences in density. Each sphere is also a different state of matter: gas, liquid, or solid.

The **atmosphere** is the outer shell of gas that surrounds Earth. The **hydrosphere** is the water of Earth. About 99 percent of this water is contained in Earth's oceans, which cover about three-quarters of the planet. The **lithosphere** is the solid rock covering Earth. (The crust is the rocky outer layer of the lithosphere.)

Table 2-1 lists the average chemical composition of each sphere. Rocks in Earth's crust represent the lithosphere because these are

Average Chemical Composition of Earth's Crust, Hydrosphere, and Troposphere

ELEMENT (symbol)	CRUST		HYDROSPHERE	TROPOSPHERE
	Percent by mass	Percent by volume	Percent by volume	Percent by volume
Oxygen (O)	46.10	94.04	33.0	21.0
Silicon (Si)	28.20	0.88		
Aluminum (Al)	8.23	0.48		
Iron (Fe)	5.63	0.49		
Calcium (Ca)	4.15	1.18		
Sodium (Na)	2.36	1.11		
Magnesium (Mg)	2.33	0.33		
Potassium (K)	2.09	1.42		
Nitrogen (N)				78.0
Hydrogen (H)			66.0	
Other	0.91	0.07	1.0	1.0

TABLE 2-1.

the rocks that are found at and near the surface. (Deep inside Earth, denser elements, such as iron and magnesium, are more common than they are near the surface.) Notice that oxygen is among the most common elements in all three parts of Earth. Elements are shown rather than chemical compounds because the crust is composed of thousands of minerals, each with a different chemical composition. However, most minerals contain roughly the same elements. Most of the atmosphere is composed of elements in the form of gases. Only the hydrosphere is made mostly of a single compound: water. Water is composed of two parts hydrogen to one part oxygen by volume.

STUDENT ACTIVITY 2-2 —PIE GRAPHS OF EARTH'S SPHERES

2: INFORMATION SYSTEMS 1

Draw three pie graphs based on the data in Table 2-1. Make one graph of the chemical composition of the crust by mass, a second of the chemical composition of the hydrosphere by volume, and the third of the chemical composition of the crust by volume. Two of these graphs represent the composition of Earth's crust. Why do these two pie graphs look so different?

The Atmosphere

A thin layer of gas, the atmosphere, surrounds the solid earth and the oceans. Most of the mass of the atmosphere, clouds, and weather changes occur in the troposphere, the lowest layer of the atmosphere. (See Figure 2-4.) Although the atmosphere accounts for a tiny part of the total mass and volume of the planet, it is in this changing environment that people and most other life-forms live.

Air is a mixture of gases that is about 78 percent nitrogen (N_2). Nitrogen is a stable gas that does not react easily with other elements or compounds. About 21 percent of the atmosphere is oxygen (O_2). Oxygen combines with many other elements in the processes of oxidation, combustion, and cellular respiration. Living things depend on cellular respiration to make use of the energy stored in food. The inert gas argon, which almost never reacts with other elements or compounds, makes up about 1 percent of the atmosphere.

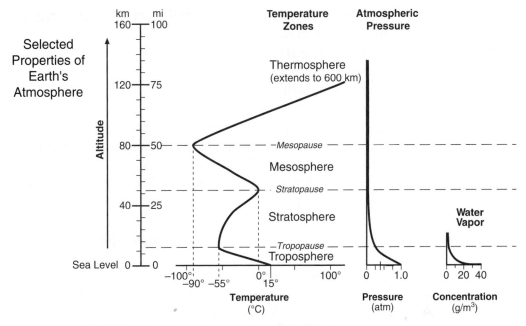

FIGURE 2-4. Selected properties of Earth's atmosphere.

The proportions of other gases in the atmosphere are not constant. The amount of water vapor, water in the form of a gas, can be as high as several percent in warm, tropical locations or a tiny fraction of a percent in deserts and cold areas. Carbon dioxide, the product of respiration and the burning of fossil fuels, makes up far less than 1 percent. However, carbon dioxide is needed by plants for photosynthesis. This gas also plays an important role in climate change, which will be explored in Chapter 25.

The paragraph above describes the composition of the atmosphere's lowest layer, the **troposphere**. Note that the names of the layers of the atmosphere end in –*sphere* because this is their shape around Earth. The names of the boundaries between layers end in –*pause,* as in stopping. Therefore, the tropopause is the place where the troposphere ends.

The atmosphere is divided into layers based on how the temperature changes with altitude, as shown in Figure 2-4. Because the layers of the atmosphere are a result of density differences, the atmosphere is most dense at the bottom of the troposphere. Actually, the troposphere contains most of the mass of the atmosphere even though it extends only about 12 km (7 mi) above Earth's surface. Nearly all the atmosphere's water vapor, clouds, and weather events occur in this lowest layer. Compared with the whole Earth, the troposphere is a very thin layer. In fact, it is the thinnest layer of our atmosphere.

Within the troposphere as altitude increases, temperature decreases. Have you ever noticed how snow lasts longer in the highest mountains than it does at lower elevations? The world's highest mountains extend nearly to the top of the troposphere. Above that height, the temperature change reverses and it actually becomes warmer with increasing altitude. The altitude at which the temperature reversal occurs is the tropopause.

The next layer of the atmosphere is the **stratosphere**, in which the temperature increases with increasing altitude. The stratosphere extends up to the stratopause, where another change in temperature trend takes place. In the **mesosphere**, the temperature falls as altitude increases. Above the mesopause, is the highest layer, the **thermosphere**, in which the air temperature rises a great deal. However, due to the atmosphere's extreme low density, the increase in temperature affects relatively few atoms. This increase in energy separates the molecules into positive and negative ions. For this reason, this layer is sometimes called the ionosphere.

The lower boundary of the atmosphere at the surface of the lithosphere or the hydrosphere is quite distinct. Because the atmosphere thins with altitude, there is no clear upper boundary of the atmosphere. The atmosphere just gets thinner and thinner as you get farther from Earth. When people refer to the atmosphere, they usually mean the relatively shallow troposphere, which actually contains about three-quarters of the atmosphere's total mass. This is layer in which we live. For information and animations about Earth's atmosphere visit the following Web site: *http://earthguide.ucsd.edu/earthguide/diagrams/atmosphere/index.html*

STUDENT ACTIVITY 2-3 —INTERPRETING REFERENCE TABLES

2: INFORMATION
SYSTEMS 1

You have probably ridden an elevator to the top of a building. What would you experience if you could ride upward in an open elevator through the atmosphere? Based on the information in Figure 2-4, write a traveler's guide to an elevator ride to a point 150 km above Earth's surface. Describe changes in temperature, air pressure, and water vapor concentration that a traveler would find on the ride. In addition, describe the protective equipment that a traveler would need to survive the various levels of the trip.

The Hydrosphere

Earth's oceans cover nearly three times as much of our planet as do the continents. People may think oceans are huge, featureless expanses of water, but oceans are not limitless and not featureless. The bottom of the oceans (the seafloor) is almost as variable as Earth's land areas. The hydrosphere is Earth's thinnest layer, averaging about 4 km (2.5 mi) in depth. Scientists think the oceans may be where life began on Earth.

The liquid hydrosphere can be divided into two parts. About 99 percent of the hydrosphere is made up of the oceans, which are salt water. Salt water is about 96.5 percent water and about 3.5 percent salt, mostly sodium chloride (common table salt). The remaining 1 percent of the hydrosphere is freshwater, which contains much smaller concentrations of dissolved solids. Freshwater is found in glaciers, streams, rivers, and lakes. However, far more liquid freshwater is in the spaces within soil and rock than on the surface. In fact, groundwater is estimated to be 25 times as abundant as the freshwater on Earth's surface.

The Lithosphere

The surface of the lithosphere, including soil and rock, is the part of Earth we see most. This natural arch shown in Figure 2-5 is part

FIGURE 2-5.
Double Arch in Arches National Park near Moab, Utah, is part of the lithosphere.

of the lithosphere, Earth's rigid outer layer. The great bulk of Earth is the geosphere. We can define the **geosphere** as the mass of solid and molten rock that extends more than 6000 km from Earth's solid surface to its center. The lithosphere, the top 100 km of the geosphere, is the most rigid (unbending) part of the geosphere. Direct explorations in mines have taken humans to a depth of less than 4 km. The deepest drill hole is about 12 km deep. Everything we know about the geosphere at depths greater than 12 km comes from indirect evidence, such as the increasing temperatures with depth, the passage of seismic (earthquake) waves, examination of meteorites, and from the determination of Earth's properties, such as its density. Therefore, scientists have directly explored or penetrated only about one-fifth of 1 percent of the distance to Earth's center.

HOW IS LOCATION DETERMINED?

4: 1.1c, 1.1d, 1.1f

How can sailors far out on the ocean determine their position so they can safely return to port? With no familiar landmarks, such as roads, cities, and geographic features, they cannot describe their location in terms of surface features the way people usually do on land. Long ago, explorers solved this problem by establishing a coordinate system that covers the whole Earth.

Terrestrial Coordinates

The grid on a sheet of graph paper is a type of coordinate system. Each point on the paper has a unique address expressed in terms of numbers along the x and y axes. Many cities are laid out in a coordinate system. Much of New York City has numbered avenues that run north-south and numbered streets that run east-west. Knowing the street address of a building can help a person quickly locate it on a map or in the city itself. However, in Earth's undeveloped areas there are no roads or street signs, and there is no way to mark the oceans' surface. Because of this, explorers used their observations of the sun and stars to find their position on Earth's surface.

The coordinate system established by early sailors and explorers is Earth's system of latitude and longitude, called **terrestrial**

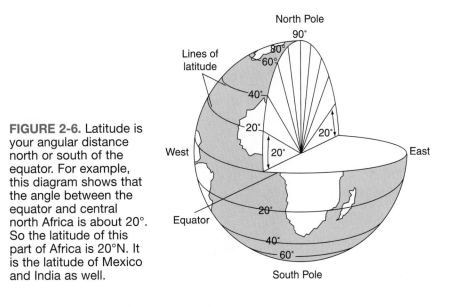

FIGURE 2-6. Latitude is your angular distance north or south of the equator. For example, this diagram shows that the angle between the equator and central north Africa is about 20°. So the latitude of this part of Africa is 20°N. It is the latitude of Mexico and India as well.

coordinates. This system is based on the spin (rotation) of Earth on its axis. The **axis** is an imaginary line that passes through Earth's North and South poles. Halfway between the poles is the equator, an imaginary line that circles Earth. The first terrestrial coordinate value is latitude. As shown in Figure 2-6, **latitude** is the angular distance north or south of the equator.

Lines of equal latitude are called parallels because they run east-west and are parallel to each other. The equator is the reference line at latitude 0°. Both north and south of the equator, latitude increases to a maximum of 90° at the poles. Parallels can be drawn at any interval of latitude from the equator (0°) to the North and South poles (90°N and S).

DIVISIONS OF ANGLES Just as meters can be divided into centimeters and millimeters, degrees of angle can be divided into smaller units. Each degree is made up of 60 minutes (60') of angle. So, $23\frac{1}{2}$ degrees is 23 degrees and 30 minutes (23°30'). Furthermore, one minute of angle ($\frac{1}{60}$ of a degree) can further be divided into 60 seconds (60"). On Earth's surface, an accuracy of a second of latitude or longitude would establish your location to within a rectangle that measures about 30 m on each side. Visit the following Web site to convert street addresses to/from latitude/longitude in one step: *http://www.stevemorse.org/jcal/latlon.php*

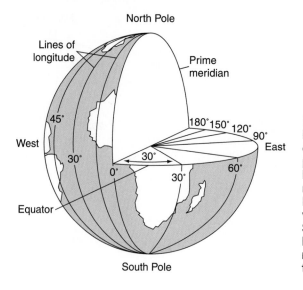

FIGURE 2-7. Longitude is your angular distance east or west of the prime meridian. At 0° longitude, the prime meridian runs through London, England, western Africa and the South Atlantic Ocean. All lines of equal longitude run from the North Pole to the South Pole.

The second terrestrial coordinate value is longitude. As shown in Figure 2-7, **longitude** measures angular distance east and west. Unlike latitude, there is no natural or logical place for longitude measurements to begin. English explorers established their reference line at the Royal Observatory in Greenwich (GREN-itch), England. Since England dominated world exploration and map-making in the 16th century, a north-south line through Greenwich became the world standard for measurements of longitude. Today, the Greenwich meridian, also known as the **prime meridian**, has become the reference line from which longitude is measured, as you can see in Figure 2-7.

Lines of equal longitude are called meridians. Meridians all run from the North Pole to the South Pole. The prime meridian has a longitude of 0°. Longitude increases to the east and west to a maximum of 180°, a line that runs down the middle of the Pacific Ocean. Meridians are not parallel because they meet at the North and South Poles. As shown in Figure 2-8, the Eastern and Western Hemispheres are the two halves of Earth bounded by the prime meridian and the north-south line of 180° longitude.

Unfortunately, some people think of latitude and longitude as only lines. For example, they confuse latitude, the angular distance from the equator, with the lines on a map that show constant latitude. If your only purpose is to read the coordinates on a map, this is not a problem. However, if you want to understand what

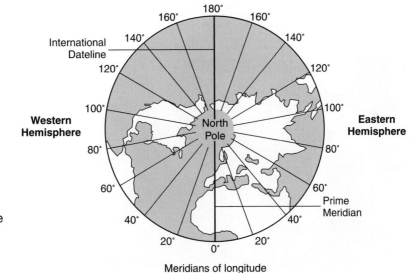

FIGURE 2-8. The prime meridian (0° longitude) and the international dateline (roughly 180°) make a circle that divides Earth into the Eastern and Western Hemispheres.

latitude and longitude are and how they are determined, you need a deeper understanding.

On a computer, Google Earth lets you "fly" anywhere on Earth to view satellite imagery, maps, terrain, 3-D buildings and even explore galaxies in the sky. Visit the following Web site to join the fun: *http://earth.google.com/*

Finding Latitude

Earlier in this chapter you read that observations of Polaris, the North Star, were used to show that Earth is a sphere. Those observations can also be used to tell how far north a person is from the equator. Earth rotates on its axis. There is no scientific reason that Earth's axis should be pointed to any particular star. In fact, the direction in which Earth's axis points moves through a 26,000-year cycle. In our lifetime the north-south axis just happens to line up with a relatively bright star called Polaris, or the North Star. (The alignment is not perfect, since Polaris is a little less than 1 degree from the projection of Earth's axis.) A navigator can also use the Sun to determine latitude. Figure 2-9 on page 44 shows a sextant being used in celestial navigation.

An observer at the North Pole sees Polaris directly overhead in the night sky. The angle from the horizon up to Polaris is therefore 90°. That observer is also located 90° north of the equator. As the

FIGURE 2-9. Precision instruments such as this mariner's sextant are used to determine latitude with remarkable accuracy.

latitude of the observer decreases, the altitude of Polaris also decreases. At the equator, Polaris is visible right on the northern horizon. Here, Polaris is 0° above the horizon, the latitude at Earth's equator is 0°. Therefore, for any observer in the Northern Hemisphere, latitude can be determined by observing the angle of Polaris above the horizon. *The altitude of Polaris equals the latitude of the observer.* Figure 2-10 illustrates how two stars in the Big Dipper can be used to find Polaris.

FIGURE 2-10. The "pointer stars" in the Big Dipper always line up with the North Star (Polaris). If you rotate your book through 90°, 180°, and 270°, you will see how the alignment of these two constellations changes over a 24-hour period. (Ignore the horizon.) The same rotation can be observed if you view these stars at the same clock time each night for one year.

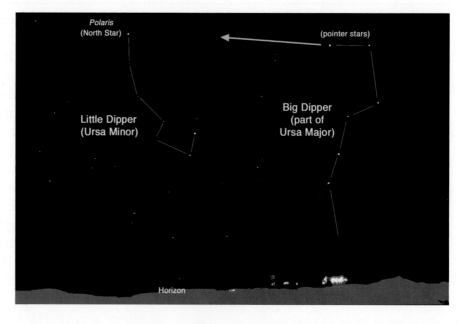

South of the equator the North Star is not visible. However, with a star map, an observer can determine the point in the night sky that is directly above the South Pole. It is near the constellation called the Southern Cross. In a procedure similar to what is done in the Northern Hemisphere, south latitude is equal to the angle of that point in the starry sky above the horizon. For people used to sighting on Polaris, it did not take long to master finding latitude in the Southern Hemisphere.

STUDENT ACTIVITY 2-4 —DETERMINING YOUR LATITUDE

1: MATHEMATICAL
ANALYSIS 1, 2, 3
6: MODELS 2
6: MAGNITUDE AND
SCALE 3

You can construct an instrument to measure your latitude using the following simple materials: a protractor, a thin string, a weight, and a sighting device such as a soda straw. This instrument is similar in principle to instruments used by mariners for hundreds of years. Figure 2-11 shows this navigation device in use.

If you use a standard protractor when you sight along the horizon, the string will fall along the 90° line. Similarly, if you look straight up, the string will line up with 0°. In these cases, you will need to subtract your angle readings from 90° to find your latitude. Your latitude is equal to the angle of the star Polaris above a level horizon. Do your results agree with the latitude given for your location on maps or other references?

FIGURE 2-11. To make a simple celestial navigation instrument, use protractor, a drinking straw, and a weight on a string.

Finding Longitude

You can determine your longitude by observing the position of the sun. If it is noon where you are, it must be midnight halfway around Earth. (A full circle is 360°, so halfway around the planet is 180°.) The sun appears to move around Earth from east to west. Therefore, when it is noon where you are, in places to your east, the local time is afternoon, and in places to your west it is still morning. Because the sun appears to move around Earth in 24 hours, each difference of 1 hour of time represents $\frac{1}{24}$ of 360°, or 15°. Each 1-hour difference in time from one location to another represents 15° of longitude.

Using the time difference of 15° per hour, you can determine the numerical value of longitude. But how can you determine whether it is east or west longitude? If local time is earlier than Greenwich Time, the observer is located in the Western Hemisphere. To observers in the Eastern Hemisphere, local time is later than Greenwich Time. To make this clearer, you can look at a globe and imagine the sun at the noon position in England. Remember that Earth spins toward the east. On your globe, most of Europe and Asia, at eastern longitudes, are in the afternoon or evening. This part of Earth east of England is turning away from the sun. At the same time, to the west of England, it is still morning. These places are rotating toward the sun.

Of course, this is based on solar time. Solar noon is the time the sun reaches its highest point in the sky. Clock time may be different from solar time by half an hour—even more if daylight saving time is in effect. If people were to set their clocks to the apparent motion of the sun across the sky in their location, clock time would be different from one place to another. This was the situation before time was standardized. In those days, towns had a clock that chimed, so the citizens would know the local time. At that time, watches were difficult to make and too expensive for most people to own. Only places on a north-south line (at the same longitude) would have exactly the same clock time.

If we still used this system and you wanted to meet someone in another town at a particular time, you could not use a clock set to the time in your town because you would probably show up early or late. Radio and television programs would not necessarily begin on the hour or half-hour. To standardize time, most of the United States is divided into four time zones: Eastern,

Central, Mountain, and Pacific Time. In each time zone, all clocks are set to the same time.

STUDENT ACTIVITY 2-5 —FINDING SOLAR NOON

1: MATHEMATICAL
ANALYSIS 1, 2, 3
1: SCIENTIFIC
INQUIRY 3

It is quite easy to measure local time by observing the sun. To determine the time of solar noon, you will need to be at a location where a tall, vertical object, such as a flagpole or the high corner of a tall building, casts a shadow onto a level surface. Throughout the middle of the day, mark the exact position of the point of the shadow, and label the positions with the accurate clock time. (To avoid making permanent marks, use a substance such as chalk that will wash away in the rain.) Call these marks the time points. Connecting the points will form a curved line north of the object casting the shadow.

The next step is to find where the curved shadow line comes closest to the base of the shadow object. (You will probably need to use a long metal tape measure to measure the distance.) Mark this point "Solar Noon." Finding the clock time of solar noon will probably require you to estimate between the marked time points to establish the precise time of "Solar Noon."

Figure 2-12 shows five positions of the sun as it travels from sunrise in the east to sunset in the west, and the solar time at each

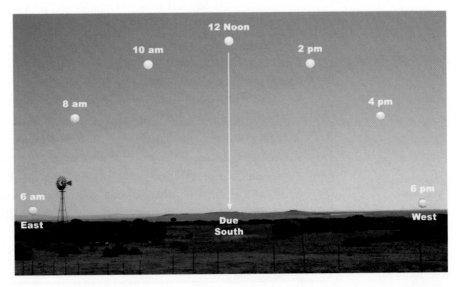

FIGURE 2-12.
Solar time in spring and autumn. Solar noon in most places does not occur exactly at clock noon.

position on the first day of spring or autumn. Solar noon occurs when the sun reaches its highest point in the sky at your location. For an observer in New York State at solar noon, the sun is highest in the sky (but not directly overhead) and due (exactly) south. An added benefit of this procedure is that it allows you to draw a line that runs exactly north to south. The line from the vertical base of the shadow object to the most distant point of the noon shadow could actually be extended all the way to the North Pole.

Modern clocks have become so precise they can measure small changes in solar noon at Greenwich throughout the year. Therefore, **Greenwich Mean Time (GMT)**, which evens out these small annual changes, is used as the basis of standard time throughout the world.

In practice, measuring longitude was not easy for the early mariners. A navigator at sea needed to know the precise time noon occurred back in Greenwich. The ships carried the most accurate clocks available at that time. However, after a long sea voyage, changes in temperature and the rocking motion of the ship caused these mechanical clocks to become inaccurate. It was easy to observe local time by observing when the shadow of a vertical object pointed exactly north. Yet, comparing local time with the time back in England depended on those mechanical clocks. Until very accurate clocks could be manufactured, measurements of longitude were poor, and maps generally showed large errors in the east-west direction.

STUDENT ACTIVITY 2-6 —DETERMINING YOUR LONGITUDE

1: MATHEMATICAL ANALYSIS 1, 2, 3

Your longitude is proportional to the time difference between local solar time and Greenwich Mean Time (GMT). This difference needs to be calculated in hours and hundredths of an hour, not hours and minutes. If you performed the solar noon activity earlier in this chapter, you can use your data to determine your longitude. In that activity, you determined the difference between clock time and solar time. For example, if you determined that solar noon occurred at 11:55 A.M., the difference between clock time and solar time is −5 minutes. If solar noon occurred at 12:09 P.M., the difference between clock time and solar time is +9 minutes. (Convert your solar time from hours and minutes to hours and hundredths of an hour by dividing the minutes by 60.)

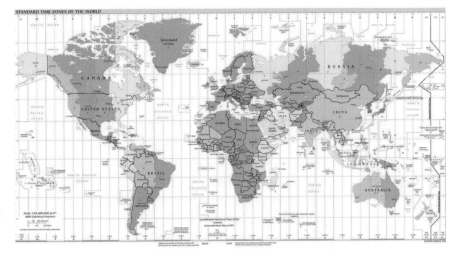

FIGURE 2-13.
The world has been divided into time zones based upon Earth's rate of rotation of 15° per hour.

Earth has been divided roughly into 24 hourly time zones, as shown in Figure 2-13. Generally, for each 15° change in longitude, clock time changes by 1 hour. Thus, when it is noon on one side of the world, on the opposite side of the world the local time is midnight (180° ÷ 15°/h = 12 h).

For locations in New York State, you can find Greenwich Mean Time by adding 5 hours to your clock time. For example, if it is 1:15 P.M. clock time, Greenwich Mean Time is 6:15 P.M. (*Note:* If it is daylight saving time in New York, you would add only 4 hours.)

Once you have calculated your longitude experimentally, you can check your results with a map of New York State, such as the *Generalized Bedrock Geology of New York State* found on page 426 or in the *Earth Science Reference Tables,* that shows local latitude and longitude. (*Note:* Mean Time and solar can differ by as much as 15 minutes. This could cause an error of as much as 4° of longitude. The following Web site explains this issue, known as the Equation-of-Time: *http://www.analemma.com/Pages/framesPage.html*)

Thanks to modern technology, finding the angle north or south of the equator (latitude) and the angle east or west of the prime meridian (longitude) has become simple and accurate. There are very accurate clocks that use the vibrations of quartz crystals to measure time. In addition, radio and telephone communications provide Greenwich Mean Time to great accuracy. Even better is the use of Global Positioning System (GPS) devices that analyze signals from orbiting satellites, allowing people to find latitude and longitude. Using a GPS device allows people to pinpoint their position to within a few meters. These devices are now small and inexpensive enough to be used by hikers and sportsmen. On land or at sea, it has become remarkably easy to find your place on the planet's terrestrial coordinates.

STUDENT ACTIVITY 2-7 —READING LATITUDE AND LONGITUDE ON MAPS

2: INFORMATION
SYSTEMS 1
6: MODELS 2

The *Earth Science Reference Tables* contain three maps (one New York map and two world maps) that can be used to read latitude and longitude. However, these world maps do not show cultural features, such as cities and political boundaries. To complete this activity you will need to use an atlas or a world map.

Your teacher may ask you to make a small "X" at each world location on a paper copy of the Tectonic Plates world map from the *Reference Tables*.

What cities are located at the following coordinates? (Please use a sheet of notebook paper. You should not mark in this book or on any reference materials.)

(1) 34°N, 118°W   (3) 41°N, 74°W
(2) 38°S, 145°E   (4) 0°, 79°W

List the latitude and longitude coordinates of the following places on a world map. Please estimate values to the nearest degree of latitude and longitude.

(5) Albany, NY   (7) São Paulo, Brazil
(6) London, England   (8) Jakarta, Indonesia

Use the *Earth Science Reference Tables* to find the terrestrial coordinates of each of these places in New York State. Round your answers to the nearest whole degree of latitude and longitude.

(9) Utica   (11) Mt. Marcy
(10) Riverhead   (12) Binghamton

(13) What city in New York State is about half way from the equator to the North Pole?
(14) There are two locations on Earth where the clock time is undefined. At these places, a clock can be set to any convenient time. Where are they?

CHAPTER REVIEW QUESTIONS

Part A

1. In Earth's atmosphere, which temperature zone contains the most water vapor?

   (1) mesosphere   (3) thermosphere
   (2) stratosphere   (4) troposphere

2. The diagram below represents Earth's rotation as observed from a position directly above the North Pole. What is the approximate time at point X on Earth's surface?

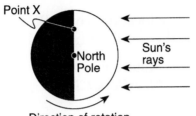

   (1) 6 A.M.   (2) 12 noon   (3) 6 P.M.   (4) 12 midnight

3. What is the approximate latitude of point X on the world map above?

   (1) 0°   (2) 45°N   (3) 90°N   (4) 180°

4. What is the most likely location of the observer shown below?

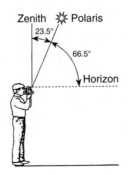

   (1) the North Pole
   (2) Northern Canada
   (3) Central New York State
   (4) Southern Florida

5. What two elements make up the largest percentage of Earth's crust?

   (1) iron and silicon
   (2) iron and oxygen
   (3) oxygen and silicon
   (4) aluminum and iron

6. What are the approximate terrestrial coordinates of Buffalo, NY?

   (1) 79°N, 43°W   (3) 79°N, 43°E
   (2) 43°N, 79°W   (4) 43°S, 79°E

7. From which New York State location would Polaris be observed to have an altitude closest to 43° above the northern horizon?

   (1) Binghamton   (3) Watertown
   (2) Utica   (4) New York City

8. Why do observers in New York State looking due south at the night sky see a different group of constellations from what they had seen six months earlier?

   (1) Constellations revolve around Earth.
   (2) Constellations revolve around the sun.
   (3) The sun revolves around the center of our galaxy.
   (4) Earth revolves around the sun.

9. The base of a cloud is located at an altitude of 2 km. The top of the same cloud is at an elevation of 8 km. In what part of the atmosphere is the cloud?

   (1) troposphere, only   (3) troposphere and stratosphere
   (2) stratosphere, only   (4) stratosphere and mesosphere

10. When the time of day for a certain ship at sea is 12 noon, the time of day at the prime meridian (0° longitude) is 5 P.M. What is the ship's longitude?

   (1) 45°W   (3) 75°W
   (2) 45°E   (4) 75°E

11. Approximately what percentage of Earth's surface is covered by water?

   (1) 100%   (3) 50%
   (2) 75%   (4) 25%

12. The arrow on the map below shows a ship's route from Long Island to Florida. As the ship travels south, the star Polaris appears lower in the sky each night.

The best explanation for this observation is that Polaris

(1) rises and sets at different locations each day
(2) has an elliptical orbit around Earth
(3) is located directly over Earth's equator
(4) is located directly over Earth's North Pole

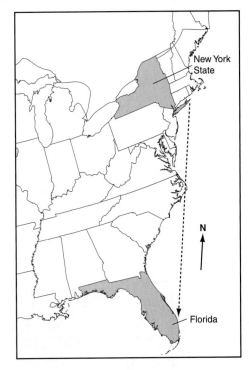

Part B

Base your answers to questions 13 and 14 on the map of Australia below.

Map of Australia

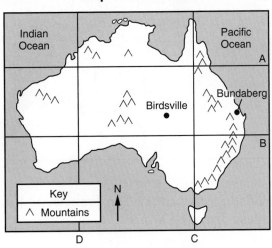

13. On the map, which straight line best represents 30° S latitude?

    (1) A   (3) C
    (2) B   (4) D

14. The map shows two cities in Australia. Explain why Bundaberg will experience solar noon before Birdsville each day.

Part C

Base your answers to questions 15 through 17 on the information and United States time zone map below. The map shows the four hourly time zones of the United States, except Alaska and Hawaii. Each time zone occupies roughly 15° of longitude. The associated lines of longitude for each time zone are shown here by dashed lines. For most locations within each, the clocks are set to the same time.

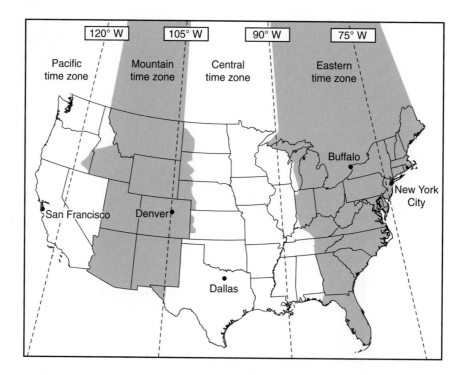

15. What is the approximate longitude of San Francisco? (Include the number, the units, and the compass direction.)

16. When it is 1:00 A.M. in New York City, what time is it in Denver?

17. Explain, in terms of Earth's rotation, why the time zones are 15° of longitude apart.

Base your answers to questions 18 and 19 on the world map below.

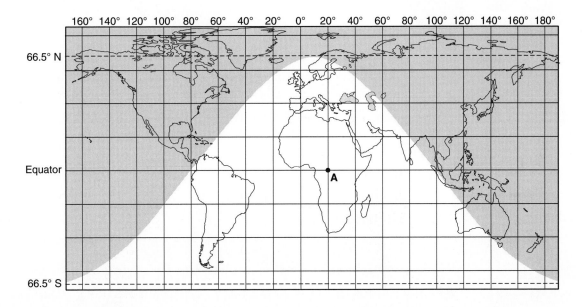

18. What are the latitude and longitude of the letter *A*?

19. At what rate (degrees per hour) is Earth rotating for a person located at *A*?

20. List the four layers of Earth's atmosphere in order from least dense to most dense.

3 Models and Maps

WORDS TO KNOW

azimuth	gradient	model	topographic map
contour line	isoline	profile	topography
field	isotherm		

This chapter will help you answer the following questions:

① What kinds of maps do we use to represent Earth's surface?

② How do maps show field quantities such as surface elevations?

③ How can maps be used to make profiles?

WHAT IS A MODEL?

6: MODELS 2
6: MAGNITUDE AND
SCALE 3

A **model** is anything used to represent something else. A photograph helps you remember people when they are not near. A photograph is an example of a physical model. There are also mathematical models. The formula for density you used in Chapter 1 is a mathematical model. This formula represents the relationship between mass and volume for any object made of a uniform substance. Memories and dreams are models of real or imagined events. In Chapter 2, you may have developed a mental model of the size and shape of planet Earth. Such models can be useful as you try to understand this planet.

Many physical models are scale models. They show the shape of the real object, but they are smaller or larger to make them easier to use. Toys and photographs are usually smaller than the real thing. In fact, a diagram showing the parts of an atom would need to be much larger than the original, perhaps by a factor of 1,000,000,000 (1.0×10^9, 1 billion).

A ratio is a convenient way to express scale. We often show a ratio as two numbers separated by a colon. A common scale used to make a small toy automobile is 1:64. This is read as "one to sixty-four." One centimeter (cm) on the toy car represents 64 cm on the real car. The model atom mentioned earlier would be at a scale of 1,000,000,000:1.

STUDENT ACTIVITY 3-1 —MODELS IN DAILY LIFE

6: MODELS 2
6: MAGNITUDE AND SCALE 3

Make a list of three models people use in their daily lives. Include physical models, mathematical models, and mental models. Organize your list into four columns: (1) Model (name each); (2) What It Represents; (3) Why It Is Used Instead of the Real Object (and if it is a scale model); and (4) Approximate Scale. Avoid different examples of the same kind of model.

Maps

Some maps are models of the whole Earth; other maps model just a part of its surface. Road maps help people to drive from one place to another. Political maps show the geographic limits of laws and government services. It is hard to imagine traveling without using maps.

Because Earth is a sphere, its surface is curved. The only kind of map that shows Earth's surface without distortion must also have a curved surface. Therefore, a globe is the most accurate model of Earth. Directions and distances are shown without distortion. However, flat maps are easier to carry. You can fold them for storage and open them on a flat surface. For small regions, the distortion of a flat map is not significant. Visit the following Web site to help you understand global maps and map distortion: *http://geology.isu.edu/geostac/Field_Exercise/topomaps/distortion.htm*

When world maps are transferred to flat surfaces, they may show increasing distortion in areas far from the equator. Compromises must be made depending how the map will be used. Compare the Ocean Currents maps (Figure 3-1) with the Tectonic Plates map (Figure 3-2). Figure 3-1 shows the Scandinavian Peninsula in northern Europe as much smaller than Australia. Figure 3-2 on page 59, the Tectonic Plates map, incorrectly shows these areas to be about the same size. Although the Ocean Currents map shows less size distortion, it distorts directions. Notice how North America seems to be slanted on the Ocean Currents map. Both maps can be used to locate places by latitude and longitude.

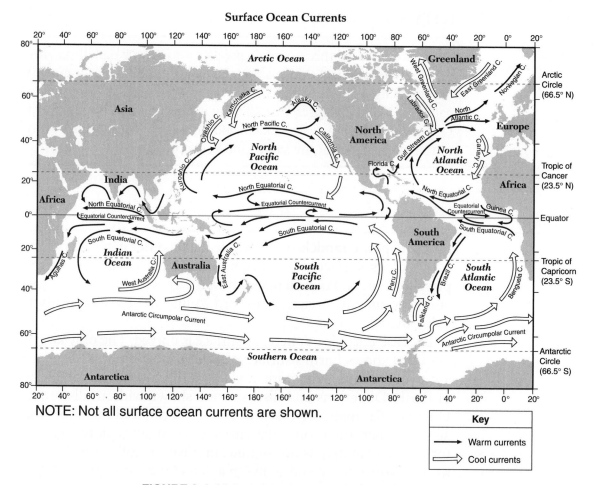

Surface Ocean Currents

NOTE: Not all surface ocean currents are shown.

Key	
→	Warm currents
⇒	Cool currents

FIGURE 3-1. Major ocean currents of the world.

Tectonic Plates

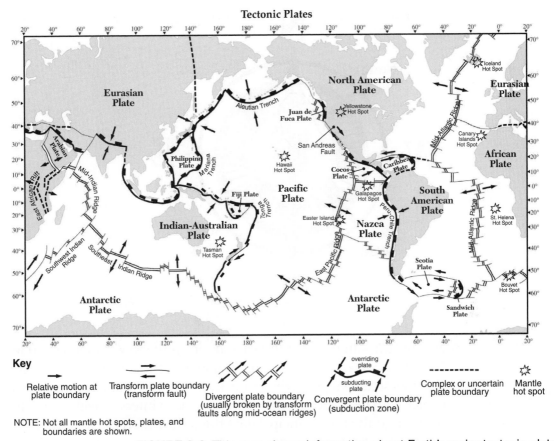

Key

Relative motion at plate boundary	Transform plate boundary (transform fault)	Divergent plate boundary (usually broken by transform faults along mid-ocean ridges)	Convergent plate boundary (subduction zone)	Complex or uncertain plate boundary	Mantle hot spot

overriding plate

subducting plate

NOTE: Not all mantle hot spots, plates, and boundaries are shown.

FIGURE 3-2. This map shows information about Earth's major tectonic plates. It can also be used to find latitude and longitude.

STUDENT ACTIVITY 3-2 —A MAP TO YOUR HOME

6: MODELS 2
6: MAGNITUDE AND SCALE 3

Use a sheet of $8\frac{1}{2} \times 11$ inch paper to draw a map that shows the most direct route from school to your home. You may use a computer-drawing program if it is available. Show the landmarks (buildings and natural features) that would be most useful in guiding a person who is unfamiliar with your community. Also include a scale of distance.

What Are Compass Directions?

Directions are a part of your daily life. Many people can give you directions, but Earth itself can also help you determine directions. In

FIGURE 3-3.

general, the sun rises in the east and sets in the west. For a person in New York State, the midday sun is always in the southern part of the sky and shadows generally point north. If you want to travel to a cooler climate, head north. To avoid cold winters, travel south. In addition to the four principal directions shown on Figure 3-3, there are intermediate directions such as northeast and southwest. Compass direction can also be specified as an angle known as azimuth. **Azimuth** starts at 0°, which is directly north, and proceeds to east (90°), south (180°), west, (270°), and back to north at 360°.

Most maps are printed with north toward the top of the map. This convention helps align the map with the area it represents. However, there may be reasons to align the map in a different way. Therefore, this rule is not always followed.

Earth as a Magnet

You may have used a magnet to pick up metallic objects, such as paper clips. In ancient times, people noticed that a large piece of the iron ore magnetite, sometimes called lodestone, attracts other iron objects. They called this attractive force magnetism. Furthermore, they noticed that a piece of lodestone floating on water tends to align itself in a consistent geographic direction. (Because this kind of rock is heavier than water, it must be placed in a flotation device.) The earliest magnetic compasses were made in this way long before anyone understood magnetism.

STUDENT ACTIVITY 3-3 —MAKING A WATER COMPASS

1: ENGINEERING
DESIGN 1
6: MODELS 2

You can construct a water compass by floating a strong bar magnet in a large container of water. Place the bar magnet in a shallow bowl that does not touch the sides of the large container. Be sure to try this activity well away from any magnetic metals or electric motors. Does your magnet tend to align consistently toward magnetic north as does the example in Figure 3-4?

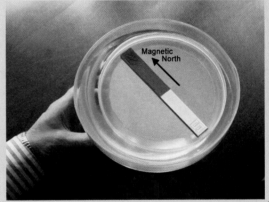

FIGURE 3-4. A simple compass can be made with a container of water and a bar magnet. Suspend the magnet on a floating object such as a lid.

A compass needle points north because Earth acts as a giant magnet. You may know that opposite magnetic poles attract each other. Then why does the north pole of a magnet tend to point toward Earth's North Pole? The reason is simple. The pole of a magnet that points north is labeled N because it is the "north-seeking pole." That end actually has a magnetic field like Earth's South Pole.

Why does Earth have a magnetic field? That question took a long time to answer. Scientists now think that Earth's outer core is made of molten liquid that circulates due to heat flow. The planet is very dense, so molten iron is a strong possibility for the major substance in the outer core. Scientists think it is this circulation guided in part by Earth's rotation (spin) that keeps the currents running generally north and south.

There are two other important things to know about Earth's magnetic field. First, the field seems to reverse itself about every 30,000 years. This is more evidence to support the outer core circulation theory. Second, the magnetic poles do not align exactly with the geographic (spin) poles. In fact, the magnetic poles wander through the Arctic and Antarctic regions.

Geographic north is determined by Earth's spin axis. There are several ways to find it. Magnetic north, which is a result of Earth's magnetic field, can be determined with a compass. Magnetic declination is the angle between geographic north and magnetic north. This angle can be measured with a protractor.

WHAT ARE FIELDS?

1: MATHEMATICAL ANALYSIS 3
6: SYSTEMS THINKING 1
6: MODELS 2
6: MAGNITUDE AND SCALE 3

A **field** is a region of space in which the same quantity can be measured everywhere. Let us look at a simple example. Have you ever noticed that when dinner is being prepared, the smell of the food drifts through the house? As you walk into the kitchen area, the smell increases. In this example, the house is the field, and smell is the field quantity that is observed.

In the previous section, you read about Earth's magnetic field. All around Earth, scientists can measure the direction and strength of Earth's magnetism in its magnetic field. Gravity is another field quantity that changes over Earth's surface. Sensitive instruments can measure very small changes in Earth's gravity from place to place. Temperature is also a field quantity. Wherever you go, you can measure or even feel the temperature and notice how it changes. Scientists often make maps to show how field quantities change over a geographic area.

Understanding Isolines

An **isoline** is a line on a field map that connects places having the same field quantity value. A weather map of the United States often shows the daily high temperatures. On some maps, there are **isotherms**, lines connecting places that have the same high temperature. These isolines generally cross the country from west to east. Other weather maps use isolines to show atmospheric pressure or the amount of precipitation.

STUDENT ACTIVITY 3-4 —CHARACTERISTICS OF ISOLINES

1: MATHEMATICAL
ANALYSIS 3
6: MODELS 2
6: MAGNITUDE AND
SCALE 3

Based on Figure 3-5, or a similar national isoline map, answer the following questions about isolines:

1. Do isolines ever touch or cross each other?
2. Do isolines usually have sharp angles or gentle curves? (Pick one.)
3. What does each point on an isoline have in common with all other points on the same line?
4. Do isolines ever end, except at the edge of the data area? (Note that the isolines in Figure 3-5 end at the edge of the continent.)
5. On a single map, is the change in value from one isoline to the next always the same?
6. Do isolines tend to run parallel as they extend around the map?
7. Does every isoline have one side where the values are higher and another side where the values are lower?

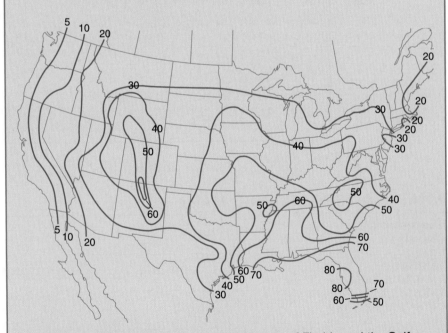

FIGURE 3-5. This isoline map shows that parts of Florida and the Gulf Coast have the largest number of thunderstorms.

To view animations that show you how to draw and check isolines, visit the following Web site: *https://courseware.e-education.psu. edu/public/meteo/meteo101demo/Examples/Section2p03.html* (Read through the text, and then click on the first green link within the text.)

Every isoline map is drawn at a specific interval. Many newspaper weather maps show temperatures at an interval of 10°F. That means when you move from one isotherm to the next isotherm, there is a change of 10°F. A smaller interval, such as 1°F, would make the map cluttered and hard to read. A larger interval would result in a map that does not give enough information. For every isoline map, an appropriate interval must be chosen to help viewers understand the data.

STUDENT ACTIVITY 3-5 —A TEMPERATURE FIELD

6: MODELS 2
6: MAGNITUDE AND
SCALE 3

In science, you sometimes collect and analyze your own data. Gathering accurate field data can be difficult, but it is an excellent way to learn about science. For example, each student in the class can use a thermometer to read and record the air temperature at the same time throughout the classroom. Individuals or lab groups can write these numbers at the appropriate location on a floor plan of the classroom and then use the numbers to draw an isotherm map of the science room. Take simultaneous temperature readings at different levels in the classroom (floor, desk level, and 1 meter above the desks). This can be a very challenging activity, but it is a good way to learn about isolines.

WHAT IS A TOPOGRAPHIC MAP?

1: MATHEMATICAL
ANALYSIS 1
4: 2.1q
6: MODELS 2
6: MAGNITUDE AND
SCALE 3
7: STRATEGIES 2

A **topographic map** is a special type of field map on which the isolines are called contour lines. These maps show a three-dimensional surface on a two-dimensional page. **Contour lines** connect places that have the same elevation (height above or below sea level). Each contour line is separated from the next by the same change in elevation, called the contour interval. A common contour interval on The United States Geological Survey (USGS) topographic maps is 10 ft, which is about 3 m. As you move your finger on a map up or down a slope, each time it crosses a contour line, the elevation has changed by the amount equal to the map's contour interval. Where the contour lines are close together, the slope of the land is steep. Where there are wide spaces between contour lines, the land is more flat.

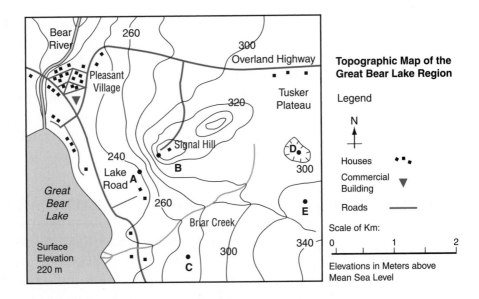

FIGURE 3-6. A topographic (contour) map of Great Bear Lake.

It may help you visualize the area represented on a topographic map if you think of each contour level as a layer on a cake. Multiple levels stack to make hills. Figure 3-6 is a topographic map of an imaginary location. Figure 3-7 is photograph of a step model constructed layer-cake style from that map. (See Activity 3-7.) A real landscape would have a more rounded shape.

Hikers and sportsmen often use topographic maps because they show the shape of the land. These maps also show where roads,

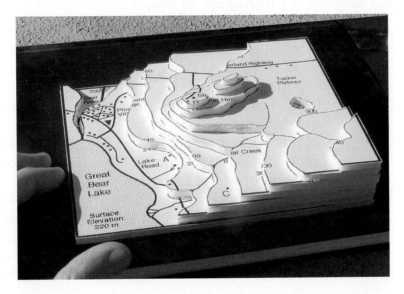

FIGURE 3-7. You can make a three-dimensional model of any topographic map or land surface using sheets of foam available at art supply stores or from supermarket food trays.

hills, lakes, and streams are, as well as a wide variety of other land-marks. Topographic maps can be used to plan the best route from one location to another. The United States Geological Survey (USGS) publishes topographic maps at various scales. Camping, sporting goods, and hardware stores often carry the local topo-graphic maps or they can be ordered from the USGS. Visit the fol-lowing Web sites to view local maps and satellite images by zip code: *http://earth.google.com*

STUDENT ACTIVITY 3-6 —MAKING A TOPOGRAPHIC MODEL

6: MODELS 2
6: MAGNITUDE AND
SCALE 3

From a simple topographic contour map selected or approved by your teacher, construct a step model of the map area similar to the model in Figure 3-7. You may select your school grounds, a por-tion of your local USGS topographic map, or the topographic map or image of a nearby landform from which to make your model. The layers can be made of corrugated cardboard, foam board from an art supply store, or the bottoms of clean foam meat/food trays. Visit the following Web site to practice making 3-D topographic models of New York's gorgeous gorges: *http://fli.hws.edu/myplace/Directions %20to%20Make%20Modelhtm.htm#Objective*

Common Features on Topographic Maps

Topographic maps use contour lines to show the **topography**, or shape of the land. In their steeper sections, small streams often cut gullies by erosion. These distinctive features show up as a sharp bend in the contour lines, which creates a V-shape that points up-stream. Figure 3-8 is a portion of the San Isabel National Forest topographic map.

Look back at Figure 3-6 on page 65. You can see that Briar Creek flows down to the west because the contour lines make V's that point upstream to the east.

Reading the elevation of a point on a contour line is easy be-cause it is the elevation of that contour line. If you want to know the elevation of a point between contour lines, you estimate the elevation with reference to nearby contour lines. For example, point *C* on Figure 3-6 is about halfway between the 260- and

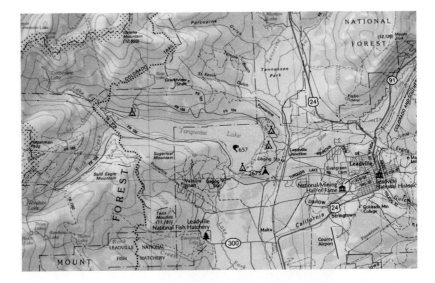

FIGURE 3-8. A topographic map of Turquoise Lake.

280-m lines. You can therefore estimate the elevation of point *C* to be 270 m.

When contour lines form an enclosed shape that is something like an irregular circle, the center of the enclosed area is usually a hill. The top of the hill is within the smallest circle. By reading the elevation of the highest contour level, you can estimate the height of the hill. A useful convention is to add half of the contour interval to the elevation of the highest contour. On Figure 3-6 the highest contour line around the house on Signal Hill is 340 m. The contour interval on this map is 20 m. So your best estimate for the elevation of the house is 350 m.

Sometimes, contour lines enclose a dip in the land surface, or a closed depression. To distinguish depressions from hills, mapmakers use contour lines with small bars that point down toward the center of the depression. Point D on Figure 3-6 is within the 300-m contour, so its elevation at the bottom of the depression is about 290 m. The first depression contour line always has the same elevation as the lower of the two adjacent contour lines.

Most maps have a legend, or key, printed outside the map area. This explains the meanings of various symbols shown on the map. The more complex the map, the greater the variety of symbols that must be explained in the legend.

As with other physical models, every topographic map has a map scale. The scale can be expressed as a ratio, such as 1:24,000.

FIGURE 3-9. To determine the length of Turquoise Lake, line up the edge of the paper with the lake and clearly mark both ends on the paper.

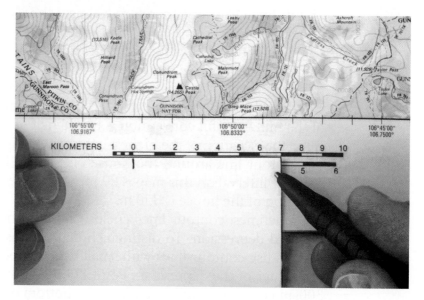

FIGURE 3-10. Holding the marked paper along the map's scale of kilometers shows that Turquoise Lake is about 7 km long.

The scale can be a translation of distances, such as 1 inch represents 2000 ft, or 1 centimeter represents 1 km. But, you will probably find the most useful scale is a line located outside the map area that is marked with divisions of distance. Figures 3-9 and 3-10 illustrate three steps in determining the distance between two map locations or the size of a geographic feature.

Step 1. To determine the length of Turquoise Lake, you will need to locate it on a topographic map.

Step 2. Hold the edge of a blank sheet of paper along the longest part of Turquoise Lake. Place the corner of the paper at one end of the lake. Make a mark along the edge of the paper where it touches the other end of the lake.

Step 3. Move the marked paper to the distance scale on the map. This indicates that Turquoise Lake is about 7 km long.

Figure 3-11 is a topographic map of a location in New York State. Bigsby Pond is located on the Minerva, New York, $7\frac{1}{2}$-minute USGS topographic map. You should be able to read the following

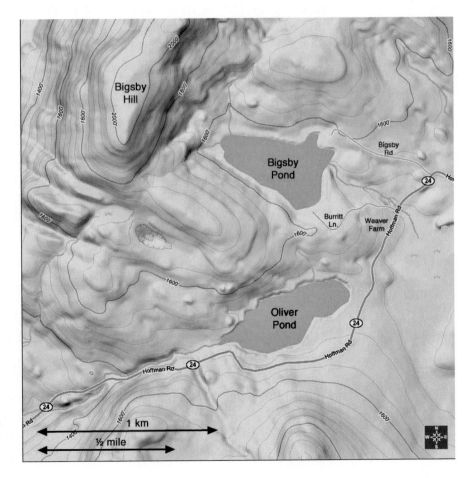

FIGURE 3-11.
Bigsby Pond map

information on this map. (1) The compass figure at the bottom right tells you that north is at the top of this map. (2) The map scales show that 1 km is a little longer than $\frac{1}{2}$ mi. (3) The scale of miles shows that both ponds are about $\frac{1}{2}$ mi long. (4) The flattest part of the map is on the eastern side, which is a swamp or bog. (5) The steepest slopes are along the eastern and western sides of Bigsby Hill and the smaller hill north of Oliver Pond. This is where the contour lines are closest together. (6) There is a closed depression (a hollow bowl shape) just to the left of the center of the map area. Note the special marks on these interior contour lines. (7) Three roads are shown on this map, Hoffman Road and two roads that lead to Bigsby Pond. (8) The contour interval is 40 feet. (9) The highest points are within small circles such as the top of Bigsby Hill and several nearby lower hills. Bigsby Hill is more than 2040 ft above sea level, but less than 2080 ft above sea level. (10) The lowest point is along the Oliver Pond outlet near the southwest corner of the map, less than 1280 ft above sea level.

This is a combination of a topographic contour map and a shaded relief map. The shading makes hills and valleys easier to see. Sometimes you can tell the direction in which a stream flows by noting where the stream flows down off the map area. But none of the three streams that meet Bigsby Pond flow to the edge of the map. However, by seeing that the contour lines make V's pointing upstream, it is clear that the stream that goes under Hoffman Road flows out of the pond to a lower elevation.

STUDENT ACTIVITY 3-7 —READING YOUR LOCAL TOPOGRAPHIC MAP

2: INFORMATION
SYSTEMS 1
6: SYSTEMS
THINKING 1
6: MODELS 2
6: MAGNITUDE AND
SCALE 3

Download free PDF images of USGS high-resolution topographic maps. In the MAPS section of the URL below, click on "Download digital scans of topo maps," then click on "Map locator and Downloader." *http://www.usgs.gov/pubprod*

Look at a copy of the United States Geological Survey topographic map that includes your school, town, or city. Do not make any marks on this map unless your teacher tells you to do so. Notice the locations of familiar features such as roads, buildings, and streams. Notice how hills and valleys are represented by contour lines. Find a symbol for your house or a friend's home, or identify its location.

1. What are the latitude and longitude coordinates of the bottom right corner of this map?
2. When was this map published? Has it been revised?
3. How many centimeters on the map represent 1 km of true distance?
4. What is the local magnetic declination?
5. What is the contour interval of this map?
6. Locate a place where contour lines become V-shaped as they cross a stream. Do the V's point upstream or downstream? In your notebook, make a sketch of the stream.
7. What and where are the highest and lowest elevations on this map?
8. State one or more things you learned about your area by looking at the map.
9. List several ways a person might find this map useful.

Gradient

If a hill changes quickly in elevation, it has a steep gradient. In fact, slope is often used as a synonym for gradient. At every location, a field value has a measurable gradient. If the field value is not changing in a particular area, we say that the gradient is zero. We can therefore define **gradient** as the change in field value per unit distance. (The field value on a topographic map is land elevation.)

You can tell where gradients are the steepest by looking at a field map. The places where the isolines are closest are the places with the steepest gradient. On Figure 3-6, the gradient is steep between points *A* and *B* where the contour lines run close together.

The following formula from the *Earth Science Reference Tables* can be used to calculate gradient:

$$\text{Gradient} = \frac{\text{change in field value}}{\text{distance}}$$

SAMPLE PROBLEMS

Problem 1 The temperature at the center of a town is 20°C, but 10 km west at the river it is only 15°C. What is the temperature gradient between these two locations?

Solution

$$\text{Gradient} = \frac{\text{change in field value}}{\text{distance}}$$

$$= \frac{20°C - 15°C}{10\,km}$$

$$= \frac{5°C}{10\,km}$$

$$= 0.5°C/km$$

Problem 2 Calculate the gradient from point C to point E on Figure 3-6

Solution   Point C, half way between the 260- and 280-m contour lines, must be at about 270 m. Similarly, point E is at about 330 m. The distance between them can be determined using the scale of kilometers in the map legend as shown in Figures 3-9 and 3-10. That distance is 2 km.

$$\text{Gradient} = \frac{\text{change in field value}}{\text{distance}}$$

$$= \frac{330\,m - 270\,m}{2\,km}$$

$$= \frac{60\,m}{2\,km}$$

$$= 30\,m/km$$

Note the following features of both solutions.

- Each solution is started by writing the appropriate formula.
- Values including units of measure are substituted into the formula.
- Each step to the solution is clearly shown.
- The units of measure are part of the solution.

Practice Problem 1
The temperature at the floor of a classroom is 21°C. At the ceiling, 3 m above the floor, the temperature is 27°C. What is the temperature gradient between the floor and the ceiling?

Practice Problem 2
The source of a stream is at an elevation of 1000 m. The stream enters a lake at 875 m, 15 km from its source. What is the gradient of the stream?

Making Topographic Profiles

A **profile** is a cross section, or a cutaway view. If you stand between a bright light and a wall, your body will cast a shadow on the wall. When you stand with your shoulder perpendicular to the wall, your shadow will include a profile of your face with features such as your chin and nose indicated clearly. As illustrated in Figure 3-12, a topographic profile shows the elevation of the land surface along a particular route. Along the profile route, the hills show as high places and the valleys as low places. Visit the following Web site to view QuickTime VR movies about topography and topographic profiles: *http://www.scieds.com/saguaro/ topo_movies.html*

A topographic map can be used to draw a profile along any straight-line route. You will need a sheet of paper that is marked with parallel lines, such as writing paper, and a blank strip of paper

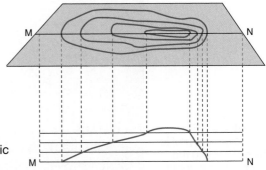

FIGURE 3-12. A profile constructed from a topographic contour map.

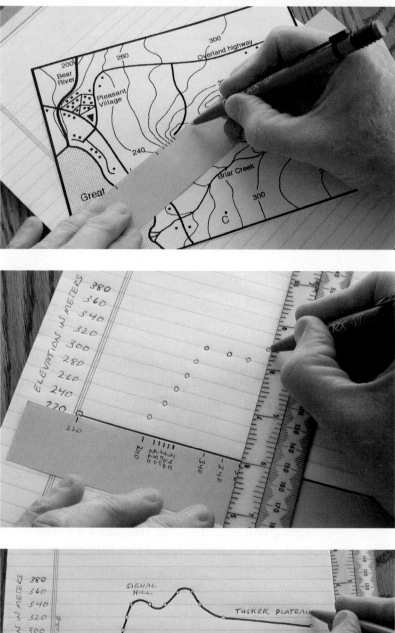

FIGURE 3-13. The first step in making a profile from a contour map is marking the blank strip where it crosses each contour line.

FIGURE 3-14. After labeling each mark with the elevation of the contour line it represents, the next step is to draw a dot at the correct elevation on the lined paper.

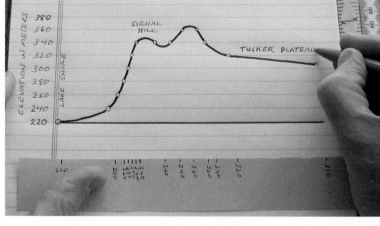

FIGURE 3-15. Complete the profile by connecting the dots with a smooth line.

a little longer than the profile route on the map. By following the steps below, you can draw a profile from a contour map.

Step 1. Place an edge of the blank strip along the profile route. Each time the edge of the blank strip crosses a contour line, make a mark at that point along the edge of the strip, as shown in Figure 3-13.

Step 2. Label each mark on the strip with the elevation of the contour line crossed. Take care with this step, because some marks will have the same elevation as others that cross the same contour line or cross another line at the same elevation.

Step 3. Along the left side of the lined paper, label the horizontal lines with the elevation of the contour lines crossed by the blank strip, as shown in Figure 3-14.

Step 4. Lay the marked blank strip along the lowest labeled horizontal line on the lined paper. Directly above the marks on the strip, make dots on the lined paper at the height indicated by the marks on the strip.

Step 5. As shown in Figure 3-15, connect the dots on the lined paper with a gently curved line. Valleys and hilltops should be rounded above the highest dots.

Step 6. Labeling features from the map line can help you visualize the curved profile. Look at Figure 3-15 again.

A similar procedure can be used to make a profile from any isoline map, such as a weather map that shows local temperatures. Some people can look at a field map and visualize its profile without following this procedure. As with making mathematical estimations, visualizing a profile is a skill that can be helpful in making accurate profiles and reading any isoline map.

STUDENT ACTIVITY 3-8 —A PROFILE ON A LOCAL TOPOGRAPHIC MAP

1: MATHEMATICAL
ANALYSIS 1
6: MODELS 2
6: MAGNITUDE AND
SCALE 3

Use the steps above to construct a profile from your local USGS contour map along a line selected by your teacher.

Using Isoline Maps for Practical Purposes

People have different uses for topographic maps. Developers and construction companies use maps to plan roads and the placement of buildings. Search and rescue teams use them when they look for lost or injured people and to plan rescue efforts. You, on the other hand, might want to find the best way to get to a fishing spot. Whether you will walk or travel by car will be an important consideration. Do you mind going over hills or crossing streams? How long will it take you to get to your destination? Considerations such as these can be important, depending on how and when you will travel.

STUDENT ACTIVITY 3-9 —INTERPRETING ISOLINE MAPS

6: MODELS 2
6: MAGNITUDE AND
SCALE 3

The map in Figure 3-16 shows two field quantities. The thinner grey isolines show surface elevations, and the thicker red isolines show the concentrations of a toxic substance in the groundwater.

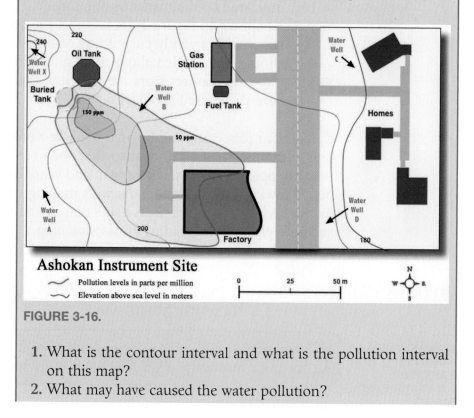

FIGURE 3-16.

1. What is the contour interval and what is the pollution interval on this map?
2. What may have caused the water pollution?

3. Which way is the contamination moving?
4. How can you tell? (There are at least two ways you can tell.)
5. Why is contamination of water well *X* unlikely?
6. In what order would you expect the remaining water wells to be contaminated?
7. If the leak occurred one month ago, at what rate has the edge of contamination (50 ppm) advanced?

STUDENT ACTIVITY 3-10 —RESCUE AND EVACUATION PLANNING

7: STRATEGIES 2

Your teacher will indicate a location on a local topographic map where a person has been injured in a fall. Figure 3-17 shows an actual rescue. Devise a detailed plan to evacuate the injured person to an ambulance waiting along a main road. In your planning, consider the cost of the evacuation and the best route to take in transporting the victim. Your plan should be written in enough detail so that the evacuation team will know how to proceed and what obstacles to avoid. Compare the plans of different groups.

FIGURE 3-17. An injured hiker in a remote area being evacuated by helicopter.

CHAPTER REVIEW QUESTIONS

Part A

Base your answers to questions 1 through 4 on the topographic map below, which shows a location in the Adirondack mountains of New York State.

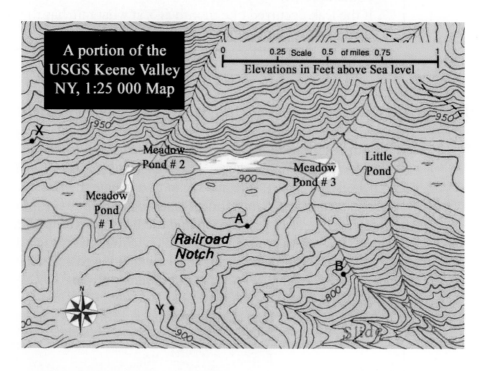

1. In what direction do the streams flowing out of Meadow Pond #1 and Meadow Pond #3 flow?

   (1) The outlet to Meadow Pond #1 flows west and the outlet to Meadow Pond #3 flows north.
   (2) The outlet to Meadow Pond #1 flows west and the outlet to Meadow Pond #3 flows south.
   (3) The outlet to Meadow Pond #1 flows east and the outlet to Meadow Pond #3 flows north.
   (4) The outlet to Meadow Pond #1 flows east and the outlet to Meadow Pond #3 flows south.

2. What is the surface elevation of Little Pond in feet?

   (1) 850   (2) 885   (3) 895   (4) 950

3. If the streams flowing out of Meadow Pond #1 and Meadow Pond #3 carry the same volume of water, which stream probably flows faster?

   (1) The outlet to Meadow Pond #1 probably flows faster.
   (2) The outlet to Meadow Pond #3 probably flows faster.
   (3) There is no way to decide if one stream flows faster than the other.
   (4) Neither Meadow Pond #1 nor Meadow Pond #3 has any stream flowing out if it.

4. Which profile below best represents the land surface from point *X* to point *Y*?

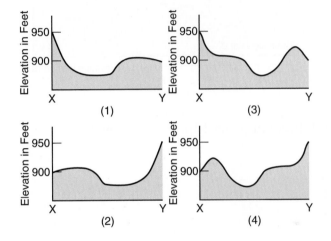

5. What is the average temperature gradient between two places in a classroom that are 10 m apart if one place has a temperature of 25°C and the other has a temperature of 23°C?

   (1) 0.2°C/m
   (2) 0.2 m/°C
   (3) 5°C/m
   (4) 5 m/°C

6. A map has a scale of 1 cm:12 km. On this map, what is the distance between two locations that are actually 30 km apart?

   (1) 2.5 cm   (3) 18 cm
   (2) 12 cm   (4) 30 cm

Base your answers to questions 7 through 9 on the topographic map below, which shows a small island in an ocean. Points *A* and *B* represent locations on the island.

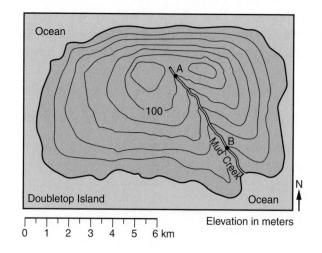

7. In what general direction does Mud Creek flow?

   (1) northeast
   (2) northwest
   (3) southeast
   (4) southwest

8. What is the greatest elevation on Doubletop Island?

   (1) 100 m   (3) 130 m
   (2) 120 m   (4) 150 m

9. What is the average gradient along Mud Creek from point *A* to point *B*?

   (1) 10 m/km
   (2) 20 m/km
   (3) 40 m/km
   (4) 80 m/km

10. Which quantity is best shown by an isoline map?

    (1) average annual rainfall at locations throughout North America
    (2) the distance from London, England, to Paris, France

(3) the current population of Binghamton, New York

(4) the position of various stars and constellations in the night sky

Part C

Base your answers to questions 11 and 12 on the Keene Valley—Meadow Ponds map from question 1 on page 78.

11. Calculate the straight-line topographic gradient between points A and B. Label your answer with the correct units.

12. How do you know that the slope is the least in the land near the four ponds?

Base your answers to questions 13 and 14 on the map below, which shows students' measurements of snow depth after a major winter storm.

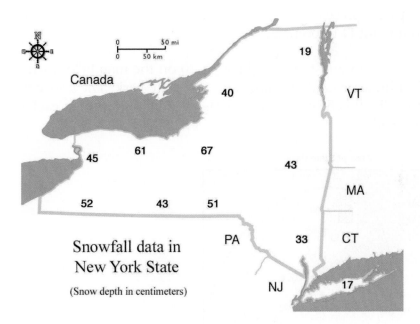

Snowfall data in New York State

(Snow depth in centimeters)

13. Use a copy of this map to draw isolines at an interval of 20 inches of snowfall. (Do not write in your book.)

14. What major city in New York State had the most snow at the end of this snow-storm?

Base your answers to questions 15 through 17 on the map below. Elevations are in meters.

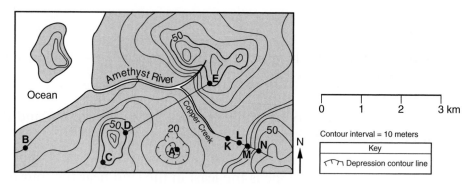

15. What is the elevation of location A?

16. Calculate the gradient between points B and C and include the correct units.

17. Use a separate piece of paper to draw a smooth, curved profile of the land surface from point D to point E.

Base your answers to questions 18 through 20 on the map below, which shows the location of an underground gasoline tank in a state park. The concentration of the gasoline shown by this field map is in parts per million (ppm) at each of the monitoring wells.

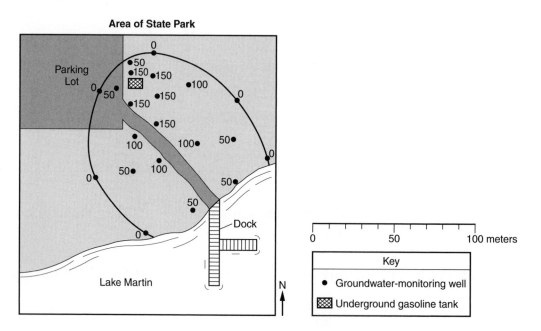

18. Use a copy of this field map to draw the 50-ppm, 100-ppm, and 150-ppm iso-lines. (Do not write in your book.)

19. State the general relationship between the distance from the gasoline tank and the concentration of contaminants in the groundwater.

20. State one action that park officials could take to prevent future groundwater contamination.

UNIT 2

Minerals, Rocks, and Resources

The natural resources we depend on are either grown or mined. The photo above shows an open-pit copper mine in Bagdad, Arizona. On average, Americans use over 100 pounds of mineral resources each day. Our homes and highways are built from Earth materials. Some of our natural resources are changed into metals and plastics. Other resources are used to provide energy for our homes and transportation. Fertilizers, mostly produced from fossil fuels, are critical to growing our food. Many medicines are made from mineral resources. There are even minerals, such as salt, that we eat.

The supplies of our natural resources are decreasing. Most mineral resources are nonrenewable. Each kilogram of natural resources we use reduces the amount left in the environment. The growing economies of the population giants China and India place even greater strain on Earth's limited resources.

The cost of a commodity such as oil is a function of supply and demand. In the 1950s, geoscientist M. King Hubbert proposed that the United States' oil supply would peak about 1970, leading to a need for more imported oil. Now, in the early part of the twenty-first century, many geoscientists think that we have reached a world peak oil production, and that future production will decrease. The slowing pace of new discoveries supports this prediction. When supply decreases while demand increases, the price of the commodity will rise dramatically. How will the decrease in oil production affect our lives? Will we face a global economic crisis? Or will we find new technologies to extract oil or otherwise meet our needs? Perhaps you will play a role in finding solutions to this growing crisis.

4 Investigating Minerals

WORDS TO KNOW

compound	element	hardness	mineral	silicate
cleavage	fracture	luster	Mohs scale	streak

This chapter will help you answer the following questions:

1 What are minerals?

2 What are the properties of minerals?

3 What are the most common minerals?

WHAT ARE MINERALS?

4: 3.1a; 4: 3.1b

If you look around the natural environment, you will probably see two kinds of things: living and nonliving. Plants and animals are parts of the living environment. The living environment is sometimes called the biosphere. The objects studied in the Earth sciences are usually nonliving things. Much of the nonliving part of the natural environment is made up of rock and soil. Just as a house is made of a variety of building materials (for example, wood, nails, concrete, and brick) so are soil and rock made primarily of minerals.

Some rocks such as granite are composed of crystals of different colored substances. (See Figure 4-1 on page 86.) The differences in the properties of the crystals identify them as different minerals.

The Most Common Minerals in Granite

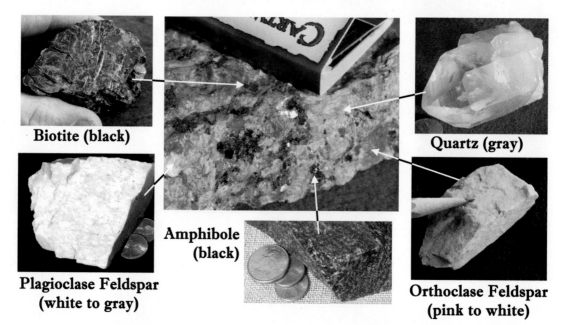

Biotite (black)

Quartz (gray)

Plagioclase Feldspar
(white to gray)

Amphibole
(black)

Orthoclase Feldspar
(pink to white)

FIGURE 4-1. Most rocks are made of a number of minerals. In granite, these five minerals are often visible.

What is a mineral? Defining minerals as the substances of which rocks are made could be acceptable in some situations, but as you will read below, there are a few exceptions to this idea. A more exact way is needed to define what a mineral is and what it is not.

Geologists have identified thousands of minerals. In fact, new minerals are discovered and named all the time. Most of the newly discovered minerals are rare and have no practical use. The wide variety of minerals makes it difficult to define exactly what a mineral is. However, geologists do have certain characteristics that identify minerals. Visit the mineral and molecule museum at the following Web site: *http://www.soils.wisc.edu/virtual_museum/displays.html*

Mineral Characteristics

Although artificial substances, such as steel or plastic, have some characteristics of minerals, they are not minerals. Synthetic diamonds, sapphires and rubies, while chemically the same as the natural gems, are not minerals either because they were not made by a natural process. The first property that characterizes all minerals is that they formed naturally.

In addition, all minerals are inorganic. This means that they were made by physical processes, and are not the result of biological activities. You have probably heard of and perhaps seen coal. Careful study shows that coal forms from layers of plant remains that were buried and compressed by the weight of more layers above. Since coal forms organically, it is not a mineral. The mineral olivine, which is very common deep within Earth, is not as common at the surface.

Gases, such as air, can be compressed. By applying enough pressure, a sample of a gas can be compressed to a very small volume. When the pressure is released, the gas expands nearly without limit. Therefore, gases do not have a definite volume. Gases and liquids take the shape of the container in which they are placed. Liquids and gases are called fluids. In general, a fluid is a substance that can flow and take the shape of its container. Solids have a definite shape and volume. A solid can be taken out of its container and it will keep its shape. You can measure the dimensions of a solid without the object being in a container. This is not true for gases and liquids.

All minerals are solids. Water pumped from the ground is not considered a mineral because it is a liquid. Petroleum, or crude oil, which is taken from deep underground, is a geologic resource but it is not considered a mineral for two reasons: first, petroleum is a liquid; second, like coal, it is of biological origin.

Solids maintain their shape and volume because of their atomic structure. Large crystals, such as those you might see in a museum or a mineral collection, usually have sharp corners and flat faces. Careful examination with powerful electron microscopes shows that these crystal shapes are a result of the way their atoms are arranged. An atom is the smallest part of an element. In crystalline substances, the atoms generally have a regular arrangement in rows and layers. The distance between the atoms in a solid changes very little with changes in temperature and pressure. Furthermore, the atoms in a solid cannot move over or around one another. Sometimes, enough atoms line up to make crystals that are visible, such as those in rock salt or granite. In rare occurrences, crystals can grow to the size seen in museums. The largest natural rock crystals can be several meters long. It is the internal arrangement of atoms that gives crystals their beautiful shapes. Visit the following Web site to learn about and see examples of Mexican selenite gypsum crystals from the "Crystal Cave of the Giants." *http://www.crystalinks.com/mexicocrystals.html*

STUDENT ACTIVITY 4-1 —SOLIDS, LIQUIDS, AND GASES

1: SCIENTIFIC
INQUIRY 1

Prepare a table with three columns. Label one column "Gas," another "Liquid," and the third "Solid." List the substances you see every day under the proper category.

Definition of a Mineral

Minerals have a definite composition, or at least a specific range of composition. Some minerals are single elements. **Elements** are the basic substances that are the building blocks of matter. Gold, as do all other elements, has just one kind of atom. The number of protons in the nucleus, or center of the atom, determines which element it is. There are 92 natural elements. Copper and sulfur are also minerals that are elements. Figure 4-2 is a photograph of copper as

FIGURE 4-2.
Native copper is a mineral that is also a chemical element.

it comes from the ground. This is called native copper. The minerals graphite and diamond are different crystal structures of the same element, carbon.

A second group of minerals with a definite composition are the chemical compounds. **Compounds** are substances that are made up of more than one kind of atom (element) combined chemically into larger units called molecules. Each molecule of a specific compound has the same number and kinds of atoms. For example, quartz (silicon dioxide, SiO_2) is a very common mineral compound in which each molecule has one atom of silicon (Si) and two atoms of oxygen (O). Every molecule in a compound is exactly the same. That is why compounds have a fixed composition, as shown in Figure 4-3.

The third group of minerals is made up of mineral families that have a variable composition. For example, the olivine family is a mixture of two chemical compounds. Olivines usually contain one compound of magnesium, silicon, and oxygen and another compound of iron, silicon, and oxygen. These compounds mix to form a single mineral family in which the properties of the mineral (color, hardness, density, etc.) are relatively consistent. These two compounds are not only similar in composition, but they are also similar in appearance and other physical properties.

The feldspars, the most common family of minerals, always contain the elements oxygen, silicon, and aluminum along with varying amounts of potassium, sodium, and calcium. All feldspar samples

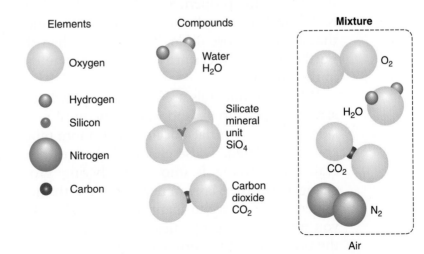

FIGURE 4-3.
All matter can be characterized as elements, compounds, or mixtures. Air is a mixture of gases.

have some common properties. Feldspars are light-colored, a little harder than glass, and have a density of about 2.7 g/cm³. Outside the laboratory, it can be very difficult to tell the different members of the feldspar family apart. For this reason, some references list feldspar as a specific mineral rather than a family of minerals.

Minerals can be defined as natural, inorganic, crystalline solids that have a specific range of composition and consistent physical properties. A surprising result of this definition is that ice is a mineral. Ice is a natural solid substance that forms crystals. Frost and snow are good examples of crystalline ice. As the solid form of water, each molecule of ice (H_2O) is composed of two atoms of hydrogen (H) and one atom of oxygen (O). But, unlike other minerals, ice has a low melting temperature, and therefore, it is not a mineral of which rocks are formed.

HOW CAN WE IDENTIFY MINERALS BY THEIR OBSERVABLE PROPERTIES?

1: SCIENTIFIC INQUIRY 1, 3
4: 3.1a

People identify things by their properties. In the identification of minerals, some properties are more useful than others. This section will concentrate on those properties that are most important in identifying minerals and putting them in groups.

Rocks and minerals change, or weather, when they are exposed to the conditions at Earth's surface. When identifying a mineral, it is important to use a fresh, unweathered sample. A fresh surface best shows the properties of the mineral. Abrasion, caused when rocks crash into each other, crushes the minerals into a powder that is much more difficult to identify than the original sample. Furthermore, as rocks are broken down, minerals are often mixed in the process.

When the minerals are broken or powdered, a greater surface area is exposed to air and water. Chemical reactions that are very slow on a fresh surface occur much more rapidly when the mineral is powdered. These chemical changes break down the mineral, transforming it into a weathering product, just as iron changes to rust. Many minerals react with moisture to form clay minerals. If you try to identify a mineral in a weathered sample, you may see the properties of a weathering product rather than the properties of the original mineral.

There are two types of mineral identification tests: those that leave the sample unchanged and those that change the sample. Observing the color of a mineral, or testing to see if it is attracted by a magnet does not change the mineral. You can get an idea of how a mineral has been broken by looking at surfaces where it broke apart. So far you have not done anything that makes the mineral more difficult to identify by the next person who sees it.

Other tests are destructive; they change the sample. If you actually break the mineral into smaller pieces to observe its properties, you have performed a destructive procedure. Although breaking the mineral may help you identify the mineral sample, this destructive procedure makes it more difficult for the next person observing the sample. Unfortunately, some useful tests are destructive. For example, to observe a mineral reacting with acid, the mineral must take part in a chemical reaction. That reaction degrades a part of the mineral into a weathering product. (You must take appropriate safety measures whenever you perform potentially hazardous tests such as this.)

When working with classroom samples, you should not perform destructive tests unless you are told to by your teacher. Rough handling or destructive procedures can change a beautiful and valuable mineral sample into useless fragments.

Color

Color is perhaps the first thing you notice about a mineral. For some minerals, color is very useful in their identification. For example, the brassy golden color of pyrite is very distinctive. No other common mineral has this color. Sulfur is one of the few yellow minerals. Almandine garnet is often identified by its red color. However, many minerals can be the same color. A black mineral might be magnetite, biotite, amphibole, pyroxene, or a number of other less common minerals.

Many minerals are colorless or white, including pure samples of quartz, halite, gypsum, and calcite. However, the color of these light-colored minerals can be changed by impurities. Impurities are small amounts of other substances that occur in the mineral. Smoky quartz is gray to black. Rose quartz is pink. Other impurities can make quartz orange, purple, or green. Agate is a banded, or striped, form of quartz in which white layers alternate with

FIGURE 4-4. Quartz is an extremely common mineral. Citrine, amethyst, rose quartz, and smoky quartz are among the colored varieties of quartz. (Note the penny for scale.)

brown or other colors. As with many other light-colored minerals, identifying quartz by its color alone can be unreliable. Dark-colored minerals are less likely to show variations in color from sample to sample because impurities usually do not cause them to change color. (See Figure 4-4.)

Streak

A different way of looking at color is **streak**, the color of the powdered form of a mineral. Figure 4-5 shows the method used to test for streak. It involves rubbing a corner of a mineral sample across an unglazed, white porcelain tile, called a streak plate. (A glaze is the glassy covering on plates, cups, and other porcelain kitchenware.) An unglazed surface is used because mineral particles do

FIGURE 4-5. The streak test reveals the color of the powder of a mineral. For many metallic minerals, the streak color is different from the color of the sample.

not rub off on a surface that is smooth. The purpose of the streak test is to identify the color of the powdered form of the mineral.

For some minerals, the color of the streak is very different from the color of a fresh surface. For example, galena, a very dense ore of lead, has a shiny gray, bright metallic surface. When galena is drawn across a streak plate, the powder that is left behind is dark gray to black. Samples of hematite, an ore of iron, can vary in color, depending upon how and where they formed. Sometimes hematite is red or brown and sometimes it is dark metallic silver. However, the streak color of hematite is always reddish-brown. Pyrite, or fool's gold, is a metallic, brassy, yellow mineral that has a green to black streak.

The streak test is most useful with minerals that have a metallic luster. These are the minerals most likely to show one color on the surface of the sample and a different color streak. Most nonmetallic minerals leave a streak that is the same color as the sample itself. A few minerals are too hard for the streak test. Samples of topaz, corundum, and diamond will scratch the streak plate rather than leaving behind a powder. (Obviously, this would damage the streak plate.)

Testing for streak is a destructive procedure. Each time you rub a mineral sample on a streak plate a little bit of the sample is lost and some of the powder remains on the mineral sample. If the streak test is done carefully, the damage to the sample is small and acceptable.

Luster

Luster can be one of the more difficult observations. However, luster is a very useful property.

Shine is a part of luster, but luster is more than how shiny a mineral appears. Luster also includes how light penetrates a fresh surface. **Luster** is defined as the way light is reflected and/or absorbed by the surface of a mineral. Is all of the light reflected? Does some of the light penetrate and some reflect? Is most of the light absorbed? The answers to these questions describe the characteristics of a mineral that determine its luster.

Luster is divided into two main categories: metallic and nonmetallic. Minerals with a metallic luster reflect light only from their outer surface. These minerals may have the hard look of a polished metal surface. Silver, copper, gold, galena, and pyrite have metallic luster. Light does not penetrate their surfaces, and you cannot see anything below the surface. Please note that luster is independent of other properties. For example, most metals are also relatively dense, but a mineral can have a metallic luster and have a density lower than most metals. Minerals that have metallic luster can be hard or soft. Like color, the only way to observe luster is with your eyes.

Minerals with nonmetallic luster can be sorted into several groups. Some minerals, such as quartz and feldspar, have a glassy luster. You have probably observed that although glass is shiny, light penetrates glass. If this were not true, windows would not let light into a room.

Few people would see a piece of glass and think that it is a metal. They just do not look the same. Metal and glass are shiny, but the glass surface does not have the hard look of a metallic surface. A porcelain dinner plate has a shiny finish, but it does not appear to be made of metal. Porcelain has a glassy luster that gives it a softer look. Figure 4-6 shows minerals with a glassy and metallic luster.

Luster can also be dull. Clay has a dull, or earthy, luster because it is not shiny at all. Dull luster is obviously nonmetallic. You would never mistake clay for a metal.

There are a few other terms used to describe nonmetallic luster. Talc has a pearly luster. Garnet has a waxy luster. These terms

FIGURE 4-6.
Selenite gypsum has a glassy luster because it is translucent. Pyrite has a nontransparent metallic luster.

are used to bring to mind substances that reflect and transmit light in the same way as the mineral sample.

1: SCIENTIFIC
INQUIRY 3
4: 3.1a

STUDENT ACTIVITY 4-2 —LUSTER OF COMMON OBJECTS

Make a list of objects in and around the classroom that can be described by the different categories of luster. For each object, record why it fits into that category of luster.

Aluminum foil is a good example. Notice how one side is more reflective than the other is. Also notice how the side that is dull still does not allow light to penetrate below its surface.

Hardness

The **hardness** of a mineral is its resistance to being scratched. The sharp corner of any mineral will scratch a substance that is softer. However, a mineral can be scratched only by a substance that has a greater hardness. The mineralogist Friedrich Mohs developed a special scale of hardness used to identify minerals.

This scale is known as **Mohs scale** of hardness. (See Figure 4-7.) The scale lists 10 relatively common minerals in order from soft to hard. The softest mineral on Mohs scale is talc. Assigned a hardness of 1, talc is so soft that any other mineral on the Mohs

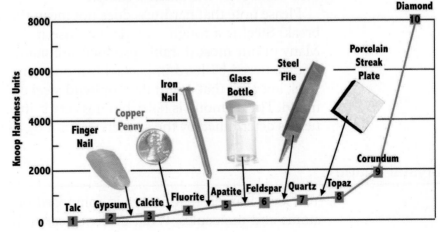

FIGURE 4-7. This graphic image of Mohs scale includes the 10 minerals selected as the standards of hardness and six common objects that can also serve as standards of hardness.

scale can scratch it. A streak plate has a hardness of about 7.5 on the scale. The hardest mineral on Mohs scale is diamond, which was assigned a hardness of 10. Diamond is the hardest known natural substance. A diamond can scratch every other mineral, but no other mineral can scratch a diamond.

Feldspar and quartz are among the most common minerals. Feldspar has a Mohs hardness of 6. The hardness of quartz is 7. This means that both minerals are relatively hard to scratch. However, quartz will scratch feldspar. On the other hand, feldspar will not scratch quartz even when a fresh edge is used because quartz is harder than feldspar. If two mineral samples have the same hardness, for example two pieces of the same mineral, each will be able to scratch the other.

Consider an example of how the hardness test works. Suppose you found a mineral that can scratch samples of talc, gypsum, and calcite. But the sample will not scratch fluorite or other minerals harder than fluorite. The hardness of the sample must be between 3 and 4 on Mohs scale. If the sample scratched fluorite and fluorite scratched the sample, the sample's hardness is 4. Perhaps the mineral you are trying to identify is fluorite.

Unfortunately, testing for hardness is a destructive test. After an object has been scratched many times, new scratches can be difficult to see. One way to make the test less destructive is to observe whether a fresh edge of the mineral will scratch a transparent glass plate. Glass is inexpensive and easy to replace. Most glass has a hardness of 5.5 on Mohs scale. Therefore, this is a good way to distinguish harder minerals from softer minerals with relatively little damage to the mineral sample.

Please note that hardness does not mean that a mineral will not break. Steel is a tough material because it does not break easily. Many of our most durable products are made from steel. Steel is not shattered by hard impacts. Diamonds are much harder than steel, but a hit that might dent or bend steel could shatter the diamond. The diamond is hard, but it is very brittle. That is why hardness is determined by the scratch test, not by any kind of collision.

Crystal Shape

You have learned that matter is composed of particles, including atoms and molecules. What evidence can you observe that solid matter is actually made of these little particles? Crystals provide visible evidence of the atomic nature of matter. The variety of crystal shapes we see is the result of the internal arrangements of atoms in different minerals.

Mineral crystals form in several different environments. Cubic crystals of halite (the mineral in rock salt) are left behind when a salty lake or lagoon evaporates. Other crystals form when water at Earth's surface or circulating underground, deposits the minerals. If the process is slow enough, layer upon layer of atoms and molecules build up to form visible crystals. Some crystals grow when molten rock cools slowly. Slow cooling allows a mineral to form large networks of ordered atoms and molecules. When lava or magma (hot liquid rock) comes to Earth's surface, dissolved gases expand and leave holes in the rock. Holes that are connected may allow water, especially heated water, to circulate through the rock. This hot water can dissolve a mineral from the large mass of rock and deposit it in the holes as crystals. Natural quartz crystals sold in some science, gift, and rock shops were formed in this way. If a large mass of magma that contains dissolved water cools slowly enough, the whole rock can be made of intergrown crystals several centimeters or more in length. Water is not required for crystal growth, but it helps form large crystals such as those shown in Figure 4-4 on page 92.

Minerals can be identified by their characteristic crystal shape. Quartz and calcite are colorless, white, or light-colored minerals with a glassy luster. Both minerals are very common. Unless you look beyond these similarities, you might get quartz and calcite confused. Quartz forms six-sided, hexagonal crystals. In cross section, they are shaped like a wooden pencil. Calcite crystals are usually four-sided with very different angles between the crystal faces. Visit the following Web site to watch snow crystal grow: *http://www.its.caltech.edu/~atomic/snowcrystals/movies/movies.htm*

Cleavage and Fracture

When minerals break, they tend to break in characteristic patterns. The term **cleavage** refers to the tendency of some minerals to break along smooth, flat planes. Minerals that display obvious cleavage break along planes (areas) of weakness where the bonding of molecules is weaker than in other directions. Cleavage surfaces can be recognized because they reflect light like a flat sheet of glass.

FIGURE 4-8. Muscovite, like other minerals in the mica family, has perfect cleavage in one direction. That is, all mica minerals split into very thin, flexible sheets.

The number of cleavage directions and the angles between them are the most important features of cleavage. Minerals in the mica family, such as biotite and muscovite, show perfect cleavage in only one direction. Large crystals of mica are sometimes called books of mica because they can be split again and again into very thin sheets, as you can see in Figure 4-8.

The feldspars have two cleavage directions that meet at nearly a right angle. Feldspar samples break into pieces that have a rectangular cross section. The ends of the pieces are not cleavage surfaces, so they are not as smooth and do not reflect light like the two cleavage planes.

Some minerals show three cleavage directions. When they break, the pieces have shiny surfaces all around. Halite, the primary mineral in rock salt, and galena, a silvery metallic mineral, split into little cubes. These minerals show three cleavage planes that meet at right angles, as you can see in Figure 4-9. Calcite is the primary mineral in limestone. When calcite crystals break, we see three cleavage directions. However, calcite's cleavage planes do not meet at right angles. Calcite breaks into rhombohedrons, which look like rectangular solids that have been pushed to one side.

Not all minerals break along definite cleavage directions. Minerals that break along curved surfaces or surfaces that are not parallel are said to show **fracture**. Natural quartz crystals have six flat sides. But when quartz crystals break, they break like window glass

FIGURE 4-9. Halite, also known as rock salt, forms cubic crystals that also break (cleave) into rectangular solids.

along curved surfaces that are not parallel to the flat sides. This is known as conchoidal fracture. Garnet breaks smoothly, and the pieces have shiny surfaces; but the surfaces are not flat and parallel. Therefore, garnet shows fracture not cleavage.

Some substances break into fragments or a powder that shows neither cleavage nor fracture. For example, dry clay disintegrates into a fine powder. Clay is among the minerals that show neither cleavage nor fracture.

To test a mineral for cleavage or fracture you must break it. Crystals are rare and beautiful. Breaking these samples destroys them. It may be better to observe the natural breakage surfaces and the angles at which they meet rather than actually testing a mineral in a destructive procedure.

STUDENT ACTIVITY 4-3 —CLEAVAGE AND FRACTURE IN HOUSEHOLD SUBSTANCES

1: SCIENTIFIC
INQUIRY 3
4: 3.1a

For this activity you will need a simple magnifying lens made of glass or plastic.

Use the magnifier to observe a variety of granular substances in your home such as sugar, salt, baking soda, and granulated laundry detergent. Check with your parents to determine which substances are safe to handle and to help you avoid making a mess.

Make a list of the substances you looked at, and tell why you think each shows cleavage, fracture, or neither.

Density

In Chapter 1, you learned that the density of an object could be calculated by dividing its mass by its volume. When mass is measured in grams and volume is measured in cubic centimeters, the unit for density is grams per cubic centimeter (g/cm^3). For example, the density of water is 1 g/cm^3. The density of the most common minerals falls between 2.5–3.5 g/cm^3. Many minerals have about the same density. Identifying a mineral by its density is helpful if the mineral is unusually dense.

Magnetite and galena are about twice as dense as the more common minerals. As mentioned previously, gold is the densest mineral substance you are likely to encounter. Gold is roughly six or seven times as dense as the most common minerals. Visit the following Web site to learn how to pan for gold: *http://www.you tube.com/watch?v=dnkg4b6EIxc*

Special Properties

Some minerals have special properties that are relatively rare. These properties can be useful in identifying these minerals. For example, graphite and talc are unusual because they feel greasy or slippery. Graphite is used as a lubricant and is the "lead" in pencils. Talc is used to make talcum powder. Minerals in the mica family (muscovite and biotite) cleave into thin, flexible sheets. Magnetite is an ore of iron that is attracted to a magnet. Sulfur has a distinctive odor and melts at a low temperature. Some minerals fluoresce (give off visible light) under invisible ultraviolet light.

It is a good rule not to taste or eat anything in a science lab. This is especially true of laboratory chemicals. (Your teacher will tell you if it is safe to taste a substance.) However, halite (the principal mineral in rock salt) can be identified by its salty taste.

Laboratory acids can be dangerous substances. If you are allowed to use them, handle them with great care because a strong acid can burn your skin and make holes in clothing. When a drop of acid is placed on calcite or a rock that contains calcite, a chemical reaction occurs that gives off bubbles of gas. A weaker acid, such as vinegar, may react visibly with powdered calcite.

A fresh surface of plagioclase feldspar may have what looks like small, parallel scratches on its surface. These scratches are striations. You are unlikely to find uranium in an Earth science lab because it is radioactive, which means that it gives off invisible rays and particles called ionizing radiation. (Ionizing radiation is a high-energy form that can damage living cells.) This radiation can be detected with special instruments such as the Geiger counter, and it is a strong indication of the presence of minerals that contain uranium or several related elements. Visit the following Web sites to learn about mineral identification data and see photographs of minerals: *http://www.minerals.net* or *http://mineral.galleries.com/default.htm*

WHAT ARE THE MOST COMMON MINERALS?

4: 3.1a
6: MODELS 2

In this chapter, you have learned about the properties used to identify a variety of minerals. Of the thousands of known minerals, just five make up about 85 percent of Earth's crust. These are the minerals that you are most likely to see.

The two most abundant elements in Earth's crust are oxygen and silicon. The group of minerals that contains both oxygen and silicon is known as the **silicate** minerals. As you might expect, silicates are the most abundant minerals in Earth's crust. Of the following minerals, only calcite is not a silicate. Figure 4-10 illustrates that the

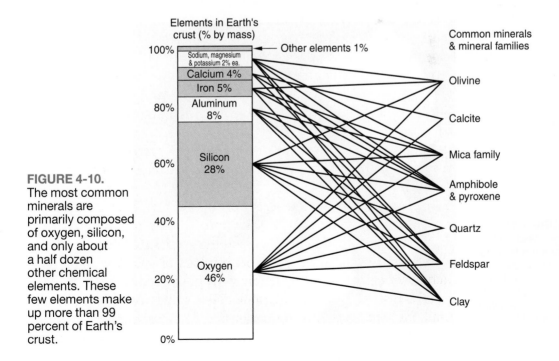

FIGURE 4-10. The most common minerals are primarily composed of oxygen, silicon, and only about a half dozen other chemical elements. These few elements make up more than 99 percent of Earth's crust.

most common minerals in Earth's crust are composed of many of the same chemical elements.

Feldspars

The feldspar family of silicate minerals makes up more than half of Earth's crust. In fact, the name feldspar comes from the Swedish words for field and mineral. The name refers to the fact that Swedish farmers had to move large amounts of feldspar-rich rocks before they could plant crops in their fields. The feldspar family of minerals is divided into groups according to composition. Plagioclase feldspars, also called Na-Ca feldspar, contain, in addition to silicon and oxygen, sodium and calcium in variable proportions. Potassium feldspar, also called orthoclase feldspar, contains potassium instead of sodium or calcium along with silicon and oxygen. Figure 4-11 shows both feldspars. Distinguishing between these groups can be difficult. In this book, they will be described by their common properties. The feldspar minerals are generally white to pink in color and cleave in two directions at nearly right angles. Feldspars have a glassy luster. With a hardness of 6 on Mohs scale, the feldspar minerals will scratch glass but will not scratch quartz.

FIGURE 4-11. Plagioclase and potassium feldspar are extremely common, both gave a glassy luster and show cleavage at nearly 90°.

Quartz

Quartz, another silicate, is second only to the feldspar minerals in abundance. If pure, quartz contains only silicon and oxygen; it has the chemical formula SiO_2. It is usually colorless or light in color and sometimes transparent. Quartz is often translucent, which means light can pass through but you cannot see objects on the other side as you could through a window. Figure 4-6 on page 94 illustrates a variety of quartz samples. Impurities can make quartz almost any color. The semiprecious gems amethyst, agate, and onyx are examples of quartz that is colored by small amounts of impurities. Like the feldspars, quartz has a glassy luster. With a Mohs hardness of 7, quartz can scratch most other minerals. Although most samples of quartz are not crystals, when quartz does form crystals, they are hexagonal, or six-sided, and can be pointed on one or both ends.

Micas

Minerals in the mica family are easy to identify because they are the only common minerals that can be split into thin, flexible sheets. We say that mica has perfect cleavage in one direction. Like quartz and feldspar, the mica minerals have a glassy luster but they are relatively soft, with a hardness of only 2.5 on Mohs hardness scale. Muscovite mica is rich in potassium and aluminum, which give it a light color. Biotite mica is dark in color due to the presence of iron and magnesium. Rocks that are rich in mica can have a reflective sheen on fresh surfaces. Figure 4-8 on page 98 is a photograph of "books" of mica surrounded by quartz.

Amphiboles and Pyroxenes

The silicates amphibole and pyroxene are the most common dark-colored minerals. Both have a Mohs hardness value between 5 and 6. They cleave into stubby splinters. Amphibole minerals, such as hornblende, can be distinguished from the pyroxene minerals, such as augite, by the angle at which their two cleavage surfaces meet. In the amphibole minerals, cleavage surfaces meet at 60° and 120° angles. The angle between cleavage surfaces in the pyroxene family is perpendicular, or 90°.

Calcite

Although calcite is not as common throughout Earth's crust as the minerals described above, it is relatively common at the surface. Calcite is also the only mineral in this section that is not a silicate. By chemical composition, calcite is calcium carbonate ($CaCO_3$). This means that calcite contains carbon and oxygen rather than the silicon and oxygen combination that defines the silicate minerals. Calcite is the most common mineral in the carbonate group. In very pure form, calcite may look transparent and colorless, but like quartz, it can have a variety of colors due to impurities. A soft mineral, calcite has a hardness of only 3. In very pure samples, breakage along cleavage directions can result in a rhombohedral shape, which looks like a rectangular solid that has been pushed toward one side, as you can see in Figure 4-12.

Olivine

Olivine is extremely common within Earth. However, it is not usually visible at the surface because it quickly weathers into clay. Olivine can often be observed in unweathered igneous rocks and it is readily identified by its olive-green color, glassy luster, and granular texture.

Clay

Clay is a family of soft, earthy minerals. Clay is the decomposition product of a wide range of other minerals, including feldspar and mica. It forms when these minerals combine chemically with water

FIGURE 4-12. Calcite crystals are easily identified by their glassy luster and rhombohedral cleavage. In addition, a transparent sample can make a single line behind the crystal look like two lines (double refraction).

in the weathering process. Clay is a major component of soil and the primary mineral in shale. The very fine particles in clay become sticky and flexible when mixed with a small amount of water. Clay often has an earthy odor.

Using a Flowchart to Identify Minerals

The simplified flowchart in Figure 4-13 can help you identify seven of the most common minerals. In this chart, you start on the left and proceed through the chart to the right. For example, if your sample is light in color, at the first branch you would take the top choice. Your sample is too soft to scratch glass so you move to the top of the next division. Your sample cleaves in three directions, which takes you to the lower portion of the final division. Finally you know that your sample bubbles when tested with acid.

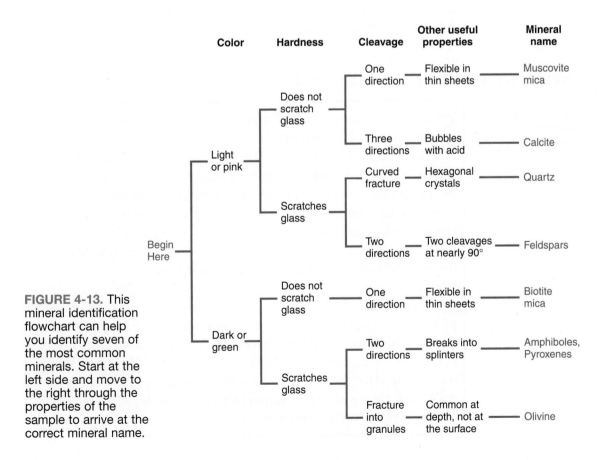

FIGURE 4-13. This mineral identification flowchart can help you identify seven of the most common minerals. Start at the left side and move to the right through the properties of the sample to arrive at the correct mineral name.

Calcite is the only common mineral that has this combination of properties.

Using the Reference Tables

The most useful properties in mineral identification are incorporated in the "Properties of Common Minerals," Figure 4-14, also found in the *Earth Science Reference Tables*. You should be able to use this chart to identify fresh samples of any of the 21 minerals listed in the "Mineral Name" column. Among the thousands of known minerals, these are the minerals you are most likely to see in the natural environment and in the science lab.

To use this table, start on the left side by identifying the luster of the sample in question. Then work your way to the right checking each property listed (hardness, cleavage or fracture, color, etc.). This table contains more than information to help you identify mineral specimens. It also includes information about the chemical composition and the principal uses of each mineral.

STUDENT ACTIVITY 4-4 —MINERAL IDENTIFICATION

1: SCIENTIFIC
INQUIRY 3
6 MODELS 2

Suggested materials: set of minerals, small magnifier, streak plate, small glass plate, magnet.

Your teacher will provide you with a kit containing several minerals. Use the "Properties of Common Minerals" chart to identify each of the samples. Record all the information you used to help you identify each mineral. But record only what you actually observed. For example, you probably have only a glass plate to determine the hardness of the minerals. Therefore, you will not be able to record the specific Mohs hardness number as listed on the chart. However, you can record whether the mineral could scratch glass.

Please handle the minerals with care and avoid damaging the samples when you perform destructive procedures. For example, rather than breaking the sample to observe fracture or cleavage, look at the surface of the sample to see how it broke when the mineral sample was prepared for the kit. Visit the following Web site to identify minerals from their properties: *http://facweb.bhc.edu/academics/science/harwoodr/GEOL101/Labs/Minerals*

Properties of Common Minerals

LUSTER	HARD-NESS	CLEAVAGE	FRACTURE	COMMON COLORS	DISTINGUISHING CHARACTERISTICS	USE(S)	COMPOSITION*	MINERAL NAME
Metallic luster	1–2	✔		silver to gray	black streak, greasy feel	pencil lead, lubricants	C	Graphite
Metallic luster	2.5	✔		metallic silver	gray-black streak, cubic cleavage, density = 7.6 g/cm³	ore of lead, batteries	PbS	Galena
Metallic luster	5.5–6.5		✔	black to silver	black streak, magnetic	ore of iron, steel	Fe_3O_4	Magnetite
Metallic luster	6.5		✔	brassy yellow	green-black streak, (fool's gold)	ore of sulfur	FeS_2	Pyrite
Either	5.5–6.5 or 1		✔	metallic silver or earthy red	red-brown streak	ore of iron, jewelry	Fe_2O_3	Hematite
Nonmetallic luster	1	✔		white to green	greasy feel	ceramics, paper	$Mg_3Si_4O_{10}(OH)_2$	Talc
Nonmetallic luster	2		✔	yellow to amber	white-yellow streak	sulfuric acid	S	Sulfur
Nonmetallic luster	2	✔		white to pink or gray	easily scratched by fingernail	plaster of paris, drywall	$CaSO_4 \cdot 2H_2O$	Selenite gypsum
Nonmetallic luster	2–2.5	✔		colorless to yellow	flexible in thin sheets	paint, roofing	$KAl_3Si_3O_{10}(OH)_2$	Muscovite mica
Nonmetallic luster	2.5	✔		colorless to white	cubic cleavage, salty taste	food additive, melts ice	NaCl	Halite
Nonmetallic luster	2.5–3	✔		black to dark brown	flexible in thin sheets	construction materials	$K(Mg,Fe)_3$ $AlSi_3O_{10}(OH)_2$	Biotite mica
Nonmetallic luster	3	✔		colorless or variable	bubbles with acid, rhombohedral cleavage	cement, lime	$CaCO_3$	Calcite
Nonmetallic luster	3.5	✔		colorless or variable	bubbles with acid when powdered	building stones	$CaMg(CO_3)_2$	Dolomite
Nonmetallic luster	4	✔		colorless or variable	cleaves in 4 directions	hydrofluoric acid	CaF_2	Fluorite
Nonmetallic luster	5–6	✔		black to dark green	cleaves in 2 directions at 90°	mineral collections, jewelry	$(Ca,Na)(Mg,Fe,Al)$ $(Si,Al)_2O_6$	Pyroxene (commonly augite)
Nonmetallic luster	5.5	✔		black to dark green	cleaves at 56° and 124°	mineral collections, jewelry	$CaNa(Mg,Fe)_4(Al,Fe,Ti)_3$ $Si_6O_{22}(O,OH)_2$	Amphibole (commonly hornblende)
Nonmetallic luster	6	✔		white to pink	cleaves in 2 directions at 90°	ceramics, glass	$KAlSi_3O_8$	Potassium feldspar (commonly orthoclase)
Nonmetallic luster	6	✔		white to gray	cleaves in 2 directions, striations visible	ceramics, glass	$(Na,Ca)AlSi_3O_8$	Plagioclase feldspar
Nonmetallic luster	6.5		✔	green to gray or brown	commonly light green and granular	furnace bricks, jewelry	$(Fe,Mg)_2SiO_4$	Olivine
Nonmetallic luster	7		✔	colorless or variable	glassy luster, may form hexagonal crystals	glass, jewelry, electronics	SiO_2	Quartz
Nonmetallic luster	6.5–7.5		✔	dark red to green	often seen as red glassy grains in NYS metamorphic rocks	jewelry (NYS gem), abrasives	$Fe_3Al_2Si_3O_{12}$	Garnet

*Chemical symbols:

Al = aluminum	Cl = chlorine	H = hydrogen	Na = sodium	S = sulfur
C = carbon	F = fluorine	K = potassium	O = oxygen	Si = silicon
Ca = calcium	Fe = iron	Mg = magnesium	Pb = lead	Ti = titanium

✔ = dominant form of breakage

FIGURE 4-14.

CHAPTER REVIEW QUESTIONS

Part A

1. Which mineral has a metallic luster, a black streak, and is an ore of iron?

   (1) galena   (3) pyroxene
   (2) magnetite   (4) graphite

2. The table below shows the density of four minerals.

   Data Table

Mineral	Density (g/cm³)
Corundum	4.0
Galena	7.6
Hematite	5.3
Quartz	2.7

   A student accurately measured the mass of a sample of one of these minerals to be 294.4 g. Its volume was 73.6 cm³. Which mineral sample did the student measure?

   (1) corundum   (3) hematite
   (2) galena   (4) quartz

3. The diagram below shows the crystal shape of two minerals.

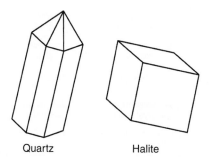

   Quartz   Halite

   Why do quartz and halite have different crystal shapes?

   (1) Light reflects from crystal surfaces.
   (2) Energy is released during crystallization.

(3) Impurities produce surface variations.
(4) The internal arrangement of atoms is different.

4. Which is a common use of the mineral graphite?
(1) a lubricant
(2) an abrasive
(3) a source of iron
(4) a cementing material

5. According to the table, which statement best describes the hardness of dolomite?

Hardness of Four Materials

Material	Hardness
Human fingernail	2.5
Copper penny	3.0
Window glass	4.5
Steel nail	6.5

(1) Dolomite can scratch window glass, but it cannot be scratched by a fingernail.
(2) Dolomite can scratch window glass, but it cannot be scratched by a steel nail.
(3) Dolomite can scratch a copper penny, but it cannot be scratched by a fingernail.
(4) Dolomite can scratch a copper penny, but it cannot be scratched by a steel nail.

6. If the four minerals shown below were placed together in a closed, dry container and shaken vigorously for 10 minutes, which material would be scratched and abraded the most?

Quartz

Amphibole

Pyroxene

Galena

(1) quartz
(2) amphibole
(3) galena
(4) pyroxene

7. The table below shows some observed physical properties of a mineral.

Physical Property	Observation
Color	White
Hardness	Scratched by the mineral calcite
Distinguishing characteristic	Feels greasy
Cleavage/fracture	Shows some definite flat surfaces

Based on these observations, the elements that make up this mineral's composition are

(1) sulfur and lead
(2) sulfur, oxygen, and hydrogen
(3) oxygen, silicon, hydrogen, and magnesium
(4) oxygen, silicon, aluminum, and iron

8. Looking at a mineral sample on a table without touching or moving it, what property could you observe?

(1) hardness
(2) atomic structure
(3) density
(4) luster

9. Which mineral is white or colorless, has a hardness of 2.5, and splits with cubic cleavage?

(1) calcite   (3) pyrite
(2) halite   (4) biotite

10. What property do nearly all minerals have in common?

(1) They are composed of a single element.
(2) They fell to Earth in the past century.
(3) They were organically formed.
(4) They are natural, inorganic, crystalline solids.

Part B

11. Base your answers to questions 11 and 12 on the graph below, which shows the conditions at which the same chemical elements crystallize to form one of three different minerals.

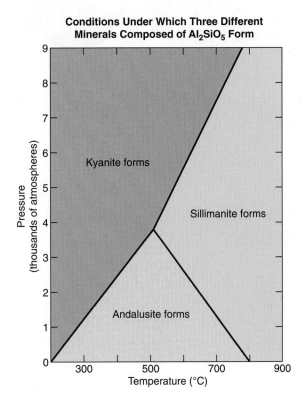

Conditions Under Which Three Different Minerals Composed of Al$_2$SiO$_5$ Form

Under what pressure and temperature conditions will andalusite form?

(1) 300°C and 6000 atmospheres
(2) 500°C and 2000 atmospheres
(3) 600°C and 4000 atmospheres
(4) 700°C and 3000 atmospheres

12. Which mineral has a chemical composition most similar to andalusite, sillimanite, and kyanite?

(1) pyrite
(2) gypsum
(3) dolomite
(4) potassium feldspar

13. Halite has three cleavage directions at 90° to each other. Which model best represents the shape of a broken sample of halite?

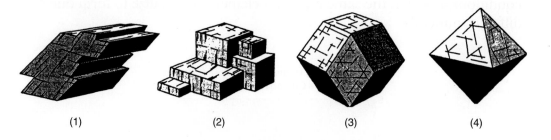

(1)   (2)   (3)   (4)

14. Quartz and calcite are both very common minerals. In what property are quartz and calcite similar?

   (1) surface luster
   (2) cleavage directions
   (3) Mohs hardness
   (4) reaction with acid

Part C

Base your answers to questions 15 through 17 on the information below.

15. Use a copy of the grid below to construct a bar graph of the hardness of these five minerals. (Do not write in this book.)

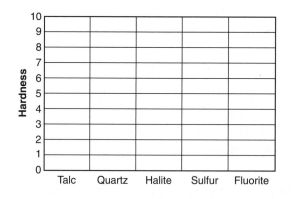

16. Which of the five minerals in the graph above would be the best abrasive? State one reason for your choice.

17. How many of these minerals could be scratched by almandine garnet?

Base your answers to questions 18 and 19 on the following information.

18. All minerals in the feldspar family contain silicon, oxygen, aluminum, and one other element. The diagram below shows the range of feldspars that contain potassium, sodium, and calcium as that last element. It is the final element or elements that determine the specific kind of feldspar mineral.

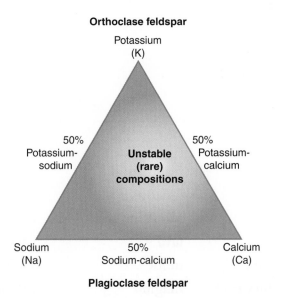

What is the most common element found in orthoclase feldspar but not in plagioclase feldspar?

19. Use a copy of this diagram to place an X to represent the position of a feldspar that has no potassium, but an equal amount of sodium and calcium atoms.

20. Based on the position of diamond on Mohs scale, state one way that a diamond can be identified if it is in a group of other similar-looking minerals.

5

The Formation of Rocks

WORDS TO KNOW

banding	fragmental	precipitation
bioclastic rock	igneous rock	regional metamorphism
classification	intrusion	rock
clastic	lava	rock cycle
contact metamorphism	mafic	sediment
crystalline sedimentary rock	magma	sedimentary rock
extrusion	metamorphic rock	texture
felsic	origin	vesicular
foliation	plutonic	volcanic
fossil		

This chapter will help you answer the following questions:

1 What are rocks?
2 How do we identify and classify igneous rocks?
3 How do we identify and classify sedimentary rocks?
4 How do we identify and classify metamorphic rocks?
5 What is the rock cycle?

WHAT ARE ROCKS?

1: SCIENTIFIC
INQUIRY 3
4: 2.1m; 4: 2.1w
4: 3.1b
6: PATTERNS OF
CHANGE 5

A **rock** is a substance that is or was a natural part of the solid earth, or lithosphere. Rocks come in many varieties. Some rocks are unusual enough for geologists and people interested in geology to collect them. Rocks can also be colorful or attractive. Landscapers often use rocks to beautify homes or parks.

Most rocks are composed of a variety of minerals. In some rocks, you can recognize the minerals as variations in color or other mineral properties within the rock. Granite is a good example. If you look carefully at a sample of granite with a hand lens, you will probably observe some parts of the rock that are transparent. This is probably the mineral quartz. Other parts are white or pink with angular cleavage. These are properties of plagioclase feldspar and potassium feldspar. Dark mineral grains that occur in thin sheets are biotite mica. Dark minerals that occur in stubby crystals are probably hornblende, the most common mineral in the amphibole family. Other minerals can occur in granite, but they are not as common as these four. (Look at Figure 4-1 on page 86.)

Other rocks are composed of a single mineral. Some varieties of sandstone, particularly if they are very light in color, can be nearly 100 percent quartz. Very pure limestone is nearly all calcite. Rock salt can be nearly pure halite.

Coal is mined in many areas of the United States, and is often used as a fuel. If we define a rock as a natural solid part of the lithosphere, coal is a rock. Many samples of coal show beautiful plant impressions.

While coal is a kind of rock, it was formed from living materials. Therefore coal is an example of a rock that is not made of minerals. (Impure coal does contain some mineral material.) Another example of a rock that is not composed of minerals is fossil limestone. This kind of limestone is the result of the accumulation of shells or coral. Coral is a colony of tiny marine animals that live in a hard external skeleton. These skeletons can be several inches or more in length and take unusual shapes. Because fossil limestone is made of the hard remains of coral, shells, and other living tissue cemented together grains of organic calcite, it is not composed of minerals. Minerals must be of inorganic origin.

STUDENT ACTIVITY 5-1 —MAKING A ROCK COLLECTION

1: SCIENTIFIC
INQUIRY1

As you read through this chapter, collect small rock samples from around your community. If you wish, you can add rocks from your travels or rocks that you have gotten elsewhere. You may recall that when you worked with minerals you tried to use fresh, unweathered samples. It is important to follow the same rule in collecting rocks. Weathered rocks tend to crumble, and they are also harder to identify than fresh samples.

This rock collection will have two purposes. First, it should help you discover the variety of rocks around you. The second purpose is to use what you learn from this chapter to classify your rocks.

Your samples should be divided into the three categories you will learn about in this chapter: igneous, sedimentary, and metamorphic rocks. You may collect as many rocks as you wish, but the final submission will consist of six small rocks, two from each category, each of which shows a property of its category. Therefore, for each rock submitted, classify it as igneous, metamorphic, or sedimentary and tell why each sample fits into its group. Your samples may be placed in a half-dozen egg carton.

You may check your progress on this assignment with your teacher. In addition, carefully observe the rock samples you use in class to identify the properties used to classify rocks. Your teacher will collect your samples after the class has completed this chapter.

Three Rock Categories

In the previous chapter, you learned that minerals are classified on the basis of their observable properties such as luster, hardness, color, and cleavage. Geologists have found it more useful to separate rocks into three groups based on how they formed. The way rocks form is their **origin**.

Igneous rocks form from hot, molten (liquid) rock material that came from deep within Earth. Only igneous rocks have this origin. Hot, liquid rock is called **magma**. At Earth's surface magma is known as **lava**. Figure 5-1 shows a geologist taking samples of molten lava. In Chapter 2 you learned that Earth's temperature increases as you go deeper within the planet. In some places within Earth, it is hot enough to melt rock. This molten rock, which is less

FIGURE 5-1. When this lava cooled, it turned to an igneous rock like the solid rock the geologist is standing on.

dense, rises to or near Earth's surface. At the surface, it is cooler, and the liquid rock material changes to solid rock. Igneous rocks are especially common around volcanoes and in places where large bodies of rock that have melted and then solidified underground have been pushed to the surface. Visit the following Web site to see a concept map that explains rock origins: *http://igs.indiana.edu/Geology/rocks/rockcycleactivities/cycleconceptmap.swf*

In Chapter 7 you will learn that most of Earth's interior is in the solid state. If temperatures underground are hot enough to melt rock, why is the interior not mostly liquid? The reason that Earth is mostly solid is the increase of pressure with depth. While the increasing temperature tends to melt the rock, the increase in pressure prevents melting.

Rocks weather and break down when they are exposed at Earth's surface. Eventually the weathered material collects as layers of **sediment**. Compression and cementing of weathered rock fragments or the shells of once living creatures is the origin of most **sedimentary rocks**, the second group of rocks.

If sedimentary or igneous rocks are buried so deeply that heat and pressure distort their structures and form new minerals, the result is **metamorphic rocks**. Metamorphic rocks are the only

FIGURE 5-2. The distortion of these pebbles shows how rock changes when exposed to the conditions that exist several kilometers within Earth.

group that forms directly from other rocks (usually igneous or sedimentary). Figure 5-2 shows a sedimentary rock that originally contained round pebbles, which forces within Earth have squashed and elongated.

But, if most metamorphic rocks form within Earth, why can they be found at the surface? The answer to this question involves two important Earth-changing processes. The first is uplift. Earth contains a great deal of heat energy. As heat escapes from the interior of the planet, it sometimes pushes up the crust to form mountains. The second process is weathering and erosion. Weathering and erosion wear down the mountains, exposing rocks at the surface that formed at depths of 10 km or more underground. Wherever you find a large mass of metamorphic rock at the surface, you are probably looking at the core of an ancient mountain range.

Classification is the organization of objects, ideas, or information according to their properties. Not all rocks fit easily into one of these three categories. Some volcanoes throw great quantities of ash into the air. The ash falls, settles in layers, and hardens. The settling part of ash layers' origin is similar to the processes that form sedimentary rock. However, because the material came from a volcano, volcanic ash is classified as an igneous rock. Another example is the gradual change from igneous or sedimentary rock to metamorphic rock. It may not be clear at what point the rock should no longer be classified as its original parent rock and when it should be called a metamorphic rock. In spite of these occasional difficulties, the classification of rocks by their origin has served geologists and Earth science students well. Figure 5-3 illustrates the way to classify rocks as igneous, sedimentary, or metamorphic by their appearance. Rocks can be classified by their appearance because the way they look offers clues to their formation. View on-line movies of Kilauea volcanic eruptions in Hawaii at the Web site *http://hvo.wr.usgs.gov/gallery/kilauea/volcanomovies/*

STUDENT ACTIVITY 5-2 —CLASSIFICATION

6: SYSTEMS
THINKING 1
6: MODELS 2

Your teacher will give you about a dozen objects to classify. Divide the objects into groups based on the properties you can see. Start by listing all the objects. Each time you divide the objects into groups, state one property that allows you to clearly decide to which group an object belongs. As you divide each group into smaller groups, limit the number of subgroups to two or three. Continue the process until each object is alone in a group. Create a flowchart that will allow someone else to separate and/or identify all the objects. Figure 4-13 on page 105 (Mineral Identification Flowchart) shows a relatively simple system for classifying minerals.

Because **igneous rocks** have formed from molten magma or lava, they are composed of intergrown crystals. Rapid cooling, however, can make the crystals too small to be visible. Igneous rocks are usually quite hard and dense, and layering is rare. Gas bubbles may give igneous rocks a frothy texture.

Most **sedimentary rocks** are composed of rounded fragments cemented in layers. In fine-grained rocks, the individual grains may be too small to be readily visible. Rocks made by chemical precipitation are composed of intergrown crystals, although these crystals are relatively soft. A rock that contains fossils is almost certainly a sedimentary rock.

Granite

Basalt

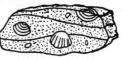

Limestone

Conglomerate

Metamorphic rocks, like igneous rocks, are usually composed of intergrown crystals. But, like sedimentary rocks, they often show layering, banding, or foliation. The layers may be bent, or distorted.

FIGURE 5-3. The appearance of a rock is often a good indicator of its origin.

HOW ARE IGNEOUS ROCKS CLASSIFIED?

4: 3.1b; 3.1c
6: MODELS 2

Igneous rocks are classified by their color and texture. The colors in rocks come from the minerals that make up the rocks. Minerals rich in aluminum (chemical symbol Al) are commonly light-colored, sometimes pink. These minerals are called **felsic** because the feldspars are the most common light-colored minerals. The word "felsic" comes from *fel*dspar and *si*licon. Minerals rich in magnesium (chemical symbol Mg) and iron (chemical symbol Fe) such as olivine and pyroxene families are called **mafic** (MA-fic). The word "mafic" comes from a combination of *ma*gnesium and *f*erric, a Latin word used to describe iron. Mafic minerals are often dark-colored.

The next characteristic used to classify igneous rocks is their texture. **Texture** describes the size and shape of the grains and how they are arranged in the rock. Texture answers the following questions. Is the rock composed of different kinds of grains? How large are these grains, and what shape are they? Do they show any kind of organization?

In igneous rocks, the size of the crystals is generally a result of how quickly the rock solidified. If the magma cooled slowly, the atoms and molecules had enough time to form crystals that are visible without magnification. Granite is a good example of a rock that cooled slowly. Granite is a popular building stone because it resists wear and weathering and because it is attractive. Granite has a speckled appearance, as you can see in Figure 4-1 on page 86. The different colors in granite come from the different minerals of which it is composed. Crystals from 0.25- to 1.0-cm long are common in granite. If the granite is pink, it probably contains a large amount of potassium feldspar, which can be pink or white.

Most granite forms in large masses within Earth. The movement of magma to a new position within Earth's crust is called **intrusion**. Intrusion occurs underground, or inside Earth. Sometimes a large quantity of hot magma rises to a place near the surface where it slowly cools to form solid rock.

In other cases, granite originates from a mass of rock buried deeply enough to melt. As the mass cools and crystallizes, it slowly forms granite. Because coarse-grained igneous rocks such as granite form deep underground, they are classified by origin as intrusive or **plutonic** rocks. (The term plutonic comes from the name of Pluto, the Roman god of the underworld.)

Basalt is also a common igneous rock, especially under the oceans. The ways that basalt differs from granite can help you understand how igneous rocks are classified. Basalt usually forms from magma that rises to or very near the surface. **Extrusion** is the movement of magma onto Earth's surface. At the surface, lava cools quickly, and the resulting crystals may be too small to see without magnification. Fine-grained igneous rocks such as basalt are therefore called extrusive or **volcanic** rocks. Basalt is rich in mafic minerals that give it a dark color, generally dark gray to black.

Figure 5-4 is a chart from the *Earth Science Reference Tables* that can help you understand and classify igneous rocks primarily by color

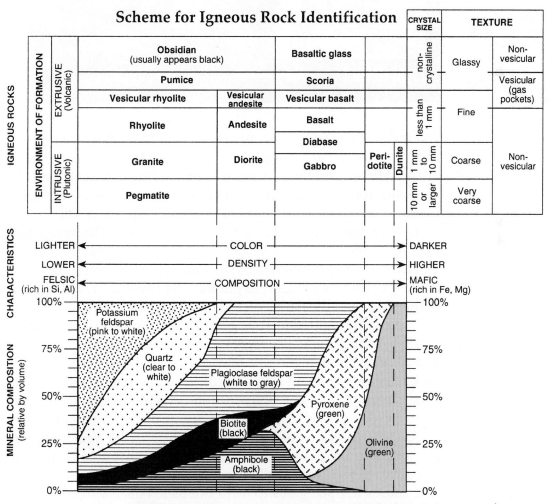

FIGURE 5-4. You can use this chart to identify the most common igneous rocks and to estimate their mineral composition.

Fine Grain

Basalt

Rhyolite

Coarse Grain

Granite

Gabbro

Light (felsic) ← Dark (mafic)

FIGURE 5-5.
Notice that these four common igneous rocks correspond to rocks identified in Figure 5-4.

and texture. The rocks with the smallest crystals (fine-grained, extrusive rocks) are at the top of this chart. In these rocks, the grains of the different minerals are too small to be seen easily without magnification. Below them are the coarse igneous rocks in which it is easy to see the different minerals. Compare this chart with images of the four major igneous rock types in Figure 5-5. This image may help you understand the variables in the *Reference Table* chart.

There are major differences in composition between the light-colored (felsic) rocks, such as granite and rhyolite on the left side of the chart, and the dark-colored (mafic) rocks, such as basalt and gabbro on the right side. Along with this change in composition (felsic to mafic) the chart shows a change in color (light to dark) and a change in density. You may not be able to feel that a mafic rock is heavier than a felsic rock of the same size, but the difference is measurable. The difference in density will become important when we consider the interior of our planet in Chapter 14.

You will find the term "vesicular" on the chart. A **vesicular** texture refers to the gas pockets, called vesicles, that are common in extrusive igneous rocks. When magma rises, the decrease in pressure causes trapped gases to form bubbles. This is similar to what happens when you open a bottle of carbonated water. These bubbles are trapped carbon dioxide. But the gas bubbles that form in

FIGURE 5-6. Pumice (right) and scoria (left) are vesicular (contain small air pockets), igneous rocks. Sometimes they have so many vesicles that these rocks will float on water.

lava are mostly water vapor, which escapes into the atmosphere. Scoria has large vesicles, and may look like cinders from a fire. Pumice has smaller gas bubbles, and it can be so light it may float on water. (See Figure 5-6.) Pumice is sometimes sold as an abrasive used to scrape the gratings of barbecue grills. If you forget what vesicular means, look at the top right portion of the chart where you will find the word "vesicular" just above the two words "gas pockets" in parentheses.

At the top of the chart is a texture called glassy. Sometimes lava cools so quickly it forms a rock that looks like a shiny, dark, glass material. This is obsidian. If obsidian (also known as volcanic glass) contains crystals, they are too small to be seen even under a microscope. The properties of the other igneous rocks listed in the Scheme for Igneous Rock Identification can be determined from the rock's position on the chart. For example, pegmatite appears at the bottom left of the chart. Like other igneous rocks on the left, pegmatite is relatively light in color. Its position at the bottom means that pegmatite is composed of very large crystals.

The bottom section of the Scheme for Igneous Rock Identification on page 121 is called Mineral Composition. This section shows the minerals that are common in igneous rocks. For example, granite usually contains potassium feldspar, quartz, plagioclase feldspar, biotite, and amphibole. If you imagine a vertical line running directly below the word granite and into this section, you will see that quartz and potassium feldspar make up about 66 percent of the volume of granite. The percent of each mineral is indicated by

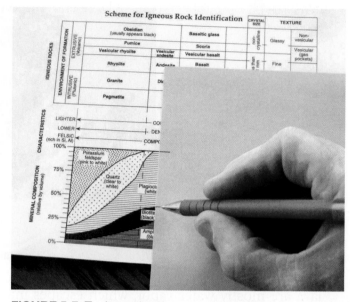

FIGURE 5-7. To determine the percentage mineral composition of a common igneous rock, mark an edge of a sheet of paper with the upper and lower limits of the proper section of the graph. (You probably will want to center the vertical edge of the paper at the center of the mineral name.) Then slide the paper over to the percent composition scale. This person is marking the limits of the plagioclase feldspar section below basalt and gabbro.

the scale that appears on each side of the Mineral Composition section of the chart. The composition of basalt is under the word "Basalt" near the other side of the chart. Basalt is mostly plagioclase and pyroxene. However, the mineral composition of igneous rocks is variable. The various compositions of each rock are enclosed by the dotted lines. Figure 5-7 shows how you can line up a sheet of paper with the mineral composition part of the chart to determine the percentage composition of various common igneous rocks.

Common Igneous Rocks

Of all the igneous rocks named on the Scheme for Igneous Rock Identification in the *Earth Science Reference Tables,* or Figure 5-4 on page 121, you are most likely to encounter just seven in your Earth science course. These seven igneous rocks are very easy to tell apart, and they show the range of properties of igneous rock.

Granite is a coarse-grained, felsic (light-colored) igneous rock. Its overall color is likely to be light gray or pink. Because of slow cooling the mineral crystals are large enough to be visible without magnification.

Rhyolite is the fine-grained equivalent of granite. Rhyolite is light-colored and felsic in composition. Rapid cooling of the magma has resulted in very small mineral grains that are unlikely to be readily visible.

Gabbro, like granite, is composed of large crystals because of slow cooling of the magma. (It is coarse-grained.) Unlike granite, gabbro is mafic in composition, which means that it is composed primarily of the dark minerals rich in iron and magnesium.

Basalt has a mineral composition similar to gabbro's, so it is also relatively dark in color. However, basalt cooled so quickly that, as in rhyolite, the individual mineral grains might be too small to see without magnification.

The next two igneous rocks share an unusual feature. Scoria and pumice are full of air pockets. This indicates that they probably formed from magma rich in dissolved gases, such as water vapor, and were ejected from a volcano during a violent eruption. The pockets in pumice are small enough that individual pockets are not obvious. Scoria has larger pockets and looks like cinders.

Volcanic glass is also called obsidian. The term "glass" describes its smooth texture, which results from rapid cooling of lava that had little dissolved water or gases. Obsidian usually breaks along curved surfaces (conchoidal fracture). It is usually black due to the even distribution of dark minerals, even though the mineral composition of volcanic glass is most often felsic. Visit the following Web site to work with a dynamic clickable igneous rocks chart: *http://csmres.jmu.edu/geollab/Fichter/IgnRx/IgnRx.html*

STUDENT ACTIVITY 5-3 —IDENTIFICATION OF IGNEOUS ROCKS

1: SCIENTIFIC
INQUIRY 1
6: MODELS 2

Obtain a set of igneous rocks from your teacher. Handle them carefully and let your teacher know if any samples are missing or damaged.

Use the information you learned in this chapter and the appropriate chart in the *Earth Science Reference Tables* to identify each of the igneous rocks in your set. List the name of each rock, such as granite or scoria, along with the characteristics you observed that helped you to identify it.

HOW DO SEDIMENTARY ROCKS FORM?

4: 2.1W
4: 3.1C

Within Earth's crust, igneous rock is the most common rock type. However, most of the surface of our planet is covered with a relatively thin layer of sedimentary rocks. Unlike igneous rocks, it is difficult to give a precise definition of sedimentary rocks. Most sedimentary rocks are made of the weathered remains of other rocks that have been eroded and later deposited as sediment in layers. Over time, the sediments are compressed by the weight of the layers above them. In addition, the layers may be cemented by mineral material left by water circulating through the sediments. The cementing material is usually silica (fine-grained quartz), clay, or calcite. All sedimentary rocks are formed at or near Earth's surface. Although this description applies only to the clastic, or fragmental, group of sedimentary rocks, these are the most common rocks of sedimentary origin. **Fossils** are any remains or impressions of prehistoric life. If fossils are present in a rock, the rock is almost certainly a sedimentary rock. The processes that create igneous and metamorphic rocks usually destroy any fossil remains.

You can recognize sedimentary rocks because they are usually composed of particles, often rounded particles, compressed and cemented into layers, called bedding planes. Shale, the most common rock on Earth's surface, is made of particles of sediment too small to be visible without magnification. Shale breaks easily into thin layers.

Clastic (Fragmental) Rocks

Clastic and **fragmental** are terms applied to the group of sedimentary rocks that are made up of the weathered remains of other rocks. These are the most common sedimentary rocks. Clastic rocks are formed by the processes of deposition, compression, and cementation of sediments. Although some sediments are deposited by wind, glaciers, or even rockfalls, most are the result of deposition in water. Seas or parts of the ocean once covered most parts of the continents. Streams and rivers carry sediments from the surrounding land into these bodies of water. The particles of sediment settle to the bottom of the water, forming fine-grained sedimentary rocks. Where deposition is rapid or currents are fast, the particles of sediment deposited are larger. Clastic, or fragmental, rocks are classified by the size of the sedimentary particles from which they are formed.

Information about the range of sizes of the various particles in sedimentary rocks is found in the sedimentary rock chart in the *Earth Science Reference Tables* and also in Figure 5-8. For example, according to this chart, sand can be defined as particles of sediment that range between 0.006 cm and 0.2 cm in size.

Conglomerate is the coarsest grained clastic rock. It is dominated by particles that are easily seen: about 0.2 cm or larger. Conglomerate sometimes looks like artificial cement with rounded pebbles embedded in it. Silica (very fine quartz), clay, and calcite (the mineral in limestone) are common cements that hold the larger particles together. There is no upper limit to the size of the particles in conglomerate, but cemented pebbles are the most common texture of conglomerate. If the particles are angular (a sign that they have not been transported very far before deposition) the term breccia (BRETCH-ee-a) is used instead of conglomerate.

Scheme for Sedimentary Rock Identification

INORGANIC LAND-DERIVED SEDIMENTARY ROCKS					
TEXTURE	**GRAIN SIZE**	**COMPOSITION**	**COMMENTS**	**ROCK NAME**	**MAP SYMBOL**
Clastic (fragmental)	Pebbles, cobbles, and/or boulders embedded in sand, silt, and/or clay	Mostly quartz, feldspar, and clay minerals; may contain fragments of other rocks and minerals	Rounded fragments	**Conglomerate**	
			Angular fragments	**Breccia**	
	Sand (0.006 to 0.2 cm)		Fine to coarse	**Sandstone**	
	Silt (0.0004 to 0.006 cm)		Very fine grain	**Siltstone**	
	Clay (less than 0.0004 cm)		Compact; may split easily	**Shale**	
CHEMICALLY AND/OR ORGANICALLY FORMED SEDIMENTARY ROCKS					
TEXTURE	**GRAIN SIZE**	**COMPOSITION**	**COMMENTS**	**ROCK NAME**	**MAP SYMBOL**
Crystalline	Fine to coarse crystals	Halite	Crystals from chemical precipitates and evaporites	**Rock salt**	
		Gypsum		**Rock gypsum**	
		Dolomite		**Dolostone**	
Crystalline or bioclastic	Microscopic to very coarse	Calcite	Precipitates of biologic origin or cemented shell fragments	**Limestone**	
Bioclastic		Carbon	Compacted plant remains	**Bituminous coal**	

FIGURE 5-8. This table can be used to identify the most common sedimentary rocks.

FIGURE 5-9.
Fossils and rounded particles are common characteristics of sedimentary rocks. The rock on the left is a fossiliferous limestone, which contains shells. The rock on the right is a conglomerate.

Figure 5-9 is a sample of conglomerate that shows obvious characteristics of a sedimentary rock.

Although sandstone is defined by a precise limit of particle sizes (0.2–0.006 cm), you can identify it by its gritty feel, like sandpaper. Figure 5-10 shows a layered sandstone. Shale feels smooth because the clay particles of which it is composed are so tiny they are invisible without strong magnification. Rocks made of particles larger than those in smooth shale but smaller than those in gritty sandstone are classified as siltstone.

Unlike igneous rocks, clastic sedimentary rocks are not classified by their mineral composition. Any clastic sedimentary rock

FIGURE 5-10.
Sandstone is composed of particles that feel gritty and are usually deposited in layers.

can contain quartz, feldspar, or clay, all of which are the remains of the weathering of other rocks. Nor does color help to tell them apart. The mineral content of the rocks influences their color. Pure quartz is usually light in color, whereas clay generally makes the rocks gray or black. Iron staining is common in sedimentary rocks, giving many of these rocks a red to brown color.

Chemical Precipitates

The next group of sedimentary rocks listed in the Scheme for Sedimentary Rock Identification is not as common as clastic rocks. This group forms as the water evaporates, leaving dissolved solids behind. When evaporation occurs, the compounds left behind become too concentrated to remain in solution. Therefore, the solids deposit as mineral crystals. This process is called chemical **precipitation**. Precipitation forms rocks known as the **crystalline sedimentary rocks**. Thick layers of underground rock salt are mined in Western New York State to be used as food additives and to melt ice on roads. These deposits as well as similar salt layers found worldwide identify places where large quantities of salt water have evaporated.

Rock gypsum and dolostone form in a similar process. However, the minerals gypsum and dolomite form from salt water with a different composition of dissolved mineral. Unlike clastic sedimentary rocks, which are classified by grain size, the chemical precipitates are classified by their chemical or mineral composition.

In general, rocks composed of crystals are not sedimentary. However, the chemical precipitates are the only sedimentary rocks made of intergrown crystals. Sedimentary rocks of chemical origin are made up of crystals that are relatively soft and often white in color. Furthermore, they are found among other layers of sedimentary rock. For these reasons, the chemical precipitates are seldom mistaken for igneous or metamorphic rocks, which are also composed of intergrown crystals. Sedimentary precipitates, such as rock salt, are usually made up of a single mineral.

Organic Rocks

Bioclastic rocks form from material made from or by living organisms. When you find a seam (layer) of coal, you are probably looking at the remains of an ancient swamp environment. This is a place where plants grew, died, accumulated layer upon layer, and were compressed and turned to stone. The green color of living plants is due to chlorophyll. But chlorophyll quickly breaks down when plants die. The carbon content of the plant remains. This gives coal its black color. Coal is mined as a fuel. In addition, it is used in making a variety of plastics and medicines. Fossil remains of extinct plants are especially common in coal, which preserves plant impressions in great detail.

A second bioclastic sedimentary rock is limestone. Limestone is usually formed by the accumulation and cementation of the hard parts of marine animals, such as the external skeletons of coral colonies and seashells. (See Figure 5-9 on page 128.) This organic material can be transformed into the mineral calcite, the primary mineral in limestone. Layers of limestone indicate the long-term presence of an active biological community in shallow seawater. This kind of active biological environment cannot be found in deep water because these ecosystems need sunlight. Sunlight cannot penetrate to the bottom in a deep-water environment. The bioclastic rocks are sometimes known as the organic group of sedimentary rocks.

STUDENT ACTIVITY 5-4 —IDENTIFICATION OF SEDIMENTARY ROCKS

1: SCIENTIFIC
INQUIRY 1, 2
4: 2.1w

Obtain a set of sedimentary rocks from your teacher. Handle them carefully and let your teacher know if any samples are missing or damaged.

Use the information you learned in this chapter and the appropriate chart in the *Earth Science Reference Tables* to identify each of the sedimentary rocks in your set. List the name of each rock, such as shale or rock salt, along with the characteristics you observed that helped you to identify it.

HOW DO METAMORPHIC ROCKS FORM?

4: 2.1m

Have you ever baked cookies? You may know that the cookie dough you put into the oven has very different properties from the baked cookies that come out. In a similar way, rocks subjected to conditions of heat and pressure within Earth are changed to metamorphic rocks. In fact, the term metamorphism means "changed in form." Metamorphic rocks are the only kind of rocks that begins as other rocks, as you can see in Figure 5-11. Heat and pressure cause changes that transform rocks from one rock type to another. This usually happens either deep underground where both the temperature and the pressure are high, or close to an intrusion of hot magma at or near the surface. It is important to remember, however, that if the heating melts the rock, cooling and solidification will form an igneous rock. Metamorphic rocks do not form from magma.

The metamorphic process causes visible changes. Minerals that are stable at the surface undergo chemical changes when they are subjected to intense heat and pressure. Figure 5-12 shows a progression of rock types that occur when clay or shale is subjected to increasing heat and pressure by being buried deeper and deeper within Earth where both temperature and pressure increase.

You learned in the last section that shale is formed by the compaction of clay-sized particles under the weight of overlying layers.

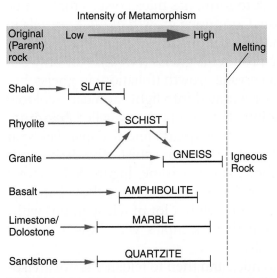

FIGURE 5-11. Metamorphic rocks are a result of changes in other kinds of rock. The black lines show the range of change in the parent rock. Metamorphism may be low or high.

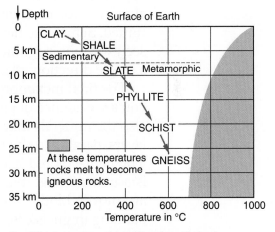

Conditions necessary to make various kinds of metamorphic rocks, or to make magma, which will cool to become igneous rock.

FIGURE 5-12. This diagram represents the transformation of clay or shale into different kinds of metamorphic rock as the rock is buried deeper and deeper within Earth. The scales of depth and temperature show the conditions at which four of the most common metamorphic rocks originate.

With deeper burial, chemical changes begin that transform shale through a series of metamorphic rocks. The clay minerals begin to change to mica, as the rock becomes harder and denser, forming the metamorphic rock slate.

At this point, a new feature of the rock starts to develop. In most slate, you can see the original bedding planes of the sedimentary rock. The parent rock, shale, usually breaks along these sedimentary bedding planes. However, the growth of mica crystals is likely to be in a direction different from that of the original layers. Mica crystals grow in response to the forces on the rock. Even though these mineral crystals may be too small to be visible, they do affect the way that slate breaks. Breakage in a direction that crosses the original bedding planes signals that mineral changes happened. This alignment of mineral crystals is called **foliation**. Foliation is a textural feature common to many metamorphic rocks.

Continued burial produces a rock called phyllite (FILL-ite). The growth of mica crystals gives phyllite a silklike sheen and may destroy the original sedimentary layering. Although the mica crystals are still too small to be visible without magnification, the shiny appearance of the rock and the even more noticeable breakage along the foliation direction indicates that mineral changes continued. Schist, the next rock to form, has mica crystals that can be seen without magnification. Continued growth of mica crystals in a single direction adds to the foliation.

The final metamorphic product is gneiss (NICE). Not only can you see evidence of parallel crystal growth (foliation) in gneiss, but also the minerals may have separated into light- and dark-colored layers, parallel to the foliation. This is a property called **banding**. The light-colored bands are mostly quartz and feldspar. Feldspar is a new mineral that may not be seen in schist. The dark bands are mostly biotite, amphibole, and pyroxene. Figure 5-13 shows banding in gneiss. Some samples of gneiss do not show banding. Gneiss may also look like granite with a lot of feldspar, but without banding. However, even these samples will show a parallel alignment of mineral crystals.

The change from clay to mica and then to feldspar is not an isolated progression. Other minerals, many of them unique to metamorphic rocks, form and disappear along the way. Red garnet is a

FIGURE 5-13. Gneiss is a metamorphic rock that is often composed of minerals separated into light- and dark-colored bands. The light bands are rich in feldspar and quartz. Amphibole and pyroxene are the primary minerals in the dark bands.

good example. Garnet can often be seen as little pods in schist or gneiss.

Red garnet (almandine) is a result of mineral changes in formerly sedimentary or igneous rocks that contain both iron and aluminum. Due to the garnet mines in the Adirondack Mountains, this is the official New York State mineral by an act of the legislature.

Changes during the formation of a metamorphic rock destroy original structures in a rock, such as sedimentary layering and fossils. Gradually, these features are eliminated by foliation and crystal growth as the rock is subjected to more heat, more pressure, and more time.

The series of metamorphic rocks explained above presents the most common examples of the foliated metamorphic rocks. Yet some kinds of metamorphic rocks do not show foliation. When limestone is subjected to intense heat and pressure, calcite crystals grow and the rock changes from limestone to marble. The growth direction of calcite crystals is not affected by the force of the overlying rock or by movements of Earth's crust. This is why foliation does not occur in marble. Sometimes marble shows a swirled layering, but this is probably due to differences in composition of the original limestone layers. When sandstone changes to quartzite, and conglomerate to metaconglomerate, these metamorphic products do not show parallel crystal alignment. Therefore, marble,

FIGURE 5-14. Marble is a light-colored metamorphic rock composed mostly of calcite. Marble forms when limestone is subjected to intense heat and pressure. Both marble, on the left, and metaconglomerate, on the right, are nonfoliated metamorphic rocks.

quartzite, and metaconglomerate are commonly *nonfoliated* metamorphic rocks, as seen in Figure 5-14.

Regional and Contact Metamorphism

Metamorphic rocks can be separated into two groups by their origin. Sometimes, large-scale movements of Earth's crust cause a huge region of rock to sink into Earth. When this occurs, a large mass of rock is changed by heat and pressure. This process is called **regional metamorphism**. As the rock is drawn deeper into Earth, chemical changes in the minerals, crystal growth, and compaction cause the original parent rock to be metamorphosed. Regional metamorphism often creates rocks that are both foliated and highly distorted, or folded, as in Figure 5-15.

If metamorphic rocks form deep within Earth, why do we find them at the surface? In Chapter 14, you will learn that large-scale movements of Earth's crust are related to heat flow from deep within the planet. Forces push rock to the depths where metamorphism occurs. Forces can also uplift metamorphic rocks and the rocks covering them to form mountain ranges. After uplift, weathering and erosion wear down the mountains to expose the

FIGURE 5-15. The complicated folding of this banded gneiss was caused by intense pressure, temperature, and mountain-building forces deep within Earth.

regional metamorphic rocks. This process may take millions of years.

The next group of metamorphic rocks occurs over a smaller area. An intrusion of hot, molten magma will change the rock with which it comes in contact. This process is called **contact metamorphism**. In this environment, rocks are not exposed to the intense pressure that is found deeper within Earth. Therefore, rocks that have undergone contact metamorphism do not show foliation. The farther you go from the heat source (intrusion), the less the parent rock has changed. In fact, it is common to find metamorphic rock grading into the original sedimentary or igneous rock within a few meters of the heat source. This change can sometimes be observed as a decrease in crystal size as you move from the intensely baked rock next to the intrusion into rock that has been altered less by the heat. Hornfels is a name often applied to contact metamorphic rock of various mineral compositions.

The *Earth Science Reference Tables* contains the Scheme for Metamorphic Rock Identification. (Figure 5-16 on page 136.) The four rock types at the right in the top half of the chart are the four foliated metamorphic rocks. They are listed in order of increasing

Scheme for Metamorphic Rock Identification

TEXTURE		GRAIN SIZE	COMPOSITION	TYPE OF METAMORPHISM	COMMENTS	ROCK NAME	MAP SYMBOL
FOLIATED	MINERAL ALIGNMENT	Fine	MICA	Regional (Heat and pressure increases)	Low-grade metamorphism of shale	**Slate**	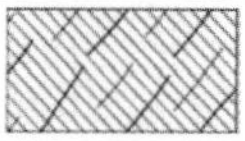
FOLIATED	MINERAL ALIGNMENT	Fine to medium	MICA, QUARTZ, FELDSPAR, AMPHIBOLE, GARNET	Regional (Heat and pressure increases)	Foliation surfaces shiny from microscopic mica crystals	**Phyllite**	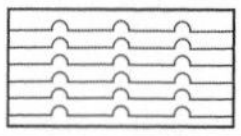
FOLIATED	MINERAL ALIGNMENT	Fine to medium	MICA, QUARTZ, FELDSPAR, AMPHIBOLE, GARNET, (PYROXENE)	Regional (Heat and pressure increases)	Platy mica crystals visible from metamorphism of clay or feldspars	**Schist**	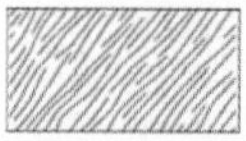
FOLIATED	BAND-ING	Medium to coarse	MICA, QUARTZ, FELDSPAR, AMPHIBOLE, GARNET, PYROXENE	Regional (Heat and pressure increases)	High-grade metamorphism; mineral types segregated into bands	**Gneiss**	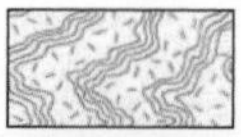
NONFOLIATED		Fine	Carbon	Regional	Metamorphism of bituminous coal	**Anthracite coal**	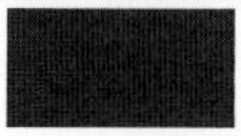
NONFOLIATED		Fine	Various minerals	Contact (heat)	Various rocks changed by heat from nearby magma/lava	**Hornfels**	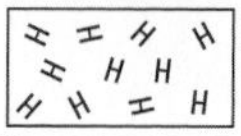
NONFOLIATED		Fine to coarse	Quartz	Regional or contact	Metamorphism of quartz sandstone	**Quartzite**	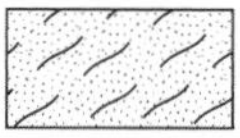
NONFOLIATED		Fine to coarse	Calcite and/or dolomite	Regional or contact	Metamorphism of limestone or dolostone	**Marble**	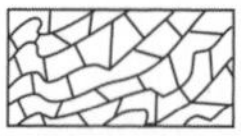
NONFOLIATED		Coarse	Various minerals		Pebbles may be distorted or stretched	**Metaconglomerate**	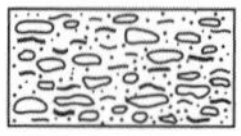

FIGURE 5-16.

metamorphic change and increasing grain size. These four rocks show the progressive metamorphism of shale that was explained earlier in this section. In the "composition" column, note the shaded bars that indicate the mineral makeup of these rocks. These bars show that minerals in the mica family are found in all four foliated metamorphic rock types. Quartz, feldspar, amphibole, and garnet are not common in slate, but are common in the three foliated rock types below slate. Of the six minerals shown here, pyroxene is the best indicator of extreme conditions of heat and pressure.

Unlike the rocks in the top half of this chart, the four rocks named at the right in the bottom half of the chart do not represent a progressive change. Each has a different mineral and chemical composition. These are the four most common metamorphic rocks that do not show foliation.

STUDENT ACTIVITY 5-5 —IDENTIFICATION OF METAMORPHIC ROCKS

1: SCIENTIFIC
INQUIRY 1
6: MODELS 2

Obtain a set of metamorphic rocks from your teacher. Handle them carefully and let your teacher know if any samples are missing or damaged.

Use the information in this chapter and the appropriate chart in the *Earth Science Reference Tables* to identify each of the metamorphic rocks in your set. List the name of each rock, such as schist or marble, along with the characteristics you observed that helped you to identify it.

WHAT IS THE ROCK CYCLE?

4: 2.1m
6: MODELS 2

There is a popular saying, "The only thing constant is change." This saying reminds us that everything around us is changing. Many geological changes occur so slowly that they are difficult to observe. Nearly everywhere, rocks are slowly changing as they adjust to the conditions and environment in which they are found. These changes are shown in a model of Earth environments and materials called the **rock cycle**. (See Figure 5-17 on page 138.)

Planet Earth receives only a very small amount of matter from outer space in the form of meteorites. At the same time, a small amount of Earth's atmosphere escapes. Therefore, in terms of mass, the planet is nearly a closed system. (A closed system is one that has no exchange with the environment outside itself.) Rocks change from one form to another.

In Figure 5-17, within the rectangles are the three major categories of rocks: sedimentary, metamorphic, and igneous. Sediments and magma are shown in ovals because, although they are important substances within the rock cycle, they are not actually kinds of rock. The lines and arrows show the processes that change materials as they go through the rock cycle. The terms printed along these lines tell you what changes are represented and the order in which they occur.

For example, magma can change to a different substance in only one way: it forms igneous rock by the process of solidification, or crystallization. However, igneous rock can change to another

Rock Cycle in Earth's Crust

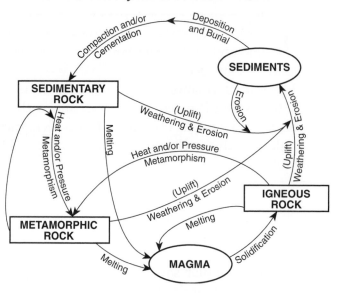

FIGURE 5-17. This diagram from the *Earth Science Reference Tables* shows that nearly all rocks are a result of geologic processes operating on other rocks. (Bioclastic and crystalline sedimentary rocks are exceptions.)

substance by any of three paths. If it is heated and melts, it can turn back into molten magma. If the igneous rock undergoes intense heating and possibly pressure, but does not melt, metamorphism will transform the igneous rock into a metamorphic rock. In the third possible path, the igneous rock is exposed to the atmosphere, probably by being pushed up to the surface. There air, water, and weather break it down and carry it away as sediments. The term "uplift" appears in parenthesis because uplift to expose the rock at the surface is likely, but it is not necessary. In the case of a lava flow onto the surface, no uplift is needed to expose the rock to weathering and erosion.

The rock-cycle diagram illustrates that nearly all rocks are made from the remains of other rocks. In the rock cycle, as you follow the arrows that show changes, notice that each begins and ends with Earth materials. None of the arrows comes in from outside the diagram. None of them take Earth materials out of the system. It is basically a closed system. To watch an animation of the rock cycle, visit the following Web site: *http://www.geolsoc.org.uk/rockcycle*

There is one group of sedimentary rocks that does not fit into this rock cycle. The organic sedimentary rocks, such as limestone and coal, are formed from the remains of plants and animals, not other rocks.

Some rocks show a complex origin. You may remember that conglomerate is made from pebbles that are held together by a cementing material, such as silica (very fine quartz), clay, or calcite. If the conglomerate contains pebbles of gneiss, granite, and sandstone, each component of the conglomerate shows a different process of rock formation found in the rock cycle.

CHAPTER REVIEW QUESTIONS

PART A

1. Which igneous rock has a vesicular texture and contains the minerals potassium feldspar and quartz?

   (1) andesite
   (2) pegmatite
   (3) pumice
   (4) scoria

2. Which set of rock drawings below is correctly labeled?

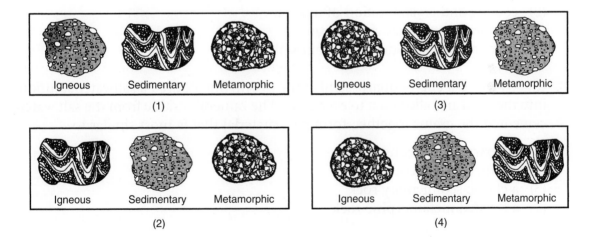

3. Dolostone is classified as which type of rock?

   (1) land-derived sedimentary rock
   (2) chemically formed sedimentary rock
   (3) foliated metamorphic rock
   (4) nonfoliated metamorphic rock

4. What is the origin of fine-grained igneous rock?

   (1) lava that cooled slowly at Earth's surface
   (2) lava that cooled quickly at Earth's surface
   (3) silt that settled slowly in ocean water
   (4) silt that settled quickly in ocean water

5. Which kind of metamorphic rock often forms from sandstone adjacent to an intrusion of magma near Earth's surface?

   (1) quartzite   (3) phyllite
   (2) marble   (4) slate

6. Which rock was formed organically and sometimes contains plant impressions?

   (1) rock gypsum
   (2) phyllite
   (3) breccia
   (4) coal

7. Wavy bands of dark minerals visible in gneiss bedrock probably formed from the

   (1) cementing together of individual mineral grains
   (2) cooling and crystallization of magma
   (3) evaporation of an ancient ocean
   (4) heat and pressure during metamorphism

8. Rita collected a cup of quartz sand at the beach. She poured a saltwater solution into the sand and allowed it to evaporate. The mineral residue from the salt water cemented the grains together, forming a material that is most similar to

   (1) an extrusive igneous rock
   (2) an intrusive igneous rock
   (3) a clastic sedimentary rock
   (4) a foliated metamorphic rock

9. Fossils of sea shells are found in a natural bedrock exposure. How did this rock probably form?

   (1) intense metamorphism of sedimentary rock
   (2) compaction and cementing of sediments
   (3) crystallization of magma deep underground
   (4) crystallization of lava at Earth's surface

10. What is the most common mineral in andesite?

   (1) quartz   (3) plagioclase feldspar
   (2) biotite   (4) potassium feldspar

11. What kind of rock is likely to form from lava extruded by a volcano?

   (1) light-colored metamorphic
   (2) dark-colored metamorphic
   (3) fine-grained igneous
   (4) coarse-grained igneous

PART B

Base your answers to questions 12 through 14 on the drawings below of six sedimentary rocks labeled A through F.

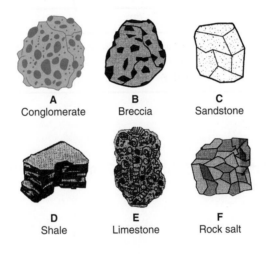

| A | B | C |
| Conglomerate | Breccia | Sandstone |

| D | E | F |
| Shale | Limestone | Rock salt |

12. Most of the rocks shown were formed by

   (1) volcanic eruptions and crystallization
   (2) compaction and/or cementation
   (3) crystal growth deep underground
   (4) melting and/or solidification

13. Which two rocks are composed primarily of quartz, feldspar, and clay minerals?

   (1) rock salt and conglomerate
   (2) rock salt and breccia
   (3) sandstone and shale
   (4) sandstone and limestone

14. Which table below shows the rocks correctly classified by texture?

Texture	Clastic	Bioclastic	Crystalline
Rock	A, B, C, D	E	F

(1)

Texture	Clastic	Bioclastic	Crystalline
Rock	A, B, C	D	E, F

(2)

Texture	Clastic	Bioclastic	Crystalline
Rock	A, C	B, E	D, F

(3)

Texture	Clastic	Bioclastic	Crystalline
Rock	A, B, F	E	C, D

(4)

PART C

Base your answers to questions 15 through 17 on the table below. Some information has been left blank.

Rock Sample Number	Composition	Grain Size	Texture	Rock Name
1	Mostly clay minerals		Clastic	Shale
2	All mica	Microscopic, fine	Foliated with mineral alignment	
3	Mica, quartz, feldspar, amphibole, garnet, pyroxene	Medium to coarse	Foliated with banding	Gneiss
4	Potassium feldspar, quartz, biotite, plagioclase feldspar, amphibole	5 mm		Granite

15. State a possible grain size, in centimeters, for most of the particles found in sample 1.

16. What is the rock name of sample 2?

17. Write a term or phrase that describes the texture of sample 4.

Base your answers to questions 18 through 20 on the flowchart below, which shows the formation of selected igneous rocks.

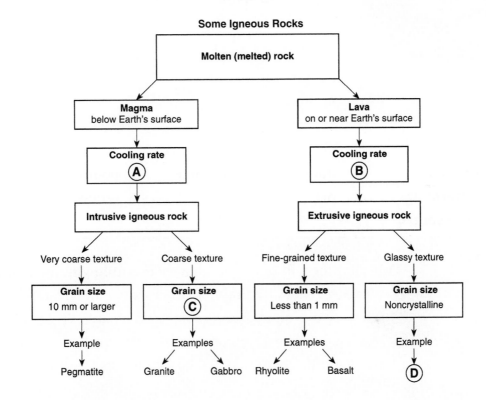

18. Compare the cooling rate at A with the cooling rate at B.

19. What is the numerical grain size at C?

20. Name one igneous rock that could be placed at D.

6

Managing Natural Resources

WORDS TO KNOW

conservation	nonrenewable resource	pollution
global warming	ore	renewable resource
natural resource		

This chapter will help you answer the following questions:

1 What are renewable and nonrenewable natural resources?

2 What are our most useful geological resources?

3 How can we conserve our natural resources?

4 What are the environmental consequences of using our natural resources?

HOW DOES THE GLOBAL MARKETPLACE WORK?

6: SYSTEMS
THINKING 1
6: EQUILIBRIUM AND
STABILITY 4
7: CONNECTIONS 1

A number of important changes in the world have dramatically changed commerce and the use of natural resources. The rise of civilization allowed individuals to work together for common goals: most notably food, shelter, security, and wealth. The Industrial Revolution added large-scale technology and mass production, allowing individuals to produce more goods. Continued progress has allowed people and products to travel worldwide. Globalization has joined national economies into a single worldwide economy, connecting humans as never before.

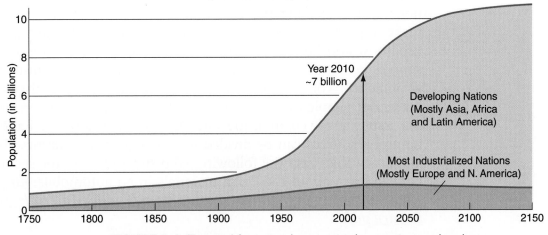

World Population Growth, 1750-2150

Year 2010
~7 billion

Developing Nations
(Mostly Asia, Africa
and Latin America)

Most Industrialized Nations
(Mostly Europe and N. America)

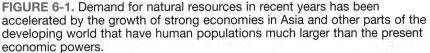

FIGURE 6-1. Demand for natural resources in recent years has been accelerated by the growth of strong economies in Asia and other parts of the developing world that have human populations much larger than the present economic powers.

But how do these changes affect you? While your great-grandparents probably got food and other needs from local suppliers, the world is very different today. For example, a car bought in your home town may be made with metals mined in Brazil or China. The car may be designed in Japan or Europe and built in the United States. Your fuel may come from Canada, Mexico, or Venezuela.

Managing our natural resources must take into account the global economy. Figure 6-1 shows that the greatest growth in world population is in the developing world. As these nations develop strong economies, they will use increasing amounts of natural resources. Visit the following Web site to see how the world population is changing and how other statistics change: *http://www.poodwaddle.com/worldclock.swf*

What is a Natural Resource?

The food you eat, the products you buy, and the clothes you wear are all produced by technology from Earth's natural resources. A **natural resource** is any material from the environment that is used to maintain people's lives and lifestyles. The word *resource* comes from an old French word that means "to arise in a new form." It is the responsibility of all people to make the best use of our natural resources, while conserving them for future generations and protecting the environment. At every step, we must make important decisions and compromises. How much will be used now and how much will be conserved? How can these resources be extracted while maintaining a clean and beautiful environment? Not everyone will agree with the decisions that are made. But people can all make better decisions if they understand the issues involved in using and conserving resources. Earth's natural resources can be divided into two groups; nonrenewable and renewable. Visit the following Web site and go to the second page to learn about the Earth materials in an automobile: *http://www.geosociety.org/educate/LessonPlans/Earth_Materials_in_Subaru.pdf*

WHAT ARE NONRENEWABLE RESOURCES?

4: 3.1c
6: SYSTEMS THINKING 1
6: EQUILIBRIUM AND STABILITY 4
7: CONNECTIONS 1

Nonrenewable resources exist in a fixed amount, or if they are formed in nature, the rate of formation is so slow that the use of these resources will decrease their availability. Rocks, minerals, and fossil fuels are the primary nonrenewable resources. In nearly every case, the use of these resources is increasing while the natural supply (reserves) is decreasing. This situation forces us to look for lower-quality reserves to meet our growing demands. Copper, which is mined from the ground, is a nonrenewable resource.

Metal Ores

Many of the minerals and rocks you studied in Chapters 4 and 5 are important resources. Rocks that are used to supply natural resources are called **ores**. The resources must be extracted, or separated, from their ores. Economically important metallic elements are obtained from minerals. Some elements, such as gold, silver, and copper, occur as native metals, that is, uncombined with other elements.

COPPER   The element copper is used in electrical wiring, plumbing, coins, and a wide variety of other applications. Gold is a better electrical conductor than is copper. However, copper is much cheaper than gold because it is more plentiful. Copper can be drawn into wire (ductile) and pressed into thin sheets (malleable). When copper is mixed with other metals such as zinc and tin, it forms alloys (mixtures of metals). These alloys, brass and bronze, are relatively hard, resist being deformed, and resist weathering.

Like most other geological resources, copper is extracted from rocks that have unusually high concentrations of copper. In some places such as northern Michigan, small amounts of native copper are found. (See Figure 4-2 on page 88) Most copper is obtained from minerals that were deposited by hot water circulating deep underground. In the early years of mining, some locations produced ores with as much as 10 percent copper. However, as the most concentrated ores were used up, new technologies were developed along with large-scale mining operations. Now, refiners can profitably extract copper from ores that contain less than 1 percent copper.

Most copper today is produced from open pit mines where giant machines scoop up tons of copper ore with each bite. The ore is transported to refineries where it is concentrated and purified.

GOLD Gold is sometimes found in sand and gravel deposits in streams and along shorelines. Due to its high density, gold transported by running water settles in places where the water slows. The pieces of gold get caught in cracks in the bedrock. (Bedrock is the solid rock that can be found everywhere under the soil and sediments.) Gold is also found in solid rock where it has been brought by hot water. Gold is valued mostly because of its mechanical and electrical properties. Gold is also attractive and does not corrode. Therefore, gold is often used in jewelry. Of the estimated 100,000 tons of gold that have ever been produced, nearly all of it is still in circulation.

IRON Today, iron is the principal metal of construction and technology. It is used in making frames for buildings, for automobiles, and even in eating utensils. Iron is found worldwide, and this metallic element is second only to aluminum in abundance in Earth's crust. Iron ore is easy to find and is inexpensive to extract.

Combining iron with small amounts of other metals makes steel. Steel is harder and more resistant to breakage and corrosion than iron. The most important iron ore deposits are the banded

FIGURE 6-2.
Banded iron ore is nonrenewable because it formed long ago when oxygen was just beginning to enter Earth's atmosphere.

iron formations that settled out of ancient seas early in Earth's history (Figure 6-2). At that time, the atmosphere contained very little oxygen. As oxygen in the atmosphere became abundant, bacteria extracted iron from the rocks and deposited concentrated iron oxide. While there are great quantities of iron ores, the fact that almost no new ores are being formed makes iron a nonrenewable resource.

Using Mineral Resources

Geologists locate ore deposits and estimate the amounts of metal that can still be extracted. Aluminum and iron are so abundant in Earth's crust that geologists are not concerned about running out of them. However, the best reserves of copper and gold in the United States have already been used up. For this reason, the amount of gold and copper we import is increasing. These rare metals are often recycled. About 25 percent of the copper produced in the United States is the product of recycling. The amount of a mineral resource that is recycled depends on its cost. The more it is worth, the more is recycled. Because of gold's high price, people do not toss away a gold object the way they discard aluminum soda cans and iron objects. Visit the following Web site to see a mineral resources map of the United States: *http://minerals.usgs. gov/minerals/pubs/mapdata/minesmap.gif*

STUDENT ACTIVITY 6-1 —ADOPT A RESOURCE

2: INFORMATION
SYSTEMS 1, 2, 3

*S*elect a specific nonrenewable resource. Then ask your teacher for permission to prepare a report on that resource. Your report should include where the resource is found, how it is refined or changed for use, and what the resource is used for. How can this resource be recycled profitably?

Fossil Fuels

Coal, petroleum, and natural gas are used for the production of electricity, for heating, and for transportation. The energy stored in these substances began as sunlight absorbed by prehistoric plants.

PETROLEUM While the United States is the world's largest user of petroleum (oil), it has only about 3 percent of the world's oil reserves. A serious interruption of the flow of foreign oil would have a devastating effect on our economy.

Petroleum formed from the remains of microscopic organisms that lived in the oceans hundreds of millions of years ago. The remains escaped complete decay when they sank to the cold, dark ocean bottom. The deposition of more layers of sediments trapped the organic remains. Over time, the organic remains changed to a complex, low-density liquid and natural gas. These substances moved upward until they were trapped by an overlying layer of fine-grained rock, such as shale. If an oil company has enough skill and luck it may drill into a layer where oil and gas are present. Then, it can pump the oil and gas out of the ground. Petroleum products are so valuable on world markets that oil is sometimes called "black gold." Figure 6-3 on page 150 predicts a decrease in petroleum production. As the world's primary source of energy, this will cause the price of petroleum products to increase rapidly.

COAL While the United States has large deposits of coal, it is used mostly in large power plants to generate electricity. In these power plants, large quantities of coal are handled efficiently and air pollution controls are used to reduce air pollution. The use of

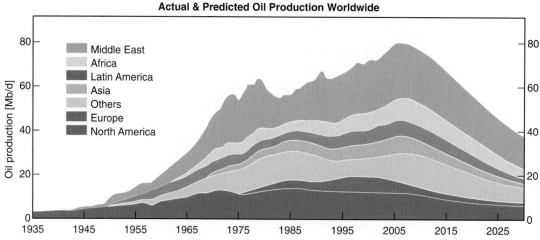

FIGURE 6-3. Some geologists predict that oil production will peak in the near future and then decline.

coal for home heating and cooking has decreased because oil, gas, and electricity are easier to use and cause less air pollution.

The fossil fuels are considered nonrenewable resources because we are using them much more quickly than natural processes can replace them. Some scientists estimate that in a few hundred years Earth could run out of fossil fuels that took 500 million years to form.

Construction Materials

By total mass, more Earth resources are used in construction than in any other application. The material used depends on the needs of the project and the materials available in the area. Sand, gravel, and crushed stone are spread as a structural base under pavements. These materials are used because they do not compress under heavy traffic, and they allow water to drain through them. A proper base layer helps the roads last longer than roads built directly on the ground. Sand and gravel are New York's most economically important mineral resources.

Concrete is made from a mixture of limestone (or dolostone) and clay that has been baked to drive off the water and carbon dioxide content of these materials. When water is added to the concrete, it "sets" to form a resistant building material. Cut stone, although it is more expensive than brick, concrete, or steel, is sometimes used for the outer facing of buildings. Like other resources, the price of concrete has been affected by economic growth in the developing countries.

WHAT ARE OUR MOST IMPORTANT RENEWABLE RESOURCES?

6: SYSTEMS
THINKING 1
6: EQUILIBRIUM AND
STABILITY 4
7: CONNECTIONS 1

Renewable resources are those resources that can be replaced by natural processes at a rate that is at least equal to the rate at which they are used. For example, sunlight is our planet's principal source of energy. Solar heat collectors make use of this energy. But, as fast as it is used, more solar energy arrives. Figure 6-4 is a map showing places in the United States that are most suited to harvesting solar energy. Sunlight also powers wind, another form of renewable energy. Electrical power from wind generators has become popular in several "wind farms" within New York State in recent years.

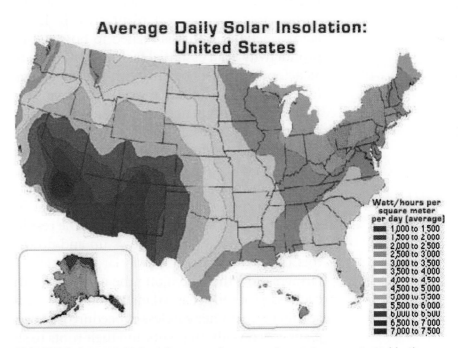

FIGURE 6-4. In the United States, solar energy is most concentrated in the desert of the Southwest. Here the sun is highest in the sky and the weather is often clear. Unfortunately, collecting solar energy is limited by high installation costs and its low concentration, requiring very large collection areas.

Although renewable resources may never run out, our use of these resources is sometimes limited by how quickly they are replaced. If your community uses a nearby river for its water supply, the amount of water used cannot be greater than the total amount of water flowing in the river.

Wood

Before people made extensive use of fossil fuels, wood was the fuel of choice. In most areas, wood could be cut locally. Even now, people burn wood to heat homes, prepare hot water, and cook food. Wood is a renewable resource. Once the trees are cut, new trees can grow back in as little as 10 years, or perhaps as long as a century. As long as usage and replacement are in balance, the supply could be unlimited.

Soil

For now, soil will be considered as the loose material at Earth's surface that can support plant growth. Later in the book, you learn a definition that is related to how soil is formed. Biological productivity is an essential feature of soil as a resource. If you look at a food chain, it begins with the organisms called producers (green plants). Plants are the only organisms that can live in an environment without needing other organisms for food. The producers use the substances they get from soil, along with air, water, and sunlight to create living tissue. All animals (including humans) depend on producers to transform the resources of the physical environment into food.

In some places, the best soil has been carried away by running water or by dust storms. This is often the result of careless farming practices. When a farmer plows up and down the slopes of hills, this allows runoff to carry away large amounts of soil. To avoid this, farmers plow the land along the contours of the hills. This makes each ridge of soil act as a dam to hold back the water and its load of sediments. Another careless procedure is leaving land without any plant cover. Without plants and their roots to slow runoff and hold soil in place, soil can be quickly washed or blown away.

The processes of weathering, infiltration of groundwater, and biological activity can restore soil, but the process could easily take hundreds of years. Although some people consider soil to be a renewable resource, the best practice is to protect soil through conservation.

Water

Water is another essential natural resource. All living things need water or they depend upon other organisms that need water. Plants use water during photosynthesis. People use water for drinking, transportation, waste disposal, industrial development, and a habitat for our food sources. Of Earth's water, 97 percent is salt water. Salt water is in the oceans and in several landlocked bodies of water, such as the Great Salt Lake in Utah. However, this water cannot be used for drinking and agriculture.

Only about 3 percent of Earth's water is freshwater. Glaciers and the polar ice caps hold 2 percent of Earth's freshwater. Of the remaining 1 percent, most is in the ground where it is not as easy to tap as are sources at the surface. Most of our water needs are fulfilled by the 0.008 percent of the total that is found as surface water in streams and lakes.

The Colorado River flows from the Rocky Mountains southwest through the desert toward the Pacific Ocean. Parts of seven states depend on Colorado River water. Through a series of agreements the states of Colorado, Utah, New Mexico, Wyoming, Arizona, Nevada, and California, as well as Mexico, have tapped into the river. Their residential, agricultural, and industrial use of the water uses the total volume of water flowing in the river. Unless there is a flood, none of the water from the Colorado River reaches the ocean. This area is the fastest-growing region in the country. How will this area find more water to support population and economic growth?

Water is usually considered a renewable resource. As water is drawn from surface sources, such as streams and lakes, more water flows in to replace it. There is an assumption here that once water is used, the water cycle replaces it. It is not always that simple.

In the 1920s, the federal government encouraged people to move into the Great Plains of the United States. People were offered loans and inexpensive land. Rich soil and good precipitation made this region seem to be a natural location for agricultural development. However, in the 1930s, there was a drought.

Meanwhile the new farm owners had plowed the land and it was exposed to wind erosion. The dust storms of the 1930s blew away much of the most productive soil. Some farmers drilled wells to bring water to the surface. That part of the United States is over a natural groundwater reservoir called the Ogallala aquifer. While farming continues today, it often depends on irrigation water drawn from the aquifer. Each year, the water level in the aquifer goes down because more water is withdrawn than nature can restore.

Lack of freshwater is a problem in parts of Africa and other arid (dry) regions of developing countries. Temporarily, new wells will help these people find clean water for drinking and household use. But what will happen when their aquifers are depleted?

HOW CAN WE CONSERVE RESOURCES?

6: SYSTEMS
THINKING 1
6: EQUILIBRIUM AND
STABILITY 4
6: OPTIMIZATION 6
7: CONNECTIONS 1

Conservation is the careful use, protection, and restoration of our natural resources. To practice conservation, people need to estimate their resources and forecast their needs. Only then can they decide what needs to be done. For example, although the use of aluminum is increasing, Earth's crust contains so much aluminum that there seems little danger of running short. However, the future for gasoline and other petroleum products is not as bright. Some people estimate that petroleum will start to run short in just a few decades. There is a popular saying "Necessity is the mother of invention." Necessity to replace petroleum fuels may be getting close. But it is not clear what "inventions" will carry us through this crisis. Conservation generally involves one or more of the three "R's" of ecology: reduce, recycle, and replace.

Reduction

If you were to find a soda can that was made 50 years ago, you might be surprised at how heavy it would be. Changes in the manufacturing process and the shape of soda cans have allowed beverage companies to use much less metal, generally aluminum, than ever before. People who carpool, drive small cars or drive hybrids not only save money, but they also use less gasoline. This helps to extend our limited and petroleum reserves.

Recycling

A large fraction of the copper we use has been recovered from old buildings, automobiles, and electrical parts. If you grew tired of a piece of gold jewelry, you would probably not put it into the trash. More likely, you would sell it for its value in gold. While aluminum is plentiful, it is costly to extract and purify from its ore. This has contributed to programs to recycle aluminum cans. (Recycling programs also help the environment by making it less likely that people will throw empty soda cans along highways.) Each time we use recycled metals we delay the inevitable time when these resources will run out.

Replacement

Replacement is often the long-term solution to decreasing resources. In the past, most people heated their homes and cooked their food by burning wood. Wood is a renewable resource. However, as the human population grew and people moved into cities, nearby wood sources could not keep up with the growing need for fuel. Coal and then petroleum fuels replaced wood as the fuel of choice. In many applications, plastics are replacing wood and leather. Necessity will certainly continue to bring about new solutions to our problems.

STUDENT ACTIVITY 6-2 —WATER USE IN THE HOME

6: OPTIMIZATION 6

Measure or estimate the amount of water that your family uses at home on a typical day. To do this you will need to identify all the household devices and activities that use water. You must also find ways to determine how much water these uses require. (You may find useful information in reference sources, on the Internet, on your water bill, or from your water meter.) After you have determined your family's daily water usage, suggest ways to reduce your water consumption.

WHAT ARE THE EFFECTS OF ENVIRONMENTAL POLLUTION?

6: SYSTEMS
THINKING 1
6: EQUILIBRIUM AND
STABILITY 4
6: OPTIMIZATION 6
7: CONNECTIONS 1

Pollution is any material or form of energy in the environment that harms humans or the plants and animals on which they depend. Pollution can be classified according to the part of the environment that has been affected: ground, water, and/or air.

Ground Pollution

The city of Niagara Falls, New York, became an industrial area because of the availability of inexpensive hydroelectric power. Chemical plants were built in the area to take advantage of this local resource. Unfortunately, the local government did not monitor or regulate the industry and its disposal of chemical wastes. When a local waterway called Love Canal was filled in, drums of toxic chemicals were buried. Eventually the land was given to the city of Niagara Falls. The city allowed low-cost housing and an elementary school to be built on the land.

Toxic liquids began to seep out of the ground. People, especially children, developed health problems. The toxic chemicals were linked to these health problems in the 1970s. The residents were shocked and angry. They did not want their families, especially the children, to be exposed to these chemical waste products.

As the tragedy became news, the value of homes in the area decreased dramatically. People could not sell their homes and they could not afford to buy homes in another area. The homeowners felt let down by their local government, which should have informed them about the chemical danger at Love Canal.

Another example of ground pollution can be found in agricultural areas. Insecticides have been sprayed to protect crops from insect damage. Over the years, these chemicals may remain in the soil and make the soil unfit for other crops. Furthermore, windblown dust can be a hazard to farm workers. In addition, dust can carry the chemicals to nearby residential areas. Children and senior citizens are especially sensitive to diseases caused by exposure to dangerous substances.

Weapons production, nuclear power generation, health services and a wide variety of other important uses have created tons of radioactive waste. Much of it is mildly radioactive and can be buried in special landfills. However, high-level radioactive waste can be isolated in places such as deep mine tunnels where it will still be very dangerous for centuries. How to dispose of or store these radioactive materials has been in debate for decades, and the debate is likely to continue for many more years.

Water Pollution

It is more efficient to send electrical power over long distances if the voltage is stepped up to levels that would be dangerous in homes. The General Electric Company (GE) needed to build transformers to change the high-voltage current sent over transmission lines into ordinary household current. Transformers will overheat if they are not cooled by a liquid that carries away excess heat. Water is a good absorber of heat, but it can evaporate or boil away. In the mid-1950s, GE began to use liquids called polychlorinated biphenyls (PCBs) in transformers. PCBs are good absorbers of heat and are less likely to boil away than is water. Because PCBs are chemically stable, the company felt that there was little danger when these substances leaked into the environment. Two manufacturing plants along the Hudson River north of Albany, New York, used the chemicals. Neither GE nor the state knew that PCB pollution would become a serious environmental issue.

After PCBs had been used for several decades, scientists discovered that PCBs could cause serious health problems if they were in drinking water. Scientists found that fish in the Hudson River had absorbed PCBs when they ate plants or other fish containing PCBs. The chemicals built up in the bodies of the fish. People learned that PCBs were not as safe as they at first appeared to be. Meanwhile, dangerous amounts of PCBs had washed into the Hudson River at the two manufacturing plants. The government warned people about the danger of eating fish from the river and looked for a long-term solution.

GE and environmental groups have disagreed on what to do about the contamination. Environmentalists (and the courts) have said that GE should remove the PCBs by dredging the most contaminated mud and dumping it on land. GE wants to leave the sediments where they are. The company says that the PCBs will eventually wash out of the river. Like many environmental controversies, all solutions have some disadvantages. The current settlement calls for GE to pay for dredging that began in 2009.

Contamination of water supplies is a problem in parts of the world where manufacturing is growing. In their efforts to compete in the world market with low-priced goods, factories may be built without adequate methods to prevent environmental pollution.

Air Pollution

Air pollution is a concern in most urban areas. Exhaust gases from cars and airborne discharge from homes, businesses, and industries can produce high levels of ozone and oxides of nitrogen. Even wind-blown dust can be a health hazard. Serious air pollution events have occurred in many cities. The response to rising air pollution is the use of cleaner fuels and pollution control devices on motor vehicles and electrical power plants that burn fossil fuels. Some dust pollution in the United States has been traced back to Asia as its source. (China is experiencing its own "Dust Bowl.")

Some air pollution issues have been dealt with quickly while other forms of pollution are more difficult to address. Scientists discovered that chlorofluorocarbon gases (CFCs) given off by some spray cans and air conditioners were weakening Earth's protective ozone layer. (Ozone is needed in the upper atmosphere because it helps protect Earth from harmful shortwave solar energy.) Government and businesses have found other chemicals to use in air conditioning systems and spray cans. These chemicals do not break down ozone in the upper atmosphere.

Our planet faces more difficult issues. By burning fossil fuels, humans have increased the carbon dioxide content of Earth's atmosphere by about 25 percent. Most of that increase has occurred in the last 50 years. The concern is that carbon dioxide absorbs heat energy that might otherwise escape into space. The increase in carbon dioxide concentration in the air is likely to cause **global warming**, an increase in the average temperature on Earth. While some locations might benefit from a warmer climate, productive farmland could be changed into desert. Another consequence could be the melting of polar ice caps and a rise in sea level, which would drown coastal cities. Unfortunately, our society depends primarily on fossil fuels for heating, electrical power, and transportation.

As the price of fossil fuels increases and the effects of global warming encourage conservation, there will be a shift to other sources of energy such as nuclear power, solar, and wind energy. Uranium for nuclear power is easily available. Nuclear energy sources do not create greenhouse gases such as carbon dioxide However, critics worry about releases of radiation and secure storage of radioactive waste products. Rising prices will also pressure consumers to use less energy. We can use less energy by constructing more efficient buildings, using smaller cars and mass transportation, and changing other habits to more carefully use heating and air conditioning.

CHAPTER REVIEW QUESTIONS

PART A

1. What is Earth's most important renewable energy source?

   (1) coal   (3) sunlight
   (2) natural gas   (4) uranium

2. Which metal is most common in Earth's crust?

   (1) iron   (3) aluminum
   (2) copper   (4) magnesium

3. Which of the following is a fossil fuel?

   (1) wood   (3) uranium
   (2) coal   (4) solar energy

4. Limestone and calcite are valuable natural resources. Which product is made from these raw materials?

   (1) gasoline   (3) cement
   (2) lubricants   (4) electrical insulators

5. Why do products made from metal ores become more expensive through time?

   (1) These resources are quickly replaced by natural processes.
   (2) Ores that are easy to mine are used first.
   (3) Exploration and manufacturing technologies do not change.
   (4) Alternative products quickly replace refined metals.

6. What property of gold is often used to separate it from other natural Earth materials?

   (1) Gold is an electrical insulator.
   (2) Gold has a glassy luster.
   (3) Gold one of the most abundant common minerals.
   (4) Gold is more dense than other minerals.

7. Which lists energy resources in the historical order of their use by humans?

   (1) wood, coal, petroleum, uranium
   (2) uranium, petroleum, coal, wood
   (3) coal, petroleum, uranium, wood
   (4) petroleum, uranium, wood, coal

Base your answers to questions 8 and 9 on the satellite image of the New York City area below.

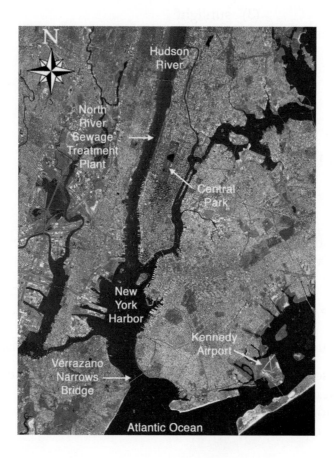

8. Which feature highlighted on the satellite image is the greatest potential source of water pollution?

   (1) Central Park
   (2) North River Sewage Treatment Plant
   (3) Verrazano Narrows Bridge
   (4) Kennedy Airport

9. As the distance from the center of New York City increases, measurable air pollution most likely

   (1) decreases
   (2) increases
   (3) does not change

10. What mineral that is common in New York State could be used as a source of the element magnesium?

    (1) coal   (3) dolomite
    (2) quartz   (4) magnetite

11. Which is the most logical first step in a long-term plan to reduce the amount of water that your family uses in the home?

    (1) Turn off the main supply of water to your home.
    (2) Measure how much water you use for various purposes.
    (3) Install an irrigation system for your lawn and gardens.
    (4) Purchase a new water softener and water heater.

12. Which group of materials does not come directly from the lithosphere?

    (1) metals such as iron and copper
    (2) fuel, including coal and oil
    (3) jewelry, including gold and silver
    (4) food, including meat and fish

13. Why is ocean water generally unfit for most human uses?

    (1) Ocean water is generally poisonous.
    (2) Ocean water has a high concentration of mineral salts.
    (3) Ocean water is a very limited water supply.
    (4) Ocean water is very expensive.

PART B

Base your answers to questions 14 through 16 on the world map below, in which the size of the nations represents their reserves of petroleum.

Who has the oil?

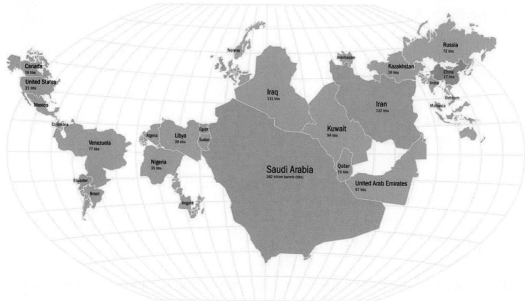

14. What part of the world can supply the most oil to world markets?

    (1) Western Europe
    (2) The Middle East
    (3) South America
    (4) The United States

15. If this map were to show by size the major users of petroleum, rather than the major sources, what part of the map would most likely increase in size the most?

    (1) Africa
    (2) The Middle East
    (3) South America
    (4) The United States

16. How could the United States most effectively reduce its use of imported oil?

    (1) Use government money to reduce the price of gasoline.
    (2) Conserve energy by using more fuel-efficient vehicles.

(3) Impose new environmental restrictions on mining and oil drilling in the United States.

(4) Eliminate all government money to mass transit projects.

PART C

17. If gold is a better conductor than copper, why do we use more copper than gold in electrical devices?

18. The fossil fuels are considered geological resources, but not mineral resources. Why?

19. Why is wood considered to be a renewable resource?

20. Why might there be public opposition to starting a coal mine in your community?

UNIT 3

Weathering and Erosion

There is a saying, "Nothing is permanent but change." This means that change is the natural state of our world. As you learned earlier, science changes. As a result of new observations, scientists come to new understandings about how the world works. A similar change occurs in your own outlook. As you grow older, not only does your world change, but your way of looking at things changes.

Even without humans or any other life forms, change is the natural state of Earth. Many geological changes occur deep within Earth or at such a slow rate that is hard to for anyone to observe them. Even natural landmarks will change. The photo on this page was taken in 1973. At the bottom of page 183, you will find another that was taken in 2000, 27 years later. Look carefully at the images. Are they identical? No. In 1975, the balanced rock in the background fell during a thunderstorm. In just a few seconds, this landmark was changed forever. How long do you think it will continue to look as it does now?

7

Weathering and Soils

7

WORDS TO KNOW

abrasion	frost wedging	residual soil
bedrock	infiltration	soil
biological activity	mechanical weathering	soil horizon
chemical change	organic matter	transported soil
chemical weathering	physical weathering	weathering
exfoliation		

This chapter will help you answer the following questions:

1 Why do rocks change when they are exposed at Earth's surface?

2 What are the two major classes of weathering?

3 How does chemical weathering differ from physical weathering?

4 What factors affect the rate at which rocks weather?

5 How do soils form?

WHY DOES WEATHERING OCCUR?

4: 2.1s
6: PATTERNS OF
CHANGE 5

In Unit 2 you learned about the solid materials that make up our planet. You were introduced to rocks, minerals, and the rock cycle. The rock cycle is a model of how these materials change. Many geological changes occur deep within Earth. Others occur at such a slow rate that is hard for a person to observe them. For example, Niagara Falls has been eroding upstream at the rate of 1 meter per year. (In recent years this upstream movement has been slowed by water diversion above the falls.) However, some geological changes occur quickly. If you look at a stream before and after a flood, the whole path of the stream may have changed. In 2003, many years of weathering and frost action led to the collapse of the Old Man of the Mountain, one of New Hampshire's most famous landmarks.

When exposed to conditions at Earth's surface rocks change. This change is called **weathering**. Weathering is influenced by exposure to wind, water, oxygen, plants, and animals. All these agents contribute to breaking down **bedrock**, the solid, or continuous, rock that extends into Earth's interior. The weathering of rock creates a loose substance known as *sediment*.

What is Physical Weathering?

Weathering processes can be classified as physical changes or chemical changes. **Physical weathering**, also known as **mechanical weathering**, breaks rocks into smaller particles. However, the chemical (mineral) composition of the particles does not change from the composition of the original rock.

ABRASION   The grinding away of rock by friction with other rocks is **abrasion**. Abrasion is a form of physical, or mechanical, weathering that occurs when pieces of rock collide or rub against one another and against the underlying bedrock. Consider a large rock that breaks off bedrock near the head of a stream. It is likely that the rock will fall into the stream and begin its journey downstream. Repeated collisions with other rocks gradually wear down the rock, grinding off any corners as it travels downstream. The farther the fragments (pieces) travel, the smaller and more rounded they become. (See Figure 7-1.) What began as a large piece of rock near the head of a stream is changed into small bits of sediment as it travels downstream. These rounded pieces are called river rocks.

FROST WEDGING   Have you ever seen the damage caused by frozen pipes in your home? Or have you left a bottle of water in

FIGURE 7-1. The three rounded river rocks may have begun their journey as larger, more angular fragments like the rock behind them.

the freezer, only to find it cracked the next day because the water inside expanded as it froze?

Water is one of the few substances that expands when it forms a solid (freezes). If this were not true, ice would sink in water. Rivers and lakes would freeze from the bottom up, killing fish and other organisms that can survive only in liquid water. Water is sometimes trapped in cracks in rock. **Frost wedging** is a kind of weathering that occurs in moist places when the temperature alternates between day temperatures above the freezing point of water and night temperatures that are below freezing. With repeated cycles of freezing and thawing, frost wedging widens the cracks, gradually forcing the rock apart. The force created when water freezes can also open new cracks for water to enter. In this way, solid rock is broken into smaller fragments.

Water can weather rocks in other ways. Figure 7-2 on page 168 is a close-up of beach sand from the island of Hawaii compared with pebbles from a beach in California. On Hawaii, molten lava flows into the ocean. The lava is cooled very quickly by contact with the sea water. The rapid temperature change causes the rock to shatter into tiny, angular fragments. Most beach sand is composed of rounded particles shaped by abrasion. But this "shatter sand" is as sharp as broken glass.

EXFOLIATION Granite forms by slow cooling and crystallization well below Earth's surface. When granite solidifies, the rock

FIGURE 7-2. Unlike the rounded pebbles found on a California beach, shatter sand is angular.

is under great pressure caused by the weight of rocks above it. When thousands of years of uplift and weathering expose granite at Earth's surface, there is a great reduction in pressure on the rock. In addition, changing temperatures in daily and annual cycles weaken the surface layer of the granite rock. The result is **exfoliation** as granite near the surface expands and cracks into slabs that break away from the solid bedrock.

There is no change in composition involved in this process, so it is a physical change. Figure 7-3 is a rock that was apparently abraded (worn down by collisions) into a rounded "river rock" before it weathered by exfoliation and was broken on one end.

BIOLOGICAL ACTIVITY   Rocks are also broken apart by **biological activity**. When the roots of a tree or other plant grow into

FIGURE 7-3. A rock that was most likely shaped by stream abrasion, exfoliation, and angular breakage.

FIGURE 7-4.
Lichens live on solid rock surfaces and help break down the rock.

a crack in solid rock, the roots apply a constant pressure that can help break the rock apart. Lichens and mosses sometimes grow on rock surfaces, as you can see in Figure 7-4. These tiny plant-like organisms also help break the rock apart. Burrowing animals such as earthworms, ants, woodchucks, and rabbits create passages through soil and allow water and air to come into contact with unweathered or partially weathered rock.

Why Do Some Rocks Last Longer Than Others?

Not all rocks wear away at the same rate. The harder a rock is, the more resistant it is to physical, or mechanical, weathering. Resistance to abrasion depends on which minerals make up the rock and how the rock is held together. For example, quartz is a relatively hard mineral (Mohs scale hardness = 7). A rock of solid quartz is likely to wear away very slowly. Although sandstone is often composed of quartz grains, sandstone will weather quickly when the quartz grains are not securely cemented. Limestone is made primarily of calcite (Mohs hardness = 3). However, a solid layer of limestone can be more resistant to abrasion and other forms of physical weathering than poorly cemented sandstone. In general, the softer parts of a rock weather more quickly than the harder parts. This differential weathering may result in a texture of harder layers sticking out more than softer layers, as you can see in Figure 7-5 on page 170.

FIGURE 7-5. In general, the softer layers of rock weather more than the harder layers, leaving the harder layers sticking out.

The Grand Canyon in Arizona is 1.5 km deep. The walls of the canyon expose more than a dozen major rock formations. The most resistant layers form the steepest rock faces because they wear away slowly and have the most strength. The weaker layers tend to form terraces because soft rocks do not have the strength to hold up as cliffs. Weak rocks can also be found as indentations or notches protected above and below by stronger rocks.

STUDENT ACTIVITY 7-1 —ROCK ABRASION

1: SCIENTIFIC
INQUIRY 1, 2, 3
6: PATTERNS OF
CHANGE 5

Materials: rock chips, mass scale, wide-mouth plastic jar with lid, sieve or strainer, plastic bucket

You can model the conditions in a fast-moving stream by placing rock chips and water in a wide-mouth plastic jar and shaking the jar. Before beginning the activity, your teacher will have soaked the rock chips in water for an hour or more. Measure out a mass of approximately 100 g of rock chips to the nearest 0.1 g. Place a few centimeters of water in the plastic jar, add the rock chips, and screw the top on tightly. (Be sure the jar does not leak.) Shake the jar vigorously for 4 minutes. Using a strainer to catch the rock chips, pour the water into a bucket. Find the mass of rock remaining after shaking and record it.

> Repeat the procedure with the same rock fragments, shaking them for two additional 4-minute intervals. Record the mass of rock remaining after each 4 minutes of shaking. Create a graph that shows the initial mass and the mass after each 4 minutes of shaking. [Label abrasion time on the horizontal (x) axis, and mass remaining on the vertical (y) axis.] If different groups use different kinds of rock, compare the data to decide which kind or kinds of rock are abraded more quickly.

What Is Chemical Weathering?

Sometimes the weathering process does more than simply break the rocks into smaller pieces. If you find a steel nail that has been exposed to the weather for a long time, it will probably be rusted. Rusting is a **chemical change**, which results in the formation of a new substance. Iron, the major ingredient in steel, can combine chemically with oxygen in the atmosphere to form rust (iron oxide). The presence of moisture (water) speeds up the rusting process.

Chemical weathering is a natural process that occurs under conditions at Earth's surface, forming new compounds. Rusting is a form of chemical weathering because a new substance (rust) is formed. Although steel is not found in nature, many minerals do contain iron. Iron is often one of the first parts of a rock to weather. When iron combines with oxygen in the atmosphere it usually forms iron oxide, which gives rock a rusty red to brown color. The chemical equation for this change is below. (As you can see on page 16 of the *Earth Science Reference Tables,* Fe is the chemical symbol for iron.)

$$4Fe \;+\; 3O_2 \;\rightarrow\; 2Fe_2O_3$$
$$\text{iron} \;+\; \text{oxygen} \;\rightarrow\; \text{iron oxide}$$

Calcite, the principal mineral in limestone and marble, is chemically weathered by water that is acidic. The chemical formula for limestone is $CaCO_3$ (calcium carbonate). Rainwater absorbs carbon dioxide as it falls through the atmosphere, making rain a weak acid (carbonic acid, H_2CO_3). This is not strong enough to hurt you or your clothing, but it can slowly break down limestone. When rainwater enters the ground, it picks up more carbon dioxide from decaying plant remains. The acid (represented by H⁺) can then react with limestone. The chemical equation for this change is written as

$$CaCO_3 + 2H^+ \rightarrow Ca^{2+} + H_2O + CO_2$$

| calcium carbonate | + acid → | calcium ions | + water + | carbon dioxide |

This process forms limestone caverns, such as Howe Caverns and Secret Caverns near Cobleskill, New York. Although the longest limestone caverns in New York State have been explored to about 10 km, Mammoth Caves in Kentucky have more than 500 km of connected underground passages. Visit the following Web site to search a database of the longest and deepest caves of the world: *http://www-sop.inria.fr/agos-sophia/sis/DB/database.html*

Limestone and marble make excellent building stones, although the calcite in them has a hardness of only 3 on the Mohs scale. These rocks are soft enough to be cut into blocks but strong enough to support the weight of a large building. Limestone and marble are also relatively easy to shape into sculptures and ornaments.

The burning of fossil fuels adds sulfur dioxide to the atmosphere. This gas forms sulfuric acid when it combines with moisture in the atmosphere. Nitrogen oxides from motor vehicles and electrical power plants form nitric acid when combined with moisture in the atmosphere. When acid precipitation (rain or snow) falls on limestone and marble it changes the mineral calcite into a chalky powder. Many historic buildings and outdoor statues are built with limestone and marble. In Europe and North America many old buildings have been damaged by acid weathering. This is a major reason there are laws to limit acid pollution. Although these measures cannot repair damaged structures, they have slowed the further chemical weathering of buildings and monuments made of limestone and marble.

Surface Area   Figure 7-6 shows that breaking a rock into smaller fragments increases the surface area of the material. Weathering occurs on surfaces. Breaking up a rock exposes more surface area and accelerates the rate of weathering. Visit the following Web site to learn about sand: *http://www.paccd.cc.ca.us/SAND/SANDHP.htm*

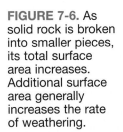

FIGURE 7-6. As solid rock is broken into smaller pieces, its total surface area increases. Additional surface area generally increases the rate of weathering.

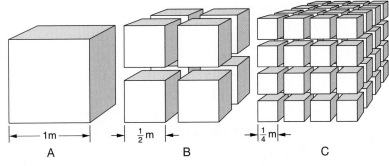

STUDENT ACTIVITY 7-2 —CALCULATING SURFACE AREA

1: MATHEMATICAL ANALYSIS 1, 2, 3

Figure 7-6 shows a single cube of rock 1 meter on each side (A) that is divided into smaller and smaller pieces. Calculate the total surface area of the samples in parts A, B, and C of the diagram. Show your work. Start with an algebraic equation, substitute numbers and units, and show the mathematical steps to each solution.

Feldspar is the most common mineral in rocks at or near Earth's surface. But feldspar is not stable when it is exposed to the atmosphere over very long periods of time. Feldspar weathers to a softer material composed primarily of clay and silica.

Figure 7-7 shows how the mineral composition of granite changes as chemical weathering takes place. The unweathered rock is composed mostly of quartz and feldspar. After a long period of weathering, the sediments are mostly clay, quartz, and iron oxide. Of the original minerals, the amount of quartz changes the least.

FIGURE 7-7. The graph shows how mineral change as granite weathers. Granite is usually composed mostly of feldspar and quartz. Exposure to Earth's atmosphere causes the minerals to change until clay, quartz, and iron oxide are the dominant minerals.

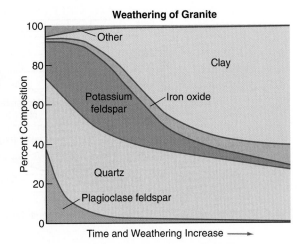

This shows that quartz is stable over a wide range of environmental conditions.

STUDENT ACTIVITY 7-3 —CHEMICAL WEATHERING AND TEMPERATURE

1: SCIENTIFIC
INQUIRY 1, 2, 3
1: ENGINEERING
DESIGN 1

Materials: small beakers (100–250 mL), hot and cold running water, thermometers, three or four small pieces of antacid tablets, stopwatch.

In this activity, you will develop your own laboratory procedure. The objective is to find out how temperature affects the rate of a chemical reaction. Once you have planned and written down your procedure, check it with your teacher. When your procedure is approved, perform the experiment and create a data table. Finally, record your conclusion about the effect of temperature on this chemical reaction. Visit the following Web site to learn about more science experiments with antacid tablets: *http://www.alkaseltzer. com/as/student_experiment.html*

What Environmental Factors Affect Weathering?

The amount and kind of weathering that takes place depends on three factors. You have read that the harder a rock is, the more it resists physical weathering. The more chemically stable its minerals are the better a rock resists chemical weathering. The final factor is climate. Figure 7-8 shows that cold climates favor physical weath-

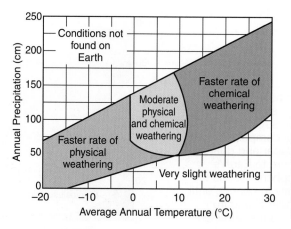

FIGURE 7-8. Cold climates favor physical weathering, especially frost action. Chemical weathering dominates under conditions of warm temperatures and abundant rainfall.

ering. Daily cycles of temperatures above and below freezing promote frost action in cold climates. Warm and moist climates accelerate chemical changes. For this reason chemical weathering is especially active in hot, humid, tropical locations.

HOW DOES SOIL FORM?

4: 2.1s

To this point in the chapter, weathering has been considered a destructive process that loosens rock and wears down the land. But weathering is responsible for one of our most important natural resources—soil. **Soil** is a mixture of weathered rock and organic matter. **Organic matter** is the remains of living organisms in which plants can grow. You may not find soil a very exciting topic. Yet, soil is a critical resource that allows us to grow and produce food. In addition, it is an absolutely critical part of the natural environment.

Figure 7-9 illustrates the development of soil on a solid rock surface. In the first column in the diagram, the bedrock is mostly unbroken, but it is exposed to the atmosphere and weather. The weathering process begins as rainwater reacts with the rock surface and water seeps into cracks in the rock. Water that seeps into crevices and fissures may change to ice and push the rock apart. Minerals soften and some minerals expand as they react with rainwater and groundwater.

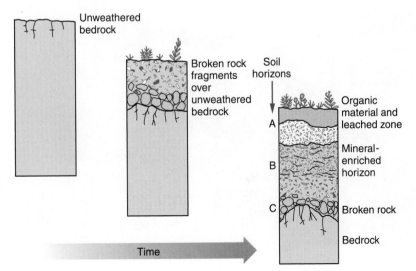

FIGURE 7-9.
Weathering, water seepage, burrowing by animals, plant growth, and decay of organic remains help form soil. This process can take thousands of years.

Many years later, the second column shows fragments of broken rock covering solid bedrock. The third column shows a mature soil in which organic remains, mostly dead plant material, have been mixed into the topsoil. **Infiltration** (water seeping into the ground) has carried some water deeper into the soil. The mature soil shows layering called **soil horizons** that are typical of well-developed soils. The topsoil is usually enriched with organic remains but may lack some dissolved minerals that water carried deeper into the soil. As a result, the soil below is enriched in soluble minerals. At the bottom of the soil profile, a layer of broken rock sits on the solid bedrock from which the soil may have formed.

The soil formed at any location depends on the composition of local bedrock, the climate, and the time for development. Warm and moist climates favor chemical weathering and usually produce thick soils, although the movement of groundwater through the soil may wash away important nutrients. Polar locations more often have thin, rocky soils with little chemical weathering.

Animals take part in soil formation as they burrow by mixing the components of soil (minerals in various states of weathering and organic remains), by loosening soil, and by allowing air and water to circulate. Figure 7-10 is a profile of soil horizons in glacial sediments along the Atlantic coast of Cape Cod, Massachusetts.

FIGURE 7-10. Soil horizons in this natural exposure range from the dark, organic rich topsoil to the less developed subsoil.

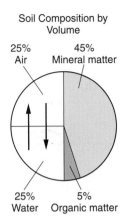

Soil Composition by Volume

25% Air

45% Mineral matter

25% Water

5% Organic matter

FIGURE 7-11.
The best soils for the growth of most plants contain a mixture of weathered minerals and organic remains. The amount of air and water in the soil depend upon the location and recent weather conditions.

Active volcanoes can be dangerous, but the soils that result from the weathering of volcanic rocks are usually very fertile. At least two cities were destroyed by the eruption of Mount Vesuvius in southern Italy in 79 CE. In spite of the danger, people soon moved back to the slopes of Vesuvius. The volcano is still active, and another major eruption is possible at any time. In spite of this, farmers are drawn back to the slopes of the mountain by the rich soil.

Figure 7-11 shows that the best soils contain air and moisture within a mix of weathered rock and organic remains. The water and air are in the pores (openings between particles) in the soil. Soil provides support for plants and the minerals provide important nutrients. Organic material holds water in the soil and holds the soil together. Water is essential for plant growth, but air is also important. Many plants cannot thrive if their roots are covered in water all the time.

Soil that is formed in place and remains there is called **residual soil**. Residual soils develop through the processes of weathering over hundreds or even thousands of years. **Transported soil** is formed in one location and moved to another location. In most areas, including New York State, transported soils, mostly by streams and glaciers, are more common than residual soils. Visit the following Web site to learn more about classification of soil types: *http://www.answers.com/topic/soil-classification?cat=technology*

Continental glaciers that repeatedly formed in Canada and moved southward pushed, carried, and dragged most of our sediment and soil from the place where it formed in Canada to New York State. The absence of a layer of broken bedrock that gradually changes with depth into solid rock in most New York locations is evidence of this transportation of soil. The soil shown in Figure 7-10 is a soil formed in a thick deposit of transported glacial debris.

CHAPTER REVIEW QUESTIONS

Part A

1. Which is a weathering process most common in a cold, moist climate?

   (1) abrasion   (3) frost action
   (2) deposition   (4) cementation

2. How does weathering usually affect rocks?

   (1) Weathering causes the mineral fragments to become larger.
   (2) Weathering makes rock harder.
   (3) Weathering occurs when sediment changes to sedimentary rock.
   (4) Weathering weakens rock so it can be carried away.

3. Which statement best describes physical weathering that occurs when ice forms within cracks in rock?

   (1) Physical weathering occurs when new minerals form.
   (2) Cracks increase in size because water expands as it freezes.
   (3) Cracks grow only because chemical reactions between water and rock dissolve the rock.
   (4) Physical weathering is most common in regions with warm, humid climates.

4. A pile of freshly broken rocks were exposed to atmospheric conditions and surface processes for a long time. Which set of graphs below best shows how the particles changed in shape and size?

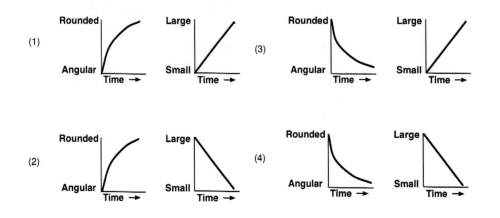

5. Which rock most quickly forms new compounds when it is exposed to air polluted with acids?

   (1) gneiss
   (2) limestone
   (3) granite
   (4) schist

6. As rock is broken apart by physical weathering processes,

    (1) its soil production decreases
    (2) its total surface area increases
    (3) new minerals form in the rock material
    (4) the mass of the rock increases

7. Marble is a metamorphic rock composed mostly of the mineral calcite. One hundred grams of marble is added to each of two identical beakers of hydrochloric acid. One marble sample is coarse marble chips. The second sample is a finely ground powder of marble. Why does the fine powder react more quickly with the acid?

    (1) Grinding changes the chemical composition of marble.
    (2) Fine particles of marble are less dense than coarser particles.
    (3) The finely ground powder has a greater total surface area.
    (4) The coarse chips have greater contact with the acid.

8. Which event is an example of chemical weathering?

    (1) rocks falling off a steep cliff
    (2) feldspar in granite being crushed into clay-sized particles
    (3) acid rain reacting with limestone bedrock
    (4) water freezing in cracks in a roadside outcrop

9. Which of the following changes does *not* directly contribute to the formation of soil?

    (1) melting of rock to make molten magma
    (2) plant roots growing into cracks in the ground
    (3) acidic rainfall reacting with the mineral calcite
    (4) rocks split apart by water freezing in cracks

10. The lowest horizon of a residual soil that formed in place is composed primarily of

    (1) organic remains
    (2) roots and twigs
    (3) broken bedrock
    (4) products of intense chemical weathering

11. The image below shows several rocks weathered by wind-blown sand in a desert environment. These angular rocks are called ventifacts.

What kind of weathering process most likely shaped these rocks?

(1) physical abrasion
(2) chemical weathering
(3) biological processes
(4) melting of rock

Part B

Base your answers to questions 12 and 13 on the figure below. The flowchart below is a model of weathering. Letters *A* and *B* identify two major kinds of weathering, and substance *X* is an important part of both kinds of weathering.

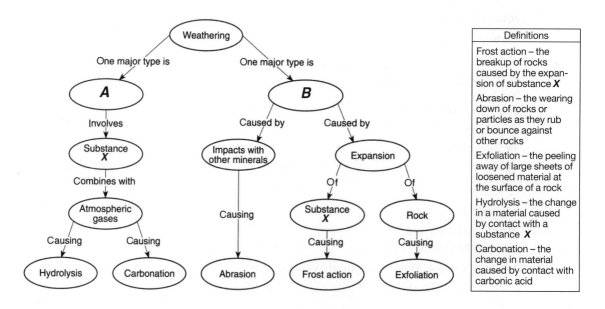

12. What kind of weathering does *A* represent?

(1) physical   (3) chemical
(2) biological   (4) glacial

13. What substance is best represented by *X* in the flowchart above?

(1) potassium feldspar   (3) hydrochloric acid
(2) air   (4) water

Base your answers to questions 14 and 15 on the diagram below, which shows how conditions of climate determine the kind of weathering that is most common in a particular location.

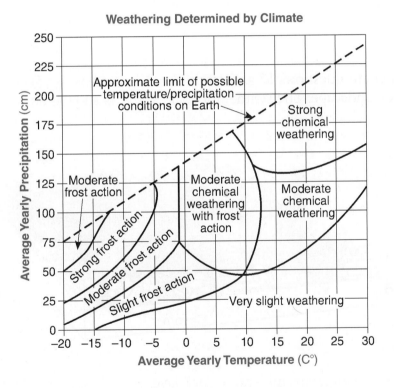

14. What kind of weathering is most common in an area with an average yearly temperature of 5°C and precipitation of 45 cm?

(1) moderate chemical weathering
(2) very slight weathering
(3) moderate chemical weathering with frost action
(4) slight frost action

15. The amount of chemical weathering will increase if

    (1) air temperature decreases and precipitation decreases
    (2) air temperature decreases and precipitation increases
    (3) air temperature increases and precipitation decreases
    (4) air temperature increases and precipitation increases

16. Look at Figure 7-10 on page 176. How did this soil form?

    (1) erosion of organic material
    (2) melting followed by crystallization
    (3) weathering and biologic activity
    (4) changes caused by heat and pressure deep within Earth

Part C

Base your answers to questions 17 and 18 on the diagram below, which represents a special kind of landscape. The symbol for rock type *A* is shown, but type *B* is not shown.

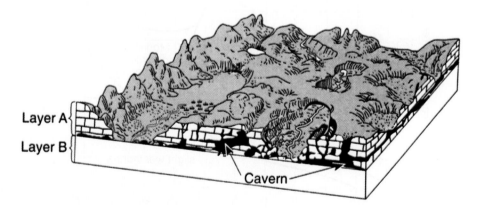

17. Name the most abundant mineral in rock layer *A*.

18. Describe how the caverns formed in rock layer *A*.

19. What is the major cause of physical weathering of big rocks transported along a large stream that has a steep gradient?

20. The image below is a fragment of broken glass found in a sand dune.

Describe the specific process that most likely changed this glass from a clear, jagged, and sharp-edged fragment to a slightly smaller object with a rounded and weathered surface.

8 Erosion and Deposition

WORDS TO KNOW

agent of erosion	erosion	mass movement
bed load	flotation	solution
deposition	glacier	sorting
dune	graded bedding	striations
dynamic equilibrium	horizontal sorting	suspension
equilibrium	landslide	vertical sorting

This chapter will help you answer the following questions:

1 How is erosion different from weathering?

2 What is the primary force that causes erosion?

3 What is the most important agent of erosion?

4 What is deposition?

5 How can we recognize particles eroded by different agents of erosion?

6 What characterizes a system in which erosion and deposition are in equilibrium?

WHAT IS EROSION?

4: 2.1t; 4: 2.1u

Have you ever visited the Grand Canyon in Arizona? Each year millions of people visit this spectacular natural wonder. The canyon (see Figure 8-1) is 10 to 20 km wide and 1.5 km deep. It is one of Earth's most inspiring examples of deposition and erosion. The walls of the canyon are mostly sedimentary rocks that represent millions of years of deposition. The Colorado River eroded the canyon in about 6 million years, as tectonic forces pushed up the Colorado Plateau more than 1000 m. Visit the following Web sites to see the Utah Slot Canyons and the Grand Canyon: *http://www.priweb.org/ed/earthtrips/s_rockies/s_rockies.htm* and *http://www.nps.gov/archive/grca/photos/*

New York State also has striking erosional features such as Letchworth Gorge on the Genesee River. This gorge is sometimes called the "Grand Canyon of the East." It has steep walls up to 120 m (394 ft) high on both sides. The Genesee River carved Letchworth Gorge after deposits from the most recent advance of continental glaciers blocked the river. Over the past 15,000 years, the river has cut a new route through thick layers of sedimentary rock. Watkins Glen (Figure 8-2 on page 186) and Enfield Glen in the Finger Lakes region are smaller gorges that have trails following streams with many waterfalls and potholes. Visit the following Web site to see the Gorges of the Finger Lakes region of New York: *http://www.citrusmilo.com/fingerlakes/iloveny1.cfm*

FIGURE 8-1.
Arizona's Grand Canyon is probably the most dramatic erosional feature on Earth's surface. Scientists estimate that the Colorado River took about 6 million years to make the canyon, and rapid erosion continues today.

FIGURE 8-2. Watkins Glen is one of the dozens of gorges that cut through the shales and other flat lying sedimentary rocks in Western New York State.

Erosion is the transportation (movement) of sediments by water, air, glaciers, or by gravity acting alone. If erosion did not occur, there would be no streams or valleys, no canyons, and no waterfalls. Mountains pushed up by tectonic forces within Earth would become higher and higher as long as uplift continued. Scientists know that the processes of weathering and erosion wear down mountains and the balance mountain-building forces within our planet.

Mass Movement

The force of gravity drives all erosional processes. Sometimes gravity acting alone transports (moves) earth materials. When weathering weakens rocks near the top of a cliff, the force of gravity may pull the rock off the cliff. In Chapter 16, you will read about the dangers to people and property caused by **mass movement** of rock and sediment.

The positions of rocks at the bottom of a landslide have little relationship to their position or organization before the landslide occurred. **Landslides** are most common in areas where tectonic forces within Earth are building high mountains. Figure 8-3 shows blocks of rock that have fallen to the bottom of a cliff.

FIGURE 8-3.
Weathering
weakens rock so
that it falls under
the influence of
gravity.

Other forms of mass movement occur more slowly. If clay-rich sediments along a slope become saturated with rainwater or snowmelt, they may move downslope in a mudflow. Sometimes blocks of sediment slide down steep slopes along weakened layers. However, the internal structure of the blocks of sediment remains unchanged, as shown in Figure 8-4. This kind of erosion is common where streams or ocean waves cut thick layers of sediment into cliffs.

Rocks eroded by gravity without being transported by water, wind, or glaciers are usually angular and rough. Recently broken surfaces show fewer signs of weathering than parts of the rock that have been exposed to weather for a longer time.

FIGURE 8-4.
Mass movement
occurred along a
slope that was
made steeper by
road construction
and lubricated by
water in the soil.

Erosion by Water

When water, wind, or ice causes erosion, they are called **agents of erosion**. Running water is the most important agent of erosion because it carries more sediment than any other agent of erosion. Each year streams and rivers carry millions of tons of sediment into lakes and oceans. The Mississippi River alone is estimated to carry about 6 tons of sediment per second into the Gulf of Mexico. Visit the following Web site to watch soil erosion animations: *http://serc.carleton.edu/NAGTWorkshops/visualization/collections/soil_erosion.html*

METHODS OF TRANSPORT Streams carry sediment in several ways. The smallest sediments are dissolved in water and are carried in **solution**. For example, when a stream flows over rock or sediment rich in the mineral halite (sodium chloride, or rock salt), the halite dissolves in water, forming a solution. The sodium and chlorine enter the solution as ions (atoms with an electrical charge). These ions are so small that they cannot be seen or separated from water by filtration. Materials carried in solution can give water a color, but you can still see through the solution; it is transparent. Natural water always has some substances in solution, even if they are present in very small amounts. Visit the following Web site to see examples of river erosion, transport, and deposition: *http://serc.carleton.edu/NAGTWorkshops/visualization/collections/erosion_deposition.html*

Sediments carried in **suspension** are small enough that they settle out of the water very slowly. Suspended sediments can be removed by passing the water through a filter. Most streams are turbulent. That is, the water does not flow smoothly downstream. When tumbling currents move faster than the speed at which suspended particles settle, the suspended load of a river can be carried indefinitely. Silt and clay in suspension give streams a muddy appearance. However, not all suspensions are in water. Clouds in the sky are actually collections of ice crystals or water droplets so small they remain suspended in atmosphere.

Some of the sediment load of a stream is too large to be carried in solution or suspension. Larger particles that are less dense than water float to the surface. Fresh leaves and other organic remains may be carried downstream by **flotation**. Large particles that are more dense than water settle to the bottom of a stream. If the stream is flowing quickly enough, these sediments will roll or bounce along the bottom of the stream as **bed load**.

SIZE AND VELOCITY   Figure 8-5, from the *Earth Science Reference Tables,* shows the relationship between the size of rock particles and the stream velocity needed to transport them. Boulders are rocks greater than 25.6 cm (almost 12 inches) in diameter. The graph shows that the smallest boulders require a stream velocity of about 300 cm/second to keep them moving.

To help you read this graph, use the corner of a sheet of paper. Carefully align one edge of the paper with the horizontal axis. Then slide the paper until the vertical edge is at the proper stream velocity. Making sure that the edge is parallel to the vertical axis, read the particle diameter from the vertical axis. (See Figure 8-6 on page 190.)

How fast must a stream be moving to keep all sand-size particles in motion? The largest sand grains are 0.2 cm in diameter.

FIGURE 8-5. The faster a stream moves, the larger the particles it can move. This graph from the *Earth Science Reference Tables* shows the stream velocity needed to keep sediment of various sizes in motion. (The stream velocity needed to start motion is usually greater.)

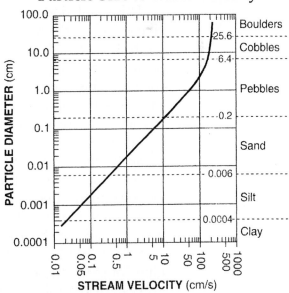

Relationship of Transported Particle Size to Water Velocity

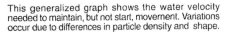

This generalized graph shows the water velocity needed to maintain, but not start, movement. Variations occur due to differences in particle density and shape.

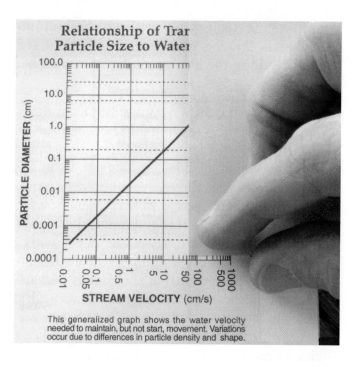

FIGURE 8-6. You can read the graph more accurately by lining up the corner and edge of a sheet of paper with the two axes of the graph.

That is the size that requires the fastest stream velocity. The line on the graph crosses the 0.2-cm sand-pebbles interface at the point where the velocity is approximately 11 or 12 cm/s.

This graph does not take into account several important factors that influence the relationship between particle size and stream velocity. For example, the stream velocity needed to start particles of sediment moving is greater than the speed needed to keep them moving. This graph applies only to particles already moving. A second issue is shape of the particles. The values from the graph apply to rocks that are neither spherical nor flat, but some average shape. Another factor is density. The denser a rock, the harder it is to move. A final factor is the nature of the bottom of the stream. The smoother and harder the bottom of the stream, the easier it is to keep sediments in motion. Therefore, this graph represents the average characteristics of shape and density of moving rock particles, as well as average conditions of the bottom of the stream.

Notice that Figure 8-5 gives the particle size for the various kinds of sediment: clay, silt, sand, pebbles, cobbles, and boulders. You can read particle size (diameter) on the vertical axis on the left side of the graph, or you can read the numbers on the dotted lines that separate each particle from larger grains above or smaller grains below. For example, sand particles are between 0.006 and 0.2 cm in diameter. Particle diameter is the typical distance across the particle. Once again, there is an assumption that the particle is partly rounded.

In this graph, clay refers to particles of sediment smaller than 0.0004 cm. Clay-sized particles are often the clay minerals, but other minerals can be reduced to the size geologists call clay.

One method to determine the relative velocity of two nearby parts of a stream is to observe the size of sediments in the stream. In the faster parts of the stream, only larger particles can settle. In the slower sections smaller particles settle out. Therefore, the speed of a stream determines the size of particles to be found in the streambed. The larger the sediments, the faster the water velocity.

Erosion by Wind

Strong winds can also transport sediments. However, because air is less dense than water, wind is generally unable to move particles larger than sand. As with stream water, the faster the wind, the larger the particles it can carry.

Wind erosion is most active in deserts and beaches. Here, there are few plants to slow the wind and hold soil in place. Wind-blown particles cause weathering by abrasion. Softer minerals in a rock are worn away, which may give wind-abraded rocks a pitted look. Ventifacts are wind-worn rocks that have flat surfaces (facets) like those of a regular geometric solid. These facets form when the rock is partly buried in sand. Abrasion by wind-blown sand wears the exposed rock face to a flat surface level with the top of the sand. Other flattened surfaces form when the rock is moved and another surface is exposed to the wind. Visit the following Web site to learn more about erosion by wind: *http://plantandsoil.unl.edu/crop technology2005/pages/index.jsp?what=topicsD&information ModuleId=1086025423&topicOrder=19&max=20&min=0*

Rocky deserts are more common than sandy deserts. Figure 8-7 on page 192 shows a wind-blasted desert surface in Arizona. This type of surface is called desert pavement. Without plants to hold the soil, fine sediment is blown away, leaving a surface covered by rocks that are too large to be carried away by the wind. Many of the remaining rocks have flat faces and straight edges, a shape typical of wind-eroded rocks.

FIGURE 8-7. Desert pavement is found where winds have carried away small sediments and left behind rocks too large to be blown away by wind alone.

Erosion by Glaciers

Geologists know that the climate of New York State has not always been the way it is now. Thousands of years ago, the climate was cold enough that winter snow did not melt during the brief summers. In eastern Canada, layers of snow built up and compressed the layers beneath them into ice. Eventually, the ice became thick enough that it flowed southward into New York under the influence of gravity. A **glacier** is a large mass of ice that flows over land due to gravity.

As the ice in a glacier flows downhill, it pushes, drags, and carries rocks, soil, and sediment. Rocks along the bottom of the glacier are worn down into partly rounded shapes. Glaciated rocks often have scratches known as **striations**, which are caused by a glacier dragging rocks along the surface of other rocks. Figure 8-8 shows rocks with shapes characteristic of different kinds of erosion.

FIGURE 8-8. Each agent of weathering or erosion produces characteristic features. (A) Wind-eroded rocks tend to be angular with broad facets. (B) Rock fragments eroded and deposited by gravity acting alone are usually rough with some very fresh surfaces. (C) Glacial erosion may leave scratches, known as striations, on the rock. (D) River-worn rocks tend to be round and smooth.

WHAT IS DEPOSITION?

4:2.1t; 4: 2.1u; 4:2.1v

Agents of erosion deposit sediments. Therefore, agents of erosion are also agents of deposition. **Deposition** is the settling, or release, of sediments carried by an agent of erosion. Gravity and glaciers transport sediment without regard to the sizes of the particles. Everything gets transported and dumped together. There is no organization in sediments deposited by gravity or by glaciers. But deposition by running water and wind is more selective. The size, shape, and density of the particles of sediment affect the rate at which they are deposited.

When streams enter the relatively calm water of a pond, lake, or ocean, sediments are deposited at the end of the stream. Sometimes the depositional feature called a *delta* forms at the mouth of the river. Figure 8-9 includes satellite images of the Nile River delta in Africa and the Mississippi River delta in North America. They are two of the world's major river deltas.

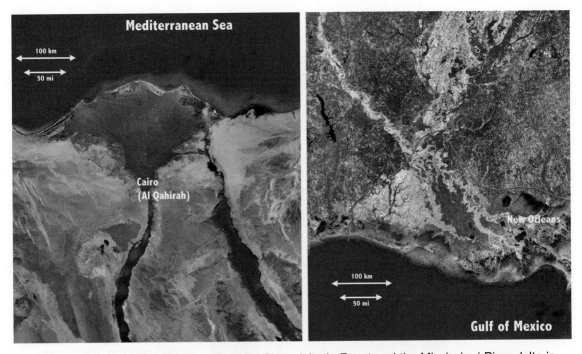

FIGURE 8-9. The Nile River delta in Egypt and the Mississippi River delta in Louisiana are among the world's most important delta regions. Both support major cities and agricultural areas. The Nile delta is easy to distinguish by the triangular-shaped green area. The more complex Mississippi Delta is outlined in red.

The Effect of Particle Size on Deposition

The size of particles transported by wind or running water determines how quickly they settle out of their transporting medium. If a landslide releases a large mass of sediment into deep water or if an underwater slide occurs along a steep slope, particles in a wide range of sizes begin to settle at the same time. The largest particles will settle to the bottom first, followed by smaller and smaller particles. Each landslide results in a layer of sediment. Within the layer, the largest particles are on the bottom and the size of particles decreases toward the top.

Figure 8-10 shows a vertical cross section of layers of sediment produced in this way. Notice that within each layer there is a gradual change in sediment size from large at the bottom to small at the top showing the order in which the particles settle to the bottom. This is called **vertical sorting**, or **graded bedding**. Note that you are not looking at alternating layers of fine sediments and coarse sediments. A single layer shows both features.

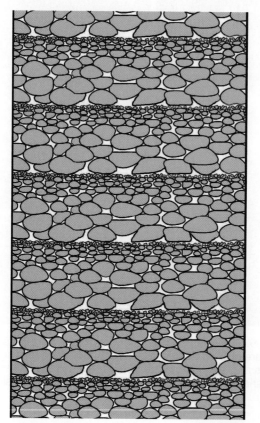

FIGURE 8-10. This diagram shows five complete layers of graded bedding. Each layer represents a single event of deposition in which the largest particles settled first. Parts of two similar layers of graded bedding are shown at the top and the bottom of the diagram.

STUDENT ACTIVITY 8-1 —WHAT IS GRADED BEDDING?

1: SCIENTIFIC
INQUIRY 1, 2, 3
6: SYSTEMS
THINKING 1
6: PATTERNS OF
CHANGE 5

The following is a procedure you can try yourself to observe how particles are deposited. Get a sample of mixed sand with particles that range from fine to coarse sand. (Very fine particles such as silt and clay will cloud the water, making the process difficult to observe.) Quickly pour about 10 mL of mixed sand into a transparent tube about 30 cm or more tall that is about half full of water. (A tall transparent jar can be substituted if necessary.) As the sand settles, look carefully at the layer that settles out. If you repeat this procedure, each layer should show this gradual change in particle size from largest at the bottom to smaller particles toward the top. The transition to new layers will be marked by a sudden change from small particles to larger particles, as you saw in Figure 8-10.

The Effect of Particle Shape on Deposition

The shape of sediment particles affects how quickly they settle. Spherical particles settle fastest because they are streamlined. Flat and irregular particles must push more water out of the way as they settle, which slows their fall. Friction is also greater for flat and irregular particles because they have a larger surface area.

The Effect of Particle Density on Deposition

Among particles of the same size and shape, the most dense particles settle first. How quickly sediments settle depends on the balance between resistance and weight. Resistance, which is determined by size and shape of the sediments, holds them back. The force of gravity (the weight of the particles) causes them to sink. Therefore if the weight (or density) of particles increases with no change in resistance, the particles will settle more quickly.

Running Water Sorts Sediments

Sorting is the separation of sediments by their shape, density, or size. Among particles of sediment transported by streams, the shape and density of particles seldom changes enough to make separation by shape or density apparent. But sediments are often

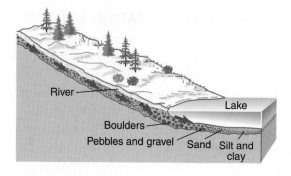

FIGURE 8-11. Horizontal sorting occurs as a fast-moving stream enters still water. The largest particles are usually deposited near shore. The finest sediments are carried the greatest distance into calm water.

sorted by their sizes. Graded bedding illustrates how water sorts sediments.

Another kind of sorting occurs when a fast-flowing river enters the relatively still water of a lake or an ocean, which causes the water to slow. The fast river current can transport sediments in a range of sizes. When the river slows, sediments are deposited. According to Figure 8-5 on page 189, the first group of sediments to settle out will be the largest particles, the boulders. By the time the current slows to about 300 cm/s, the boulder-sized particles have been deposited and cobbles will settle. If the current continues to decrease, pebbles, sand, silt, and finally, clay particles are deposited. Flatter, less dense, and smaller particles are carried farther and settle more slowly. The pattern that results from the decrease of current velocity is different from the vertical separation of graded bedding. **Horizontal sorting** is a decrease in the size of sediment particles with increasing distance from the shore that is produced as a stream enters calm water. (See Figure 8-11.)

Deposition by Wind

Like running water, wind sorts sediments. The primary difference is that wind cannot pick up large rocks. Sediments deposited by wind tend to be sand size (0.006 cm to 0.02 cm) and smaller. In areas where wind is the primary agent of erosion and deposition, wind blows sand into hills or ridges of sediment called **dunes**. The wind blows particles of sand up the gentle slope of the windward side of sand dunes and deposits them on the steeper, protected downwind face. Dunes form in some desert regions where there are few plants to hold the sand in place and some locations along lakes and oceans where sand is plentiful. Figure 8-12 shows

FIGURE 8-12.
The shape of these dunes makes it clear that the wind blows left to right. Sand blows up the gentle left side of these dunes and settles on the right side where it forms a steeper slope. There are many shapes of sand dunes, but all active dunes are steeper on the lee (downwind), depositional side.

a sand dune in a desert area of southern Utah. It is clear that the wind has been coming from the left.

Have you been to a place along the ocean where land areas are separated from the water by sand dunes? The line of dunes protects inland areas from storms and erosion by waves. There may be no plant cover on the dunes because sand lacks important nutrients, the sand is unable to hold onto water, and the sand has a loose consistency. Sand dunes are fragile ecological features that are important to the beach environment. Only recently have people become aware of the importance of protecting dunes from motor vehicle and foot traffic as well as residential or commercial development.

Deposition by Ice

In some ways, glaciers are like running water. They flow from higher areas to lower areas. Glaciers often occupy valleys that they form by erosion. The rate of flow of a glacier is usually a meter or less per day. Glaciers are not able to separate, or sort, different sizes of particles. Ice is not fluid enough to allow sediments to settle through the ice. As a result, sediments carried by moving ice are not deposited until the glacier melts. Giant boulders, fine clay, and every sediment size in between are deposited in irregular mounds with no separation or sorting. Figure 8-13 on page 198 shows glacially deposited sediments in New York State. Notice the range

FIGURE 8-13.
This construction site in Westchester County, New York, contains unsorted sediments deposited by glacial ice.

of particle sizes with no layering or apparent sorting. You will learn more about glaciers in Chapter 12.

Sorted and layered sediments can be found in areas where glaciers are or were active. However, these sorted sediments were deposited by streams running out of the glaciers. Sediments deposited directly by ice are unsorted.

STUDENT ACTIVITY 8-2 —WHAT'S IN SEDIMENT?

6: SYSTEMS THINKING 1
6: PATTERNS OF CHANGE 5

Materials: containers of sand and other sediments, magnifiers, metric rulers

Your teacher will set out containers of sediments collected from a variety of locations. Compare the sediments by overall color, colors of the grains, shapes of the grains, average particle size, sorting, range of sizes, and any other unusual characteristics you may observe.

WHEN IS THERE EQUILIBRIUM BETWEEN EROSION AND DEPOSITION?

6: EQUILIBRIUM AND STABILITY 4

Consider a section of a river that is carrying sediment. You may observe that the river looks about the same over a long period of time. If the riverbed is not filling with sediment and it is not cutting deeper into its bed, the river is in equilibrium. **Equilibrium** is a state of balance. In the case of the river, the sediment washed into this stretch of river must be equal to the sediment that is carried away.

There are two ways in which this can occur. It is possible that the sediment entering this part of the river is carried along without any erosion or deposition. The particles of sediment that are carried in are the same particles that are carried out. The river does not change and the sediment that makes up the bottom of the river does not change. In this condition there is no erosion and no deposition.

It is more likely, however, that some new sediment is deposited, and some sediment from the bottom of the river channel is carried away by erosion. This is especially likely if the volume of water and the velocity of the water change through time, as they do in nearly all rivers. Flooding causes erosion of the streambed and deposition occurs when the flow is reduced. If equilibrium is reached over the course of the year, erosion and deposition are equal. However, some sediment from the bed of the river has been carried away, while some new sediment has been deposited. This is **dynamic equilibrium**. In a dynamic equilibrium, opposing processes are taking place, but they balance out because they take place at the same rate.

CHAPTER REVIEW QUESTIONS

Part A

1. The largest particles that a stream deposits as it enters a pond are 0.008 cm in diameter. What is the minimum velocity of the stream at this location?

    (1) 0.1 cm/s   (2) 0.5 cm/s   (3) 1.0 cm/s   (4) 5.0 cm/s

2. What process was primarily responsible for the formation of a delta?

    (1) glacial erosion
    (2) cementation of sediment
    (3) deposition of sediment
    (4) mass movement

3. Which event is the best example of erosion?

    (1) shale breaking apart as water freezes in its cracks
    (2) dissolving of limestone to make a cave
    (3) a pebble rolling along the bottom of a stream
    (4) bedrock crumbling to form a residual soil

4. A stream flowing at a velocity of 75 cm/s can transport

    (1) clay, only
    (2) pebbles, only
    (3) pebbles, sand, silt, and clay, only
    (4) boulders, cobbles, pebbles, sand, silt, and clay

5. A stream transports particles *W, X, Y,* and *Z* as shown below. They are shown actual size.

Density = 3.8 g/mL

Density = 3.8 g/mL

Density = 2.4 g/mL

Density = 2.4 g/mL

   Which particle is likely to settle most quickly as the velocity of the stream decreases?

   (1) *W*   (2) *X*   (3) *Y*   (4) *Z*

6. Look at the photograph of the sand dunes in Figure 8-12 on page 197. How did the dunes most likely form?

    (1) water flowing from the left
    (2) water flowing from the right
    (3) wind blowing from the left
    (4) wind blowing from the right

7. What process best describes the downward slide of some rock layers from a cliff?

    (1) tidal changes   (3) mass movement
    (2) glacial erosion   (4) lava flow

8. The diagram below is a profile of a sediment deposit. In what type of event were these sediments most likely deposited?

    (1) a landslide into deep water
    (2) a slowly moving glacier
    (3) a delta deposited at the end of a river
    (4) sand blown into a dune

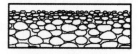
(Drawn to scale)

9. When small particles settle through water faster than large particles, the small particles are probably

   (1) lighter in color   (3) better sorted
   (2) less rounded   (4) more dense

10. Which property of a mineral sample does *not* influence how quickly it settles in water?

    (1) density   (3) shape
    (2) hardness   (4) size

11. Where the stream velocity decreases from 300 to 200 cm/s, which size sediment will be deposited?

    (1) cobbles   (2) sand   (3) silt   (4) clay

12. The particles of sediment at the top of a single layer of graded bedding are usually those that are

    (1) most dense, most rounded, and smallest
    (2) least dense, most flattened, and smallest
    (3) most dense, most rounded, and largest
    (4) least dense, most flattened, and largest

Part B

13. The diagram below shows sediments along the bottom of a stream.

    What can we say about the water velocity in this stream at the time these sediments were deposited?

    (1) The stream velocity decreased to the left.
    (2) The stream velocity increased to the left.
    (3) The stream velocity was constant in this section of the stream.
    (4) There is no relationship between sediment size and stream velocity.

14. These diagrams represent a stream flowing into an ocean. Which diagram best shows the distribution of large and small particles?

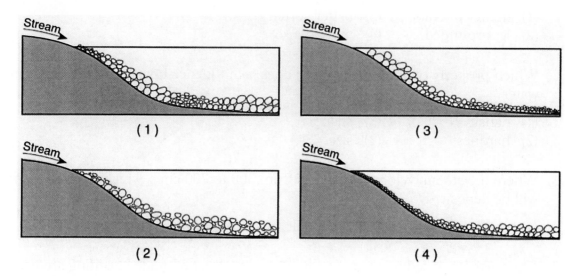

15. The diagram below shows four identical columns of water. Four sizes of spherical object made of the same uniform material are dropped into the columns where they settle to the bottom as shown below.

Which graph best shows the relative settling times of the four objects?

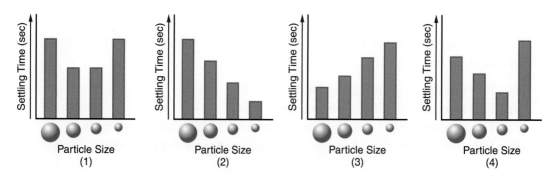

Part C

Base your answers to questions 16 through 19 on the block diagram below. The diagram shows streams flowing over bedrock. In the cross-sectional view on the left, the particles are drawn to actual size and the arrows show their motion in the water. The rock symbols match those shown in the *Earth Science Reference Tables*.

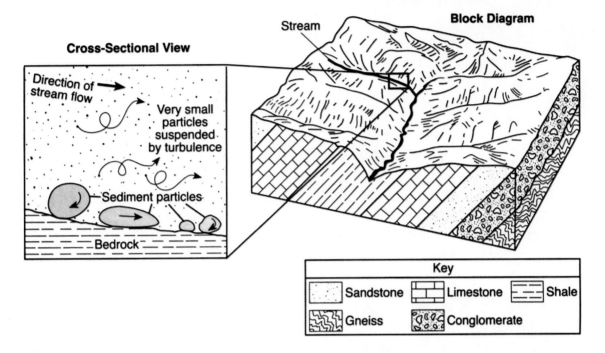

16. Measure the particles and determine the name of the largest particles in the cross section that are shown.

17. What process is responsible for the rounded shape of the large particles in the cross section?

18. What kind of rock in the diagram appears to be most easily eroded?

19. What size sediment is classified as clay?

20. At a brick factory along a river, stacks of identical bricks were left outside the building for shipment. A major flood destroyed the factory and washed away the bricks. Some bricks were carried several miles downstream; others were deposited close to the factory site. State two characteristics of the brick recovered far downstream that are likely to be different from the brick found near the factory site.

UNIT 4

Water Shapes Earth's Surface

As you learned in Chapter 8, water is a major agent of erosion. Water carries sediments and carves canyons, changing the shape of Earth's surface. Rivers also shape our history. The trade routes of Native Americans and the routes of European explorers followed rivers to reach the interior of North America. Travel by water was easier than travel by land, especially where the forests were thick and travelers were carrying food and supplies. The earliest colonial settlements were along the coastline. Here they had access to shipping and manufactured goods. From coastal cities, settlements and commerce followed the rivers to the interior of the country. Before settlers built roads, rivers were the highways they used for travel, trade, and communications.

The Colorado River of the American Southwest, while not especially large in terms of water flow, is one of the world's most important rivers in terms of its regional influence. The Colorado River starts in the Rocky Mountains and forms the boundary between Arizona and California before it enters Mexico. Due to human water use, the river has not reached the sea for many years. This picture shows a portion of the Colorado River near Page, Arizona.

9 Stream Dynamics

WORDS TO KNOW

delta	levee	stream
discharge	meander	stream system
drainage divide	overland flow	tributary
drainage pattern	runoff	watershed
floodplain		

This chapter will help you answer the following questions:

1 Why are rivers important to us?

2 Why are streams sometimes called stream systems?

3 What is a drainage basin?

4 What are the common features of streams?

5 How can you measure properties of streams such as stream speed?

WHAT DOES THE HUDSON RIVER MEAN TO NEW YORK STATE?

7: CONNECTIONS 1

The Hudson and Mohawk Rivers played a crucial role in the history of New York State. The Hudson-Mohawk lowlands provided the only low-level route from New England and the Atlantic coastal colonies to the interior of the growing nation. The Hudson River is actually a sea level passage that can be traveled by oceangoing vessels all the way to Albany. It was the Hudson River and the deep-water harbor that led to New York City becoming the major center for commerce in the colonies. The Hudson River corridor played a central role in the American Revolution. The British tried to divide the colonies by taking control of this inland passage. Later, the Erie Canal was built to connect the Great Lakes to New York City and the East Coast. These valleys also provided a path for the New York State Thruway, one of the first links in the Interstate Highway System. Today, highways, railroads, and air routes have replaced streams as our primary arteries for travel, settlement, and commerce. Rivers and lakes still carry some commerce while they supply freshwater and serve as recreational areas.

WHAT IS A RIVER SYSTEM?

1: SCIENTIFIC
INQUIRY 1, 3
4: 1.2g
4: 2.1 p, 2.1v
6: SYSTEMS
THINKING 1

A system is a collection of components that work together to perform a function. A river system, or **stream system**, consists of all the streams that drain a particular geographic area. A **stream** is any flowing water, such as a brook, river, or even an ocean current. The function of a river is to transport water and sediments from a specific land area to an ocean or a lake.

Water and sediments have potential energy at the beginning of their journey. The amount of energy depends on how high they are above the end, or mouth, of the stream. As water and sediment flow downhill, potential energy changes to kinetic energy. At the end of the stream where water flows into the calm water of a lake or ocean, potential energy has decreased because water and sediments are at their lowest elevation. The kinetic energy also decreases as these materials stop moving. For these materials, the transporting function of the stream system has been accomplished. Figure 9-1 shows a stream that is transporting a large load of sediment.

Watersheds

The geographic area drained by a particular river or stream is its **watershed**, or drainage basin. All the rain, snow, and other precipitation that falls into the watershed and does not escape by infiltration, evaporation, or transpiration must exit the watershed through its principal river, stream, or other body of water. (Sometimes a lake or an ocean rather than a river defines a watershed.)

FIGURE 9-1. This braided river in Alaska carries so much sediment that the river breaks into many gravel-clogged channels. Much of the sediment load comes from glacier fed streams in the mountains. Visit the following Web site to see current water flow conditions in the United States: _http://waterdata. usgs.gov/nwis/rt_

Drainage divides separate one watershed from the next. These are often ridges from which water drains in opposite directions.

You can identify watershed boundaries and trace the perimeter of a watershed by drawing a line that separates all the streams draining into one watershed from streams that flow into neighboring watersheds. The Continental Divide is a geographic line that separates North America into watersheds that drain into different oceans.

Watersheds are important because they show the region drained by a particular stream system. Communities that draw water from a nearby stream or river depend on rain and snow that falls within the watershed. The availability of water for such a community depends only on the amount and the quality of precipitation in the watershed upstream from where the community takes its water. When a water-soluble form of pollution is released into the environment, it flows downhill, and is carried into the nearest stream. Salt used to melt ice on roads during the winter is a good example. As smaller streams join larger streams, the pollution affects only downstream locations in the watershed. Like water pollution, flooding is also confined to a particular watershed. Visit this Web site to see high-resolution images of rivers and watersheds compiled by the United States Geological Survey and the World Wildlife Fund: _http:// hydrosheds.cr.usgs.gov_

Most of the precipitation that falls over the continents reaches solid ground. If this water does not infiltrate the ground or evaporate, it flows downhill under the influence of gravity as **overland flow**, or **runoff**. The amount of runoff depends on the slope of the land, the permeability of the surface, and the amount of precipitation. The steeper the slope, the greater the runoff. More water runs off a hard surface, such as concrete, than off a permeable surface, such as soil. Grasses and shrubs in the soil also decrease runoff because they absorb water. The greater the amount of precipitation, rain or snow, the greater the runoff. Overland flow continues until the water reaches a stream.

Names such as brook or creek are often used to label small streams that flow into larger streams such as rivers. A stream that flows into another larger stream is called a **tributary**. In large watersheds, small tributaries join to form larger tributaries which themselves may be a tributary of even larger streams. The Bedrock Geology of New York State map in the *Earth Science Reference Tables* shows some of the major rivers of New York State. Figure 9-2 shows the watershed of the Hudson River.

FIGURE 9-2.
The watershed of the Hudson River lies mostly within New York State, although it also includes small areas of nearby states.

STUDENT ACTIVITY 9-1 —DRAINAGE OF THE SCHOOL GROUNDS

6: MODELS 2
6: MAGNITUDE AND
SCALE 3

Make a map of your school grounds to determine how water drains off different parts of the property. On the map, identify potential sources of water pollution and show what parts of the grounds would most likely be affected by these sources of pollution. Also show where runoff could cause erosion problems and suggest ways to prevent these problems.

Features of Streams

As most streams flow from their source to their mouth, the *slope,* or *gradient,* of the stream decreases and the shape of the valley becomes wider. Streams that form steep, V-shaped valleys in mountain areas move into regions where the gradient is smaller. Here the valleys become wider with floodplains. (See Figure 9-3.)

FLOODPLAINS   Most of the time, a stream is confined to a relatively narrow and winding path along the bottom of the valley. In

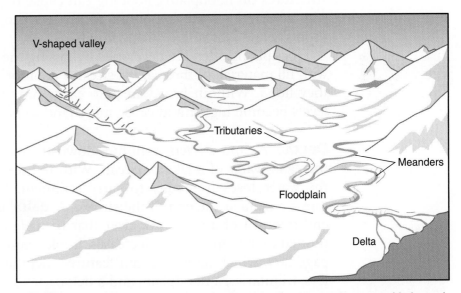

FIGURE 9-3. Many rivers start as small mountain streams in narrow V-shaped valleys. As they move into a lower gradient and are joined by tributary streams, they develop a broad floodplain and curves called meanders. Finally, the river may deposit a delta at its mouth.

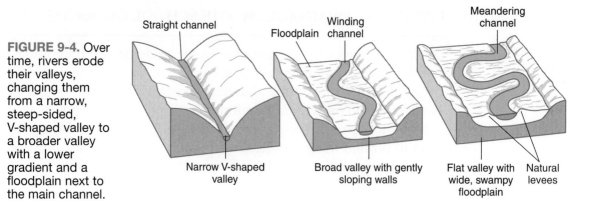

FIGURE 9-4. Over time, rivers erode their valleys, changing them from a narrow, steep-sided, V-shaped valley to a broader valley with a lower gradient and a floodplain next to the main channel.

times of flood, streams overflow their banks and spread over a floodplain. A **floodplain** is a flat region next to a stream or river that may be covered by floodwater. Floodplains are good for agricultural land because sediments brought by floods enrich the soil with important minerals and nutrients for plant growth. Figure 9-4 shows a stream as it changes through time from being in a narrow, steep-sided valley to wandering over its floodplain in a broad valley.

People may agree that the occasional flooding of farm land is an acceptable risk. However, when people build houses and other structures on floodplains, flooding can cause huge losses. This is why zoning laws are important to protect people from loss of property and loss of life during floods.

Floodplains often are developed by raising the level of the land with fill material. In these areas, streams are more restricted when they flood. Without broad areas to carry away floodwater, flood levels become higher and cause more land to be flooded.

DELTA As most rivers continue downstream, they empty into the calm water of a lake or ocean. With the decrease in velocity, the water loses its ability to carry sediment. Deposition often forms a delta at the end of the stream. A **delta** is a region at the end (mouth) of a stream or river that is made of sediments deposited as stream velocity decreases. Look again at Figure 9-3 on page 209, which shows stream features including V-shaped valleys, tributaries, a floodplain, and a delta.

MEANDERS As a stream flows over relatively flat land, its path develops curves called **meanders**. Builders of irrigation canals have discovered that when the channel has a soft streambed and little slope, the path of the canal tends to meander. Even when the path of the water is initially straight, meanders develop, through time, unless the banks are lined with a hard material such as concrete. The curves of a meandering stream are the natural shape of streams and rivers that have a low gradient and flow over a broad valley.

Where the stream is straight, the fastest current is near the center. The current at the sides and bottom is slowed by friction with the banks and river bed. When the water flows through a meander, the fastest current is at the outside of the curve, and the slowest current is at the inside of the curve.

Figure 9-5 illustrates a meandering stream highlighting where deposition occurs and where erosion occurs. These processes cause meanders to change and move downstream. The longer arrows indicate the fastest current. Notice that deposition occurs where the stream slows on the inside of the meanders. Along the outside, where the water flows fastest, is where erosion takes place. The three cross sections to the right show that where the water flows fastest (see the red circles) and near the eroding banks is where the channel is also the deepest.

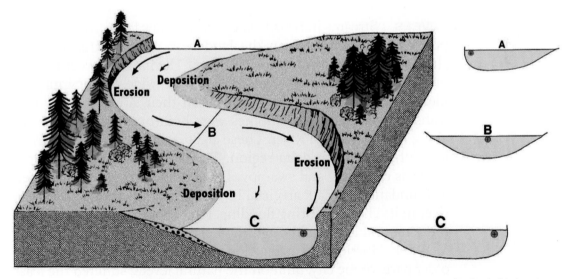

FIGURE 9-5. Erosion tends to occur where the stream velocity is highest along the outsides of meanders. On the inside of these curves, where the water slows, deposition takes place. Also notice that where the stream velocity is the greatest, the stream is deepest.

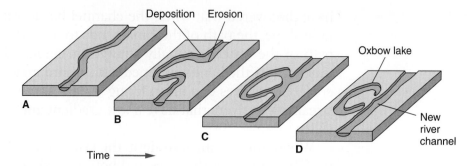

FIGURE 9-6. Meanders form where stream gradient is low and there is a broad floodplain. Diagram A shows a meander beginning to form in a stream. In B, erosion and deposition are starting to cut through the meander. In C, the cutoff is complete. In D, deposition along the edge of the river leaves the meander isolated, or cut off, as an oxbow lake.

Once meanders have formed, they do not remain in place through time. If you could see a greatly speeded-up view of a meandering stream, the meanders would shift like the slithering motion of a snake. How do they do this? Streams change their course as a result of erosion and deposition. Erosion occurs where the water flows fastest, and deposition takes place where the water slows. If a stream has a low gradient, it will tend to form meanders and then cut off its meanders as shown in Figure 9-6.

LEVEES Streams in broad valleys sometimes flood and leave deposits of sand and silt on the land next to and parallel to the streams. These ridges are called **levees**. The natural levees may be only a few feet high and hundreds of feet wide, so they may be identified by their dry land rather than their slope. In New Orleans, Louisiana, the first homes were built on the natural levees along the Mississippi River. These areas are the highest and driest locations in the flat delta region. The land away from the riverbanks is often low and swampy. The natural levees are often used as foundation for tall, artificial levees, which are built to keep the river in its channel during flooding. Figure 9-4 on page 210 shows these natural levees, although the height has been exaggerated.

When Hurricane Katrina flooded New Orleans in 2005, the older parts of the city, built on the natural levees, suffered the least flooding. Newer neighborhoods have been built on lower, drained swampland farther from the river. These areas were under as much as 5 m of water, flooding some homes to their roofs. The long-time residents understood this contradiction very well; the closer you lived to the river, the *less* likely your house was to be flooded.

When people change natural systems, we often get unintended results. Tall levees may force a river to deposit its sediment load in the river channel. In some locations, the Mississippi River is now higher than the surrounding land. This increases the danger of flooding. The artificial levees may need to be built higher and higher. However, the new and higher levees often shift the danger to older levees. They protect one area from flooding while putting another in danger.

STUDENT ACTIVITY 9-2 —MODELING A STREAM SYSTEM

6: MODELS 2

Materials: stream table, and fine sand or coarse silt
You can buy or build a table-top model including a small running stream system. Observe and list characteristics and common features of streams that develop on a stream table. How does the path of the stream change through time? Where do erosion and deposition occur?

HOW DO WE MEASURE STREAMS?

1: MATHEMATICAL
ANALYSIS 1, 2, 3
6: SYSTEMS
THINKING 1

Scientists measure the velocity and size of streams. Several factors affect the velocity of a stream. Size is more than just the length of the stream.

Stream Velocity

How quickly water flows in a stream is a function of three factors: shape of the stream channel, gradient (slope), and volume of water. If the stream channel is straight and smooth, water can flow quickly. However, if the stream is flowing over large rocks, the rocks slow the water as it bounces from rock to rock. Many mountain streams are filled with coarse sediment such as large cobbles and boulders, which slow the velocity of the stream.

The gradient of a stream affects how quickly water moves. The force of gravity maintains flow of water. Other factors being equal, the steeper a stream channel, the faster water will flow.

The third factor is the volume of water flowing in a stream. When the discharge (water flowing) in a stream increases, the weight of water usually increases faster than the resistance of the stream channel. Most streams are steepest at their beginning, with the gradient decreasing as they flow downstream. But many rivers flow faster as they move downstream because the increase in the volume has a larger effect on its speed than the reduction in slope.

In Figure 9-7 you see profiles along the length of the Hudson River and the Colorado River. The concave shape of these profiles is typical of many streams. These streams are steeper near their source than they are near their end.

The vertical lines on the profile of the Colorado River in Figure 9-7 represent two major dams. The lower dam is Hoover Dam. Lake Mead is the reservoir that formed behind this dam. The Hoover and the Glen Canyon dams prevent flooding on the Colorado River, provide hydroelectric power, and store several years' supply of fresh water. The water is used, for example, to grow crops, for drinking water, for laundry, and for watering lawns in the cities of the desert Southwest. Without these dams, economic development of this area would have been impossible.

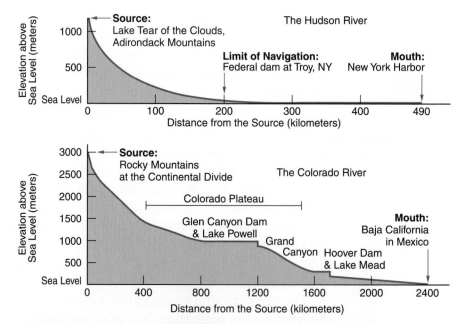

FIGURE 9-7. The longitudinal profiles of the Hudson and Colorado rivers are concave as their slope decreases downstream. The double concave shape of the Colorado River profile indicates uplift of the Colorado Plateau.

STUDENT ACTIVITY 9-3 —WATER VELOCITY

1: MATHEMATICAL
ANALYSIS 2
1: SCIENTIFIC
INQUIRY 3

Materials: Stream table or running water and tilted trough to represent a stream bed, meterstick, stopwatch or timer

Determine the influence of gradient and stream discharge on stream velocity. Measure the velocity of the water as the gradient and the discharge are changed. Use your data to explain how gradient and discharge affect stream velocity. (May also be a teacher demonstration.)

You can measure the speed of a stream by selecting a relatively straight section of a small stream. You will need a device to measure distance, such as a meterstick, a timing device such as a watch with a second hand, or a stopwatch, and an object to float downstream. (In case there is wind, it may be best to use an object that floats, but is mostly submerged.) Measure the length of the stream section in units such as meters. Then place the floating object in the water above the measured section and time how long it takes for the object to float through the measured distance. The stream velocity can be calculated using the following formula:

$$\text{Velocity} = \frac{\text{distance}}{\text{time}}$$

SAMPLE PROBLEM

Problem Two students stand 53 m apart along a straight portion of a small stream. One student places a floating marker in the stream and immediately begins timing it with a stopwatch. If the marker passes the second student in 50 s, what is the average velocity of this section of the stream?

Solution
$$\text{Velocity} = \frac{\text{distance}}{\text{time}}$$
$$= \frac{53 \, \text{m}}{50 \, \text{s}}$$
$$= 1.06 \, \text{m/s}$$

Practice

1. A floating marker takes 30 s to travel 105 m along a straight portion of a stream. What is the average velocity of this section of the stream?

2. The average velocity of a stream is 0.4 m/s. How far will a marker travel in 1 minute?

Stream Size

The size of a stream can be measured in several ways. One measure is the area of its watershed. In general, the larger the drainage basin, the larger the stream. However, some locations receive more precipitation than others. In a dry region, a large watershed may supply water only to streams that are dry most of the year. Some watersheds receive so little rain and snow that none of the water in the stream flows out of the watershed. Streams in these areas run into bodies of water that lose their water by evaporation, such as the Great Salt Lake in Utah, or the stream water may seep into the ground before the stream reaches its lowest level.

Stream size is more often measured by finding discharge. **Discharge** is the amount of water flowing in a stream past a particular place in a specified time. For example, a small stream may have discharge of a fraction of a cubic meter per second. However, the Amazon River in South America has the greatest discharge of any river on Earth. It discharges about 200,000 m³ of freshwater into the Atlantic Ocean each second. That's about 6 times as much water as the Mississippi and about 150 times the discharge of the Hudson River.

To measure the discharge of a stream, you can measure the area of its cross section at a particular location, then multiply that value by the velocity of the stream. You can estimate the area of the cross section by multiplying the average depth of the stream by its width at that point. Area is measured in square meters. Velocity is expressed in meters per second. The product of these values is cubic meters per second. The following formula shows how to calculate discharge volume:

$$\text{Discharge} = \text{area of cross section} \times \text{stream velocity}$$

STUDENT ACTIVITY 9-4 —MEASURING STREAM DISCHARGE

1: ENGINEERING DESIGN 1

Materials: Device to measure distance, watch, or timer
Use the method described in the text above to measure the discharge of a stream near your school or home. Devise a different method to find the discharge in order to verify the first value that you obtained.

The size of a stream determines how it responds to rainfall. As you can see in Figure 9-8, large streams are slower to respond than small streams. This is because the water in a large drainage basin has a greater distance to flow before it reaches the river. Streams in smaller watersheds generally respond quickly because the precipitation does not flow far to reach a small stream.

When land is covered by buildings and paved areas (urbanized), all streams in the area respond more quickly and dramatically to rainfall. The curve for the river may become similar to the shape to the small stream, but with a far higher curve and larger discharge volume. Therefore, cities need runoff channels that can manage much larger stream flows than their former, natural streams.

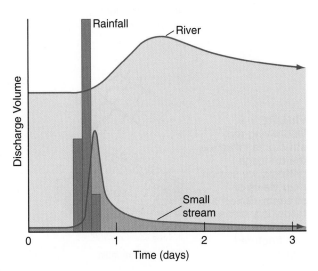

Stream Response to Rainfall

FIGURE 9-8. Hydrographs of a river and a small stream. The dark-blue bars show rainfall in a summer thunderstorm. Note that the river responds slowly and does not return to its former discharge volume even after three days. The smaller stream responds quickly, growing to many times its original discharge volume, but it also returns to its original state quickly. What is the lag time (from maximum rainfall to maximum water discharge) for each stream?

WHAT IS A DRAINAGE PATTERN?

4: 2.1t, 2.1u

Streams seek the lowest path as they move downhill, and they tend to erode their beds in places where the ground is weak. Therefore, both topography and geologic structure affect the **drainage pattern**, the path a stream follows through an area. By looking at a map view of a stream, you can often infer the underlying bedrock structures. Figure 9-9 shows the relationship between stream pattern and rock structure.

The most common stream pattern is dendritic drainage (see Figure 9-9 A). Dendritic streams flow downhill in the same general direction and they join to make larger streams. As a result, they have a branching appearance. This pattern is common where the bedrock is uniform, without faults, folds, or other major structures or zones of weakness to capture the streams. Dendritic drainage is

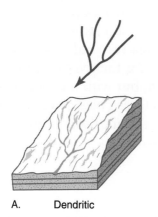

A.   Dendritic

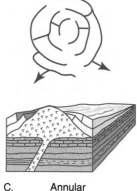

C.   Annular

FIGURE 9-9.
Underlying rock structures influence the drainage patterns we see on map views of streams. Streams follow low areas, weak rock types, and fractured rocks.

B.   Rectangular

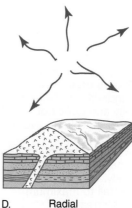

D.   Radial

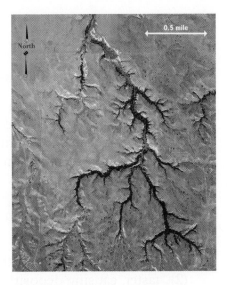

FIGURE 9-10. Dendritic drainage pattern in the flat-lying sedimentary layers of the Great Plains near Hisle, South Dakota.

also common where the rock layers are horizontal. Much of the region of western New York State north of the Pennsylvania border has dendritic drainage because rock layers are flat and there are few faults or folds to divert streams. The satellite view of stream drainage in Figure 9-10 shows dendritic drainage pattern in South Dakota.

A region that has prominent parallel and perpendicular faults, repeated folds, or a strong rectangular jointing pattern will display a rectangular drainage pattern (see Figure 9-9 B). (Joints are cracks in bedrock along which no significant movement has occurred. They may be related to expansion or regional forces acting on bedrock.) Streams run to the lowest areas of folds, fractured rocks along faults, or the weakest surface bedrock locations.

Annular drainage is a pattern of concentric circles that are connected by short radial stream segments (see Figure 9-9 C). This type of drainage occurs in an eroded dome.

A radial drainage pattern resembles the spokes of a wheel (see Figure 9-9 D). Streams flow away from a high point at the center of the pattern. Radial drainage may develop on a smooth dome or a volcanic cone. The Adirondack Mountain region of New York displays radial drainage, although rock structures such as faults and folds in the Adirondacks change the regional pattern and may make radial drainage hard to observe.

The important point is that the underlying rock types and geologic structures influence streams, and that different structural features produce different patterns of drainage.

CHAPTER REVIEW QUESTIONS

Part A

1. Which condition would cause surface runoff to increase in a particular location?

   (1) covering a dirt parking area with pavement
   (2) reducing the gradient of a steep hill
   (3) planting grasses and shrubs on a hillside
   (4) having a decrease in the annual rainfall

2. A stream bank is steeper on the outside of a meander because the water there is flowing

   (1) slower, causing deposition
   (2) faster, causing deposition
   (3) slower, causing erosion
   (4) faster, causing erosion

3. What unit is used to measure the discharge volume of the stream?

   (1) kilometers per hour
   (2) cubic meters per second
   (3) centimeters per minute
   (4) meters per square meter

4. Why are streams and rivers with the features of their watersheds called systems?

   (1) All streams and rivers are the same size.
   (2) Streams and rivers form straight channels.
   (3) Different parts contribute to the same outcome.
   (4) The features of one stream system are not found in other stream systems.

5. Which river is a tributary of the Hudson River?

   (1) Mohawk River
   (2) Susquehanna River
   (3) Delaware River
   (4) Genesee River

Part B

Base your answers to questions 6 and 7 on the two graphs below. The graphs show the discharge of storm water into a nearby stream following two equal rain events in the same urban location. Graph II was constructed from data obtained after several major construction projects.

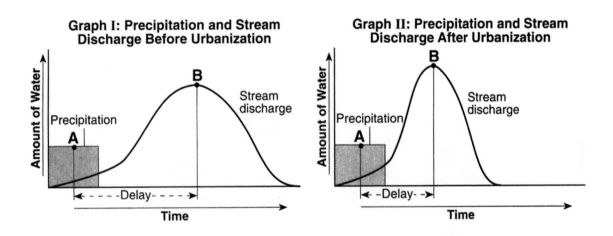

6. The delay time between A and B on both graphs is primarily due to the time needed for

   (1) groundwater to evaporate
   (2) precipitation water to move into the drains
   (3) green plants to absorb precipitation
   (4) rainfall to slow from its maximum rate

7. How did urbanization affect the maximum rate of discharge and the delay time between rainfall and stream discharge?

   (1) The maximum discharge rate decreased, and the delay time decreased.
   (2) The maximum discharge rate decreased, and the delay time increased.
   (3) The maximum discharge rate increased, and the delay time decreased.
   (4) The maximum discharge rate increased, and the delay time increased.

8. The diagram below shows a coastal region that is drained primarily by a single stream that flows from location *X* to *Y* and *Z*. The stream is not shown in this diagram.

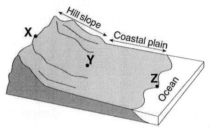

Which diagram below shows the most probable path of the stream?

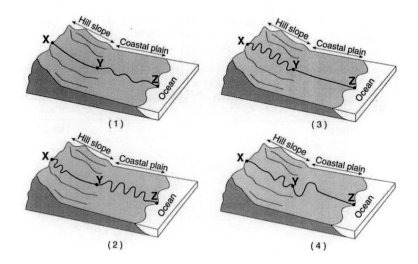

Base your answers to questions 9 through 11 on the diagram below. The stream table models a meandering stream over a land surface with a gentle slope.

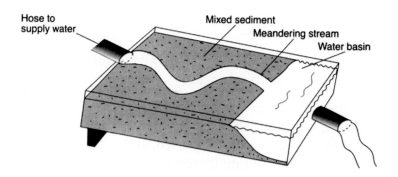

9. Which diagram below best represents where erosion, *E*, and deposition, *D*, are occurring?

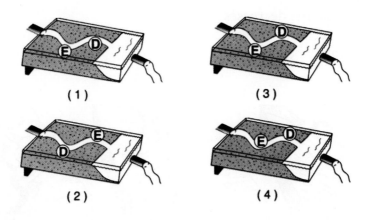

(1)   (3)

(2)   (4)

10. As the stream enters the still water at the bottom of the slope, where are the larger, *L*, and smaller, *S*, particles deposited?

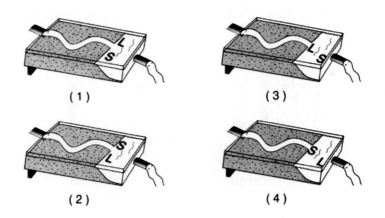

(1)   (3)

(2)   (4)

11. How can the stream table be changed to increase the amount of sediment transported by the stream?

(1) decrease the temperature of the sediment
(2) decrease the slope of the stream
(3) increase the size of the sedimentary particles
(4) increase the rate of the water flow

12. For most streams, as you travel from the source of the stream toward the mouth, the

    (1) gradient increases
    (2) discharge volume decreases
    (3) floodplain becomes wider
    (4) force of gravity becomes weaker

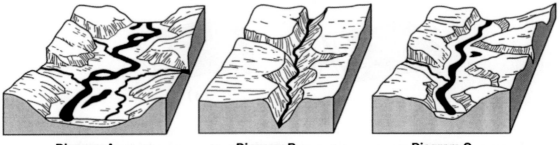

Diagram A   **Diagram B**   **Diagram C**

13. The block diagrams above represent three river valleys. Which bar graph below best represents the relative gradients of the principal rivers shown in diagrams *A, B,* and *C?*

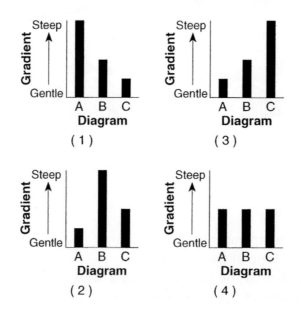

Part C

Base your answers to questions 14 and 15 on the data below. The table shows the average monthly discharge for a stream in New York State. Below the data table is a sample graph that you can use as a model to construct your own graph. (Please do not write in this book.)

Data Table

Month	Jan	Feb	Mar	Apr	May	Jun	Jul	Aug	Sept	Oct	Nov	Dec
Discharge (ft³/s)	48	52	59	66	62	70	72	59	55	42	47	53

14. On a piece of graph paper, construct a graph of the discharge data. Choose an appropriate scale for the x and y axes. The x axis is the horizontal axis, the y axis is the vertical axis. Write the months of the year along the x axis and Discharge (ft³/s) along the y axis. Plot an X for the average stream discharge for each month shown in the data table. Then connect the X's with a line.

15. Give one likely reason that the stream discharge volume in April is greater than the discharge volume is in January.

16. This is an illustration of a portion of the Susquehanna River in south-central New York State. Describe the structure of the rocks that most likely underlie this location.

17. Why are low areas near rivers in New York State better suited for farmland and growing crops than they are for use as home sites?

18. A student decided to measure the speed of a stream by floating apples down a straight section of the stream. Describe the steps the student must take to determine the stream's rate of surface movement (speed) by using a stopwatch, a 4-m rope, and several apples. Include the equation for calculating velocity.

19. The diagram below shows a stream with a constant flow running through an area where the land around the stream is made of uniform sand and silt. Make a sketch copy of this map view. Show the future path of this stream resulting from erosion and deposition as it usually occurs along a meander by drawing two dotted lines (one on each side of the river) to show the future stream banks.

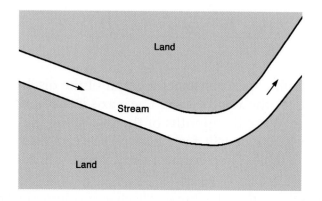

20. On a separate sheet of paper, draw a map view of water's likely drainage pattern on the landform below.

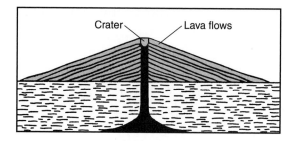

10 Groundwater

WORDS TO KNOW

aquifer	groundwater	spring
capillarity	hydrologic cycle	transpiration
condensation	infiltration	water table
convection	permeability	zone of aeration
dew point	porosity	zone of saturation
evaporation	precipitation	

This chapter will help you answer the following questions:

1. Where is most of Earth's water?
2. How does water circulate through the environment?
3. How can we find water underground?
4. What are the properties of a good aquifer?
5. What problems are associated with groundwater?

WHERE DOES THE WATER IN STREAMS COME FROM?

4: 1.2g

When it rains hard or when snow melts you can see water running over the ground toward streams. This is easy to see in cities where the ground is covered by pavement and buildings. The link between precipitation and stream flow is easy to see. But there can be weeks between rainstorms, and the streams still flow. So where

FIGURE 10-1. Even when it has not rained for weeks, streams can flow because they are fed by groundwater that comes to the surface in springs. Deep within the Grand Canyon water gushes from the canyon wall at Vasey's Paradise. Note the people on the raft to the left.

does the water come from? In some places, water seeps out of the ground. These features are called **springs** such as the spring in Figure 10-1. Even in desert climates where rain is infrequent, springs can run, permanently feeding small streams.

In the desert environment, many streams run overland and then disappear into the ground. Sometimes water that seeps into the ground at one location along a stream shows up downstream as it comes back to the surface. Where water disappears and where it comes back to the surface depend on conditions underground.

WHERE IS EARTH'S WATER?

4: 1.2g

Earth is sometimes called the "water planet," or the "blue planet." After all, about 71 percent of Earth's surface is covered by oceans and another 3 percent by glacial ice. Earth is the only planet known where it is neither too hot nor too cold for liquid water to exist on the surface. In fact, water is the only common substance that exists on Earth in three states: solid, liquid, and gas. Water is necessary for life. If people were to settle on another planet, one of their most important considerations would probably be the availability of water.

Water that enters the ground filling open spaces in soil and sediment as well as openings in bedrock, including cracks and spaces between the grains, is **groundwater**. For humans, groundwater is an important part of Earth's water supply. Groundwater is usually freshwater. It is available nearly everywhere on the continents, and it is generally free of organic pollution, germs, and dangerous contaminants.

However, groundwater is endangered. Once groundwater is polluted or used up, recovery will be very slow. In some locations, it may be too late. But in most locations understanding groundwater and the *hydrologic cycle* could allow us to conserve it for future use.

The Hydrologic (Water) Cycle

Earth has a limited supply of water. However, water is considered a renewable resource because it circulates through various parts of the environment. Figure 10-2 shows that about 97 percent of Earth's supply of water is salt water in the oceans, and another 2 percent is ice, mostly in polar ice caps. The remaining 1 percent is the freshwater that supports life. As the human population has expanded and the use of water in our homes, for farming, and for industry has increased, there are growing concerns about the availability of usable water. Perhaps it is time to consider what water has in common with nonrenewable resources and plan more carefully for the present and the future. Visit the following Web site to learn more about the hydrologic cycle: *http://ga.water.usgs.gov/edu/watercycle.html*

Distribution of Earth's Water

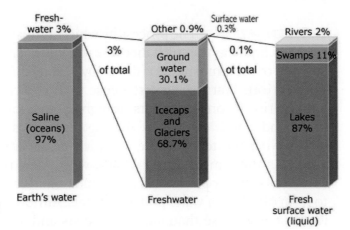

FIGURE 10-2. Most of Earth's water is in the oceans. Of the remaining (fresh) water, about a third is groundwater.

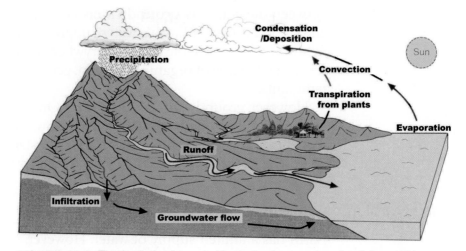

FIGURE 10-3. The hydrologic cycle illustrates how water circulates within Earth's lithosphere and atmosphere. The labels show eight processes that allow water to circulate.

EVAPORATION   Water is constantly being recycled through the oceans, atmosphere, and land in the **hydrologic cycle**, as shown in Figure 10-3. As with any cycle, a description of water circulation can begin at any point in the cycle. Most of the water that enters the atmosphere comes from the oceans through evaporation. **Evaporation** is the process by which a substance changes from a liquid to a gas as the substance absorbs energy. Evaporation in nature is powered by solar energy, which lifts water into the air. On land areas, evaporation from open water and from the soil is extended by **transpiration**, which is the release of water vapor by plants.

CONDENSATION   The process by which a substance changes from a gas to a liquid as the substance loses energy is **condensation.** Circulation of the atmosphere is driven by density differences that cause **convection**. Warm air is less dense than cooler air and therefore tends to rise. If convection lifts air high enough into the atmosphere, it expands and cools. As cooling continues, air reaches the **dew point**, the temperature at which air is saturated with water vapor. As the temperature falls below the dew point, water vapor condenses on tiny particles of dust, smoke, and salt from ocean spray to form droplets of water and ice crystals. Although liquid water and ice are more dense than air, the droplets and ice crystals are small enough to remain suspended in the atmosphere. We see these water droplets and ice crystals when they form a cloud.

STUDENT ACTIVITY 10-1 —OBSERVING CONDENSATION

1: MATHEMATICAL ANALYSIS 1, 2, 3
1: SCIENTIFIC INQUIRY 1

On a humid summer day, take a very cold can of soda from the refrigerator and measure its mass with an electronic balance. Allow moisture to condense on the cold can and weigh it again. Calculate the percentage increase in the mass of the soda can.

PRECIPITATION   As some clouds develop, the ice crystals and water droplets grow and combine until they are too large to be held up by the air. Then they fall back to Earth as **precipitation**: rain, snow, sleet, or hail. Most precipitation falls into the ocean where it completes the water cycle. Precipitation that falls onto land can accumulate on the surface, it can run overland into streams, or it can infiltrate the ground.

Infiltration is the process by which water soaks into the ground under the influence of gravity. Whether water sits on the surface, runs off, or infiltrates depends on many factors. If the ground is saturated with water, it will not be able to hold any more water. Frozen ground also stops infiltration. Precipitation that falls as snow may remain where it fell until it melts. The ground can absorb more water when rainfall comes as a long, steady rain. Rapid, intense rainfall may be more likely to run into streams than seep into soil. Slope is also important. The steeper the surface gradient, the more likely water is to run off into streams and the less likely it is to infiltrate. Water is more likely to infiltrate a permeable soil with large pores.

Ground cover is also important. Grass, trees, and other vegetation slow runoff and give surface water more time to soak in. However, when the ground is covered with pavement and buildings, water run off is rapid and infiltration may not take place at all.

Permeability is the ability of rock, soil, or sediment to allow water to flow down through it. A loose or sandy soil is more permeable than clay or a soil with mineral deposits that block infiltration. Permeable sandstone makes an excellent aquifer to hold groundwater.

DOES GROUNDWATER OCCUR IN SPECIFIC ZONES?

4: 1.2g

Gravity pulls water into the ground until it reaches the **zone of saturation**, the part of the rock and soil where all available spaces are filled with water. Below the zone of saturation, the layers of rock or other material (such as clay) that do not have pores that water can penetrate. (These layers are impermeable.) Above this zone of saturation is the **zone of aeration**, the region in which air fills most of the available spaces in the rock and soil. As groundwater infiltrates, it moves through the zone of aeration and enters the zone of saturation.

The upper limit of the zone of saturation is the **water table**, as shown in Figure 10-4. Therefore the zone of saturation extends from impermeable layers at the bottom to the water table at the top. The height of the water table changes depending on infiltration, horizontal flow of groundwater, and usage. Sometimes there is a period of plentiful precipitation. At this time, infiltration is greater than the amount of water taken from wells or that flows out through the ground. In response, the water table will rise and groundwater will move closer to the surface. However, a dry spell is likely to result in a drop of the water table, as inflow is less than usage from wells and water that flows away.

The depth of the water table is important to anyone who uses a well to tap groundwater. Unless the well reaches below the water table, water will not flow into the well. As water is drawn from a

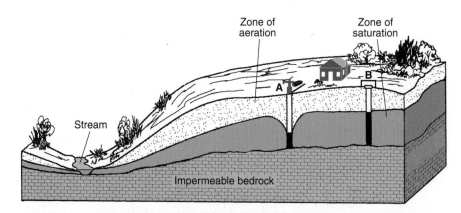

FIGURE 10-4. The zone of saturation is between impermeable layers below and the zone of aeration above. The top surface of the zone of saturation is the water table. The height of the water table depends on the balance between water entering as infiltration, water leaving as underground flow, and usage from wells. Well *A* is active and it has depressed the water table nearby. The well at *B* does not affect the water table because it is not being used.

well, the water table falls near the well, as shown in Figure 10-4. It is important to limit the usage of water from a well. The amount of water taken out should not be greater than the amount of water that can flow into the well.

If the water table falls below the bottom of the well, the well is said to run dry. Unfortunately, a well is most likely to run dry when there is a lack of surface water and the need for well water is the greatest. There are several ways to correct this problem: reduce the pumping of groundwater, allowing the water table to rise back into the well; find other sources of water to reduce the use of well water; and dig the well deeper.

The water table is often an irregular surface that has high and low places where there are hills and valleys on the land. However, the water table usually has less relief, or change in elevation. The water table is usually deeper below the surface at hilltops, and the water table may come to the surface in valleys. If the water table comes to the surface, it can feed ponds or lakes or it may flow out of the ground onto the surface.

Figure 10-5 shows two streams in different climate regions. The top stream is in a humid location where it is fed by groundwater. The other stream is in a dry area. It loses water through its bottom into the ground. Visit the following Web site, How Our Rivers Run, to learn how precipitation affects our rivers: *http://www.bigelow.org/virtual/water_sub2.html*

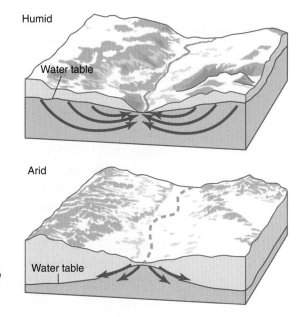

FIGURE 10-5. In humid climates, groundwater feeds into streams to maintain flow between rainstorms. But in arid climates, streams may run dry as they lose water into the ground through their streambed.

STUDENT ACTIVITY 10-2 —GROUNDWATER MODEL

1: SCIENTIFIC
INQUIRY 1, 3
6: MODELS 2

You can construct a model of groundwater zones in a watertight container such as a fish tank. The bottom of the tank can represent the impermeable zone, although you may be able to place a layer of clay in the bottom of the tank to represent this layer. Well-sorted sand, such as beach sand, is a good choice to represent the part of the ground in which water can circulate. Soda straws can represent wells of various depths. Note the flow of groundwater and changes in the position of the water table. Make a list of your observations as water infiltrates the model and as water is drawn from the wells.

HOW DOES GROUNDWATER MOVE?

4: 2.1g

The ability of soil and bedrock to hold and transfer groundwater changes from place to place. In some places, groundwater is plentiful while other places have little or no groundwater. Even if there is groundwater, it is easier to remove water from some materials than from others. Several factors affect where groundwater occurs and how it moves. Visit the following Web site to learn more about groundwater: *http://ga.water.usgs.gov/edu/mearthgw.html*

Permeability

If the ground is very permeable, infiltration will be quick, and water can flow freely within it. Therefore, permeable ground prevents or reduces flooding. Permeable soils allow water to infiltrate before it reaches streams. *Permeability* affects the recharging, or replacement, of groundwater. The more permeable the soil, sediment, and bedrock are, the more quickly water can flow down into the zone of saturation to replace groundwater lost to outflow or pumping from wells.

Large openings that are connected make rocks or soil very permeable. Even bedrock can be permeable if it is composed of particles with spaces between them, such as sandstone. Bedrock can also be permeable if it has large and connected cracks, such as a fractured granite, or limestone that has underground passageways.

Among sediments, the most permeable are well-sorted sediments with the largest, roundest particles. Soils made of uniform large particles have large spaces between the particles. This is especially true for rounded grains, which cannot be packed as tightly as flat or rectangular particles. If smaller particles are mixed with the large particles, the smaller grains fill in spaces between the large particles and reduce the permeability.

Even though a fine-grained soil may have the same total space between the grains as a coarse-grained soil, water cannot flow as quickly through these smaller spaces. Water clings to the surfaces of grains of sediment. Because fine-grained sediments have more total surface area, water cannot pass through fine-grained sediments easily. Therefore materials like silt and clay have low permeability. Compact clay can be nearly as impermeable as solid bedrock.

Porosity

Porosity is the ability of a material to hold water in open spaces, or pores. It is an important property of a soil, sediment, or bedrock because porosity determines how much water the ground can store. If the porosity is low, there cannot be much groundwater stored. Rock and sediment with a high porosity generally result in a good supply of groundwater. Porosity can be calculated using the following equation:

$$\text{Porosity} = \frac{\text{volume of pore space}}{\text{total volume of the sample}}$$

Well-sorted sediments with round grains have a high porosity because of the large spaces between the particles. Sand that is rounded and well sorted can have porosity as high as 50 percent. However, unlike permeability, the size of the particles does not affect porosity. Figure 10-6 on page 236 shows three containers of spherical particles. The container on the left holds a sample of small particles, the container in the middle holds medium particles, and the container on the right holds large particles. In each container, the particles are packed in the same way. Each sample has the same porosity.

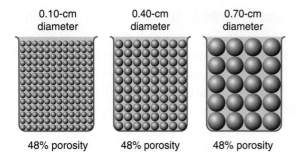

0.10-cm diameter   0.40-cm diameter   0.70-cm diameter

48% porosity   48% porosity   48% porosity

FIGURE 10-6. Changing the size of the particles does not change the porosity of a sample. This is true as long as there is no change in the sorting, shape, and packing of grains. Note that each of the samples shown has the same porosity, 48 percent.

STUDENT ACTIVITY 10-3 —COMPARING THE POROSITY OF DIFFERENT MATERIALS

1: SCIENTIFIC INQUIRY 1, 2, 3
1: ENGINEERING DESIGN 1
6: MODELS 2

Determine the porosity of several different samples of sand and gravel by comparing the volume of each sample with the volume of water needed to fill each sample to its surface with water.

To remember that the size of the particles does not affect the porosity, imagine a hollow cube that encloses the largest possible solid sphere. The solid sphere takes up about 52 percent of the volume of the cube. Therefore, the percent of open space is the remaining 48 percent. This is true of any size of cube in which the length of the side of the cube and the diameter of the enclosed sphere are equal. Although smaller particles leave smaller spaces between particles, the increase in the number of spaces balances the decrease in the sizes of the spaces.

Porosity does depend on the shape and the packing of the particles. Flat or rectangular particles can pack more closely than spherical particles. Figure 10-7 shows how tighter packing can even reduce the porosity of sediment composed of identical spherical particles.

Two other factors affect the porosity of a soil. In a mixture of sediment sizes, small particles fit into pore spaces between larger particles. Figure 10-8 illustrates how mixing different particle sizes

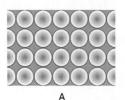

A

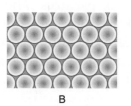

B

FIGURE 10-7. Sample *A* has a greater porosity than sample *B* because the particles are more tightly packed in sample *B*. Tighter packing reduces the porosity.

FIGURE 10-8. The porosity of a soil sample is reduced when small particles occupy spaces between the larger particles.

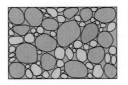

reduces porosity. A final factor is mineral cement. When particles are held together by a substance such as calcite, clay, or silica cement, the cementing substance reduces the porosity of material.

Capillarity

The property of adhesion causes water to stick to surfaces. Adhesion is related to the surface tension of water, which can be a remarkably strong force. This force draws water up into tiny spaces; this is called **capillarity**, or capillary action. Sometimes, this action can pull water from the water table upward to where it reaches the roots of plants. Capillarity also allows trees to draw water from the soil into their leaves tens of meters above the ground. Tree trunks contain narrow passageways that draw water toward the leaves. Capillarity is also the way that a towel soaks up water or the way wax moves up through a wick to vaporize and serve as fuel for the flame of a candle. Figure 10-9 illustrates capillary action in small openings in glass tubes.

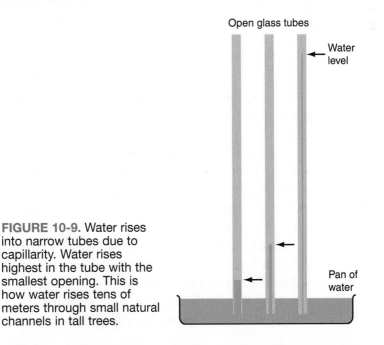

FIGURE 10-9. Water rises into narrow tubes due to capillarity. Water rises highest in the tube with the smallest opening. This is how water rises tens of meters through small natural channels in tall trees.

STUDENT ACTIVITY 10-4 —CAPILLARITY OF SEDIMENTS

6: MODELS 2

> Find several transparent plastic or glass tubes that are about 0.5 cm in internal diameter. (Transparent drinking straws may work.) Cover the bottom of each tube with cotton fabric held in place by a rubber band. The fabric will allow water to pass through the opening but prevent the sediments from falling out. Fill each tube with a different size of sediment from sand to silt. Place the tubes of sediment in a pan of water. Observe how high the water rises into each sample of sediment.

Capillary action occurs only in rock or sediments with very small, connected pore spaces. When openings become wider, the weight of water in the openings increases without a corresponding increase in the surface area of the openings. Adhesion cannot hold or draw water into large openings.

WHERE IS GROUNDWATER AVAILABLE?

4: 1.2g

In places where surface water evaporates, especially in the summer, groundwater may be the only reliable supply of freshwater. Unless there is a source of freshwater in an area, people cannot live there. People need freshwater to drink and prepare food, to clean and for other household purposes, and to grow food. In most places, groundwater is more reliable and cleaner than surface water.

In most places, groundwater can be found within 100 m of Earth's surface, but the depth of solid, impermeable rock varies greatly from place to place. In some places, solid rock without open spaces is exposed at the surface, and the impermeable rock extends into Earth's interior. In other places, water can be brought to the surface from many kilometers underground. Digging or drilling wells can be a major expense for a property owner. The deeper the well goes, the greater the cost. Knowing where to place a well, how deep to dig, and when to stop drilling can be difficult decisions.

About 98 percent of Earth's supply of freshwater is within the ground. Lakes, rivers, and streams are more visible than groundwater. However, there is far more freshwater stored in the ground than on the surface. The best supplies of groundwater are in underground aquifers. An **aquifer** is a zone of porous material that contains useful quantities of groundwater. Farming in the western United States depends on well water drawn from aquifers. The circular patterns familiar to people who have flown over the Great Plains are created by center-pivot irrigation systems. In some cases, water has been taken out of the aquifers so much faster than it can be replenished by rainfall and infiltration that there is danger of completely depleting the aquifer. This is sometimes called "mining" water because farmers are taking water that has been in the ground for hundreds of years. Although scientists usually consider freshwater a renewable resource, the rate of usage of some aquifers far exceeds the rate at which they receive new water.

In some places, groundwater is trapped between impermeable layers. If the recharge area is higher than the outlet of an aquifer, spring water can gush quickly out of the ground. Artesian flow can be started by drilling a well through an impermeable layer and into an aquifer under pressure. In some cases, the well water just rises above the level of the surrounding groundwater. In other cases, water may flow out of the well without being pumped.

WHAT ARE SOME GROUNDWATER PROBLEMS?

7: CONNECTIONS 1
7: STRATEGIES 2

Just as many streams have been contaminated by careless disposal of wastes, some aquifers are also being seriously polluted. Wherever soluble waste materials are left on or in the ground, infiltration can carry them to an aquifer. Several factors make the contamination of groundwater even more difficult to deal with than the pollution of surface waters. People cannot see or smell buried wastes. They cannot see them contaminating an aquifer. In addition, it may take a long time for waste materials to reach the aquifer. By the time scientists know an aquifer is becoming polluted it might be impossible to prevent a very serious problem. It will also take far more time to rid an aquifer of contamination than it would to flush surface water of organic or inorganic toxic materials.

Sewage

Many people use septic systems to dispose of waste water and human waste. In a septic system, liquid wastes and water seep into the ground from an underground tank. Solid wastes remain in the tank. There are nutrients in sewage; organisms that live in the ground use these nutrients. As a result they clean sewage-filled water. However, this cleaning is effective only as long as nutrients are not present in amounts too large to be consumed by the organisms in the ground. When communities conclude that increasing population and density of housing threaten to create a health problem, they install public sewage systems. These systems pipe sewage to a central location and treat it to speed up the rate at which organisms use and therefore remove toxic substances. Keeping waste materials out of the ground helps preserve the quality of groundwater. Visit the following Web site to learn more about how septic and sewer systems work: *http://home.howstuffworks.com/sewer3.htm*

Saltwater Invasion

The increasing use of groundwater in coastal locations has allowed salt water from the ocean to invade some aquifers. Because salt water is more dense than groundwater, the salt water flows in and under the freshwater. This has occurred in eastern Long Island, New York, as shown in Figure 10-10. The problem has been dealt with in several ways. Some parts of Long Island use water piped in from upstate rivers to reduce groundwater withdrawal. Returning treated wastewater to the aquifer has also been used to reduce salt-

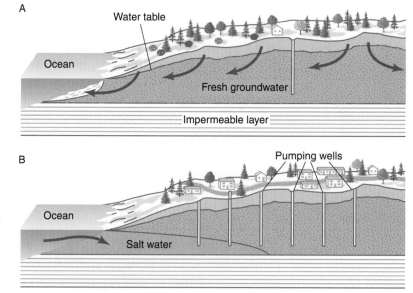

FIGURE 10-10. Pumping groundwater from an aquifer in a coastal region such as Long Island, New York, can let salt water flow into the aquifer and contaminate wells. Part *A* is a profile before extensive pumping from the aquifer. Part *B* shows salt water invading the emptied aquifer.

water invasion. Visit the following Web site to learn more about the endangered Long Island, New York, aquifer. (You must sign up for a password to access the *New York Times* archives.) *http://www.nytimes.com/2006/12/02/nyregion/02water.html?fta=y*

A related problem occurs in arid parts of the United States. In these areas, the soil is often salty. Irrigation of salty soil carries the salt deeper into the ground where it pollutes underground water supplies.

Subsidence

Taking water from the ground can cause the land to sink, or subside. Figure 10-11 shows a crack in the ground created by the use of groundwater. Usually, the sinking is a slow process that is not noticeable. However, cracks sometimes appear near the edge of the valley where deep sediments meet shallow bedrock. Sediments held up by bedrock are unable to sink with the central parts of the valley. This causes cracks to open. While the movement itself is not a danger, it can cause problems in the foundations of buildings and it can damage roads. The sinking ground level can also increase the danger of flooding when rivers run full. Perhaps more important, this is a sign that water is being mined too quickly. In a location such as this, groundwater is a nonrenewable resource.

FIGURE 10-11. In Arizona, this crack opened in the ground in a matter of days. Water taken out of the ground nearby caused the land level to sink several meters.

CHAPTER REVIEW QUESTIONS

Part A

1. The water table usually rises when there is a(n)

   (1) decrease in the amount of infiltration
   (2) decrease in the amount of surface covered by vegetation
   (3) increase in the amount of precipitation
   (4) increase in the slope of the land

2. By which process do plants add water vapor to the atmosphere?

   (1) precipitation   (3) condensation
   (2) transpiration   (4) absorption

3. Water was poured into the container under a clay flowerpot. The water level in the bottom container dropped to level *A* as the wet part of the clay pot moved upward to *B*.

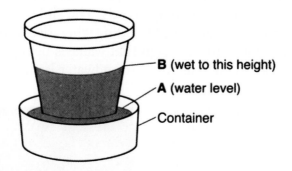

   Water level *B* is higher than the water at *A* because water

   (1) is less dense than the clay pot
   (2) is more dense than the clay pot
   (3) traveled upward in the clay pot by capillary action
   (4) traveled downward in the clay pot by capillary action

4. Which sediment size would allow water to flow through at the fastest rate?

   (1) clay   (3) sand
   (2) silt   (4) pebbles

5. Why is much of Earth's water unfit for use by humans?

   (1) Most of Earth's water is frozen.
   (2) Most of Earth's water contains natural salt.
   (3) Most of Earth's water is contaminated with sewage.
   (4) Most of Earth's water is too deep underground to pump out.

6. How deep must a well be dug to yield a constant supply of freshwater?

   (1) to the bottom of the soil
   (2) to the top of the zone of aeration
   (3) below the water table
   (4) several meters into the bedrock

7. During a heavy rainstorm, runoff is most likely to occur if the surface of the soil is

   (1) firmly packed clay-sized particles
   (2) loosely packed sand-sized particles
   (3) covered by trees, shrubs, and grasses
   (4) unsaturated and has a gentle slope

Part B

8. Which soil type has the slowest permeability rate and is most likely to cause flooding?

   (1) clay   (2) silt   (3) sand   (4) pebbles

9. Tubes *A* and *B* below with their drain valves are identical. Both are partly filled with spherical plastic beads of uniform size. However, the beads in *A* are smaller than in *B*. Each tube was filled with water to the same level; to the top of the beads. The information in the data table was recorded for tube *A* only.

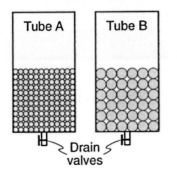

Data Table 1: Tube A	
Water required to fill pore spaces	124 mL
Time required for draining	2.1 sec
Water that remained around the beads after draining	36 mL

If the same experiment was performed with tube *B*, which data table below best represents the expected results?

Data Table 2: Tube B	
Water required to fill pore spaces	124 mL
Time required for draining	1.4 sec
Water that remained around the beads after draining	26 mL

(1)

Data Table 2: Tube B	
Water required to fill pore spaces	124 mL
Time required for draining	3.2 sec
Water that remained around the beads after draining	36 mL

(3)

Data Table 2: Tube B	
Water required to fill pore spaces	168 mL
Time required for draining	3.2 sec
Water that remained around the beads after draining	46 mL

(2)

Data Table 2: Tube B	
Water required to fill pore spaces	168 mL
Time required for draining	1.4 sec
Water that remained around the beads after draining	36 mL

(4)

Base your answers to questions 10 through 12 on the diagram below, which shows selected processes of the hydrologic cycle.

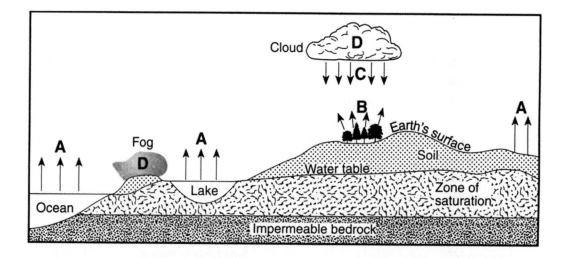

10. Which two letters represent processes in the water cycle that usually cause a lowering of the water table?

(1) A and B   (3) B and D
(2) A and C   (4) C and D

11. What are two water-cycle processes *not* represented by arrows in this diagram?

    (1) transpiration and condensation
    (2) evaporation and melting
    (3) precipitation and freezing
    (4) runoff and infiltration

12. What change in state is represented by *D* and how does this change affect the temperature of the air?

13. Which of the following events will upset people who use well water in their homes?

    (1) building a community park
    (2) constructing a gas station
    (3) installing pipes under the roads to carry away storm water
    (4) passing a law to prevent real estate development

14. Columns *A, B, C,* and *D* below are partly filled with different sediments. Each column contains particles of a uniform size. A fine wire-mesh screen covers the bottom of each column to prevent the sediment from falling out. The lower part of each column has just been placed in a beaker of water.

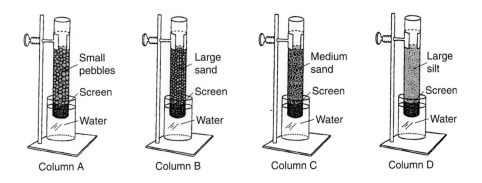

In which column would capillary action cause the water from the beaker to rise the highest?

    (1) *A*   (3) *C*
    (2) *B*   (4) *D*

Part C

15. Base your answer to question 15 on the diagram below. Name the process of the hydrologic cycle that occurs at each of the letters on the diagram: *A, B, C, D,* and *E.*

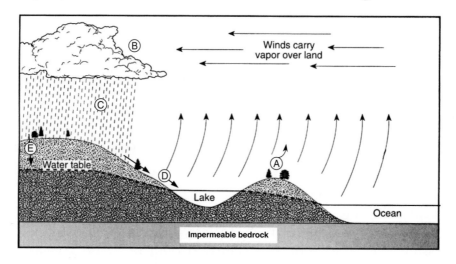

16. Describe the process of evaporation.

17. State one land-surface condition that would allow runoff to occur.

18. Explain one role of plants in the water cycle.

19. The diagram below is a profile view of a working water well. If the owner stopped using the well for many weeks, the position of the water table would probably change. On a sketch or a copy of this diagram, draw a dotted line to show the most likely new water level after this long period of no withdrawal of groundwater.

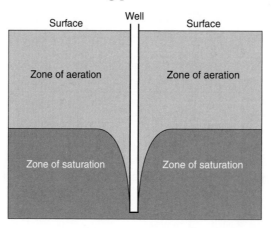

20. By what process do plants affect the moisture content of the atmosphere?

11 Oceans and Coastal Processes

WORDS TO KNOW

barrier island	longshore transport	sandbar	tidal range
Coriolis effect	neap tide	spring tide	tide
El Niño	ocean current	surf zone	

This chapter will help you answer the following questions:

1. Why are the oceans salty?
2. How do scientists explore the oceans?
3. How does ocean water circulate?
4. What causes the tides?
5. Why do coastlines change?

WHY IS EARTH CALLED THE BLUE PLANET?

4: 1.1i
4: 1.2f; 1.2g

Scientists' ability to study other planets has increased greatly over the past few decades. Through studies of other planets, it has become clear that Earth is unique among all known planets. Earth is the only planet known to have liquid water on its surface. So much of our planet is covered by oceans that Earth has a unique blue color when seen from space.

Early forms of life thrived and evolved in Earth's oceans. The oceans protected them from harmful ultraviolet rays. The water circulating in the oceans carried oxygen and food to organisms that did not move. Other organisms developed that could move through the oceans in search of food.

WHAT MAKES OCEAN WATER DIFFERENT?

4: 1.2c; 1.2f
6: EQUILIBRIUM AND
STABILITY 6

Our planet probably began as a rocky mass without surface water. Although scientists do not know how long it took for oceans to form on Earth, evidence of surface water can be found in rocks that date back to very early in Earth's history.

The Origin of Earth's Water

There are several possible sources of the water now in the oceans. Perhaps most of the water came from magma that formed deep within Earth's molten interior. Most of the water vapor from the earliest eruptions may have remained in the atmosphere until the surface cooled enough for liquid water to collect. Even today, erupting magma contains large amounts of water vapor.

Some of the water could have come from outer space. Comets are composed mostly of ice. They are sometimes described as dirty snowballs. Comets striking Earth probably added some of the water in the oceans. Even rocky meteorites, which showered Earth much more frequently early in its history, contain water. There are few remains of Earth's earliest rocks; therefore, details of the formation of oceans will probably remain unknown for many years.

The Composition of Ocean Water

You may have read stories of people stranded at sea who suffered from a lack of water. Surprisingly, they were surrounded by more water than they could ever need. However, ocean water contains about 3.5 percent dissolved salts. Figure 11-1 shows the average composition of ocean water.

Our bodies use much of the water we drink to absorb and remove waste products. Drinking ocean water would add unwanted salts rather than help the body get rid of them. That is why people cannot drink ocean water unless most of the salts have been removed. Other than water, the most common substance in seawater

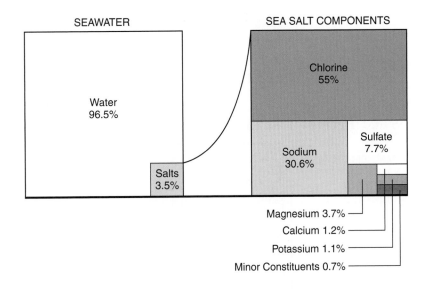

FIGURE 11-1. On average, about 3.5 percent of the mass of salt water is dissolved salts. The most common salt is sodium chloride, table salt.

is sodium chloride, or table salt. Also present are magnesium, calcium, and potassium compounds, which are also called salts by chemists.

If the oceans receive mostly freshwater, why are they salty? The oceans are part of the hydrologic cycle. The water that enters the oceans will eventually evaporate into the atmosphere. The average time a molecule of water stays in the ocean is about 4000 years. However, water cannot take along its load of dissolved solids when it evaporates; the salts are left behind.

You might think that through time the oceans would become more and more salty. However, the salinity of ocean water has been in a state of dynamic equilibrium for millions of years. Processes that take dissolved salts out of the oceans balance the dissolved salts that enter the ocean. Some animals that live in the ocean remove salts to make bones and shells. In addition, some salts leave the water as precipitates, forming salt deposits.

Source of Salts

Where did all that salt come from? The water given off by volcanoes is freshwater with few dissolved salts. The salts found in ocean water come from the land. Chemical weathering of rocks releases salts. Overland flow (runoff) and groundwater dissolve the salts in bedrock and soil. This amounts to adding about 4 billion tons of dissolved solids to the ocean each year. In spite of those salts, most of the water entering the oceans is considered to be freshwater. Additional dissolved substances, including salts, enter the oceans through deep-sea vents, which release water that has circulated through the rocks that make up the ocean bottom.

STUDENT ACTIVITY 11-1 —THE DENSITY OF SEAWATER

1: MATHEMATICAL
ANALYSIS 1
1: SCIENTIFIC
INQUIRY 1, 2, 3

Collect a few cups of clean ocean water, or mix your own in a ratio of 3.5 g of table salt per liter of freshwater. Carefully pour the water into a balloon. Be sure that there is no air bubble in the balloon, and then tie the end of the balloon. Gently place the balloon in a large container of freshwater. Does it sink or float? What does this tell you about the density of ocean water? Find a way to measure the volume and mass of the salt water to calculate its density. Visit the following Web site to try some oceanography activities: *http://www.msc.ucla.edu/oceanglobe/investigations.htm*

Salinity Changes with Latitude

The balance between inflow of freshwater and evaporation of water depends on latitude. At about 25° north and south of the equator there are regions where the oceans are a little saltier than average. This is because the climate at these latitudes is generally dry. Consequently, there is relatively little rainfall and more evaporation of ocean water in these regions.

Near the equator, precipitation is plentiful and rivers such as the Amazon dilute the salt water of the oceans. Ocean-water salinity is also lower at high latitudes where temperatures are cool and evaporation is low.

HOW CAN WE INVESTIGATE THE OCEANS?

4: 2.1l, 2.1o
7: STRATEGIES 2

Until the middle of the twentieth century, finding the depth of the oceans was a time-consuming process. Rolling out miles of steel cable to reach the ocean bottom took a long time. Today, oceanographers can measure the depth of the oceans by bouncing sound waves off the seafloor. Figure 11-2 shows the distribution of land elevations and ocean depths over Earth. From this figure it is clear

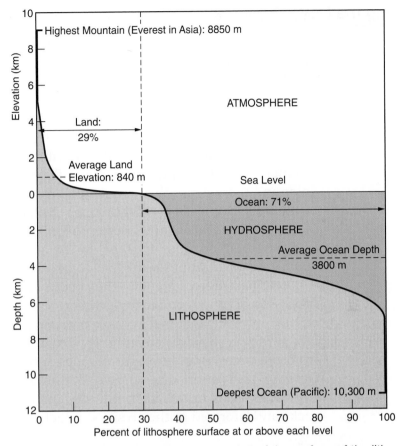

FIGURE 11-2. This graph shows the portion of the surface of the lithosphere that is found at various elevations over Earth's surface. Notice the large portion of the surface just above sea level and about 4 km below sea level. This is not a profile of any location, but a variation of a bar graph.

that the depth of the oceans ranges from sea level at the shore to a maximum depth of more than 10 km.

Exploring the Shallow Ocean

Scientists know the most about the shallowest parts of the oceans. Here they can observe the ocean bottom most easily. Where the ocean is shallowest, light can reach the bottom, and life is abundant. Divers using air tanks can dive down a few hundred meters. The greatest danger to humans in ocean exploration is the extreme pressure caused by the weight of overlying water, which limit the depths of dives.

Exploring the Deep Ocean

To explore deeper parts of the oceans, scientists use special diving chambers known as submersibles. However, most exploration of the deepest parts of the ocean is done with remote-controlled diving devices. Figure 11-3 is a model of a remote controlled deep-ocean exploration vehicle.

Exploration of the ocean bottom has found igneous rocks of mafic composition, such as basalt and gabbro, usually underlie the sediments covering the ocean floor. These rocks are darker in color and more dense than granite and rocks of similar composition that are found in the continents. The two relatively flat parts of the line seen in Figure 11-2 on page 251, one just above sea level and another about 4 km below the ocean's surface, are a result of this division of Earth's crust into two basic rock types.

Geological forces renew the ocean bottom through the processes of plate tectonics, which you will learn about in Chapter 15. Upwelling material from deep within Earth reaches the surface at the ocean ridges creating new crust. The crust moves away from the ocean ridges toward trenches and zones of subduction carrying the continents with it. At the zones of subduction, oceanic crust is drawn back into the interior while continental rocks are deformed as they resist subduction.

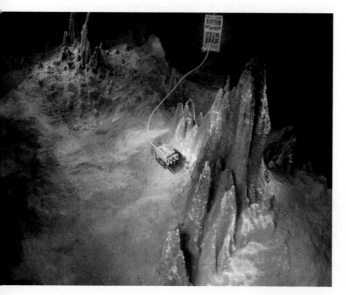

FIGURE 11-3. Due to the extreme pressure in the deepest part of the oceans, widespread exploration by humans is nearly impossible. Remotely operated vehicles such as the one shown in this museum model are the best way to explore the deepest parts of the oceans. Unlike this unusual location, most of the ocean bottoms are as flat and featureless as any land area.

WHY DOES THE WATER IN THE OCEAN CIRCULATE?

4: 2.1a; 2.1b
4: 2.2b; 2.2c, 2.2d

The water of the oceans is moving constantly. The primary cause of deep currents is differences in density. Dense water sinks to the bottom and forces water that is less dense to the surface. Near Earth's poles water is cooled, becomes more dense, and sinks to the bottom. Water reaches its greatest density at a temperature of 4°C. Therefore, over the entire planet, deep-ocean water is near freezing.

Surface temperatures vary considerably with latitude. It is warmer near the equator and colder near the poles. The sinking of cold water at the poles must be balanced by upwelling that brings deep water back to the surface. Cold water can hold more oxygen and support more marine life than warm water. For this reason, upwelling, cold currents bring nutrients to the surface in some of the world's best fishing grounds.

Turbidity Currents

In Chapter 8 you learned about the vertical sorting of sediments that occurs when particles of mixed sizes settle in deep water. Graded bedding occurs on the ocean bottom when sediment-rich streams, sometimes called turbidity currents or density currents, flow down under water slopes. The largest particles usually settle first. These sudden underwater "landslides" move at surprising speeds, sometimes reaching more than 100 km/h (60 mph). The primary factors affecting these currents are their size, flow density, and the slope, or gradient of the ocean bottom.

The Coriolis Effect

The circulation of surface water follows wind circulation. Both are affected by Earth's rotation. Winds and ocean currents generally curve as they travel long distances over Earth's surface. This curving is called the **Coriolis effect**. Actually, the winds and ocean currents are going as straight as they can, but Earth's rotation makes them appear to curve to the right in the Northern Hemisphere and to the left in the Southern Hemisphere.

Figure 3-1 on page 58 is a map of the world showing the most common surface current directions. Notice that most of the currents in the North Atlantic follow a circular path curving constantly to the right in a great clockwise circle. The currents in the northern part of the Pacific Ocean also follow this clockwise (to the right) pattern. Currents in the South Atlantic and southern parts of the Pacific Ocean curve to the left in a counterclockwise pattern.

STUDENT ACTIVITY 11-2 —OBSERVING GYRES

USING *THE EARTH SCIENCE REFERENCE TABLES*

A gyre is a large, curving pattern of circulation in the ocean. Use the surface ocean current map in the *Earth Science Reference Tables,* or Figure 3-1 on page 58, to locate gyres in the Northern Hemisphere and in the Southern Hemisphere. For each gyre, list the ocean or part of an ocean it occupies, the names of the surface currents that form the gyre, and the direction in which it circulates (clockwise or counterclockwise). What is the most common direction of circulation in each hemisphere?

Currents and Climate

Ocean currents influence the climates of coastal locations. The temperature of ocean water does not change as quickly as the temperature of rock and soil. Therefore, coastal locations usually have a smaller range of temperature than do inland locations. Cold and warm currents also affect coastal temperatures. For example, people who live along the coast of California are not as likely to swim in the ocean as people who live along the Gulf of Mexico or the Atlantic coastline of the United States. Cold ocean currents along the California coast keep the water temperature too chilly for most swimmers, even in the summer. The cool ocean water also prevents summer temperatures from getting too hot along the coast of southern California.

The climate along the southern coast of Alaska is influenced by relatively warm ocean water. Summer and winter temperatures are fairly mild in this part of Alaska. Palm trees grow in some areas along the west coast of Great Britain where the warm currents of the Gulf Stream and the North Atlantic Current regulate winter temperatures. These coastal locations have winters far less severe than central European cities that are far from the ocean. By looking at the arrows on Figure 3-1 you can tell where warm or cool ocean currents affect coastal areas. The black arrows show warm currents and the white arrows show cool currents.

El Niño

In recent decades, scientists have become more aware of how changes in ocean currents affect the climate over large areas. Most of the time, cold ocean currents and nutrient-rich water are found off the western coast of South America. Good fishing in this region provides food and employment in ocean-side villages. However, in some years the upwelling of cold water is replaced by warm water, which reduces fish production. This usually happens about the time of the Christmas holidays. Local people call it **El Niño**, a Spanish term for the Christ Child, although it is an unwelcome "Christmas present." However, this event affects more than the local fishing industry. A strong El Niño can cause increased winter rain and flooding along the coast of California and drought in the western Pacific. The relationship between ocean currents and regional climatic changes is giving scientists new methods to predict weather and prepare for its consequences. Visit the following Web site to learn about El Niño and coastal change: *http://coastal. er.usgs.gov/hurricanes/elnino/*

WHAT CAUSES THE TIDES?

4: 1.1a, 1.1i

People who live along ocean coastlines are familiar with the periodic rise and fall of the oceans. The twice-daily cycle of change in sea level is the **tides**. Currents associated with tides can affect fishing and the ability of boats to sail in some places. If a storm strikes a coastal area at high tide, wave and water damage is likely to be greater than from a storm that comes ashore at low tide. Visit the following Web site to find current tides data for the USA: *http://tidesandcurrents.noaa.gov/station_retrieve.shtml?type= Current+Data*

Tidal range is the difference between the lowest water level and the highest water level. Most locations have two high tides and two low tides each day. However, some places have only one

FIGURE 11-4 *A* AND *B*. (*A*) Parrsboro Harbor, Nova Scotia at high tide. Note the boy diving into the water. (*B*) The harbor at low tide. The water level has dropped to expose the bottom of the harbor, leaving the boat on land. In addition, the water has receded quite a distance. Notice the people walking along the land near the end of the pier.

daily cycle. In some locations, the change in sea level is too small to be noticeable. On the other hand, in the Bay of Fundy along the eastern coast of Canada, the tidal range can be as much as 15 m. This is about the height of a four-story building. Figure 11-4 *A* and *B* were taken along the Bay of Fundy. They show the same scene at high tide and about 6 h later at low tide.

STUDENT ACTIVITY 11-3 —EXTREMES OF TIDAL RANGES

2: INFORMATION
SYSTEMS 1

Prepare a report about places around the world that have an un-usually large or small range of ocean tides. Plot these locations on a world map. Why do some locations have higher tides than others, and how do these extreme tides affect the local people and economy?

Gravity

Gravity is the force of attraction between objects. The strength of the force is determined by the masses of the objects and the distance between them. The force of gravity holds Earth in its path around the sun. It also keeps the moon in orbit around Earth.

People do not feel the gravitational attraction between their body and most of the objects around them because the objects are too small. People certainly do feel the attraction between their body and Earth. That force is weight. It is a strong force because Earth is so massive and because we are so close to it. If you climb a tall mountain, you move a little farther from the center of Earth. This decreases your weight, although the change is too small to observe without careful measurement. (Your mass, however, remains the same.) If you could move far enough above Earth into space you would actually notice a decrease in your weight. Astronauts in orbit around Earth feel completely weightless as the result of their distance above Earth and their orbital motion.

The tides are caused by the gravitational pull of the moon and the sun. Although the moon is much smaller than the sun, it is much closer to Earth. Therefore, the moon has a greater gravitational effect on Earth than does the sun.

The Moon and Tides

The moon's gravity affects the solid Earth and the oceans. The moon pulls most strongly on the part of Earth closest to it. When the moon is directly over the ocean, this part of the ocean experiences a high tide. The moon has a smaller affect on the solid Earth, so it pulls Earth away from the water on the far side. This causes a high tide on the side of Earth away from the moon. That is why most locations have two high tides each 24-h day. Figure 11-5 shows how the moon pulls more strongly on the ocean water on the side of Earth closer to the moon.

The Sun and Tides

The sun also influences ocean tides. When Earth, sun, and moon are in a line with one another, the highest, or **spring tides**, occur. At spring tides, the sun and moon do not need to be on the same

FIGURE 11-5. The main cause of the tides is the gravitational attraction of the moon. The moon pulls water on the near side of Earth away from the solid Earth. It also pulls Earth away from water on the far side. This is why most locations have two high tides every 24 h. (The length of the arrows represents the affect of the moon's gravity.) Distances are not to scale.

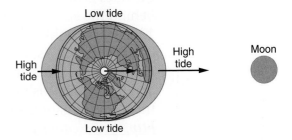

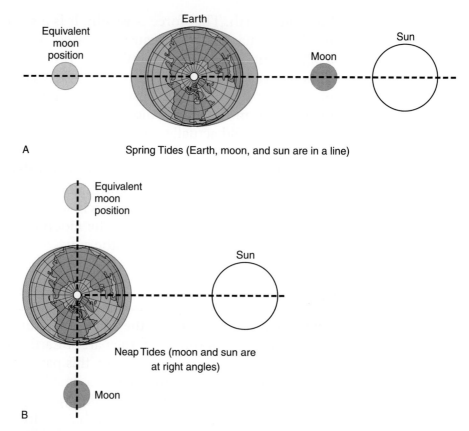

A   Spring Tides (Earth, moon, and sun are in a line)

B

FIGURE 11-6 *A* AND *B*. The tidal range at any location depends on the angle between the moon, the sun, and Earth. When they are in a line (*A*), the greatest range, or spring tides, occurs. The smallest tidal range happens when the sun and moon are at right angles to Earth (*B*). (Neither distances nor size of the sun are to scale.)

side of Earth, as you see in Figure 11-6 part *A*. The range of the tides is the lowest when the sun and moon are at right angles to Earth, and **neap tides** occur. Figure 11-6*B* shows the configuration of Earth, sun, and moon at neap tides.

The moon orbits Earth every 27 days. Therefore, the period of the tides is not exactly 12 or 24 h. Each day the moon seems to fall behind the sun by about an hour. A full cycle of the tides is about 12.5 h, or roughly 25 h in places that only experience one cycle per day. Figure 11-7 shows the cyclical nature of the tides. The graph clearly shows the daily cycle of the tides and the cycle of the spring and neap tides. Visit the following Web site to see an animation of the tides animation showing the daily cycle, spring and neap tides:
http://www.edumedia-sciences.com/a475_12-tides.html

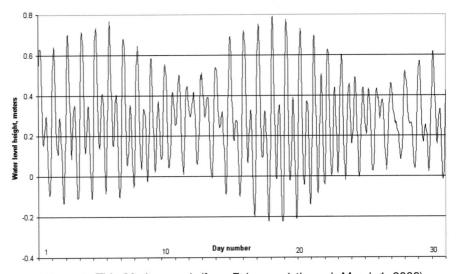

FIGURE 11-7. This 30-day graph (from February 1 through March 1, 2000) records the change of the water level at Hilo Bay, Hawaii. Notice how the cycle repeats.

STUDENT ACTIVITY 11-4 —GRAPHING THE TIDES

6: PATTERNS OF
CHANGE 5

Graph the height and time of tides. (Graph Time on the horizontal axis and Water Height on the vertical axis.) If you live on the coast, you may be able collect your own data, or you can use data from a local newspaper. If you live inland, you can use data from the Internet.

HOW DO COASTLINES CHANGE?

4: 2.1u

In earlier chapters, you read about erosion caused by glaciers, wind, running water, and gravity acting alone. It is now time to consider coastal erosion and the movement of sediments along coastlines. When you think of visiting an ocean beach you may picture a broad strip of sand where you can play, rest, get a suntan, or enjoy the water. You may not realize that the beach is a dynamic part of a system that transports sediment. The sediment making up the beach is on a journey that transports weathered rock from the land into the ocean. A beach is one of Earth's most active environments of deposition and erosion.

There are two primary sources of sediment for beaches. Waves, particularly in storms, erode the coast and cause the shoreline to

FIGURE 11-8. The sand on this beach came from erosion of the cliffs behind the beach. There are no rivers nearby to supply sediment from inland areas.

erode (move inland). Rock and sediment fall or are washed onto the beach. Streams and rivers sweep other material into the ocean. Beaches are zones of transport where sediments move along the shore by wave action and currents. Figure 11-8 shows a wide beach composed of sediment eroded from the cliff behind the beach.

Waves

The energy of most waves comes from wind. The greater the wind's velocity and the greater the distance it blows over open water, the larger the waves it creates. Because winds can blow for greater distances over the ocean than over a lake, ocean waves are usually larger than waves on lakes. Friction between moving air and the surface of the water sets up waves that move forward in the direction of the wind.

The waves you observe can be deceiving. It may look as if the water is moving forward with the waves. However, energy not water is transferred by waves. Figure 11-9 shows that as the energy of the wave moves forward, surface water moves in circles. Deep water is not affected by waves. When the wave enters shallow water near shore, the crest moves faster than the bottom of the wave, and a breaker forms. As the wave breaks, it gives up its energy along the shore. This energy can do three things along the beach:

1. Wave energy breaks up sand and rock in the surf zone by causing abrasion.
2. Wave energy can erode the beach, including sediments and rock on the beach.
3. Wave energy transports sand and sediment parallel to the shore.

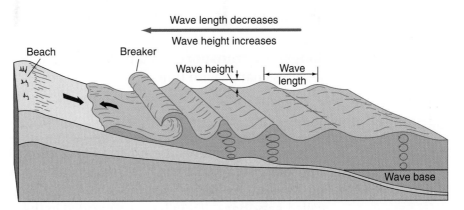

FIGURE 11-9. Ocean waves are driven by winds. In deep water, waves make the surface water move in circles as they carry their energy forward. Waves break in shallow water, giving up their energy to abrade and transport beach sediments.

Longshore Transport

Most beaches have a region called the surf zone. The **surf zone** extends from where the base of the waves touches the bottom (a depth of about half the distance between wave crests) to the upper limit the waves reach on the beach. The surf zone along most beaches is like a river. Waves cause sand to wash onto the beach with the breakers and then wash back into the water with the return flow. Most waves approach the beach at an angle. The result is a zigzag motion that carries sand (or whatever sediment the beach is made of) downwind along the beach, as shown in Figure 11-10.

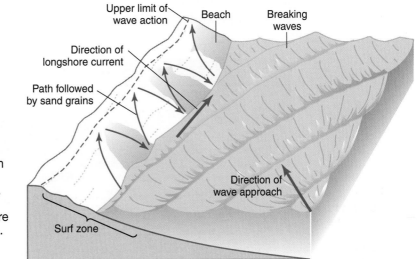

FIGURE 11-10. As waves wash onto a beach, beach sediment moves forward and back in the surf zone. Beach sediment is also carried parallel to the shore in the downwind direction. The result is longshore transport.

The resulting motion of the water along the shore is called a long-shore current, and the motion of the sediment is known as **longshore transport**. As a result of these processes, oceanfront features change with time.

Depositional Features

Many coastal features are related to wave erosion and longshore transport. Sometimes the back-and-forth motion of the waves deposits sand that forms low ridges along the shore. These ridges are called **sandbars**. If you have ever waded in the ocean along a sandy beach and discovered a shallow area separated from the shore by deeper water, you found an underwater sandbar.

A spit is a sandbar that forms a continuation of a beach into deep water. Spits sometimes grow across bays, forming a baymouth bar. Similar offshore features that rise above sea level are **barrier islands**. A shallow bay called a lagoon separates barrier islands from the mainland.

Figure 11-11 shows the series of islands that separate the south shore of Long Island from the Atlantic Ocean. Jones Beach and Fire Island are a part of this series of barrier islands. These features are common on gently sloping coastlines with a large amount of sand.

BEACHES Figure 11-12 *A, B,* and *C* illustrates a sequence of events in a shore area with a sandy beach. Part *A* shows a shoreline in balance. Beach sand originates from sediment carried by the river on the right and eroded from the bluffs along the shore. Waves from

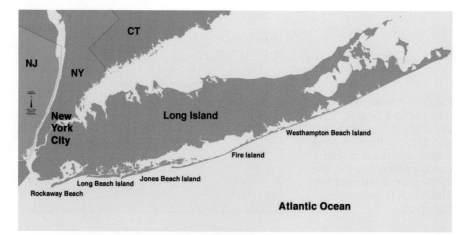

FIGURE 11-11. The south shore of Long Island is protected by a series of barrier islands.

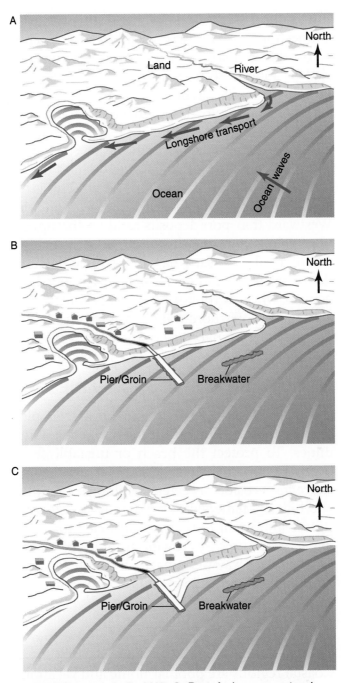

FIGURE 11-12 *A, B,* AND *C.* Part *A* shows a natural shoreline in dynamic balance. The construction of a breakwater and a pier/groin are shown in Part *B.* Part *C* shows how these structures cause the beach and shoreline to change as sand is deposited in some places, such as upwind from a groin. The downwind side usually experiences increased erosion with a narrowing beach.

the southeast bend as they enter shallow water near the shore, and a longshore current carries sand westward. The sand spit growing across the bay makes it clear that the principal direction of sand carried by longshore transport is toward the west. Part *B* shows a breakwater built parallel to the shore to protect boats from large waves. A pier/groin has been built from the shore out into the ocean. The structures are new in Part *B* and no changes in the beach are visible. Part *C* shows how the beach changes in response to these two barriers. The beach gets wider behind the offshore breakwater as sand builds outward from the beach. This is because wave energy has been reduced behind the breakwater and deposition increases. Westward transport deposits sand on the upwind side of the solid pier/groin. However, the beach shrinks on the downwind side where the flow of sand has been stopped. Even the sand spit is reduced because sand movement was stopped by the pier/groin. In general, when a groin or solid pier is constructed into the ocean in a region of longshore transport, the beach becomes wider on the upwind side and narrower in the downwind side.

HOW SHOULD WE MANAGE ACTIVE SHORELINES?

2: INFORMATION
SYSTEMS 1
7: STRATEGIES 2
6: SYSTEMS
THINKING 1
7: CONNECTIONS 1
7: STRATEGIES 2

Humans affect shorelines in many ways. People increase shoreline erosion by trampling on protective vegetation, especially in sand dunes. To protect the beach or unstable features, people build breakwaters, groins, and jetties. This is common in areas where shoreline erosion threatens buildings or other property. Dunes and hills are flattened to make building sites and parking areas.

It is important to understand that shore areas are delicate and dynamic features. A growing number of citizens are recognizing that the best way to manage changing coastal regions is to limit development and allow natural processes to continue without human interference.

Legal Issues

Coastal regions are popular home and vacation sites. Recreational opportunities, including swimming, boating, and fishing, make beachfront property highly attractive. However, there is discussion of whether there should be private ownership of beaches. In addition, people debate the wisdom of building on unstable areas near shorelines.

This is especially true around New York City and near other urban areas. Should the beaches of Long Island be playgrounds only for the wealthy, or should they be available to everyone? Should anyone be allowed to walk along ocean beaches? Should people be allowed to construct homes in unstable, sandy areas and low areas that are subject to storms and flooding? Do roads and buildings seriously affect the natural resources of oceanfront property? Beachside communities constantly deal with these issues. There are probably no solutions acceptable to everyone. We usually try to balance the factors and select the best policies from a wide range of controversial solutions. Visit the following Web site to access the Environmental Atlas of the Great Lakes: *http://www.epa.gov/glnpo/atlas/index.html*

STUDENT ACTIVITY 11-5 —ZONING FOR COASTAL PRESERVATION

6: SYSTEMS
THINKING 1
7: CONNECTIONS 1
7: STRATEGIES 2

In a cooperative group, develop a set of policies to guide both public and private development of ocean coastal areas. Prepare a document that could be given to coastal communities to help them develop zoning regulations for their oceanfront areas.

CHAPTER REVIEW QUESTIONS

Part A

1. Swimmers notice that it is easier to float in ocean water than it is to float in freshwater. Why is it easier to float in salt water?

   (1) Salt water is more dense than freshwater.
   (2) The ocean has larger waves than lakes and rivers.
   (3) Ocean water is usually deeper than freshwater.
   (4) Ocean water has more dissolved gases than freshwater.

2. Which ocean current transports warm, equatorial waters into the cooler regions of Earth?

   (1) California Current
   (2) Guinea Current
   (3) Falkland Current
   (4) Brazil Current

3. A tombolo is a sandy spit that connects a formerly unconnected island to the shoreline. Which process would form a tombolo?

   (1) coastal weathering
   (2) Coriolis effect
   (3) shoreline deposition
   (4) spring tides

4. Which location has a coastal climate that is made warmer by the influence of a nearby ocean current?

   (1) Southern California
   (2) Peru in South America
   (3) The east coast of Australia
   (4) Northwestern Africa near the Canary Island

5. Most of the Gulf Stream ocean current is

   (1) warm water that flows southwestward
   (2) warm water that flows northeastward
   (3) cold water that flows southwestward
   (4) cold water that flows southeastward

6. What process most likely formed the barrier islands along the south shore of Long Island, New York?

   (1) mass movement
   (2) wave action
   (3) stream erosion
   (4) glacial deposition

7. According to Figure 11-11 on page 262, in what direction are the longshore currents moving along the south shore of Long Island?

   (1) northeast   (3) southeast
   (2) northwest   (4) southwest

8. For most oceanfront locations, what is the usual period of time between one high tide and the next high tide?

   (1) about 1 h   (3) about 1 week
   (2) about 12 h   (4) about 2 weeks

9. Sand from beaches near recent Hawaiian lava flows is angular. Most other beaches, such as those on New York's Long Island, are composed of more rounded particles.

How can you account for the difference in the shape of sediment particles in these two places?

(1) The ocean temperature is colder at the Hawaii beach.
(2) The Long Island sediment is derived from harder rocks.
(3) The Hawaii beach has larger waves and more storms.
(4) The Long Island sediments have been on the beach longer.

10. Homes built in coastal sand dunes are more likely to suffer storm damage than homes located several miles from the shore. Which is *not* a factor in making the shorefront homes more vulnerable to storm damage?

(1) nearness to the ocean
(2) foundations built on sand
(3) a cooler summer climate
(4) changes in sea level

Part B

11. To preserve the wide, sandy beaches in a coastal resort, the local government built rocky groins from the top of the beach 50 m straight out into the ocean. After the construction of these groins, how did the beach change?

(1) The size of the sand particles increased near the groins.
(2) The beach became narrower everywhere near the groins.
(3) The beach became wider everywhere near the groins.
(4) The beach became wider in some places and narrower in others.

12. A student visited the seashore at high tide and at low tide to record her observations of the sun, the moon, and the planet Venus. Which observation was most likely recorded at high tide?

(1) The moon was low in the sky.
(2) The moon was high in the sky.
(3) The bright planet Venus was visible.
(4) The bright planet Venus was not visible.

13. A student collected several gallons of unpolluted ocean water. Using a clean metal pot, he boiled away all the water. Which statement best describes what he saw on the bottom of the pan after the water boiled away?

    (1) The pan was as clean as it was before the water was boiled away.
    (2) A film of calcite was left in the bottom of the pan.
    (3) A substance resembling table salt was left in the bottom of the pan.
    (4) The pan contained a transparent film of quartz along the bottom.

14. A research ship in the middle of the Pacific Ocean took measurements of ocean water at the surface and near the ocean bottom 6 km below the surface. In what way is most water from deep in the oceans different from the water they observed near the surface?

    (1) The water near the bottom of the ocean is warmer than surface water.
    (2) Water near the bottom of the ocean is more dense.
    (3) Surface water is salty but bottom water is freshwater.
    (4) Water near the bottom receives more light than water near the surface.

Part C

Base your answers to questions 15 through 17 on the diagram below, which represents a part of the Atlantic Ocean seafloor. When an earthquake occurred at *A,* a sediment flow was released. The times indicated at positions *B, C,* and *D* show when the sediment flow arrived at each position.

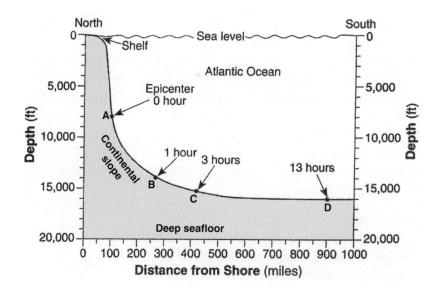

15. Calculate the average gradient of the ocean floor from *A* to *D*. Be sure to start with the appropriate formula and label your answer with the proper units.

16. How did the velocity of the sediment flow change from the time it was first caused until it arrived at *D*?

17. Why did the speed of the sediment flow change as it moved from *A* to *B, C,* and *D*?

18. In California, the prevailing winds come from the west. Therefore, how does the California Current affect the climate of coastal locations in California?

19. The double image below is of a coastal area before and after a hurricane hit the area. What can local governments do to reduce storm damage to homes in coastal areas without constructing protective structures or changing the natural beach processes?

20. Most surface ocean currents in the Northern and Southern hemispheres do not curve in the same direction. Describe the direction in which most surface ocean currents curve in the Northern Hemisphere.

12

Glaciers

WORDS TO KNOW

continental glacier	grooves	outwash	till
drumlins	kettle	terminal moraine	valley glacier
erratics	moraine		

This chapter will help you answer the following questions:

1 What evidence of glaciation can be observed in New York State?

2 What are the two major kinds of glaciers?

3 How can we recognize the erosional and depositional features caused by glaciers and by meltwater?

4 What are ice ages?

HOW DID SCIENTISTS DISCOVER THAT GLACIERS ONCE COVERED NEW YORK STATE?

6: MODELS 2

The beauty and variety of landforms in North America fascinated the first European settlers. In New York, settlers also found mysterious features they could not explain. For example, many of the rocks in the soil were kinds not found in local bedrock. Some of these foreign rocks were large and perched on hilltops. These boulders were much too high and too large to have been moved there by nearby streams. These features could not be explained by the geological processes at work in the area at that time.

In the nineteenth century, European geologists began to understand that prehistoric, continental glaciers played a major role in the evolution of their landscape. American geologists also found that the idea of a great ice age helped them understand puzzling aspects of North American geology. Visit the following Web site and search for glacial legacies to watch a Power Point slide show featuring glacial features in New York State: *http://www.newyork scienceteacher.org/earth_science/es_shareathon.php*

WHAT IS A GLACIER?

4: 2.1t, 2.1u

Snow falls and sticks to the ground everywhere in New York State. In the highest parts of the Adirondack Mountains, winter snow often lasts until early summer. If these mountains were 1000 or 2000 m higher, the reduced warmth of summer would not melt the previous season's winter snow. Each year more snow would collect and exert pressure on the underlying snow. This pressure would change the snow to ice, and gravity would make the ice begin to flow downhill. This is how glaciers form. The reason that no glaciers exist today in New York State is that there are no places where the snow does not completely melt before the following winter.

Snow and ice exist as crystals. When snow falls, the flakes are usually light and feathery. After the flakes reach the ground, they are buried under more snow. The delicate crystals gradually change to solid ice over a period of time that depends on such factors as speed of burial and temperature.

Is ice a solid or a liquid? Ice is the crystalline solid form of water. It fits the definition of a solid. Under short-term stress, ice behaves as a solid. An ice cube in an environment below freezing has a fixed shape. Hit it with a hammer and it breaks into smaller pieces. Yet, ice in a glacier flows. Glaciers do not flow because the ice is melting. They flow because solid ice responds to long-term stress by bending and changing shape. This is similar to the behavior of solid rock within Earth's mantle. If you could strike the rock deep in Earth's mantle with a hammer, it would shatter. Forces applied over long periods of time result in fluid behavior of both rock and

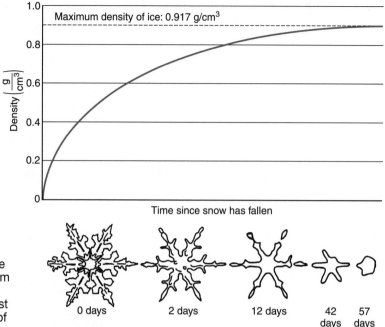

FIGURE 12-1. Delicate snow crystals transform into compact ice crystals. This is the first step in the formation of glaciers.

ice. The differences between solids and liquids are not as clear as you might have thought in the past. Figure 12-1 shows how the increasing density of the snow pack relates to changes in ice crystal shape.

STUDENT ACTIVITY 12-1 —SNOW TO ICE

6: MODELS 2
6: PATTERNS OF CHANGE 5

This activity must be done in the winter when there are piles of snow outside. Get two 200-mL beakers and measure the mass of each. Find a location where the snow has been pushed into a deep pile. Take the snow samples from the pile at different, measured depths. Without packing the snow, dig out the same volume and place each sample in a different 200-mL beaker. Also, gather fresh, surface snow that has not been moved. Determine the mass of snow in each beaker. Draw a graph of snow depth versus the mass of snow in a given volume.

Valley Glaciers

Valley glaciers, which begin high in mountain areas, flow from the high ice fields through valleys to lower elevations. In some locations, valley glaciers spread over flat lowlands as a piedmont glacier. Some glaciers flow into a lake or ocean where the ice breaks off, forming icebergs. The Portage Glacier flows out of permanent snow and ice fields high in the mountains near Anchorage, Alaska. The advancing glacier enters a freshwater lake.

Other valley glaciers move down to an elevation where it is warm enough to melt the ice as quickly as it advances, or moves forward. If the ice in the glacier moves forward at 1 m per day while the ice also melts back 1 m per day, the front of the glacier will not seem to move. This is a *dynamic equilibrium* because the rate of flow and the rate of melting are in balance. Even though the end of the glacier may be in the same place from year to year, the ice is moving downhill constantly.

The rate at which ice moves varies from glacier to glacier. Ice in a small glacier may move only a few millimeters a day. Large valley glaciers commonly move a few centimeters a day. Some glaciers experience periodic surges during which the glacier advances more than 30 m per day over a period of several months. Surges are generally caused by water accumulating under the glacier and serving as a lubricant.

Just as gradient and volume affect the speed of water in rivers, they also influence the speed of ice flow in glaciers. Also, like the water in most rivers, ice moves fastest near the center and near the surface of a glacier. Figure 12-2 illustrates ice flow in a glacier. The long arrows indicate fast flow, whereas the short arrows indicate slower flow.

FIGURE 12-2. Three glaciers meet below this ridge in Alaska. The arrows show that, like streams and rivers, each glacier flows fastest near the center.

Continental Glaciers

A **continental glacier** flows outward from a zone of accumulation to cover a large part of a continent. If the process could be speeded up, it might resemble what pancake batter looks like as it is poured onto a griddle. Figure 12-3 shows grooves on bedrock surfaces in New York City's Central Park. These grooves show that area was once completely covered by ice. The Catskills are well over 1000 m above sea level. They also show evidence that they were once covered by ice. Furthermore, ice flowed southward over the Catskills from eastern Canada where the ice must have been even thicker. Yet this part of Canada does not have mountains as high as New York's Catskills, so the ice in the zone of accumulation must have been very thick. Even the highest mountains in the northeastern United States were completely covered by the great thickness of ice covering eastern Canada.

Ice sheets several kilometers thick now cover most of Greenland and Antarctica. Scientists have studied both regions to understand ice caps and how they flow outward and down to the oceans. Ice flowing into an ocean breaks away from the main body of the glacier to float away as icebergs. Because ice is less dense than water, icebergs do not sink. Icebergs from Greenland and Antarctica are large enough to threaten shipping. A number of vessels, including the *Titanic,* have sunk because they ran into

FIGURE 12-3. The wide grooves in this rocky outcrop in New York City's Central Park were made by rocks carried under the great continental glaciers. These grooves show that glaciers moved at least as far south as this part of the island of Manhattan.

icebergs. Fortunately, icebergs break up and melt before they can invade most ocean areas. Visit the following Web site to learn about glaciers of the world: *http://swisseduc.ch/glaciers/earth_icy_planet/glaciers02-en.html*

STUDENT ACTIVITY 12-2 —A MODEL OF A GLACIER

6: MODELS 2

You can make a model of a glacier with something called "oobleck." You make oobleck by mixing cornstarch with water in a mass ratio of about 1 part water to 1.2 parts cornstarch. Like ice, this substance will shatter when hit hard. But, placed on a slope, oobleck will flow downhill.

HOW DO GLACIERS CAUSE EROSION?

4:2.1t, 2.1u

When a glacier advances down a valley or over a continent, the ice pushes, carries, and drags large amounts of soil and sediment. These loose materials have little chance of remaining in place when a mass of ice hundreds or even thousands of meters thick moves over them. Ridges and knobs of bedrock are pried loose or rounded by the moving ice. Although ice is much softer than most bedrock, the rocks and sediment dragged along the bottom of a glacier scrape and scour the bedrock over which the glacier passes.

Valleys

In mountainous or hilly terrain, advancing glaciers seek the lowest passages and move through valleys first. Stream valleys often have a V shape in profile, especially in mountain areas. Streams and the sediment they carry occupy only the bottom of the valley and do not erode the sides of the valley above flood levels. The sides of a stream valley in mountainous terrain collapse under the influence of weathering and gravity, which often gives them a steep but uniform slope of the V profile such as the stream valley in Figure 12-4*A* on page 276.

When a glacier moves down a mountain valley filling it with ice, the erosive action of the glacier and its load of sediment pluck, scrape, and scour the sides of the valley changing its profile to a

FIGURE 12-4. Valleys made by mountain streams often have a narrow, V-shape. This is the Yellowstone River in Wyoming. (*A*) This U-shaped valley in the Western Finger Lakes was shaped by glacial action. (*B*)

broader U shape. U-shaped valleys are strong evidence of glacial erosion.

In the western part of New York, north-south-aligned river valleys that drained the Allegheny Plateau were eroded into soft shale, siltstones, and limestones. The continental ice sheets that advanced southward into New York State followed the lowest passages southward. The ice sheets advanced into these valleys first, making them wider and deeper. Dry valleys in the Finger Lakes region show the U shape especially well. (See Figure 12-4*B*.) A number of large U-shaped valleys running north-south in this part of the state were blocked at their outlets by glacial sediments to form the Finger Lakes. (See the map of New York State in the *Earth Science Reference Tables*.)

Striations and Grooves

Parallel scratches called glacial *striations* can be found on hard bedrock surfaces throughout New York State, but they are best preserved in the hard metamorphic and igneous rocks of the Adirondacks and New York City regions. These scratches indicate the direction in which the glaciers moved. **Grooves** are deeper and wider cuts formed by glaciers. Grooves are often found in softer bedrock than where there are striations. Central Park in New York City has both grooves and striations in exposed bedrock.

HOW CAN WE RECOGNIZE DEPOSITION BY GLACIERS?

4: 2.1u, 2.1v

There are several differences between sediments deposited by ice and sediments deposited by water or wind. Water and wind sort sediments. Moving ice transports and deposits sediment without regard to particle size. Therefore, sediments deposited directly by glaciers are unsorted and do not show layering. This unsorted glacial debris is sometimes called **till**, as you saw in Figure 8-13 on page 198.

Stream sediments are deposited where streams flow, usually in the bottom of a valley. But a glacier can move its debris anywhere the ice covers, even to the highest parts of New York State. Glacial sediments often cover the whole land surface with an uneven blanket of till composed of mixed particle sizes.

Glaciers carried their load of sediments, including boulders of granite and gneiss from Canada, hundreds of kilometers southward into New York State. In western New York State where the local bedrock is sedimentary, most often shale, siltstone, and limestone, these foreign rock types are especially noticeable. New York soils have a greater variety of minerals and they are more fertile than they would have been if they contained only local rocks. Large rocks that were transported from one area to another by glaciers are known as **erratics**. Figure 12-5 shows a large erratic located in New York City.

Moraines

A **moraine** is a deposit of unsorted glacial sediment (till) pushed into place by an advancing glacier. Where a glacier stopped advancing, the moraine takes the form of irregular hills. This kind of

FIGURE 12-5. Many large boulders in New York City's Central Park were left there by the continental glaciers. Streams could not have deposited such large rocks at this elevated location.

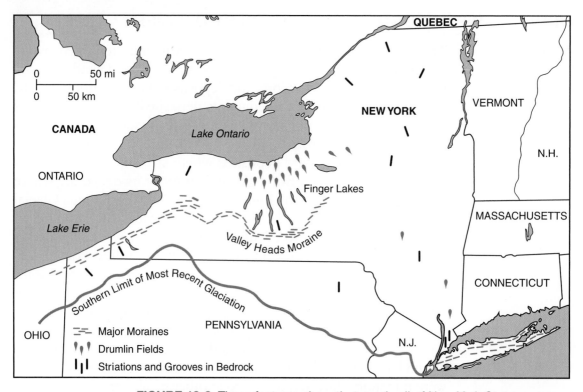

FIGURE 12-6. These features show that nearly all of New York State was under continental glaciers approximately 15,000 years ago. At this time, about half of North America was covered by glacial ice. Evidence suggests that there were several glacial advances and retreats during a period that lasted tens of thousands of years.

deposit is a **terminal moraine**. Even though the front of the glacier was nearly stationary, ice continued to carry sediment to its front where the sediment was deposited as the ice melted. Figure 12-6 shows the parts of New York State and neighboring areas that have moraines and other glacial features. Most of these are places where the ice front stalled and dumped its sediment load over a period of time.

The Valley Heads Moraine dammed the southern ends of the valleys that became New York's Finger Lakes. This allowed many of them to fill with water that now drains northward. Long Island was built on two principal end moraines that extend into New England.

Among the irregular hills in a moraine are depressions called kettles. A **kettle** is a small closed basin with no low-level outlet. Some kettle holes form when a block of ice within the till melts, leaving a closed depression. Rainwater that runs to the bottom of the kettle hole can escape only by evaporation or by infiltrating the

FIGURE 12-7. Kettle holes are often found in glacial moraines. This kettle pond near Laurens, New York, was partly filled with water that froze on top.

ground. Some kettle holes fill with water and become kettle lakes or ponds. Figure 12-7 shows a kettle pond in central New York State.

Distinctive glacial landforms in New York State include eskers. These are long, narrow, winding ridges most likely deposited by streams in tunnels under a glacier. There are many eskers in the western foothills of the Catskills.

Drumlins

Drumlins are streamlined hills as much as 100 m high and 1 km long. Most are aligned north to south. They have steep sides, a relatively blunt north slope with a gentler slope to the south. Scientists have different explanations of how drumlins formed. Some think that a glacier forms the hills by riding up and over sediment it is pushing forward. Thus, drumlins show the direction in which the ice was moving. However, others say that the unsorted and unlayered nature of till within the drumlins supports the idea that drumlins were deposited by ice. Figure 12-8 shows a

FIGURE 12-8. Drumlins are long, narrow hills that are often aligned north-south. Most geologists think that the drumlins of New York State were created when the advancing glacier rode up and over some of the sediment, pushing southward.

drumlin in New York State. A drumlin is usually shaped like a long, narrow tear drop.

HOW CAN WE RECOGNIZE DEPOSITION BY MELTWATER?

4: 2.1t, 2.1u, 2.1v

At the end of a glacier there are usually large quantities of water. Some flows from under the ice and more comes from the melting ice front. Sediments deposited by water from melting ice are **outwash**. The principal difference between till and outwash is that outwash deposits, like other sediments laid down by water, are sorted and layered. Outwash is generally less hilly than moraine. Kettle holes and kettle lakes are common in outwash plains where blocks of ice caught in the outwash melted leaving depressions.

The difference between ice deposits and meltwater deposits can be seen very clearly on Long Island. (See Figure 12-9.) Most of the island is made of sediments that can be traced back to glacial origin. The only bedrock on Long Island is at the far western end of the island. High bluffs of unsorted sediment in the Harbor Hill Moraine dominate the north shore of Long Island. The beaches of the north shore are composed of pebbles, cobbles, and even large boulders washed out of the moraine.

The south-shore beaches of Long Island are made of sand washed out of the glaciers by meltwater. A few kilometers inland

FIGURE 12-9. The beaches on the north and the south shores of Long Island, New York, illustrate the difference between deposits by glacial ice and meltwater deposits. Many north-shore beaches have boulders eroded out of unsorted till moraine deposits. South-shore beaches are uniform, fine sand transported by meltwater.

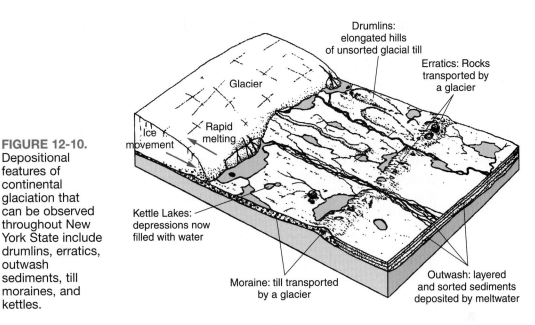

FIGURE 12-10. Depositional features of continental glaciation that can be observed throughout New York State include drumlins, erratics, outwash sediments, till moraines, and kettles.

from the southern beaches are deposits of sand that show layering and sorting: strong evidence that the southern part of Long Island is made of sediment deposited by meltwater, not by ice. Figure 12-10 shows features of continental glaciation that can be observed throughout New York State.

STUDENT ACTIVITY 12-3 —INVENTORY OF GLACIAL FEATURES

7: STRATEGIES 2

Make a list of glacial features found near where you live. This list should include the name of the feature (moraine, large erratic, drumlins, etc.), how each formed, and its location. If possible, add photographs of the features. A geology textbook can help you identify features of glaciation not listed in this book. Visit the following Web site to access the USGS Glossary of Glacier Terminology: *http://pubs.usgs.gov/of/2004/1216/index.html*

WHAT ARE ICE AGES?

4: 2.1t, 2.1u
6: PATTERNS AND CHANGE 5

The most recent ice age in the Pleistocene Epoch ended only 10,000 to 15,000 years ago. At that time, the ice front melted back northward from Long Island and Pennsylvania. The Geologic History chart in the *Earth Science Reference Tables* shows this most recent ice age (the Pleistocene) near the top of the chart. In terms of Earth's history, this is very recent.

During the past 2 to 3 million years, there have been many warm and cold periods. A drop in Earth's average surface temperature of only about 5°C is enough to cause a major ice age. The time between ice ages is variable. It appears that there may have been 10 or more major ice advances just in the past 1 million years. Figure 12-11 shows how much of North America was covered by ice during the most recent major advance of continental glaciers. Today, both Greenland and Antarctica are nearly covered by continental glaciers.

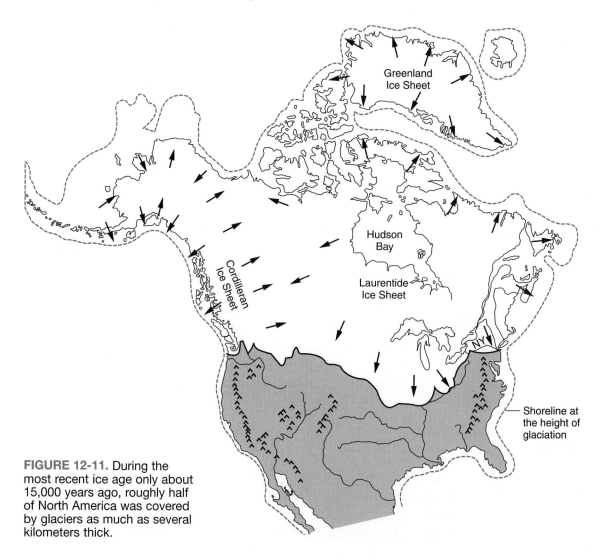

FIGURE 12-11. During the most recent ice age only about 15,000 years ago, roughly half of North America was covered by glaciers as much as several kilometers thick.

Scientists have debated the cause of the ice ages since they first accepted ice ages over a century ago. Movements of the continents, changes in the tilt of Earth's axis, changes in the shape of Earth's orbit, and changes in the carbon dioxide and dust content of Earth's atmosphere have all been suggested as contributors to global climate change. Understanding the reasons for changes in global climates is more important than simply unraveling the past. Future climate changes of this magnitude would certainly have a major effect on human civilization.

CHAPTER REVIEW QUESTIONS

Part A

1. On Mohs scale, ice has a hardness of 1.5. Why can a glacier cause the grooves and parallel scratches (striations) found in bedrock of granite and gneiss?

   (1) The moving ice wears away at the rock surface.
   (2) The ice carries rocks that abrade the rock surface.
   (3) Alternating freezing and thawing wear the rock smooth.
   (4) The weight of the ice pushes on rock to make it smooth.

2. The diagram below represents a profile of New York's Catskill Mountains. The diagram includes four major valleys. Which valley was most likely carved by glacial ice?

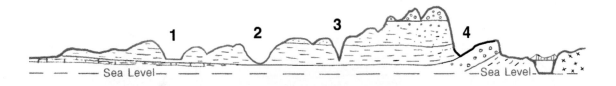

3. Which property would best distinguish sediment deposited by a river from sediment deposited by a glacier?

   (1) mineral composition of the sediments
   (2) thickness of the layers of sediment
   (3) age of fossils found in the sediment
   (4) layering and sorting of sediments

4. The photograph below was taken from an airplane flying over the Rocky Mountains. Winter snow covers most of the higher land surface and forests can be seen in the lower elevations.

Which agent of erosion appears to have done most of the excavation to shape these valleys?

(1) wind
(2) running water
(3) valley glaciers
(4) continental glaciers

5. Which features of the New York landscapes were created by glacial processes?

(1) the Finger Lakes and Long Island
(2) the Catskills and the Adirondack Mountains
(3) Niagara Falls and the Mohawk River Valley
(4) the Hudson Highlands and Mt. Marcy

6. The diagram below represents three stages in the formation of a post-glacial feature.

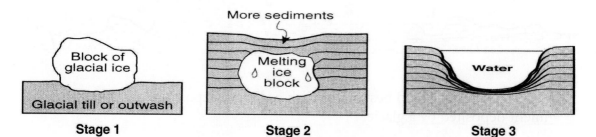

What name do geologists give to the feature in stage 3?

(1) kettle
(2) drumlin
(3) finger lake
(4) parallel scratches

Part B

Base your answers to questions 7 and 8 on the diagram below, which shows landscape features formed as the most recent glacier retreated from New York State.

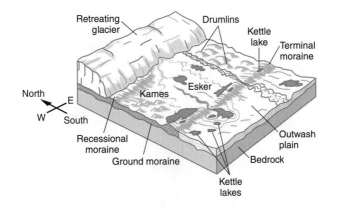

7. The moraines pictured in the diagram above were deposited directly by the glacier. The sediments within these moraines are most likely

   (1) sorted by size and layered
   (2) sorted by size and unlayered
   (3) unsorted by size and layered
   (4) unsorted by size and unlayered

8. At the stage shown in the diagram above, the ice in the glacier is probably moving toward the

   (1) north   (2) south   (3) east   (4) west

9. The diagram below shows estimated trends in the temperature of North America over the last 200,000 years.

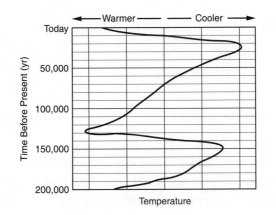

According to this graph, how many glacial periods that have occurred in North America in the last 200,000 years?

(1) 5   (3) 3
(2) 2   (4) 4

Base your answers to questions 10 and 11 on the shaded relief map below, which shows many hills north of Manchester, New York.

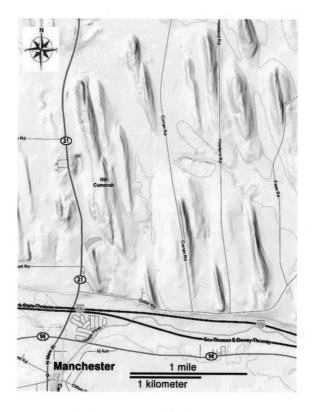

10. What do geologists call the hills shown on this map?

(1) grooves   (3) kettles
(2) drumlins   (4) erratics

11. How can you tell that the glacier that shaped these hills moved from the north?

(1) The hills are aligned east-west.
(2) The hills on the northern part of the map are the highest.
(3) The roads run mostly north-south and few roads cross the hills.
(4) Most of the hills are steeper on the north ends than on the south ends.

12. The graph below shows the snow line (the elevation above which glaciers form at different latitudes in the Northern Hemisphere).

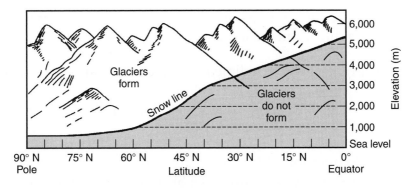

At which location would a glacier most likely form?

(1) 0° latitude at an elevation of 6000 m
(2) 15° N latitude at an elevation of 4000 m
(3) 30° N latitude at an elevation of 3000 m
(4) 45° N latitude at an elevation of 1000 m

13. On a field trip 40 km east of the Finger Lakes, a student observed a boulder of gneiss on the sedimentary bedrock. Which inference is supported best by his observation?

(1) The local sedimentary bedrock was weathered to form a boulder of gneiss.
(2) The gneiss boulder was transported from its original area of formation.
(3) The gneiss boulder was formed from sediments that were compacted and cemented together.
(4) The local sedimentary bedrock melted and solidified to form a boulder of gneiss.

Part C

Base your answers to questions 14 through 16 on the passage below.

Antarctic Ice Sheet

The size and shape of the West Antarctic Ice Sheet depends on many factors, including melting and freezing beneath the glacier, the amount of snowfall, snow removed by wind, iceberg formation, and the rate of ice flow. Glacial moraines are found in the Executive Committee Mountains. Moraines are located up to 100 m above the present ice surface, which indicates that a thicker ice sheet existed 20,000 years ago. The world's oceans and climate are influenced by Antarctica's ice. Even a small increase in sea level from melting glaciers would be a disaster for the nearly 2 billion people who live near coastal areas.

14. List one form of evidence that might be found near the present glaciers that would indicate a much thicker ice sheet existed in the past.

15. Describe the arrangement of sedimentary particles found in a glacial moraine.

16. If Earth's climate became colder, more ice might pile up on Antarctica. How would this affect the level of Earth's oceans?

Base your answers to questions 17 through 20 on the map below, which shows several lobes (sections) of the Laurentide Ice Sheet that advanced into New York State thousands of years ago. The arrows show the direction of movement of the ice. The terminal moraine represents the maximum advance of this ice sheet.

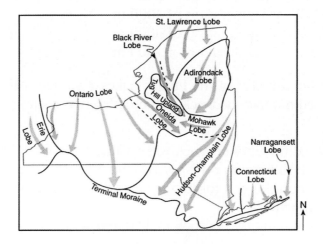

17. During which epoch did the Laurentide Ice Sheet advance over New York State?

18. Ice ages led to deposition of sediments directly by ice and by meltwater from the ice. How do meltwater deposits differ from sediment moved and deposited directly by the ice?

19. According to the map above, in what direction did the ice advance over the Catskills?

20. What evidence might be found on bedrock surfaces in the Catskills that would indicate the direction in which the glacier moved?

13 Landscapes

WORDS TO KNOW

escarpment	mountain landscape	plateau
landform	plains	relief
landscape		

This chapter will help you answer the following questions:

1. Why are many places in New York State popular sightseeing attractions?
2. What are the three major landscape types?
3. Why are there different landscapes in different geographic regions?
4. What are the major landscape regions of New York State?

WHAT ARE NEW YORK'S NATURAL WONDERS?

4: 2.1r

No matter where you live in New York, unique features of geologic interest that your family can enjoy are only a short drive away. For example, Niagara Falls (see Figure 13-1 on page 290) is a broad sweep of thundering water that carries the outflow of four of the world's largest freshwater lakes over a 60-m (197-ft) plunge.

The beaches of Long Island offer miles of white sand and pounding surf. The freshwater beaches in the Finger Lakes and the Adirondack Mountains are famous for their scenic and recreational

FIGURE 13-1. Niagara Falls is one of the most powerful and impressive waterfalls in the world. The outflow of four Great Lakes tumbles nearly 60 m (200 ft) over a rim about 900 m (0.5 mi) wide.

value. Letchworth Gorge on the Genesee River south of Rochester has impressive cliffs hundreds of meters high. The Hudson is one of America's most scenic rivers. Visit the following Web site to learn about New York State Parks (click on the desired box): *http://nysparks.state.ny.us/gmaps*

WHAT ARE LANDSCAPES?

4: 2.1r

A **landscape** is a region that has landforms related by similarities in shape, climate, and/or geologic setting. Landscapes include a variety of topographic features related to the processes that shaped the surface. For example, New York's glaciated landscapes include such landforms as U-shaped valleys; rounded, grooved, and polished bedrock surfaces; drumlins; and moraines. A **landform** is a feature of a landscape. Landscapes are generally made of a variety of related landforms such as mountains, valleys, and river systems. Geological structures are major influences on landscapes. Most land areas can be divided into regions that have similar landforms. Visit the following Web site to see how landforms are shown on topographic maps: *http://www.csus.edu/indiv/s/slaymaker/Archives/Geol10L/landforms.htm*

STUDENT ACTIVITY 13-1 —INVESTIGATING LANDFORMS

2: INFORMATION SYSTEMS 1

Collect original photographs, travel brochures, or images from the Internet of landforms in and near where you live. For each landform, write a brief description of the processes that formed it.

Most geologists think of a landscape as the product of geological events that took place over hundreds, thousands, or even millions of years. These events include building processes such as uplift, volcanic eruptions, and glacial deposition. Other equally important events, such as weathering and erosion, wear down the land surface. Therefore, landscapes are the surface expressions of opposing geological processes.

Plains

Most landscape regions can be classified as plains, plateaus, or mountains. **Plains** are usually relatively flat. That is, the range of elevations is small. Hill slopes are gentle and streams commonly meander over broad floodplains. Figure 13-2 is a view of the Great Plains of eastern Colorado.

The bedrock under plains can be any kind of rock that has been in place long enough to be eroded to a low level. Shale and other layered sedimentary rocks are especially common in plains landscapes. Soils tend to be thick and lasting due to the low gradients. Much of the Mississippi Valley and central part of the United States is a plains landscape with a base that is mostly flat layers of relatively soft sedimentary rock.

Plateaus

Plateau landscapes have more relief than plains but they are not as rugged as mountains. **Relief** is the difference in elevation from the highest point to the lowest point on the land surface. Some use the term tableland as a synonym for plateau. Indeed, some

FIGURE 13-2. The flat-lying sedimentary rocks of the Great Plains make this a classic plains landscape.

FIGURE 13-3. Like many other plateau regions, the base of the Colorado Plateau is flat-lying layers of sedimentary rocks. In the mid-distance the Little Colorado River has eroded a deep canyon.

plateaus look like relatively level areas that are nearly as flat as plains but at a much higher elevation.

Parts of the Colorado Plateau of Arizona and Utah (see Figure 13-3) look like a plains landscape that has been pushed up more than 1000 m (0.6 mi). The Columbia Plateau of eastern Oregon was built up by lava flows that spread over great distances. Flat-lying, sedimentary rocks form the base of most plateaus, including the Colorado Plateau.

Landforms in the eastern part of the United States are generally more rounded than those in dryer areas of the West. Therefore, the Appalachian Plateau, the largest plateau in the eastern United States, is mostly a region of rolling hills as you can see in Figure 13-4.

Mountains

Mountain landscapes have the greatest relief. Figure 13-5 shows a rugged mountain region in Alaska. Variations in rock types, which are common in mountain areas, often include hard metamorphic and igneous rocks as well as weaker kinds of rock.

FIGURE 13-4. The Appalachian Plateau in southwestern New York State is more rounded than plateaus in the arid Southwest.

FIGURE 13-5.
Mount McKinley
(Denali) in the
Alaska Range is a
classic mountain
landscape with
high and rugged
peaks that
dominate the
landscape.

Forces deep within Earth cause deformed rock structures such as folds and faults that contribute to the complexity and relief of mountain landscapes. A close look at the rocks in any major mountain area, such as the Rockies, the Alps, or the Himalayas, is likely to reveal a complex geologic history and a wide variety of rock types. Figure 13-6 shows the major landscape regions of the United States.

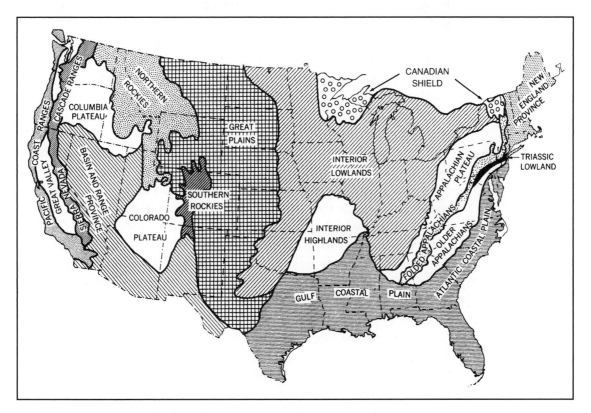

FIGURE 13-6. Landscape regions of the United States include plains, plateaus, and mountains.

WHAT FACTORS INFLUENCE LANDSCAPE DEVELOPMENT?

4: 2.1r
6: SYSTEMS
THINKING 1

The factors that influence landscapes are geology and climate. Geologic factors include uplift, rock type, and geologic structures. The major climate factor in landscape development is the annual precipitation.

Geologic Factors

Among geologic factors, upward and downward movement of Earth's crust is very important. Where Earth's crust is being pushed up, high peaks and deep valleys form young mountain ranges, such as the Himalayas of Asia or the relatively young mountain ranges of the western United States. Rapid uplift and steep slopes also lead to rapid erosion. Rivers that drain these areas carry far more sediment into the oceans than rivers in older landscapes.

The characteristics of rock types, especially their resistance to weathering and erosion, are very important in landscape development. The hardest rocks of any area usually are found at the highest land elevations. In places where some bedrock is more resistant to weathering and erosion than others, ridges tend to follow the hardest rock types. Valleys in these areas generally follow the softer and/or fractured rock. Visit the following Web site to investigate the USGS Tapestry map—a union of geology and topography: *http://tapestry.usgs.gov*

The Niagara Escarpment is a major landform that runs from the Niagara River south of Lake Ontario across much of western New York State. An **escarpment** is a steep slope or a cliff of resistant rock that marks the edge of a relatively flat area. The Niagara Escarpment, which separates the lowlands along Lake Ontario from higher land to the south, is made of a particularly hard sedimentary rock known as the Lockport dolostone. (Dolostone is a sedimentary rock similar to limestone but harder, due to a slightly different mineral composition.)

Geologic structures also influence landscapes. The Adirondack Mountains have been pushed up into a broad dome structure. The waves of hills of the Appalachian Plateau north of Harrisburg, Pennsylvania, are the result of large folds in the bedrock caused by compression of the crust. Faults are zones of crushed bedrock along which movement has occurred. Faulting also controls the

FIGURE 13-7. The Hudson Highlands are a landscape region of rugged hills about 80 km (50 mi) north of New York City. The Hudson River has eroded a path through the Highlands along zones of weakness created by movement of the crust at geologic faults.

direction of many northeast-flowing streams in the high peaks region of the Adirondack Mountains.

The Hudson Highlands, about 30 km (19 mi) north of New York City, is a block of Earth's crust that has been pushed up. Figure 13-7 is an aerial view looking north to where the Hudson River passes through the Hudson Highlands. The Highlands are bounded by geologic faults to the north and south. In addition, the zigzag route of the Hudson River through the Hudson Highlands is probably a result of erosion along geologic zones of instability.

Climatic Factors

Climate is the other major factor in landscape development. A humid climate favors chemical weathering, which produces rounded landforms. Most of the hill slopes of New York State are rounded and gentle because of the relatively humid climate. A moist climate also allows plants to grow and protect soil from erosion. Desert areas sometimes have a steplike profile, with flat hilltops and terraces on a base of softer rock layers separated by steep escarpments of more resistant rocks. Figure 13-8 on page 296 shows the contrast between the rounded shapes of a humid landscape and the angular features of many desert landscapes. All of New York has a moist, temperate, mid-latitude climate. Therefore the landscape differences in New York State are caused by geological factors.

FIGURE 13-8. The rounded landforms of New York State are caused by the action of the continental glaciers, the humid climate, and extensive plant cover that holds soil in place. Contrast this Hudson River landscape (left) with Monument Valley in the high desert along the Arizona-Utah border (right).

WHAT ARE THE LANDSCAPES OF NEW YORK STATE?

4: 2.1r

Figure 13-9 is the Generalized Landscape Regions map from the *Earth Science Reference Tables* that shows the landscape regions of New York State. New York State is a limited geographic region with a relatively uniform climate. However, the state has a remarkable variety of landscapes including plains, plateaus, and mountains. The boundaries between landscape regions are sometimes sharp and easy to see. For example, the eastern front of the Catskills is marked by a dramatic change in elevation from the lowlands along the Hudson River to a landscape nearly 1000 m (0.6 mi) higher to the west. The Hudson Highlands are an uplifted block of land bounded by fault scarps to the north and to the south. However, the long boundary between the Allegheny Plateau and the Erie-Ontario Lowlands in western New York is less visible.

The Generalized Landscape Regions map can be used along with the Generalized Bedrock Geology map (Figure 18-10 on page 298), also in the *Reference Tables,* to locate cities and other geographic features on the landscape map. For example, looking at the landscape region map and the bedrock geology map you can see that Watertown, Oswego, and Rochester are located in the Erie-Ontario Lowlands.

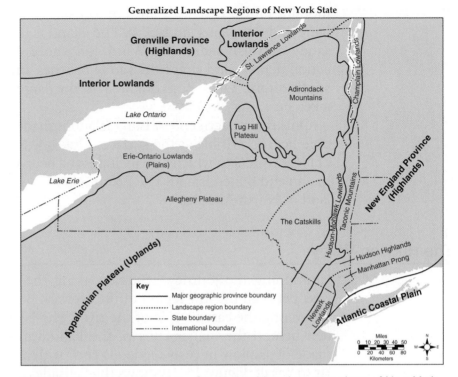

FIGURE 13-9. This map of the Generalized Landscape regions of New York includes plains, plateaus, and mountains. In addition, there are several regions difficult to classify. Boundaries between landscape regions often follow changes in elevation and rock types.

STUDENT ACTIVITY 13-2 —LANDSCAPE BOUNDARIES

2 INFORMATION
SYSTEMS: 1
6 SYSTEMS
THINKING 1

If you live near a landscape boundary, draw that boundary on a local road map. In what ways does the land look different on the two sides of the boundary? How do the landscape features and the underlying soils affect land use?

Long Island

We will begin our tour of New York State at the eastern end of Long Island. Glacial moraines give this part of the island the split, fish-tail shape seen on maps. The two moraines mark the most southerly advance of the recent continental glacier. These moraines form east-west lines of low hills on an otherwise low-lying and flat plains landscape. The island is largely composed of sandy outwash sediments. Wind, longshore currents, and wave action deposit and shape the sandy beaches and dunes along the south shore. Lake Ronkonkoma is the largest of many kettle lakes on Long Island.

Only the western end of Long Island is made of hard metamorphic bedrock visible at the surface. This is within the boundaries of New York City, which also includes the bedrock islands of Manhattan and Staten Island. The Bronx is built on the bedrock mainland of North America.

The Hudson Valley

Continuing the journey upstream along the Hudson River, we pass between the low region of the Newark Basin and the rolling hills along the eastern side of the river. The Newark Basin is hidden behind the higher cliffs of the Palisades. The Palisades are made of shallow intrusive igneous rock with a composition similar to basalt. Figure 13-10 shows the Palisades and the distant skyline of New York City.

Mostly on the east side of the Hudson River, the Taconic Mountains, the Hudson Highlands, and the Manhattan Prong are all supported by very old metamorphic rock. This old rock gives them a complex landscape, but mostly without the steep slopes and high peaks of a true mountain landscape.

FIGURE 13-10. You can see the cliffs of the Palisades, the Hudson River, the George Washington Bridge, and the skyline of New York City in this photograph.

The Catskills-Allegheny Plateau

The Catskills rise above the western edge of the Hudson Valley. Sometimes called the Catskill Mountains, they include some of the highest elevations in New York State. However, the base of the Catskills is made of flat layers of sedimentary rock with rounded and even relatively flat summits. The Catskills are the higher, eastern end of the Allegheny Plateau, a large region of rolling hills that extends nearly to Lake Erie at the western end of the state. The Allegheny Plateau, an extension of the Appalachian Plateau, contains the Finger Lakes as well as other valleys deepened and U-shaped by the advancing continental glaciers. Letchworth Gorge cuts a deep canyon through the flat-lying sedimentary layers of the Appalachian Plateau of Western New York State.

Plains

Lakes Erie and Ontario, two of North America's Great Lakes, are bordered by a plains landscape that includes drumlins and other features of the ice ages. Local soils have a good mix of minerals due in part to the large amount of glacial till carried south from Canada. The lakes moderate the climate of this agricultural region, extending the growing season. This plains landscape extends along the St. Lawrence River and around the Adirondack Mountains to join the valleys of the Hudson and Mohawk rivers.

Tug Hill Plateau and Adirondacks

Two landscape regions remain on our tour of New York. The Tug Hill Plateau is a remote and thinly populated highland that receives the greatest winter snowfalls in New York State.

The Adirondack Mountains are New York's best example of a true mountain landscape. The land here was pushed up into a broad dome, exposing at the surface ancient metamorphic rocks that were formed deep underground. Faulted stream valleys add to the mountain relief. The highest point in the state, Mount Marcy, and some of the most scenic lakes are located in this region. Tourism supports the economy since the area is too rugged and cold for good farming and it is too remote for business. The following Web site will take you to virtually any place in the world using Landsat satellite imagery: *http://worldwind.arc.nasa.gov/java/*

STUDENT ACTIVITY 13-3 —LANDFORMS OF NEW YORK STATE

2: INFORMATION
SYSTEMS 1

As a class, collect photographs from friends and relatives, travel brochures, or the Internet, and attach them at their appropriate place on a large wall map of New York State. Select photographs that show landforms characteristic of the various landscape regions of New York. Visit the following Web site to learn about New York State Parks (click on the desired box): *http://nysparks.state.ny.us/gmaps*

A similar project can be conducted collecting photographs of landforms in the United States and placing them on a map of the United States.

CHAPTER REVIEW QUESTIONS

Part A

1. Which landscape region of New York State would *not* be considered a part of a larger landscape region that extends beyond the New York borders?

   (1) Allegheny Plateau   (3) Newark Lowlands
   (2) Tug Hill Plateau   (4) St. Lawrence Lowlands

2. According to the map below, what landscape region of New York State usually has the most thunderstorms?

   (1) Adirondack
       Mountains
   (2) Erie-Ontario
       Lowlands
   (3) Taconic Mountains
   (4) Allegheny Plateau

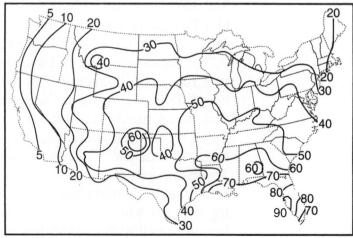

Average Number of Thunderstorms Each Year

3. The Catskill region is often classified as a plateau primarily because the Catskills have

    (1) V-shaped valleys   (3) horizontal rock layers
    (2) jagged hilltops   (4) folded metamorphic rocks

4. In which New York landscape is Rochester, New York?

    (1) Adirondack Mountains
    (2) Erie-Ontario Lowlands
    (3) Allegheny Plateau
    (4) Atlantic Coastal Plain

5. New York State's generalized landscape regions are identified primarily on the basis of elevation and

    (1) bedrock structure   (3) geologic age
    (2) climate zones   (4) latitude

6. From 1817 to 1825 the Erie Canal was built from the Albany area through Utica and Syracuse, then just south of Rochester to Buffalo. What landscapes did the canal follow?

    (1) The canal was built primarily through valleys in mountain landscapes.
    (2) The canal was built mostly through plateau landscapes.
    (3) The canal was built mostly across plains and lowlands.
    (4) The canal connected all the landscape regions of New York State.

7. Which block diagram below best represents a portion of a plateau landscape?

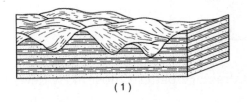
(1)

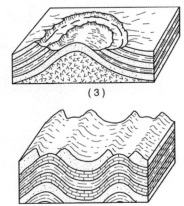
(3)

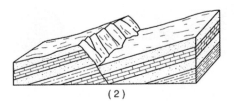
(2)

(4)

8. What is the smallest distance from the southern boundary of the Adirondack Mountains landscape to the northern boundary of the Allegheny Plateau?

   (1) approximately 1 km
   (2) approximately 10 km
   (3) approximately 50 km
   (4) greater than 100 km

9. Which two locations are in the same New York State landscape region?

   (1) Albany and Old Forge
   (2) Massena and Mt. Marcy
   (3) Binghamton and New York City
   (4) Buffalo and Watertown

10. What landscape region is located at 44°N latitude, 74°W longitude?

    (1) Allegheny Plateau
    (2) Erie-Ontario Lowlands
    (3) Adirondack Mountains
    (4) Atlantic Coastal Plain

11. The map below shows a stream drainage pattern. Arrows show the direction of stream flow.

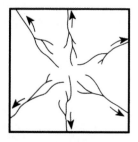

On which landscape region did this drainage pattern most likely develop?

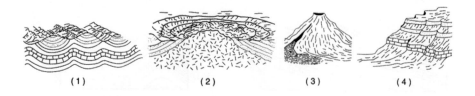

(1)　　　　　(2)　　　　　(3)　　　　　(4)

Part B

12. The table below describes three landscape regions of the United States.

Landscape	Bedrock	Elevation/Slopes	Streams
A	Faulted and folded gneiss and schist	High elevation, steep slopes	High velocity, rapids
B	Layers of sandstone and shale	Low elevation, gentle slopes	Low velocity, meanders
C	Thick, horizontal layers of basalt	Medium elevation, steep to gentle slopes	High to low velocity, Rapids and meanders

Which choice below best identifies landscapes A, B, and C?

(1) A–plain, B–mountain, C–plateau
(2) A–mountain, B–plateau, C–plain
(3) A–plateau, B–mountain, C–plain
(4) A–mountain, B–plain, C–plateau

13. What popular New York State tourist destination is located in a plateau landscape?

(1) Mt. Marcy
(2) The Finger Lakes
(3) Niagara Falls
(4) Lake Champlain

14. Scientists think that global warming may melt the polar ice caps and cause sea level to rise. Which part of New York State would most likely be flooded if this were to happen?

(1) Long Island
(2) Tug Hill Plateau
(3) Catskills
(4) Mt. Marcy

Part C

15. The aerial photograph below shows a well-known region of New York State where long, north-south aligned lakes dominate the landscape. Where in New York State was this photograph taken?

16. The map below represents New York State. Make a copy or a sketch of this map and write an X at the position of the Tug Hill Plateau.

17. In what landscape region of New York State is Kingston, New York, located?

18. What kind of landscape has a base of folded and faulted bedrock of many different kinds of rock?

19. If you journey overland from Binghamton, New York, to Oswego, New York, what change would you see in the general topographic relief as you travel northward?

20. If over a period of a million years climate change were to cause New York State to become a desert, how would hill slopes most likely change?

UNIT 5

Earth's Internal Heat Engine

The churning currents within Earth's mantle are responsible for earthquakes and the movement of lithospheric plates. In addition, mountains such as California's Mt. Whitney (shown here) rise as the land nearby sinks.

Among short-lived natural disasters, earthquakes are probably responsible for the greatest number of human deaths. If scientists could find a way to give people just a few hours advanced notice, many lives could be saved. However, scientists have not had much success in predicting earthquakes.

The United States Geological Survey places scientific instruments across the San Andreas Fault in Parkfield, California, to detect signs of future earthquakes. These instruments measure twisting and very small stretching of the land surface. Geologists can identify sections of faults where stress must be building. However, this information is most useful for making predictions about what will happen over the next decades or longer. Can reliable indicators of seismic events be found that will allow people to make short-term preparations? The need to save lives and property will drive continuing interest in this line of research.

14 Earthquakes and Earth's Interior

WORDS TO KNOW

conduction	Mercalli scale	seismic moment
convection	Moho	seismograph
convection cell	origin time	seismologist
earthquake	primary waves	seismology
epicenter	radiation	strain
fault	refraction	stress
focus	Richter scale	travel time
logarithmic	secondary waves	

This chapter will help you answer the following questions:

1. Why do we have earthquakes?
2. How do seismologists measure earthquakes?
3. How can the location of an earthquake be determined?
4. What is Earth's interior structure?

WHAT CAUSES EARTHQUAKES?

4: 2.1a, 2.1b, 2.1k
4: 2.2b
6: SYSTEMS
THINKING 1

You might think that the rocks of Earth's crust cannot move. After all, "steady as a rock" is a common expression for someone who is unbending in his or her attitude. But if you look around, you

FIGURE 14-1. Folded rock layers show that forces within Earth can move and bend solid rock. This kind of layer that is bent downward in the middle is a syncline.

can see evidence that rocks have moved. For example, Figure 14-1 shows a reservoir where a layer of rock that was originally laid down flat has been bent by forces within Earth. The ability of rock to bend before it breaks depends on such factors as its temperature, time for deformation, and mineral composition.

Have you ever felt an earthquake? An **earthquake**, or seismic event, is a sudden movement of Earth's crust that releases energy. The mild motion of a small seismic event can be exciting. But what if you felt that your life was in danger? Large earthquakes can be very frightening.

Early in the morning on Friday, March 27, 1964, one of the largest seismic events on record struck the coast of Alaska. For the residents of Anchorage, it started as a gentle rolling motion. Then, for the next 5 minutes the ground shook and heaved. One side of Fourth Avenue in the main shopping district moved downhill 3 m (about 10 ft), causing large cracks to open. Windows broke and material fell from buildings onto the sidewalks. A large housing development and a school were destroyed in similar land failures in another part of the city. The nearby port of Valdez was badly damaged by the shaking. Later, Valdez was completely destroyed by a series of gigantic waves that left a large ship sitting on a city street. In this chapter, you will learn why earthquakes occur, how they are measured, and what they tell us about Earth's interior. Visit the following Web site to design a building, make your own earthquake parameters, and see the results: *http://www.tlc.discovery.com/convergence/quakes/interactives/ makeaquake.html*

STUDENT ACTIVITY 14-1 —ADOPT AN EARTHQUAKE

2: INFORMATION
SYSTEMS 1

Choose a historic earthquake to write a report about. Before you begin your report, let your teacher know which earthquake you chose. Your teacher may require that each student or group select a different seismic event. After your choice is approved, use library resources or the Internet to gather information about your seismic event. Your report must include a bibliography of your sources of information. Your teacher will specify what information will be needed for each entry in the bibliography.

Earth's Internal Energy

You may recall from physical science that energy always flows away from the hottest places toward cooler places. Lava erupting from volcanoes, walls of deep gold mines that are hot enough to burn your hand, and hot springs such those in Yellowstone National Park in Wyoming remind us of the heat energy stored deep within our planet. In fact, at Earth's center the temperature is greater than 6000°C. This is the energy that causes earthquakes as well as the slow bending of the crust.

There are two principal sources of Earth's internal heat energy. Some of the energy is left over from the formation of the planet about 4.6 billion years ago. In addition, there is the energy of radioactive decay. Some large atoms, such as uranium, are unstable. They spontaneously break down into more stable substances, such as, lead. This slow process releases heat energy.

Heat energy can travel in three ways. Some heat travels through Earth and to the surface by conduction. **Conduction** occurs when heated molecules pass their vibrational energy to nearby molecules by direct contact. Most metals are good conductors of heat energy. This is one of the ways that heat can travel through Earth toward the surface. **Radiation** is the movement of electromagnetic energy through transparent materials. The sun radiates light and other forms of electromagnetic energy into space. Earth absorbs some of that energy and changes it to heat. Earth radiates heat energy into the atmosphere. The third form of heat flow is convection. In **convection**, heated matter circulates from one place to another due to differences in temperature and density. On Earth's surface the wind circulates energy by convection. Even within Earth, heated material rises toward the surface in some places. In other places, cool material sinks into Earth. Horizontal currents complete the circulation by connecting the areas of vertical heat flow. This pattern of circulation is known as a **convection cell**. Although convection motion within Earth occurs at only a few centimeters per year, over time heat flow by convection is capable of causing catastrophic changes such as earthquakes.

Stress and Strain

You might think of Earth's mantle and crust as solid and unbending. However, forces applied to solid rock at high temperatures and over a long time can bend, or deform, rocks. This slow bending builds up forces on rocks near the surface. This kind of force is called **stress**. Surface rocks do not deform as easily as the rocks inside Earth that are under intense heat and pressure. Surface rocks respond to stress by elastic bending called **strain**. Therefore, the stress on rocks near the surface builds until the rocks break, and an earthquake occurs. What you feel as an earthquake is the sudden movement of the ground, which releases stored energy. This sequence is shown in Figure 14-2.

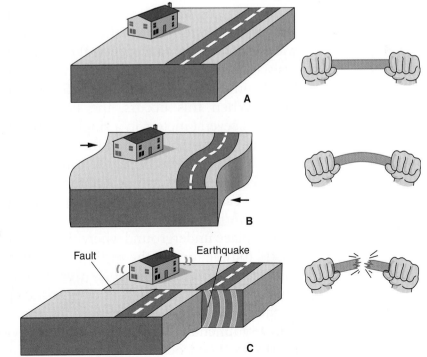

FIGURE 14-2.
Earthquakes occur when stress (force) causes strain (bending) that exceed the ability of the rocks to resist breakage. A sudden release of energy (breaking) generates energy waves that radiate as an earthquake.

FIGURE 14-3. Earthquakes occur when the ground suddenly shifts along geologic faults such as this small fault near Kingman, Arizona. The displacement here is small, but major faults can offset many miles.

Earthquakes do not occur randomly. Earthquakes occur at cracks in Earth's crust along which there is movement. These cracks are **faults**. (See Figure 14-3.) Faults are usually zones of weakness. The places most likely to break are those that have broken in the past.

Perhaps the most active fault in the United States is the San Andreas Fault system, which extends from the Gulf of California north past Los Angeles and through the San Francisco area. The land on the western side of the San Andreas Fault is moving northward at an average rate of several centimeters per year. Due to the resistance of Earth's crust, this motion is not uniform. Stress builds until it is greater than the friction that holds the rocks in place. When this occurs, the Pacific side jolts northward as much as several meters in a single earthquake. Large earthquakes occur at various places along the San Andreas Fault roughly once every 10 to 20 years. Visit the following Web site to watch the PowerPoint—Transforming California, A Journey Along the San Andreas Fault: *httP://www.Priweb.org/ed/earthtriPs/transforming_ca_home.htm*

The place underground where the rock begins to separate is the **focus** (plural, foci) of the earthquake. Directly above the focus, at the Earth's surface, is the **epicenter**. (See Figure 14-4.) An earthquake is felt most strongly at or near the epicenter. The intensity of the shaking is also influenced by the nature of the ground. Loose sediments usually experience more violent movement than nearby solid bedrock.

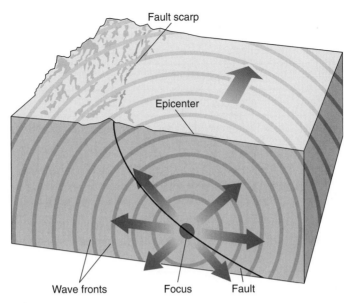

FIGURE 14-4.
The focus of an earthquake is the place underground where the rock breaks and shifts. Energy waves (shaking) radiates from the focus. Directly above the focus is the epicenter, where the energy waves first reach the surface.

HOW ARE EARTHQUAKES MEASURED?

6: MAGNITUDE AND SCALE 3

Some seismic events are stronger and do more damage than others. The most violent earthquake in modern times was probably a 1960 event centered in the Pacific Ocean off the coast of Chile. But this earthquake did not have the human toll of two earthquakes that occurred in China, one in 1556 and the other in 1976. During these earthquakes, buildings were destroyed and hundreds of thousands of people died.

Earthquake Scales

The study of earthquakes is a branch of geology and engineering called **seismology**. Seismologists use two different methods to measure earthquakes. Table 14-1 on page 312 is an abbreviated form of the **Mercalli scale**, which is based on reports of people who felt the earthquake and observed the damage it caused. The 12 levels of intensity on the Mercalli scale are designated by Roman numerals. Intensity I or II is felt by only a few people who are at rest but awake and alert. Everyone feels intensity VI, but there is little damage. An earthquake of intensity XII causes great damage to all buildings, and

Table 14-1. An Abbreviated Mercalli Scale

Intensity	Effects
I–II	Almost unnoticeable.
III–IV	Vibrations are noticeable; unstable objects are disturbed.
V–VI	Dishes can rattle; books may be knocked off shelves; damage is slight.
VII–VIII	Shaking is obvious, often prompting people to run outside; damage is moderate to heavy.
IX–X	Buildings are knocked off foundations; cracks form in the ground; landslides may occur.
XI–XII	Wide cracks appear in the ground; waves seen on ground surface; damage is severe.

the shaking is so violent that large objects can be thrown into the air. The major advantage of the Mercalli intensity scale is that no instruments are required and information can be gathered after the earthquake. Furthermore, everyone who felt the event or observed the damage becomes a virtual measuring instrument.

The **Richter scale** of earthquake magnitude is based on measurements made with an instrument called a **seismograph**, as shown in Figure 14-5. Inside a seismograph, a weight is suspended on a spring. As the ground shakes, the weight tends to remain still. A pen attached to the weight traces a line on paper attached to a rotating drum. The larger the earthquake the more the line appears to move from its rest position. Mechanical seismographs can measure a very limited range of ground vibrations. Using electronic amplification systems, modern seismometers can record and measure events over an extremely wide range of magnitudes. Visit the following Web site to download a Mac OS X application that turns your MacBook or MacBook Pro into a seismograph: *http://www.suitable.com/tools/seismac.html* Visit the following Web site to download Seismic/Eruption software from Dr. Alan Jones at SUNY Binghamton (PC only): *http://bingweb.binghamton.edu/~ajones/*

Richter magnitude is based on the size of the waves recorded on the drum of a seismograph. The advantage of the Richter scale is its precision based on measurements. There is no human bias or need to adjust for places that have different kinds of soils or

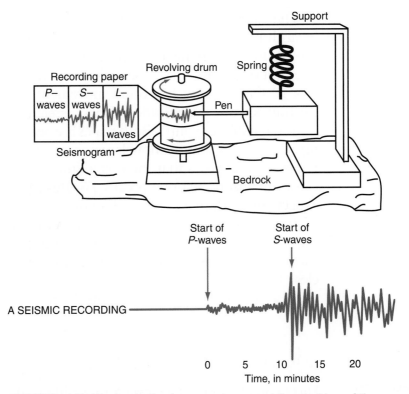

FIGURE 14-5. Mechanical seismographs record the shaking of the ground with a large mass suspended on a spring. As the ground shakes, a pen makes a record on a rotating drum. The recording that is produced is called a seismogram. Today, most seismometers use electronic amplification in place of mechanical systems.

building construction. One disadvantage is that measurements are taken only in those places that have an instrument. In addition, the Richter scale was developed for earthquakes in California, and is based on measurements made with a seismograph that no longer represents the latest in technology.

In recent years, seismologists have developed a magnitude scale called **seismic moment**. This scale is based on the total energy released by the earthquake. For small and moderate events, the seismic moment and Richter scales give very similar numbers. Both scales are based on the use of instruments, although the seismic moment scale is usually based on modern electronic instruments. These electronic instruments are better than mechanical seismographs at measuring low-frequency vibrations and large movement of the ground. Low-frequency motions are an important part of the largest seismic events.

The Richter and seismic moment scales are **logarithmic**. In a logarithmic scale, an increase of one unit on the scale translates to a 10-fold increase in the quantity measured. Therefore, an increase of one unit on the Richter or seismic moment scale means that there was 10 times as much ground movement. It also means that an earthquake one magnitude number smaller caused only $\frac{1}{10}$ the movement of the ground. Both scales are open-ended scales. In theory, there is no limit to how large or small an earthquake can be. However, it appears that there is a limit to how much energy the crust can absorb before it breaks. For this reason, scientists do not expect any earthquakes larger than magnitude 10. At the other end of the scale, small earthquakes get lost in vibrations caused by passing cars and trucks and people walking nearby. However, the largest earthquakes release about 1000 times the energy of the largest nuclear bomb ever tested. This is a gigantic range of energies. Even worldwide, very large seismic events do not occur very often.

HOW DO EARTHQUAKES RADIATE ENERGY?

1: MATHEMATICAL
ANALYSIS 1
4: 2.1j

Have you ever tossed a stone into a still pond and seen the waves it makes? Starting from the point at which the stone hit, water waves carry energy away in all directions. If the stone hitting the water represents an earthquake, it would be the arrival of those waves that are felt and measured as an earthquake.

Seismic waves are vibrations of the ground generated by the break that started the earthquake. The energy released is carried away in all directions by these vibrations. Three kinds of seismic waves carry earthquake energy. **Primary waves**, or P-waves, travel the fastest. (They are called primary because they are the first waves to arrive at any location.) P-waves cause the ground to vibrate forward and back in the direction of travel. This direction of vibration makes them a type of *longitudinal* wave. P-waves are like sound waves because they alternately squeeze and stretch the material through which they pass. This is the kind of motion that you might observe if a speeding car ran into a line of parked cars and the energy of the collision traveled along the line of cars. P-waves can travel through any material—solid, liquid, or gas.

Secondary waves, also known as S-waves, are the next to arrive. They travel at about half of the speed of P-waves. As the

S-waves pass, the ground vibrates side-to-side, the way that waves move along a rope. Like light, S-waves are *transverse* waves, which cause vibration perpendicular to the direction in which they are moving. S-waves can travel through a rigid medium (solids) but they cannot pass through a fluid, which includes liquids or gases.

Collectively, P- and S-waves are known as body waves because they pass *through* Earth rather than along its surface. Figure 14-6 illustrates how P-waves and S-waves travel as vibrations in different directions relative to their forward motion. You can use a long spring toy to demonstrate P- and S-waves.

When P- and S-waves reach the surface, they become surface waves, which are slower and have a circular motion like waves on water. Surface waves usually do the most damage because they include both vertical and horizontal motion.

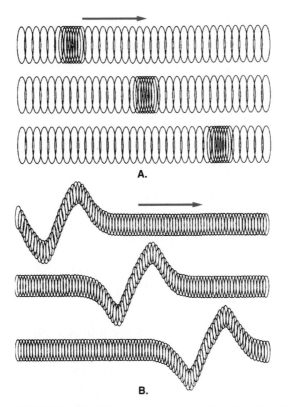

FIGURE 14-6. Seismic waves are classified by the direction of vibration. As P-waves carry energy from the focus, they cause the ground to vibrate forward and back parallel to the direction of motion as shown in *A*. However, S-waves cause the ground to move perpendicular to the direction in which they move as shown in *B*.

STUDENT ACTIVITY 14-2 —MODELING SEISMIC WAVES

6: MODELS 2

Make a line of 5 to 10 students across the front of the classroom. Students should stand with their hands on the shoulders of the person in front of them. Model P-, S-, and surface-wave motions passing along the line. Seismic waves may also be demonstrated with a long spring toy. The spring can be stretched and released quickly to show how P-waves travel. It can be moved side-to-side to show how S-waves travel.

Adding and Subtracting Time

To find earthquake origin times you will need to subtract one time from another. Adding and subtracting time takes practice. Most numbers you work with are in base 10. However, there are 60 seconds in a minute and 60 minutes in an hour. Furthermore, the hour is never greater than 12. For these reasons, calculations with time are not the same as most calculations you do.

Time can be written in the form of hours:minutes:seconds. For example, the time 06:15:20 means 6 hours, 15 minutes, 20 seconds. (In this chapter it will not be important whether the time is A.M. or P.M.) The time 12:45 can be expressed to the nearest second as 12:45:00.

To add 2 hours:10 minutes:15 seconds to 9:45, the addition can be set up as follows:

$$
\begin{array}{r}
02:10:15 \\
+\ 09:45:00 \\
\hline
11:55:15
\end{array}
$$

But suppose the numbers are a little larger:

$$
\begin{array}{r}
08:50:15 \\
+\ 09:45:00 \\
\hline
17:95:15
\end{array}
$$

You may have noticed that there is no time written as 17:95:15. The hour can be no larger than 12 and the minutes' and seconds' columns cannot be greater than 60. It is therefore necessary to add a minute and/or an hour to the next column on the left. Instead of 95 minutes, it becomes 1 hour and 35 minutes. Adding an hour changes the hour number from 17 to 18 (18:35:15). You are not done yet. Most clocks do not read 18 hours. So 12 hours is subtracted (a full cycle of the clock), making it 6 hours (06:35:15):

$$
\begin{array}{rr}
08:50:15 & 18:35:15 \\
+09:45:00 & -12:00:00 \\
\hline
18:35:15 & 06:35:15
\end{array}
$$

This subtraction problem is quite simple because none of the numbers are greater than 60 seconds, 60 minutes, or 12 hours:

$$
\begin{array}{r}
11:34:50 \\
-06:27:10 \\
\hline
05:07:40
\end{array}
$$

In subtraction problems you may also need to borrow from the column to the left. Consider the next calculation.

$$
\begin{array}{r}
11:14:20 \\
-06:27:50 \\
\hline
\end{array}
$$

If you subtract a larger number from a smaller number, you get a negative number. However, negative numbers are not used to express time. So you need to change the top number by borrowing from the units to the left while keeping the same value. When you borrow 1 minute and add 60 seconds to the right column, 20 seconds becomes 80 seconds. Having added a minute to the seconds' column, you need to subtract 1 minute from the minutes column. So, 14 minutes becomes 13 minutes (11:13:80).

But now you have a problem in the minutes' column because 27 minutes is greater than 13 minutes. It is now necessary to borrow 1 hour (60 minutes) from the hours' column. The 13 minutes becomes 73 minutes, and 11 hours becomes 10 hours (10:73:80). This problem now can be written as:

$$
\begin{array}{r}
10:73:80 \\
-06:27:50 \\
\hline
04:46:30
\end{array}
$$

Practice Problem 1
Add 3 hours:10 minutes:55 seconds to 8 hours:40 minutes:10 seconds.

Practice Problem 2
Subtract 6 hours:45 minutes:50 seconds from 9 hours:30 minutes:30 seconds.

Using this system of borrowing from the column to the left, you should be able to solve any problem of this type and determine earthquake origin times.

HOW ARE EARTHQUAKES LOCATED?

1: MATHEMATICAL
ANALYSIS 1
ESRT PAGE 11
4: 2.1j

Seismologists know an earthquake has occurred when they feel the ground shaking or they see their instruments detect energy waves. In this section, you will learn how seismologists can locate an earthquake with data recordings from as few as three seismic stations.

The first step is to determine how far away an earthquake is. The fact that P-waves travel much faster than S-waves allows seismologists to determine the distance to an earthquake's epicenter. When an earthquake occurs, energy waves move away in all directions. P- and S-waves start out from the epicenter at the same time. A person at the epicenter would feel both waves immediately. There would be no separation between them. But the farther you are from the epicenter, the longer the delay between the arrival of P-waves and the arrival of S-waves. **Travel time** is the time inferred between the breaking of the rocks that causes an earthquake and when the event is detected at a given location. The greater the time delay between the arrival of the first P- and S-waves, the greater the distance to the epicenter. For a specific time delay, you can establish a distance.

Figure 14-7 is a graph from the *Earth Science Reference Tables* that shows how long it takes P- and S-waves to travel as far as 10,000 km. Note that the distance from the epicenter is on the horizontal axis and travel time is on the vertical axis. Along the

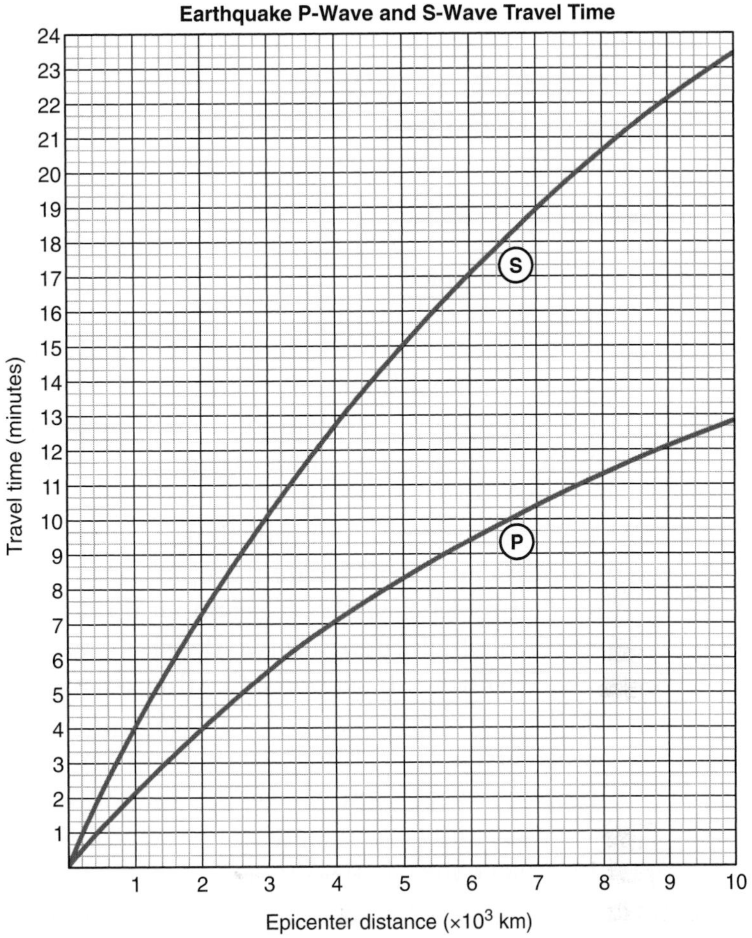

Earthquake P-Wave and S-Wave Travel Time

FIGURE 14-7.
This graph from the *Earth Science Reference Tables* allows you to determine the distance to an earthquake epicenter based on arrival times of P- and S-waves.

horizontal scale, the distance from one dark line to the next is 1000 (10^3) km. Each 1000-km division is split into five subdivisions. Therefore each thin line represents 1000 kms ÷ 5 = 200 km. On the vertical scale, each thick line represents 1 minute and there are three divisions. What is the value of each thin line? Remember there are 60 seconds in each minute. Therefore each vertical division is one-third of a minute, or 20 seconds: 1 minute (60 seconds) ÷ 3 = 20 seconds.

For example, this graph tells you that if an earthquake epicenter is 4000 km away, the P-wave will take 7 minutes to get to you. That is, at your location you will feel the first vibrations 7 minutes after the rupture occurs at the epicenter. The graph shows that the S-waves need about 12 minutes:40 seconds to travel the same distance. Therefore, if the time delay between the arrival times of the P- and S-waves is 5 minutes:40 seconds, the epicenter must be 4000 km away. An observer closer to the location of the epicenter would notice a smaller delay between the arrival of the P- and S-waves. A more distant recording station would register a greater delay between the arrival of the waves.

Imagine that you observed a 5-minute:40-second time delay. What does this tell you about the location of the epicenter? It tells you that the epicenter of the earthquake is 4000 km from your location. Therefore, on a map you could draw a circle centered on your location with a radius of 4000 km and know that the epicenter is somewhere on that circle. However, you do not know where on the circle the epicenter is, but you do know it has to be that distance away from your station.

The text and photographs that follow will demonstrate a procedure you can use to locate the epicenter of an earthquake. Notice that Figure 14-8 shows a seismogram recorded at Denver, Colorado, on which there are two divisions in each minute. Therefore each division is 30 seconds. On this seismogram you can see that the P-wave arrived at 8 hours:16 minutes:0 seconds, and the S-wave arrived at 8 hours:19 minutes:30 seconds. The time separation between P- and S-waves is therefore 3 minutes:30 seconds.

There is only one distance at which this time delay can occur. To find it, place the clean edge of a sheet of paper along the vertical distance scale. Make one mark at 0 minutes and another at 3 minutes:

FIGURE 14-8. By reading the seismogram you can see that P-waves arrive 3 minutes: 30 seconds before the S-waves arrived at this recording location.

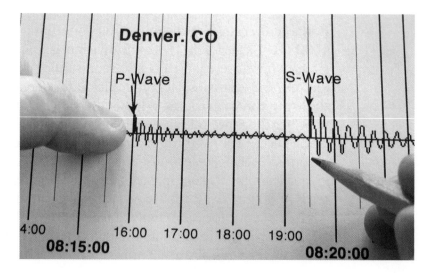

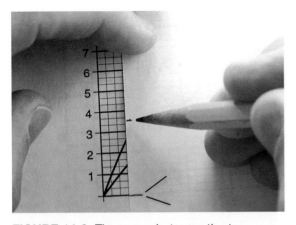

FIGURE 14-9. The space between the two marks on the edge of the sheet of paper represents the time difference between the arrival of the P- and S-waves.

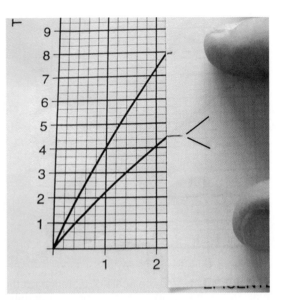

FIGURE 14-10. The time-delay marks are moved up and to the right until they line up with P- and S-wave graph lines. The time delay indicated on the edge of the paper will occur at a specific distance, which can be read below on the distance scale.

30 seconds as shown in Figure 14-9. Perform this step carefully to avoid problems further along in the procedure.

Next, as you keep the edge of the edge of the paper vertical, move the paper to the position at which each marks rests on (matches) a graph line. As shown in Figure 14-10, the 0 mark will be on the P-wave line and the mark for 3 minutes:30 seconds will be on the S-wave line. Notice that this time delay occurs only if the epicenter is 2200 km away.

The next objective is to draw a circle around the Denver seismic station with a radius equivalent to 2200 km. Figure 14-11 on page 322 shows a drawing compass stretched along the map scale to 2200 km. Do this carefully because it is another step where errors are common.

On the map, place the pivot of the compass on the Denver seismic station. Be sure the compass opening has not moved since it was set at 2200 km. Draw a circle around Denver at a distance equivalent to 2200 km, as you see in Figure 14-12 on page 322.

Now you have drawn a circle on a map that shows how far the epicenter is from Denver. The epicenter can be anywhere on that

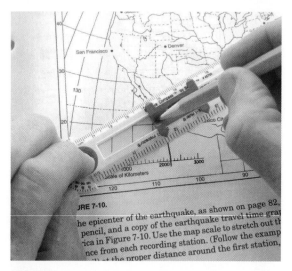

FIGURE 14-11. Use the scale of distance on the map to set the compass to the correct epicenter distance.

FIGURE 14-12. Place the pivot of the compass on the recording station and draw a circle at the epicenter distance. The epicenter is somewhere on this circle.

circle. You need two more seismic stations to locate the epicenter. Following the same steps with another seismogram of this earthquake recorded at Miami would give you a second circle. Most likely those two circles will intersect at two points. The epicenter must be at (or near) one of those points. Using the same procedures with a seismogram from a third location draw another circle. These three circles should meet at one location on the map. That marks the epicenter, as you can see in Figure 14-13.

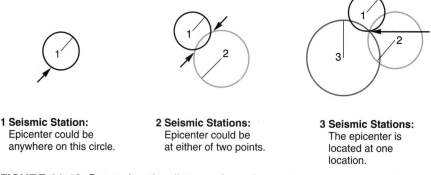

1 Seismic Station:
Epicenter could be anywhere on this circle.

2 Seismic Stations:
Epicenter could be at either of two points.

3 Seismic Stations:
The epicenter is located at one location.

FIGURE 14-13. Determine the distance from the earthquake epicenter for three recording stations. Draw a circle on a map around each station. The epicenter is located where the three circles meet. (In practice, the three circles seldom meet exactly.)

If you have the data, you can draw more than three circles. That is a good way to check your work. Is it really that neat and simple? The theory is, but in practice the circles may not meet exactly. More often, the three circles make a small triangle, and the epicenter is located inside that triangle. When seismologists working with real seismic recordings draw these circles to locate actual seismic events, the circles seldom meet at a single point. Differences in the composition of Earth's interior and a wide range of other variables can contribute to errors that prevent a perfect meeting of the circles. Visit the following Web site from California State University System to investigate a virtual earthquake: *http://www.sciencecourseware.com*

Finding the Origin Time

The **origin time** is the time at which the fault shifted, producing the earthquake. Knowing when the P- or S-waves arrived at your location and how far away the epicenter is, you can calculate the origin time. Suppose that a friend arrived at your house at 10:30 and told you that the trip took him 20 minutes. You could determine that your friend left home at 10:10. In the same way, if you know the time that a seismic wave arrives and you know how long it traveled, you can tell when the earthquake began at the epicenter.

When a P- or S-wave arrives, how can you tell how long it was traveling? The key is knowing how far away the epicenter is and using the travel time graph. You have already learned that the distance to an epicenter is determined from the time delay (difference) between the arrival of the P- and S-waves. Once you know the distance to the epicenter, you can use the travel time graph to discover how long it took for the P- and S-waves to travel from the epicenter to your location: the travel time.

Let us continue with an example in the "How are Earthquakes Located?" section. In the first example in that section, the separation between P- and S-waves determined that the epicenter was 4000 km from your location. You also need to know when the P- and S-waves arrived. Let us work with a P-wave arrival time of 02:10:44 and an S-wave arrival time of 02:14:04. By reading the graph you can tell that a P-wave takes 7 minutes to travel a distance of 4000 km. Subtract the 00:7:00 travel time from the 02:10:44 P-wave arrival time to find the origin time.

$$02:10:44 \text{ (P-wave arrival time)}$$
$$- \underline{00:07:00} \text{ (travel time)}$$
$$02:03: 44 \text{ (origin time)}$$

You can check this time by using the S-wave data. To travel 4000 km an S-wave takes 12 minutes:40 seconds:

$$02:16:24 \text{ (S-wave arrival time)}$$
$$- \underline{00:12:40} \text{ (travel time)}$$

Borrowing 1 minute and adding it to the seconds' column makes this

$$02:15:84 \text{ (S-wave arrival time)}$$
$$- \underline{00:12:40} \text{ (travel time)}$$
$$02:03: 44 \text{ (origin time)}$$

This is one earthquake, so you would expect the two origin times to be the same. If they are not at least close in time, you probably need to go back and look for an error in your work.

WHAT IS INSIDE EARTH?

4: 2.1j, 2.1k, 2.1l

For a long time people have wondered about what is inside Earth. In Chapter 2 you learned that Earth has a radius of more than 6000 km (3800 mi). In South Africa, the world's deepest mine shaft reaches about 5 km (3 mi) below the surface. To learn about rock properties far below the surface, the Russians have drilled to a depth of 12 km (7.5 mi), but even this is much less than 1 percent of the distance to Earth's center.

If a doctor needs to see inside your body without cutting you open, he or she can use x-rays and CAT scans. However, these techniques cannot penetrate thousands of kilometers of solid rock. Scientists have found another way to examine Earth. Earthquake waves do travel through Earth's interior, and they can be used to get information about its structure. From observing which earthquake waves travel through the planet, where they are detected, and the speed at which they travel, scientists make inferences about Earth's structure. Based on direct observations and analysis of seismic waves, scientists have divided the planet into four layers: crust, mantle, outer core, and inner core.

The Crust

The crust is the outermost layer of Earth. In all the mining and drilling humans have done, this is the only layer that has been observed directly. In most places, a thin layer of sedimentary rocks covers metamorphic and igneous rocks. The deeper you dig, the greater the chance of finding igneous rocks. Within the continental crust, the overall composition is close to the composition of granite. Under the sediments and sedimentary rocks of the oceans, the oceanic crust is more dense. This is verified by observing that the lava from most ocean-basin volcanoes hardens to form basalt. Geologists therefore know that the crust under the oceans is mostly basalt or has a similar composition. Furthermore, from studying very deep mines, scientists know that the temperature within the crust increases with depth at a rate of about 1°C per 100 m. Based on this rapid rate of temperature change, it seems logical that Earth's interior is very hot.

The bottom of the crust is inferred from observations of earthquake waves. Seismic waves that travel through just the crust are slower than those that dip below the crust and enter the mantle. This change in seismic wave speed occurs at depths as shallow as 5 km (3.1 mi) under the oceans to as deep as 60 km (37 mi) under some mountain ranges on land. This indicates that the continental crust is thicker than the oceanic crust. The Croatian geophysicist Andrija Mohorovičič first noticed this boundary, which is now named in his honor. However, most people shorten the name of the boundary between the crust and the mantle to the **Moho**.

The Mantle

As you can see in Figure 15-6 on page 344, the mantle extends from the bottom of the crust to a depth of about 2900 km (1800 mi). In fact, it contains more than half of Earth's volume. However, because we cannot reach below the crust with mines or drills, investigations of the mantle must be conducted by other means.

Most magma originates within Earth's crust beneath mountains where the crust is thick and felsic in composition. However, magma that originates below the crust is rich in dense, mafic minerals such as olivine and pyroxene. These are the minerals on the right side of the Scheme for Igneous Rock Identification in the *Earth Science Reference Tables.*

Most natural diamonds originate in particularly violent eruptions of material from deep within Earth. These gas-charged intrusions from deep within the mantle are called kimberlites. They are named for Kimberley, a city in South Africa, where diamonds have been mined for many years. Diamonds can form only under conditions of extreme heat and pressure, such as the conditions within the mantle. The fact that diamonds are found in these volcanic rocks indicates that the magma came to the surface from deep within the mantle. There are a few small and isolated exposures of kimberlites in Central New York State. However, these intrusions do not seem to be from deep enough in the crust to contain diamonds. Visit the following Web site and select the geology animation to learn more about diamond formation: *http://www.debeers canada.com/files_2/victor_project/victor_animations/animation_ 2007/main_menu.html*

The stony meteorites that reach Earth are composed of dense, mafic minerals such as pyroxene and olivine. These are the same materials found in Earth's mantle. Scientists think that these meteorites are material left over from the formation of Earth and the other planets billions of years ago, or they are the remains of a planet that was torn apart by a collision with another object. If this is the case, you would expect meteorites to have a composition similar to the planets, including Earth.

The diameter of Earth has been known for centuries. From this value, it is easy to calculate Earth's volume. The mass of Earth has been determined based on its gravitational attraction. Knowing Earth's mass and volume, scientists have calculated that Earth's overall density is about 5.5 g/cm^3. That is about twice as dense as most rocks in the crust. Therefore, scientists expect the mantle, which includes most of Earth's volume, to be composed of minerals that are more dense than those in the crust.

Both P- and S-waves travel through the mantle. Because S-waves will not travel through a liquid, the mantle is known to be mostly in the solid state. Furthermore, earthquake waves travel faster in the mantle than they do in Earth's crust. This indicates that the rock in the mantle is more brittle than crustal rocks. Again, olivine and pyroxene fit the observations.

All these observations provide strong evidence that Earth's mantle is composed primarily of iron- and magnesium-rich, mafic silicate minerals.

The Core

Starting at 2900 km (1800 mi) and extending to Earth's center is the next layer. Seismologists noticed that P-waves can penetrate the whole planet, but S-waves do not travel through Earth's core. Therefore the core must be liquid.

Starting from the atmosphere, each of Earth's layers is more dense than the layer above it: atmosphere, hydrosphere, crust, and mantle. You might therefore expect the core to be even more dense than the mantle.

Meteorites come in two basic varieties. You read earlier that the more abundant stony meteorites are thought to be composed of the material found in the mantle. The second group of meteorites is mostly iron. Therefore, scientists infer that Earth's core is also mostly iron with some nickel.

The behavior of seismic waves, including reflection and refraction within the core, has led scientists to infer that the inner part of the core is a solid due to extreme pressure at Earth's center. The solid state of the inner core may also result from different elements mixed with iron. Geologists therefore infer that a liquid outer core surrounds a solid inner core, both composed mostly of iron.

The conditions of heat and pressure deep within Earth prevent scientists from any direct observations. But they have been able to duplicate these conditions in laboratories. Very small amounts of the substances geologists believe to be present in the mantle and core can be squeezed under extreme pressure between the flat faces of two diamonds. The instrument is called a diamond anvil. Heat is applied with a laser beam. Then x-rays are used to probe the crystal structure of the substance. These studies confirm, for example, that pressure at Earth's center can cause the iron-rich inner core to solidify.

6: MODELS 2
6: MAGNITUDE AND
SCALE 3

How can you see the actual proportions of Earth's inner layers? Select a location in a long hallway or on the school grounds to show the relative thickness of Earth's primary features. This will be your base line from sea level to Earth's center. Establish a scale to place one student at sea level and another at Earth's center. Then use this scale to position other students at the Moho, the bottom of the lithosphere, and the boundaries of both cores. Finally, have other students show the scale elevation of Mt. Everest above sea level, and the world's deepest bore hole.

Earthquake Shadow Zones

When a large earthquake occurs, seismic waves travel through Earth's layers to recording stations around the world. The behavior of these waves provides important information about Earth's structure. For example, scientists know that S-waves will not travel through the outer core because it is liquid.

The zones where direct P-waves are recorded are more complicated. These P-waves arrive at locations as far as half way around Earth. These locations also receive S-waves. P-waves are also received in the region directly opposite the epicenter. Those are the P-waves that travel through the liquid outer core (and the solid inner core). But there is a shadow zone that does not receive any direct P- or S-waves. Figure 14-14 shows that beyond the reach of S-waves there is a zone that circles Earth where no direct P- or S-waves are received. This shadow zone circles Earth like a donut with a hole in the middle. While you might expect this region to receive at least direct P-waves, the bending of seismic waves prevents even them from reaching the shadow zone.

The bending of energy waves is known as **refraction**. It can be observed whenever energy waves enter a region in which they speed up or slow down. From Figure 14-14 you can see how seismic waves curve as they travel through the mantle and core. You can also see the P-waves bend when they move from the mantle into the core as well as when they move from the core back into the mantle. The result of this bending (refraction) of seismic waves is the shadow zone where no direct P- or S-waves are received.

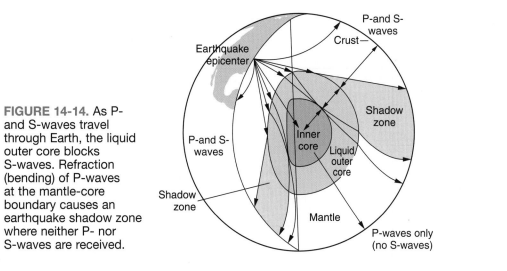

FIGURE 14-14. As P- and S-waves travel through Earth, the liquid outer core blocks S-waves. Refraction (bending) of P-waves at the mantle-core boundary causes an earthquake shadow zone where neither P- nor S-waves are received.

In short, the primary methods scientists use to investigate the deep Earth rely on remote sensing. Remote sensing is the gathering of data without physical contact. In spite of how difficult it is to make direct observations, geologists have constructed a logical understanding of Earth's interior. There is probably no other place where scientists have never been, and where they probably never will go, about which they know so much.

CHAPTER REVIEW QUESTIONS

Part A

1. Which statement correctly compares seismic P-waves with seismic S-waves?

   (1) P-waves travel slower than S-waves and pass through Earth's liquid zones.
   (2) P-waves travel slower than S-waves and do not pass through Earth's liquid zones.
   (3) P-waves travel faster than S-waves and pass through Earth's liquid zones.
   (4) P-waves travel faster than S-waves and do not pass through Earth's liquid zones.

2. A seismograph station recorded the arrival of the first P-waves at 7:32 P.M. from an earthquake 4000 km away. If it is in the same time zone, when did the earthquake occur at the epicenter?

   (1) 7:20 P.M.   (3) 7:32 P.M.
   (2) 7:25 P.M.   (4) 7:39 P.M.

3. The map below shows the use of data from three seismic stations to locate an earthquake epicenter.

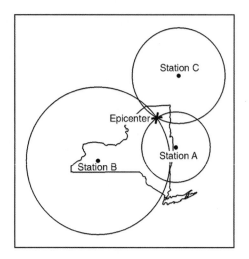

The seismogram recorded at station A would show the

(1) arrival of P-waves only
(2) earliest arrival of P-waves
(3) greatest difference in the arrival times of P-waves and S-waves
(4) arrival of S-waves before the arrival of P-waves

4. An earthquake's first P-wave arrives at a seismic station at 12:00:00. This P-wave has traveled 6000 km from the epicenter. At what time did the first S-wave from the same earthquake arrive?

(1) 11:52:20   (3) 12:09:20
(2) 12:07:40   (4) 12:17:00

5. The magnitude of an earthquake can be determined by

(1) measuring seismic waves
(2) calculating the depth of the faulting
(3) analyzing the origin time
(4) finding the speed of P- and S-waves

6. Compared to Earth's continental crust, Earth's oceanic crust is

(1) thinner and more dense
(2) thinner and less dense
(3) thicker and more dense
(4) thicker and less dense

7. How far away is the epicenter of an earthquake if the P-waves arrive at your location 9.5 minutes before the first S-waves?

   (1) 1600 km   (3) 6000 km
   (2) 2800 km   (4) 8000 km

8. Which object is likely to have the same composition as Earth's mantle?

   (1) a granite rock   (3) an iron meteorite
   (2) a rhyolite rock   (4) a stony meteorite

Part B

9. What is the approximate speed of P-waves traveling though Earth's crust?

   (1) 4 km/minute   (3) 400 km/minute
   (2) 40 km/minute   (4) 4000 km/minute

10. The diagram below is a profile of Earth. An earthquake with an epicenter at X is recorded at seismic stations A, B, and C.

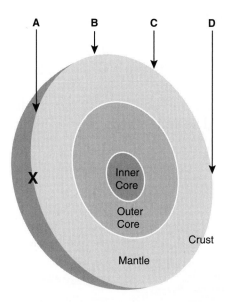

    Which statement below describes how direct P-waves and S-waves would be observed at stations A, B, and C?

    (1) A, B, and C will all record both P-waves and S-waves.
    (2) A, B, and C will all record only P-waves.
    (3) A will record both, B will record only P-waves, and C will record neither.
    (4) A will record both, B will record neither, and C will record only P-waves.

Base your answers to questions 11 and 12 on the diagram below, which shows the inferred internal structure of the planets Mercury, Venus, Earth, and Mars.

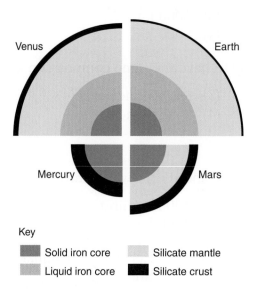

11. How are the crusts of the four planets similar in composition?

12. Identify the two planets that would allow S-waves from a crustal quake to be transmitted through the core to the opposite side of the planet

Base your answers to questions 13 through 15 on the diagram below, which shows a seismograph that recorded seismic waves from an earthquake located 4000 km from this seismic station.

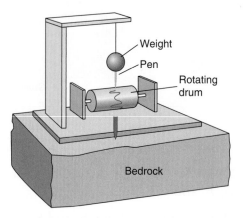

13. State one possible cause of the earthquake that resulted in the movement of the bedrock detected by this seismograph.

14. Which type of seismic wave was recorded first on the rotating drum?

15. How long does the first S-wave take to travel from the earthquake epicenter to this seismograph?

Part C

Base your answers to questions 16 through 18 on the diagram below, which shows a vertical profile of Earth. Layer X is part of Earth's interior.

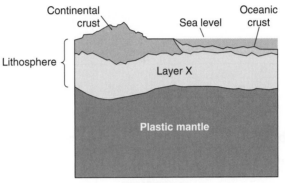

Continental crust   Sea level   Oceanic crust

Lithosphere

Layer X

Plastic mantle

(Not drawn to scale)

16. Compare the texture and relative density of the granite bedrock of the continental crust to the basaltic bedrock of the oceanic crust.

17. Name three minerals that are abundant in both oceanic and continental crust.

18. What name do geologists generally apply to the layer labeled X in the diagram above?

19. A seismic station receives P-waves and S-waves from 12 earthquakes both nearby and at great distances. How can a person tell which of the 12 seismic recordings came from the earthquake with the closest epicenter?

20. Explain why earthquakes are common in California.

15

Plate Tectonics

WORDS TO KNOW

asthenosphere	lithospheric plates	polarity
convergent plate boundary	mid-ocean ridges	seafloor spreading
divergent plate boundary	ocean trenches	subduction zone
fluid	paradigm	tectonics
hot spot	plastic	transform boundary
island arc	plate tectonics	

This chapter will help you answer the following questions:

1. How do we know that the continents travel over Earth's surface?
2. Why do the continents move?
3. How do we know about Earth's deepest layers?
4. How does the escape of Earth's internal heat affect Earth's surface?
5. How will the geography of Earth change in the future?

DO CONTINENTS MOVE?

1: SCIENTIFIC INQUIRY 1, 2, 3 4: 2.1k, 2.1l, 2.1m, 2.1n

This is a story of changing scientific ideas, or a paradigm shift. A **paradigm** (PARA-dime) is a logical set of principles and understandings. Until the late 1960s, most scientists thought of Earth's continents as motionless features attached to the interior by solid rock. When the evidence for moving continents became so strong that scientists could no longer dismiss it, the geological community experienced a paradigm shift. They changed from thinking of Earth's surface features as fixed in position to proposing a planetary surface composed of moving pieces.

Continental Drift

When explorers and mapmakers crossed the Atlantic Ocean in the 1600s, some of them noticed an interesting coincidence. The more complete their maps became, the more the Atlantic Ocean shorelines of the Americas and the Old World continents of Africa and Europe looked as if they would fit together like pieces of a jigsaw puzzle, as you see in Figure 15-1 and Figure 15-2 on page 336. Initially this was simply noted as a curiosity.

In 1912, Alfred Wegener (VEYG-en-er)—a German explorer, astronomer, and atmospheric scientist—proposed that the continents we see are the broken fragments of a single landmass that he called Pangaea (pan-JEE-ah), which means *all Earth* in Greek. He hypothesized that over millions of years, the continents slowly moved to their present positions. Wegener supported his

FIGURE 15-1. The earliest maps of "The New World" (the Americas) revealed the roughly matching shorelines across the Atlantic Ocean.

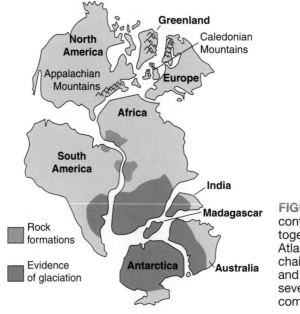

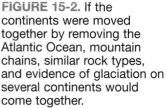

FIGURE 15-2. If the continents were moved together by removing the Atlantic Ocean, mountain chains, similar rock types, and evidence of glaciation on several continents would come together.

theory by pointing out that a particular fossil leaf (*Glossopteris*) is found in South America, Africa, Australia, Antarctica, and India. Bringing the continents together would put the areas where these fossil organisms are found next to each other, as shown in Figure 15-2.

The same placement would also line up similar rock types and mountain areas. Wegener also noted glacial striations in places such as Australia and Africa, where now the climate is too warm for glaciers. He suggested that these places were once located near the South Pole. On the basis of this evidence, Wegener proposed his hypothesis that came to be known as continental drift.

The idea that continents could move was so radical that few geologists took it seriously. How could the landmasses move through the solid rock of the ocean basins? The causes suggested by Wegener could not provide the force needed to move continents. As another way to challenge Wegener's idea of continental drift, scientists pointed out that he was not a geologist. Alfred Wegener died in 1930 on an expedition to Greenland. At that time, very few geologists supported his ideas. His most enthusiastic supporters were in the Southern Hemisphere, where the most convincing evidence for continental drift had been observed.

STUDENT ACTIVITY 15-1 —MATCHING SHORELINES

6: MODELS 2
6: PATTERNS OF
CHANGE 5

Y ou can reconstruct the supercontinent, Pangaea. Get an outline map of the world. Cut out the continents and lay the pieces on your desk in their present locations. Then, slide them together to show how they might have fit to make a single landmass. Paste or tape them onto a sheet of paper in the position they might have been in before they separated.

Seafloor Spreading

The rocks of the ocean bottoms are very different from most rocks on land. Rocks from the continents have a composition similar to granite. That is, the continental rocks are generally felsic (rich in feldspar), light-colored rocks with a relatively low density. The low-density material that makes up the continents tends to rise higher than the ocean basins. Most oceanic rocks have a composition similar to basalt. They are mafic. You may recall from Chapter 5 that mafic rocks are darker in color and relatively dense. This is why the ocean basins tend to ride lower on the rocks of Earth's interior and are covered by the oceans.

At the time Wegener worked on his hypothesis, the oceans were vast expanses of unknown depth. After his death, the strongest evidence for his theory of moving continents came from the oceans. During World War II, there were important advances in technology and oceanic exploration. A moving ship could make a continuous record of the depth of the ocean by bouncing sound waves off the ocean bottom. After the war, scientists used this technology to map the world's ocean bottoms.

Scientists discovered an underwater system of mountain ranges that circles Earth like the seams on a baseball. These are the **mid-ocean ridges**. Most of these 64,000-km-long features are underwater. However, they rise above sea level in Iceland and on several smaller islands. By looking at the landscape of Iceland, scientists can investigate processes that occur hidden from view at the bottom of the ocean.

When magma solidifies quickly as it comes in contact with the cold ocean water, it forms pillow-shaped rocks. Wherever scientists find pillow lavas, they can be sure that magma moved into water and solidified quickly. Pillow lavas are common along the mid-ocean ridges but are uncommon elsewhere in the oceans.

In 1960, Princeton University geologist Harry Hess and his colleagues reconsidered Wegener's ideas. During World War II, Hess had commanded a ship that measured ocean depths. He had seen the newly discovered shape of ocean bottoms including the mid-ocean ridges. Hess suggested that molten magma from the mantle rises to the surface at the mid-ocean ridges and erupts onto the ocean bottom. In many places, the ocean ridges are like two mountain chains separated by a valley. It is in the valley that the most active eruptions take place, forming a strip of new ocean floor.

Hess suggested that the mid-ocean ridges are the places where new lithosphere is made and adds on to older material that moves away from the ridges on both sides. Hess called this process **seafloor spreading**. Visit the following Web site to watch a USGS animation of seafloor spreading: *http://www.uwsp.edu/geo/faculty/ ritter/glossary/S_U/sea_flr_spread.html*

MAGNETIC EVIDENCE   New technology also allows scientists to make continuous records of the magnetism of the ocean floor. In Chapter 3, you learned that Earth is a giant magnet. Scientists are not certain what causes Earth's magnetic field. They think that the likely cause is currents of molten iron in Earth's outer core. These currents tend to line up with Earth's spin axis and therefore make Earth's magnetic poles close to the geographic north and south poles. When magma cools to make rock, iron forms the magnetic mineral magnetite. Within molten magma, iron is free to align itself with Earth's magnetic field and change its alignment as the magma flows or as Earth's magnetic field changes. When magma rises to the surface and solidifies, the alignment of the iron in the rock can no longer change. The new rock holds a record of Earth's magnetism at the time it crystallized.

Earth's magnetic poles have reversed many times. The magnetic field weakens and then builds in the opposite direction. In other words, the north magnetic pole becomes the south magnetic pole, and the south magnetic pole becomes the north magnetic pole. The time from one reversal to the next is not equal, but most reversals occur after 100,000 to 1,000,000 years. The last reversal was nearly a million years ago, so some scientists speculate that a reversal may be overdue. A record of these reversals is preserved in igneous rocks that solidified at different times. In Chapter 17, you will learn how scientists use radioactive substances to find the age of igneous rocks.

If the whole ocean floor formed at the same time, all the igneous rocks would be magnetized in the same direction. The direction is determined with an instrument that measures Earth's magnetic field. However, scientists have found bands of polarity. **Polarity** is the direction of a magnetic field. The polarity of some of the bands is the way they would be if they solidified with the present alignment of the magnetic poles. These areas have normal polarity. Other parts of the ocean floor show reversed polarity. These rocks crystallized when the magnetic north and south poles were opposite to the way they are now. These bands of normal and reversed polarity are the strongest near the ocean ridges. In addition, they show a symmetrical (balanced) pattern on each side of the ocean ridges. That is, the striped pattern of normal and reversed polarity on one side of an ocean ridge is almost a mirror image of the pattern on the other side, as you see in Figure 15-3. These bands of normal and reversed polarity provide evidence that new crust is forming as the plates move away from the ridges.

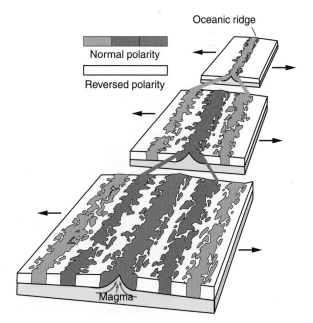

FIGURE 15-3. Earth's crust is created as magma upwells, then cools, and solidifies at the spreading ocean ridges. Cooling allows the new ocean floor make a permanent record of Earth's magnetic field. The three drawings represent growth of the ocean floor at a single location.

AGE OF ROCKS Geologists can determine the age of rocks recovered from the ocean floors. In the 1950s, the oceanographic research ship *Glomar Challenger* drilled into the basaltic rock of the ocean floor to collect pieces of the solid bedrock of the oceans. When laboratory work on these samples revealed their age, a pattern emerged. Geologists discovered that the rocks recovered close to the ridges were younger than those collected far from the mid-ocean ridges. As the distance from the mid-ocean ridges increases, so does the age of the ocean floor. Figure 15-4 shows that the age of rocks changes as you move away from the mid-ocean ridges that circle Earth. The youngest rocks would be at the ridge; the oldest, near the continents.

These changes in polarity and age of the rocks support the idea that ocean floor is formed continuously at the mid-ocean ridges and moves like a conveyor belt toward the continents. But how can new lithosphere be formed if Earth is not getting larger? In Chapter 6, you read about recycling to conserve resources. Earth recycles its lithosphere. Old lithosphere moves back into Earth's interior at places called **subduction zones**. As lithosphere is created by magma moving to the surface at the mid-ocean ridges, the older lithosphere moves away from the ridges. The rate of relative

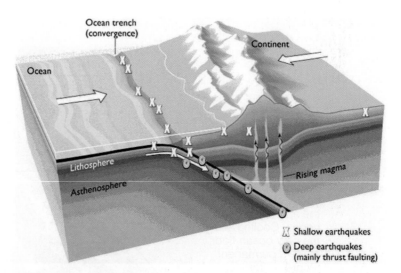

FIGURE 15-4. Scientists trace subducting oceanic crust as it dives into the lithosphere by determining the depths of earthquake foci. For example, shallow earthquakes are common in the Pacific Ocean off South America. However, as the Pacific Plate dives under the continent, foci depths increase.

motion varies from one plate boundary to another. Some move as fast as 10 cm/year. Typically, the plates move about as fast as your fingernails grow. Figure 15-4 illustrates that at the subduction zones sections of the lithosphere slowly sink into Earth. Geologists use seismology to trace subduction. Earthquake foci are generally deeper when the quakes occur in subducting ocean floor.

At **ocean trenches,** the more mafic parts of the lithosphere are absorbed into the mantle. However, the felsic continental lithosphere resists being drawn into Earth in the process of subduction. This lighter material forms intensely folded and faulted mountains near the subduction zones. Felsic rocks drawn into Earth often melt and come back to the surface in volcanic eruptions. Due to the water in these continental rocks changing to steam, volcanic eruptions of felsic rocks are often violent and explosive. This is similar to the rapid formation of bubbles and eruption of liquid caused by the sudden release of pressure that occurs when you open a bottle of seltzer. Both the highest mountains and the deepest parts of the oceans occur near zones of subduction and both are related to the process of subduction.

Plate Tectonics

You may recall that one of the most important objections to Wegener's idea of continental drift was that there was no mechanism to explain how continents could move through the ocean basins. Seafloor spreading provided an explanation for the motion of Earth's lithosphere. When the discoveries of seafloor spreading were added to Wegener's ideas about continental drift, geologists proposed the more comprehensive theory of **plate tectonics**. (The word **tectonics** means large-scale motions of Earth's crust.) Tectonic forces cause the rocks of the continents to become folded, such as you see in Figure 15-5 on page 342. This is evident in uplift and mountain building.

In 1965, Canadian geologist J. Tuzo Wilson proposed that Earth's surface is made of about a dozen rigid sections called **lithospheric plates**. The plates include the crust and the rigid upper mantle. Plates are approximately 100 km (60 mi) thick, and ride over a portion of the mantle that is less rigid than the upper part of the mantle. Compared with Earth's 6000-km (3700 mi) radius, the plates are relatively thin. The lithospheric plates are

FIGURE 15-5. These large S-shaped folds are evidence of forces within Earth strong enough to fold rock layers. These folds are along Interstate 84 in eastern New York State.

formed at the ocean ridges and destroyed at the subduction zones. Some plates contain only ocean floor, but most major lithospheric plates have oceanic and continental lithosphere. Zones of geological activity, such as the mid-ocean ridges and subduction zones, form the boundaries of the plates. In addition to the major plates, a number of smaller plates exist where the larger plates do not meet precisely.

The motion of the plates is driven by Earth's internal heat. Conduction does allow some of Earth's energy to reach the surface. However, conduction is too slow to account for most of the heat flow from Earth's center. The major form of heat flow outward from Earth's interior is convection. At the mid-ocean ridges, convection brings heated matter to the surface. There the heat energy warms the ground and radiates into space.

Convection occurs only in fluids. A **fluid** is any substance that can flow. Generally, scientists consider gases and liquids to be fluids, but not solids. But how can solid rock flow like a fluid? If you could visit the mantle, the rocks there would seem just as solid as similar rocks at the surface. If you were to hit mantle rock with a hammer, the hammer would bounce off the rock and make a high-pitched clink. The hammer does not deform the rock because it creates only a short-term force. Although there are some places where rock has melted, forming magma, nearly all the mantle is in the solid state.

You may recall that S-waves do not pass through liquids. The passage of S-waves through Earth's mantle indicates that the mantle is solid. Earth's mantle is a solid, but it can flow to carry energy by convection. How can this be? The key to resolving this contradiction is the extreme heat and pressure inside Earth. Under extreme pressure and high temperature, solid rock can flow slowly like a fluid. In addition, the rate of movement of the plates is only a few centimeters per year. This is slow enough to allow convection in a material that otherwise behaves like a solid.

Convection is driven by differences in density. Most materials expand when they are heated. Expansion increases volume and makes matter less dense. In places where a fluid is heated, it becomes less dense and rises. Cooling usually causes matter to become more dense. This makes the substance heavier than an equal volume of the surrounding material. When fluids are cooled they usually sink. A heat source in one place, such as Earth's interior, and cooling in another place, such as Earth's surface, can set up a circular motion known as *convection*. In convection, the motion of the fluid occurs when heated material moves away from the source of heat. This is what is happening in Earth's plastic mantle. Visit the following Web site to download animations of plate tectonics: *http://www.scotese.com/newpage13.htm*

STUDENT ACTIVITY 15-2 —EXPERIMENTING WITH OOBLECK

1: SCIENTIFIC
INQUIRY 1
6: MODELS 2

You know that water has three forms: solid (ice), liquid (water), and gas (water vapor). However, the three states of matter are not always so easy to tell apart. Remember oobleck from Chapter 12? It has the properties of a liquid and a solid. Mix cornstarch or potato flour and water in a volumetric ratio of about two parts powder to one part water. (The specific ratio to get the best consistency may take some adjustment.) Place a clump of the mix on a slope to see it flow. Toss a ball of it into the air. Will a ball of oobleck bounce on a hard surface like a rubber ball? Hit it gently with a hammer to see how it responds to quick stress. What happens if you leave it in a bowl to partially dry out on a dry, sunny day? Is it a liquid or is it a solid? Find the most interesting YouTube video with oobleck. Visit the following Web site to watch an experiment with oobleck: *http://ellen.warnerbros.com/2008/02/all_steve_spangler_do_not_try_this.php*

WHAT IS EARTH'S INTERNAL STRUCTURE?

4: 2.1j

In Chapter 14, you read about the four layers within Earth as determined by seismic waves: crust, mantle, outer core, and inner core. Motion of the plates leads us to consider a different way to divide the outer layers of Earth into lithosphere, asthenosphere, and stiffer mantle. Note that Figure 15-6 is from the *Earth Science Reference Tables*. The following paragraphs explain this chart.

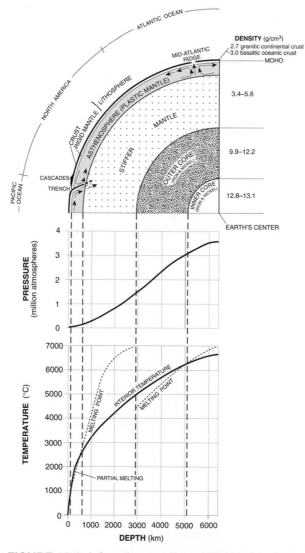

FIGURE 15-6. Inferred properties of Earth's interior.

Temperature

An *inference* is an interpretation or a conclusion based on indirect observations. The word "inferred" tells you that this diagram is based on laboratory simulations and other investigations rather than on direct observations. This diagram has three parts: a temperature graph, a pressure graph, and a profile view inside Earth. The three parts of this figure describe Earth's interior. The temperature graph shows how scientists think Earth's temperature changes with depth. It is easy to find an average temperature at Earth's surface because people can travel over the surface. You may have noticed that you feel cooler when you go down into a basement or into a cave. However, scientists have observed that beyond a relatively shallow depth measurements show that the temperature increases the deeper they go below Earth's surface. The temperatures in deepest mines and wells are much higher than the temperature at the surface. Beyond the depths of mines and wells, scientists must use laboratory instruments to simulate conditions inside Earth.

This graph has two lines: the dark Actual Temperature and the dotted Melting Point. At the surface, both lines are close together. Since the rocks at the surface are solid, the surface temperature is clearly below the melting point of these rocks. At a depth of about 100 km, the dark temperature line is very close to the dotted melting point line. Much of the magma that erupts from volcanoes comes from this depth. This is also where the lithosphere, the crust and the rigid part of the mantle that moves with the plates, ends.

Have you ever played with Silly Putty? Silly Putty shatters like a solid when it is hit with a hammer. However, leave a ball of Silly Putty on a flat surface and it will slowly collapse into a puddle. A material is said to be **plastic** if it has the properties of a solid under short-term stress, but flows like a liquid when stress is applied over a long period of time. Below 100 km, there is partly melted rock that is close to the melting temperature of mafic silicates. The plastic nature of that layer allows rigid lithospheric plates to slowly move over Earth's surface. The **asthenosphere** is the plastic part of the mantle. Plastic refers to its state of matter.

Below the asthenosphere, the melting point increases due to higher pressure, and the mantle loses its plastic nature. This happens because the temperature of this part of the mantle is well below the melting temperature. Recent research has shown that some magma does seem to originate within the stiffer mantle. This might indicate that variations in composition occur in this part of the mantle. Some parts of the stiffer mantle might melt at lower temperatures than other parts.

There is a sudden change in melting temperature at a depth of about 2900 km. Geophysicists, scientists who study the physical properties of Earth materials, believe that this is where the silicate composition of the mantle ends and the mostly iron outer core begins. Iron has a lower melting temperature than mafic silicate minerals.

The composition of meteorites and the density of Earth support the inference that Earth's core is mostly iron. If scientists subject iron to the pressure estimated to exist at Earth's center, it melts at just over 6000°C. Therefore, the outer core is a molten liquid. Finally, there is a transition from the liquid part of the outer core to the solid phase of the inner core, even without a major change in composition. This indicates that the temperature at Earth's center is again below the melting point, and therefore the inner core is probably solid.

Pressure

The graph at the center of Figure 15-6 illustrates how the pressure within Earth changes with depth. This pressure is caused by the weight of the layers above. Therefore, as you might expect, pressure increases with depth. Note that the scale on the vertical axis is marked in millions of times normal atmospheric pressure.

The top diagram illustrates Earth's internal structure. By looking at the labels and surface features, you can tell that the diagram represents a region of Earth from the middle of the Atlantic Ocean, across North America, and into the Pacific Ocean. The height of the Cascade Mountains is greatly exaggerated. It would be difficult to see the mountains if they were actually drawn to scale. Notice the arrows that show upwelling at the Mid-Atlantic Ridge and subduction at the trench. This is a part of the internal convection explained earlier.

The diagram also shows two ways to divide Earth's interior. From studies of earthquakes, scientists have learned much about Earth's crust, shown here by the dark line at the surface. (The thickness of this line has also been exaggerated.) The white, gray, and dotted sections are within Earth's mantle. Below the mantle are the outer and inner cores.

Another way to show the interior is based on motions of the lithospheric plates. In the plate tectonic model, Earth's top layer is the lithosphere. The lithosphere includes the crust, shown in black, and the upper part of the mantle, shown in white. The asthenosphere and the stiffer, more solid part of the mantle are above the outer and inner cores.

WHAT HAPPENS AT PLATE BOUNDARIES?

4: 2.1m, 2.1n

A tectonic plate moves over Earth's surface as a single unit. The rigid nature of the plates generally transfers force applied anywhere in the plate to the edges of the plates. Although large earthquakes can and do occur within the plates, earthquakes are more common where one plate meets another. When seismologists were able to record and locate earthquakes all over the globe, they noticed that there are distinct zones of crustal activity that stretch around the world. These zones of earthquake activity are also regions in which there are many volcanic eruptions and where tectonic mountain building is occurring. Compare Figure 15-7 with the map of plate boundaries in the *Earth Science Reference Tables*. (See Figure 3-2 on page 59.)

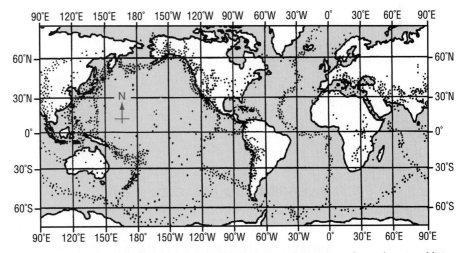

FIGURE 15-7. Each dot on this map represents an earthquake epicenter. Note that the greatest concentration of epicenters is in distinct bands of crustal activity, while in other regions earthquakes are relatively rare.

STUDENT ACTIVITY 15-3 —ZONES OF CRUSTAL ACTIVITY

1: SCIENTIFIC
INQUIRY 1
2: INFORMATION
SYSTEMS 1

Use the Internet to find lists of active volcanoes and earthquakes that provide latitude and longitude. Use this information to plot the position of at least 25 volcanoes or earthquake epicenters (choose one) on an outline map of the world. Compare the location of the features on your map with the features that others have plotted from different lists. Can you see similarities? Do the maps show any kind of pattern? Why do earthquakes and volcanoes often occur in the same places?

At the ocean ridges where new crust is forming, the plates are relatively thin. Deep earthquakes do not occur in this region because the upwelling material is relatively plastic, so it tends to flow and bend rather than break suddenly. Although there are many earthquakes at the mid-ocean ridges, their foci are shallow. However, where plates descend into the asthenosphere, the earthquake foci are deep. As they sink into Earth, plates remain rigid until they are heated enough to become plastic. Figure 15-4 on page 340 shows the concentration of shallow seismic foci near a mid-ocean ridge and the deep foci where subduction occurs near an ocean trench. Seismologists can locate sinking plates by plotting deep-focus earthquakes.

Types of Plate Boundaries

Plate boundaries are some of the most active zones of earthquakes and volcanic eruptions. These boundaries can be classified into three types based on the relative plate motions. Figure 15-8 illustrates the three types of plate boundaries. Refer to it and Figure 3-2, page 59, as you read the following sections.

DIVERGENT BOUNDARIES   The double lines on the map in Figure 3-2 on page 59 indicate rift boundaries. These are places such as the Mid-Atlantic Ridge where heated material rises toward the surface. At the same time, the lithosphere is spreading away from the ridge allowing magma to come to the surface, cool, and make new lithosphere. At the mid-ocean ridges new ocean floor spreads

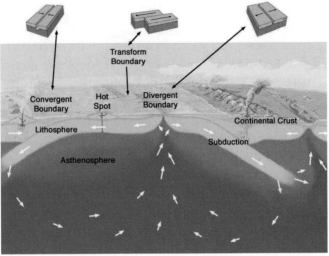

FIGURE 15-8. This diagram shows how the three plate interactions (divergence, convergence, and transform) are related to and driven by convection currents within Earth.

away from the plate boundary. For this reason these areas are called **divergent plate boundaries**. Recall that this is also a region of many small and shallow earthquakes. Ocean ridges are where heat energy is escaping from Earth's interior most rapidly.

CONVERGENT BOUNDARIES   The lines with black rectangles indicate subduction zones. These are the places where old, cool lithosphere sinks into Earth. Because the motion of the lithosphere is toward the plate boundary, zones of subduction are also known as **convergent plate boundaries**. Scientists often find continental and oceanic lithosphere meeting near subduction zones. You may recall that the rock of the continents is relatively low-density granitic rock and therefore resists subduction.

A good example of a convergent plate boundary is the Peru-Chile Trench along the western coast of South America. The Nazca Plate is ocean floor and mafic in composition. It is therefore easily drawn into Earth. As it descends, the edge of the plate bends and shifts causing earthquakes. Seismologists observe many earthquakes in these regions. Subduction zone earthquakes can be large because the brittle rock can absorb a great deal of energy before it breaks. The earthquakes can also have deep foci because the descending slab of lithosphere is cooler and less bendable than other rocks at the same depth. Descending plates stay rigid until they have absorbed heat. Observation of deep-focus earthquakes has given geologists an important tool that allows them to locate the plates as they move into Earth at subduction zones. Descending plates are the only place where earthquakes can originate from deep within Earth.

On the eastern side of the Peru-Chile Trench, most of the continental, felsic rocks are too light to be pulled down by subduction. They therefore bend, break, and pile up to make intensely folded and faulted mountains. In fact, the Andes Mountains along the western side of South America are second only to mountains of Asia in height.

You may recall from Chapter 5 that felsic minerals generally melt at a lower temperature than do mafic minerals. Therefore, as they absorb heat from their surroundings and from friction caused by plate movements, felsic rocks are the first to melt. The change to magma makes the rock fluid. Due to the low density of felsic magma, it rises toward the surface. For this reason, subduction zones, such as the one near the west coast of North America, are regions of volcanic activity. Unlike the mafic volcanoes of Hawaii, which generate long-lasting streams of lava, felsic volcanoes, tend to be more violent. Felsic magma is likely to contain water, which expands explosively when it reaches the surface. The 1980 eruption of Mount St. Helens in Washington State is a good example of the eruption of less-mafic lava.

The edges of two converging plates of continental crust resist subduction. This type of collision can produce a great mass of jumbled rock that builds the world's highest mountains. The Himalaya Mountains of Asia are the result of a collision between the north-moving Indian-Australian Plate and the giant plate that contains most of Europe and Asia. Measurements conducted in the Himalaya Mountains have shown ongoing uplift of several centimeters per year.

Volcanoes are not restricted to land areas. Volcanoes are common where the oceanic portion of one plate dives under another ocean floor segment. Partial melting of the descending plate may result in a curved line of volcanic islands known as an **island arc**. The Aleutian Islands, which extend westward from Alaska, and the islands of Japan are island arcs created at subduction zones.

TRANSFORM BOUNDARIES   Some plates do not converge or diverge. A **transform boundary** occurs when two plates slip past each other without creating new lithosphere or destroying old lithosphere as shown in Figure 15-8, on page 349. If you could straddle a transform fault for a long enough time, you would see one foot heading in one direction parallel to the fault and your other foot going in the opposite direction. The San Andreas Fault system in California is an excellent example of a transform boundary. In this area, the Pacific Plate is moving northwest with respect to the North American Plate. Motion along the fault is not continuous. At any place along the fault, the plates may be locked together by friction. When the force on the fault becomes great enough to overcome friction, the fault breaks suddenly and the plates move. This motion generates an earthquake. Visit the following Web site to watch animations of plate tectonic activity: *http://www.nature.nps.gov/geology/usgsnps/animate/pltecan.html*

Hot Spots

A **hot spot** is a long-lived source of magma within the asthenosphere and below the moving lithospheric plates. If the plates were not moving, we would observe repeated eruptions in a single location. Instead, we usually see a line of volcanoes in which the oldest volcanoes are at one end and the youngest are at the other end. Scientists interpret this pattern as evidence of the motion of a plate over a stationary hot spot under the lithosphere.

The Hawaiian Islands (Figure 15-9 on page 352) are an excellent example of volcanic eruptions over a hot spot. The oldest volcanoes in the Hawaiian chain are at the northwestern end of the chain. The volcanoes that produced the island of Kauai were active 3.8 to 5.6 million years ago. The Kauai volcanoes have not erupted since then. The major islands of Oahu, Molokai, and Maui were built by successive eruptions of volcanoes approximately 2.5, 1.5, and 1 million years ago, respectively. The youngest of the Hawaiian Islands is Hawaii, which gave its name to the whole chain of islands. Kilauea volcano on the island of Hawaii has been erupting continuously for nearly half a century.

If you did not know about plate motions, you would think that the source of lava was moving southeast. However, scientists now realize that the northwestward motion of the Pacific Plate has transported the ocean floor over the hot spot. Successive eruptions created new islands, while the older islands were carried northwest on the moving Pacific Plate.

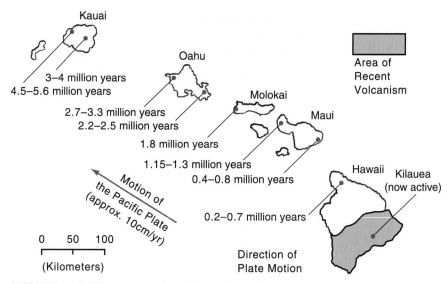

FIGURE 15-9. The progression of ages in the Hawaiian Islands shows that the Pacific Plate is moving to the northwest over a deep hot spot below the lithospheric plate. Kilauea volcano on the largest island (Hawaii) is currently active because it is above the hot spot.

STUDENT ACTIVITY 15-4 —GRAPHING HAWAIIAN VOLCANOES

1: SCIENTIFIC
INQUIRY 1
6: MODELS 2

Use the information in Figure 15-9 to draw a graph that shows ages of the Hawaiian Islands compared with their distance from Kilauea. The position of Kilauea is shown by a dark dot on the island of Hawaii. Then, from your graph, determine the rate at which the Pacific Plate is moving over the stationary hot spot.

The line of Hawaiian volcanoes did not begin at Kauai. Ocean-ographers (scientists who study Earth's oceans) have analyzed basalt from the Emperor Chain of islands and seamounts that extends from the Hawaiian Islands northwestward nearly to Alaska. The age of these islands and seamounts steadily increases in that direction to a maximum age of approximately 60 million years. It seems clear that the Hawaiian hot spot below the lithosphere has been supplying magma for at least the past 60 million years. Even at a pace of just a few centimeters per year, over a period of millions of years the lithospheric plate can move great distances. (See Figure 15-10.)

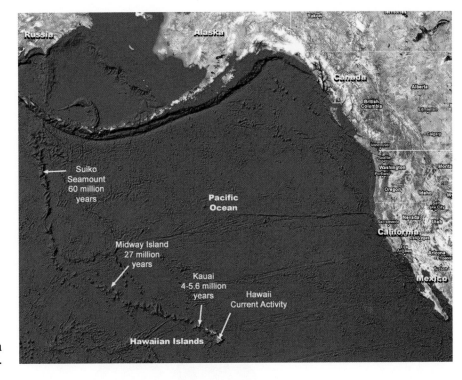

FIGURE 15-10.
A chain of islands and mountains on the seafloor extends from Hawaii toward the Aleutian Islands of Alaska. The increasing ages of volcanic rocks shows the movement of the Pacific Plate over a stationary hot spot.

DOES EARTH'S GEOGRAPHY CHANGE?

4: 2.1l, 2.1m, 2.1n

Scientists have used a wide variety of evidence to show how the continents have moved over millions of years. Geologists know that ocean basins are younger than the continents. This is because as new seafloor is created at the ocean ridges, old parts of the seafloor are drawn back into Earth's interior at the ocean trenches. The dense basaltic rock of the ocean floor is constantly recycling. However, the continents resist subduction due to their lower density. Large parts of the continents are composed of rock much older than the rocks found anywhere on the ocean floor.

Figure 15-11 on page 354, from the *Earth Science Reference Tables*, shows the evolution of Earth's surface over more than 300 million years. Notice that 359 million years ago North America was located along the equator. As time passed, North America moved north along with Africa and South America. In the past 200 million years, North America separated from Africa and Europe, which formed the North Atlantic Ocean. Africa and South America split apart more recently, forming the South Atlantic Ocean.

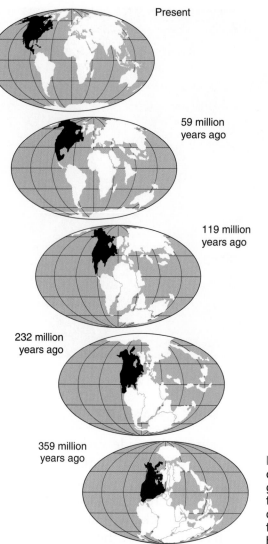

Present

59 million years ago

119 million years ago

232 million years ago

359 million years ago

FIGURE 15-11. Based on physical evidence, geologists have been able to reconstruct the positions of the major continents through most of Earth's history.

Scientists cannot be sure how the plates will move in the future. But, if present trends continue, the Atlantic Ocean will become wider as separation continues at the Mid-Atlantic Ridge. As the American Plates continue to push into the Nazca and Pacific plates, the mountains near the western edges of the Americas could grow higher. (Mountain heights also depend on how fast erosion takes place.) In a few tens of millions of years, movement along the San Andreas Fault will carry Los Angeles northward to a position near San Francisco. Visit the following Web site to watch animations of a variety of plate tectonic processes, including the motions of the continents through geologic time: *http://whs.moodledo.co.uk/mod/resource/view.php?id=972*

One important principle in geology is sometimes stated as, "The present is the key to the past." This means that events in the prehistoric past are likely to be similar to changes that you can observe today. It also means that if scientists understand the geological processes that are happening in modern times, they can better predict what will happen in the future.

CHAPTER REVIEW QUESTIONS

Part A

1. The edges of most lithospheric plates are characterized by

   (1) reversed magnetic orientation
   (2) low rates of weathering and erosion
   (3) frequent volcanic activity
   (4) S-waves that travel faster than P-waves

2. Why does oceanic crust sink beneath continental crust at subduction boundaries?
   (1) Oceanic crust has greater density.
   (2) Magnetic forces pull on the oceanic crust.
   (3) Continental crust is more mafic.
   (4) Continental crust is pulled by tidal forces.

3. The diagram below shows a portion of the boundary between the African and Arabian tectonic plates. Arrows show the directions of plate motions.

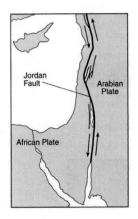

According to this diagram, which kind of plate boundary is the Jordan Fault?

(1) divergent   (3) convergent
(2) subduction   (4) transform

4. Which two tectonic plates are separated by a mid-ocean ridge?

(1) Indian-Australian and Eurasian
(2) Indian-Australian and Pacific
(3) North American and South American
(4) North American and Eurasian

5. Which graph below best shows the inferred density of Earth from the upper mantle to the lower mantle?

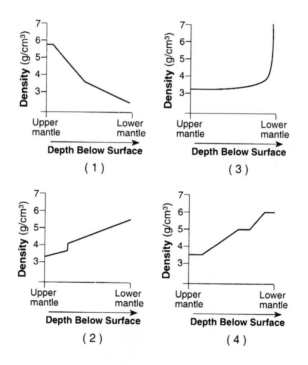

6. What is Earth's inferred pressure at a depth of 3500 km below the surface?

(1) 1.9 millions of atmospheres
(2) 2.8 millions of atmospheres
(3) 5500 millions of atmospheres
(4) 6500 millions of atmospheres

7. Which temperature is inferred to exist within Earth's plastic mantle?

(1) 2000°C   (2) 3000°C   (3) 5000°C   (4) 6000°C

8. Earth's outer core is best inferred to be

   (1) liquid, with an average density of approximately 4 g/cm³
   (2) liquid, with an average density of approximately 11 g/cm³
   (3) solid, with an average density of approximately 4 g/cm³
   (4) solid, with an average density of approximately 11 g/cm³

Part B

Base your answers to questions 9 and 10 on the diagram below. The diagram is centered on the South Pole and shows the movement of the continent of Australia. Each position is labeled with the period of the geologic past when Australia was located in various places.

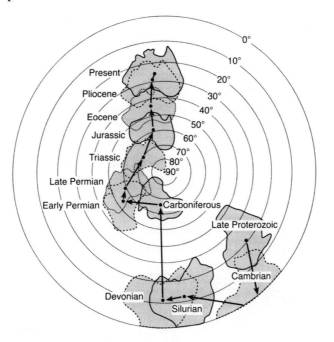

9. Why has the geographic position of Australia changed?

   (1) The gravitation of the moon pulls on landmasses of Earth.
   (2) The escape of Earth's internal heat energy drives convection.
   (3) Earth's rotation has moved the continent of Australia.
   (4) Changes in the tilt of Earth's axis moved Australia.

10. During which geologic period did Australia most likely have a tropical climate?

   (1) Cambrian   (3) Permian
   (2) Carboniferous   (4) Eocene

11. The diagram below is a profile of a part of the Mid-Atlantic Ridge. The motion of the crust, shown by the arrows, is occurring at the same rate on both sides of the ridge.

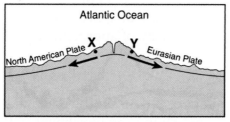

(Not drawn to scale)

Which statement about the magnetic orientation and age of the ocean floor at points *X* and *Y* is correct?
(1) The basalt at *X* is younger than the basalt at *Y*. Both locations have the same magnetic orientation.
(2) The basalt at *X* is older than the basalt at *Y*. Location *X* has reversed magnetic orientation, whereas *Y* has normal magnetic orientation.
(3) The basalt at *X* is the same age as the basalt at *Y*. Both locations have the same magnetic orientation.
(4) The basalts at *X* and *Y* are the same age. Location *X* has normal magnetic orientation, whereas *Y* has reversed magnetic orientation.

Part C

Base your answers to questions 12 and 13 on the map below, which shows the locations of four major earthquakes. (Two of them occurred at San Francisco.)

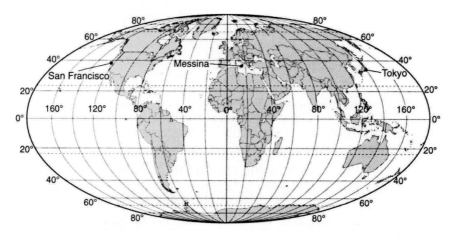

12. Explain how the locations of these earthquakes are related to the tectonic plates.

13. What process within Earth's asthenosphere is responsible for plate motions?

Base your answers to questions 14 and 15 on the diagram below. The diagram shows a plate boundary between Africa and North America 300 million years in the past. Arrows at *A, B, C,* and *D* represent relative crustal movements. *X* is a volcanic eruption.

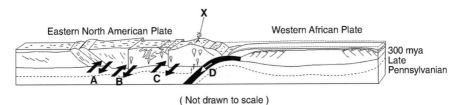

(Not drawn to scale)

14. Identify the type of plate motion represented by the arrow at *D.*

15. Identify the type of plate motion represented by the arrows at *A, B,* and *C.*

Base your answers to questions 16 through 20 on page 360 on the passage below and map on page 360. The passage describes the Gakkel Ridge found at the bottom of the Arctic Ocean. The map shows the location of the Gakkel Ridge.

The Gakkel Ridge

In the summer of 2001, scientists aboard the U.S. Coast Guard icebreaker *Healy* visited one of the least explored places on Earth. The scientists studied the 1800-km-long Gakkel ridge at the bottom of the Arctic Ocean near the North Pole. The Gakkel Ridge is a section of the Arctic Mid-Ocean ridge and extends from the northern end of Greenland across the Arctic Ocean floor toward Russia. At a depth of about 5 km below the ocean surface, the Gakkel Ridge is one of the deepest mid-ocean ridges in the world. The ridge is believed to extend down to Earth's mantle, and the new seafloor being formed at the ridge is most likely composed of huge slabs of mantle rock. Bedrock samples taken from the ocean floor at the ridge were determined to be the igneous rock peridotite.

The Gakkel Ridge is also the slowest-moving ocean ridge. Some ridge systems, like the East Pacific Ridge, are rifting at a rate of about 20 cm/year. But the Gakkel Ridge is rifting at an average rate of less than 1 cm/year. This slow rate of movement means that there is less volcanic activity along the Gakkel Ridge than along other ridge systems. However, heat from the underground magma slowly seeps up through cracks in the rocks at the ridge at structures scientists call hydrothermal (hot water) vents. During the 2001 cruise a major hydrothermal vent was discovered at 87°N latitude, 45°E longitude.

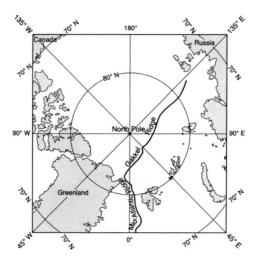

16. Make a copy of the diagram above and place an *X* at the location of the major hydrothermal vent described in the passage on page 359.

17. Describe the relative motion of the two tectonic plates that meet at the Gakkel Ridge.

18. What are the two tectonic plates at the boundary of Gakkel Ridge?

19. Identify one other feature, other than hydrothermal vents, often found at mid-ocean ridges that shows the escape of Earth's heat energy.

20. State two minerals that were most likely found in the rocks recovered from the Gakkel Ridge.

16 Geologic Hazards

WORDS TO KNOW

avalanche	hazard	mass movement	vent
caldera	landslide	tsunami	volcano
crater	liquefaction		

This chapter will help you answer the following questions:

1. How do geologic hazards put us in danger?
2. What are the dangers of earthquakes?
3. How can we prevent damage and injury in earthquakes?
4. What threats do volcanoes present?
5. What are the dangers of land failure?
6. How have humans contributed to geologic hazards?

WHAT IS A GEOLOGIC HAZARD?

4: 2.1l, 2.1t

People who live near the coast of southern California enjoy mild temperatures and beautiful mountain scenery. The climate is usually dry, but rain and snow in the nearby mountains provide freshwater and recreational opportunities. However, other aspects of life in California are not as favorable. If it rains hard, there can be flooding and landslides. This is because the mountains are young with steep, unstable slopes. In addition, the dry summer climate results in relatively few plants to hold back surface water and keep soil in place. California also has frequent forest fires and more destructive earthquakes than any other part of the United States except Alaska. Visit the following Web site to enter the USGS Geologic Hazards Team web site: *http://geohazards.cr.usgs.gov*

Many residents of coastal California do not know that the mild climate and mountain scenery are a result of active geologic forces. The San Andreas Fault system and other faults along which earthquakes occur are responsible for the mountains. If it were not for earthquakes, the mountains would not be there. The mild climate benefits from winds off the ocean that are stopped by the mountain barriers. If it were not for the mountains, the climate would not be as mild. It all relates to geology.

Most geologic events, such as uplift, weathering, and erosion, occur over a long time. Such slow events rarely cause danger to people. However, some geologic changes happen quickly, such as earthquakes, volcanic eruptions, landslides, and avalanches. **Hazards** are unsafe conditions that pose a threat of property damage, injury, or even loss of life. Potential events, such as earthquakes that might cause damage are natural hazards. Visit the following Web site to see the USGS Interactive Seismic Hazards Maps: *http://gldims.cr.usgs.gov/nshmp2008/viewer.htm*

Earthquakes

Plate boundaries are active earthquake zones. The only plate boundary within the continental United States is the western edge of the North American Plate. The boundary between the Pacific and North American plates runs through California and then off the Pacific coast from Oregon to Alaska. Although Alaska is one of the most seismically active regions in the world, earthquakes in California threaten more people.

Large earthquakes also can occur within continental plates at places where the plates seem to be breaking apart. Two of the strongest seismic events in American history took place in the nineteenth century, one in Missouri and the other in South Carolina. Figure 16-1 shows the parts of the United States that are considered most likely to experience major seismic events. Visit the following Web site to learn about the seismic zones in the New York City area: *http://www.sciencedaily.com/releases/2008/08/080821164605.htm*

Scientists know that when seismic waves pass from solid rock into loose sediment, the shaking becomes stronger. In 1985, there

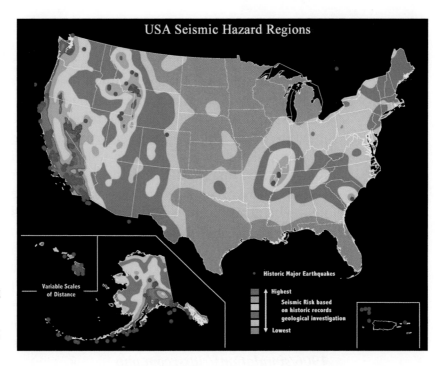

FIGURE 16-1. Alaska and California have the greatest risk of damaging earthquakes. However, New York State does occasionally have seismic events that cause damage.

was a major earthquake in Mexico. The quake did more damage in Mexico City than in Acapulco, even though Acapulco was closer to the epicenter. Acapulco is built on or close to solid rock. On the other hand, Mexico City is built on sediments of an ancient lake bed. Not only do thick sediments multiply the shaking of an earthquake, they are a less secure foundation for large buildings.

Seismologists sometimes say, "Earthquakes don't kill people, buildings do." People are rarely hurt by motion of the ground. The collapse of buildings is the major cause of death and injury in most earthquakes. This is especially true if the buildings have no reinforcement to hold them together. Wood-frame houses are among the best at withstanding earthquakes. In this kind of house, walls are held in place by other walls. The foundation, floors, walls, and roof of the house can be secured with bolts. Wood-frame structures absorb energy by bending and can go back into their original shape. Steel-frame buildings are also good absorbers of ground motion.

In January 2010, a magnitude 7 earthquake struck the Caribbean nation of Haiti, causing widespread devastation. At least 230,000 people lost their lives. Several factors contributed to the human impact of this event. The epicenter was very close to Haiti's largest city and the focus was relatively shallow. Furthermore, most of the buildings in Haiti were poorly constructed and could not withstand seismic shaking. Just a month later, an 8.8 magnitude earthquake struck Chile on the west coast of South America. Although the Chile event released about 500 times more energy than the Haiti earthquake, the death toll was much lower. Lives were spared because Chile has laws that require stronger buildings with reinforced walls. Visit the following Web site to watch USGS videos and animations: *http://education.usgs.gov/common/video_animation.htm*

In populated areas, fires often follow earthquakes. The fire that followed the magnitude-8 San Francisco earthquake of 1906 caused far more damage than was caused by the collapse of buildings. The ground broke and shifted as much as 6 m (20 ft), causing gas lines to break. Damaged and sparking electrical lines ignited the highly flammable natural gas escaping from broken gas lines. Water pipes, which could have provided water to fight fires, were also broken, making it impossible to save burning buildings. The fire burned for several days, destroying most of the city. Visit the following Web site to watch video simulations of shaking during the 1906 earthquake: *http://earthquake.usgs.gov/regional/nca/1906/simulations/classroom.php*

Fortunately, some builders and communities have learned from their mistakes. Studying cities that have experienced major earthquakes has helped engineers understand how to avoid future damage. When the Alaska oil pipeline was constructed in the 1970s, special bends and joints were added where the pipeline crosses known fault lines. Should an earthquake occur causing movement along the fault, these bends should prevent the pipeline from breaking.

Earthquakes can also trigger ground failure, such as landslides. Sediments that hold groundwater are another hazard. Saturated sediments can turn into a material like quicksand in the process of **liquefaction**. Strong shaking allows water to surround the particles of sediment, changing the sediment into a thick fluid. Buildings can tilt where the ground is weakened by liquefaction.

STUDENT ACTIVITY 16-1 —MAKE YOUR SCHOOL SAFER

1: ENGINEERING DESIGN 1

Prepare an engineering report for your school building. In it identify damage that might occur in a destructive earthquake. How much would it cost to repair the damage? Suggest ways to make the school safer and estimate the costs of these measures.

Dam failure is also a hazard associated with earthquakes. The shaking of the ground or a landslide can break dams. If a reservoir of water is held back by the dam, people who live downstream will be in danger from flooding. Table 9-1 lists some of the world's

Table 9-1. Selected Historic Earthquakes

Location	Date	Magnitude	Notes
AROUND THE WORLD			
Shensi, China	1556	~8 (est.)	Most deadly natural disaster known. Most people died in the collapse of homes.
Valdivia, Chile	1960	9.5	Occurred in Pacific Ocean, causing tsunamis. Largest earthquake measured with seismographs.
Haicheng, China	1975	7.5	Predicted by scientists; city was evacuated, saving many lives
Tangshan, China	1976	7.6	Worst modern natural disaster. Prediction failed.
Mexico City, Mexico	1985	8.1	Worst damage in Mexico City, 200 miles from epicenter.
Sumatra, Indonesia	2004	9.1	Tsunamis displaced more than 1 million people and killed more than 280,000.
Port-au-Prince, Haiti	2010	7.0	Killed more than 230,000 people
Concepción, Chile	2010	8.8	Spawned a small tsunami
UNITED STATES			
New Madrid, MO	1811–1812	~8 (est.)	Mississippi River reversed flow briefly.
Charleston, SC	1886	~7.3 (est.)	Destroyed most of the buildings in the city. Largest event along the East Coast.
San Francisco, CA	1906	~8.3	Water lines broken. Fires caused most of the damage.
Prince William Sound, AK	1964	9.3	Large tsunamis affected entire Pacific Basin. Slope failure in Anchorage.
Los Angeles, CA	2008	5.4	Shaking lasted 5 to 10 seconds. Little damage reported.
NEW YORK STATE			
Massena	1944	~5.8 (est.)	Largest in NYS. Chimneys fell; water lines broke.
Au Sable Forks	2002	5.1	Felt throughout state. Damaged some local roads.

greatest earthquakes as well as United States and New York seismic events.

Some earthquakes, such as the Chile quake of 2010, cause a series of waves called a **tsunami** (sue-NAHM-ee). The most destructive tsunamis are caused by a sudden motion of the ocean bottom or an underwater landslide caused by an earthquake. In the open ocean, tsunamis may travel 1000 km/h (600 mph) as a gentle swell that would be hard to notice on board a ship at sea. At this speed, a tsunami can cross a major ocean in a few hours. When a tsunami moves into an open but shallow bay, its energy becomes more concentrated and the water can build high waves. On shore, the first sign of these waves may be a sudden drop in sea level. People have been drawn to the shore by the sudden withdrawal of the sea. They may not realize that this is a sign that a giant wave is coming that could sweep them away. Tsunamis cause great damage and loss of life in coastal locations. The city of Hilo in Hawaii has experienced several destructive tsunamis. Nations around the Pacific Ocean now have a tsunami early warning system to alert people who live in coastal areas of approaching danger.

Unfortunately, no such warning system existed in the Indian Ocean when a magnitude 9.2 earthquake occurred about 160 km (100 mi) off the Indonesian island of Sumatra in 2004. (See Figure 16-2.) Damage from collapsing structures and other land-based events was small compared with the devastating tsunamis that rolled across the Indian Ocean killing an estimated 230,000 people.

FIGURE 16-2. The Indian Ocean (Great Sumatra–Andaman) earthquake of December 26, 2004 affected areas such as this on the west coast of Indonesia. The magnitude 9.2 event caused a tsunami that flooded nearby coastal towns in Sumatra to a depth of as much as 30 m (100 ft).

Tsunami-related deaths occurred from coastal Indonesia to Africa, some 6000 km (4000 mi) from the epicenter.

The Indian Ocean earthquake was generated as the Indian-Australian tectonic plate was suddenly pulled approximately 15 m (50 ft) under the Eurasian Plate. A 10-year-old British student vacationing along the coast of Thailand is credited with saving as many as 100 lives. Tilly Smith had studied tsunamis in her suburban London school. She recognized the ocean withdrawing from the beach as a sign that a massive tsunami was approaching. Unfortunately, in other locations people did not understand this warning of a tsunami.

Earthquake Preparedness

Reducing the risk of injury or loss of property is especially important in places that have a history of damaging earthquakes. There are things people can do to prevent injuries and property loss. The following measures can be split into two categories: preparing for an earthquake and what to do during an earthquake. The following are examples of advanced preparations:

- Select a home site on or close to solid bedrock.
- Select a location that is not near a steep hill, ocean or ocean-bay shoreline, or downstream from a reservoir.
- Be sure your home meets local building codes.
- Know how to shut off gas, electricity, and water.
- Avoid storing heavy objects on high shelves.
- Store some food and freshwater in your home, as well as a battery-operated radio.
- Keep emergency telephone numbers in handy locations.
- Know where to find medical supplies and the location of the nearest doctor or hospital.
- Learn how to help any family member with special needs.
- Plan and rehearse what to do in case of an earthquake.

After a major earthquake, your help may be needed. Below is a list of what to do during and after an earthquake:

- Find safe shelter, undamaged food supplies, and clean water.
- Listen to the radio for information and instructions.

- Until help arrives, respond to the needs of people who are injured.
- Assist official emergency services as requested, but do not get in their way.
- Prevent further injuries by identifying unsafe structures, broken objects, chemical spills, and similar hazards.
- If possible, turn off gas, water, and electric supplies.
- Establish a way to contact or meet family members.

In an earthquake, finding protection depends on where you are. If you are outside, it is best to stay away from buildings or trees, since they may fall. If you are in a building, the best protection is probably under a strong object, such as a desk or a table that can protect you from falling debris. Small earthquakes last only a few seconds, giving you little time to move very far. Large earthquakes last longer, but the shaking can make it impossible to walk. Furthermore, there is immediate danger of building collapse or falling objects. So relocating is likely to be more dangerous than just ducking under nearby furniture. Hold on to keep yourself protected as things bounce around and cover your head to shield it from falling objects. Remember that aftershocks often occur for many days following a major earthquake. Visit the following Web site to learn about the Seven Steps to Earthquake Safety: *http://www.earth quakecountry.info/roots/seven_steps.htm*

STUDENT ACTIVITY 16-2 —DEVISING AN EARTHQUAKE PREPAREDNESS PLAN

1: ENGINEERING
DESIGN 1
4: 2.1l
7: STRATEGIES 2

Work with your family or classmates to devise an emergency plan to use in the event of a damaging earthquake. Brainstorm with them to take into account unique characteristics of where you live and the needs of people around you. Create a list of things you can change to make your home more earthquake-safe.

Volcanoes

A **volcano** is a landform on Earth's surface where molten magma has erupted. The source of magma is deep within Earth where the temperature is above the rocks' melting temperature. The temperature at which rock melts depends on its mineral composition and pressure on it. Felsic rocks melt at a lower temperature than mafic rocks. As pressure increases, so does the melting temperature. Fluid magma moves toward the surface through cracks or zones of weakness in the overlying rock. When the magma reaches the surface and releases gases into the atmosphere, it is called lava.

Figure 16-3 is a cross section, or profile view, of a volcano. In this diagram you can see the magma chamber with vents that lead to the surface. Note the layering of ash and lava inside the mountain. These layers are built up by repeated eruptions. Explosive eruptions of some volcanoes leave a bowl-shaped depression at the top of the mountain called a **crater**.

TYPES OF VOLCANOES   The cooling of magma at the surface builds volcanic features around the **vent** where the lava comes to the surface. Scientists recognize four types of volcanoes based on their shapes: shield volcanoes, cinder cones, composite volcanoes, and lava plateaus.

Some volcanoes erupt quietly. Kilauea on the island of Hawaii has continuously vented rivers of lava for several decades. The basaltic lava that feeds Kilauea is very hot and contains little gas, making the lava very fluid. Repeated eruptions of hot, fluid, basaltic magma build a broad structure with gently sloping sides

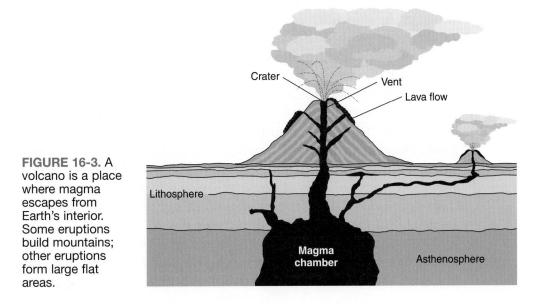

FIGURE 16-3. A volcano is a place where magma escapes from Earth's interior. Some eruptions build mountains; other eruptions form large flat areas.

known as a shield volcano. The Hawaiian Islands contain shield volcanoes as much as 100 km (60 mi) across such as Mauna Kea on the Big Island of Hawaii.

Cinder cone volcanoes are usually small features. They are built by cooler lava that was blown into the air, fell back to Earth, and hardened into a pile around the vent.

Composite volcanoes are mounds built up by alternating lava flows and layers of ash. Mount St. Helens is a good example of a composite volcano. Volcanoes such as Mount St. Helens vent lava that is felsic in composition. Felsic lavas are recycled continental rocks. They contain more silicate minerals as well as dissolved gases such as water vapor, carbon dioxide, and sulfur dioxide. As felsic lava comes to the surface, decreasing pressure causes the gases to expand explosively. Such volcanic eruptions release large quantities of volcanic ash and toss larger objects known as blocks and bombs into the air. These are considered the most dangerous eruptions.

Before it erupted in 1980, Mount St. Helens in Washington State was a nearly perfect volcanic cone almost 3000 m (10,000 ft) high. The mountain was known to be an active volcano. In fact in 1975, scientists from the United States Geological Survey predicted that it would probably erupt before the turn of the century.

Signs of activity began in late March of 1980 with many small earthquakes generated by underground movement of magma. A bulge in the northern slope of the mountain grew to about 100 m (300 ft). On May 18, it broke loose and rushed down the northern slope of the mountain in a great, gas-charged cloud. (See Figure 16-4.) The loss of pressure within the mountain resulted in a blast of hot ash and volcanic gases that filled Spirit Lake, a popular fishing resort on the north side of the mountain. About 1 km^3 (35 million ft^3) of the mountain was blown away. Hundreds of miles away, ash fell several centimeters deep. The mountain lost nearly a quarter of its height, and 60 people lost their lives. Since the 1980 eruption, a small dome has grown inside the new crater. This dome could grow to the original height of the mountain, or it could lead to another eruption in the future.

The eastern part of the state of Washington is covered by hundreds of meters of flat layers of successive fluid lava flows that created lava plateaus. The lavas that formed these plateaus were so

FIGURE 16-4. The eruption of Mount St. Helens in 1980 was the most dramatic eruption in the continental United States in recorded history.

hot that they flowed over the surface almost like water before they hardened into basalt.

Some volcanoes form a **caldera**. This is a large, bowl-shaped depression that forms when the top of the volcano collapses into the emptied magma chamber. Crater Lake in Oregon is a very large caldera that has filled with water, making a large, round lake where the top of the mountain used to be.

Volcanoes are sometimes classified as active or dormant. If scientists see evidence of recent activity or if they see steam rising out of a volcano, it is considered active. However, dormant volcanoes can suddenly erupt, showing how difficult it is to classify them reliably.

Volcanoes as Hazards

Figure 16-5, on page 372, shows that volcanoes, like earthquakes, tend to occur near plate boundaries. Notice the way that active volcanoes nearly surround the Pacific Ocean. That region is called the Ring of Fire. Many of these volcanoes occur near ocean trenches. Subduction zones are especially dangerous because this is where granitic (felsic) rocks are pulled into Earth's interior. With their low melting temperature, low-density, high-viscosity magma, and considerable gas content, subduction zone volcanoes are the most dangerous volcanoes.

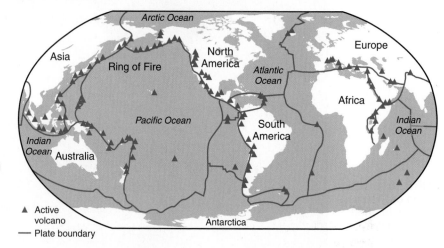

FIGURE 16-5. Like earthquakes, most volcanoes are concentrated along plate boundaries, such as the "Ring of Fire" that surrounds the Pacific Ocean.

▲ Active volcano
— Plate boundary

STUDENT ACTIVITY 16-3 —ADOPT A VOLCANO

2: INFORMATION SYSTEMS 1

Select a famous volcano. Be sure that your teacher approves the volcano you have chosen. Prepare a report about your volcano's activity. Please give your source(s) of information in the form of a bibliography.

In real estate, it is said that the value of a property depends on three factors: location, location, and location. The same factor(s) will determine your vulnerability to volcanic hazards. The first question is, are there volcanoes or volcanic rocks in your area? If not, the likelihood of danger is low. If there is local evidence of volcanoes, answer the next question. How close are you to a volcano that is or might become active? If the volcano is within a few tens of kilometers (10 to 20 mi), a third question comes into play. Is your home, school, or place of work in a valley connected to the volcano?

Volcanoes cause damage in several ways. Sometimes, lava flows out of a volcano and runs downhill into valleys destroying anything in its path. However, most lava flows are slow enough that people can usually escape them although buildings may be destroyed. A greater threat is gas-charged flows of hot ash that can descend from a volcano at 100 km/h (60 mph) or more. It was a flow of hot ash that killed observers and flattened forests around Mount St. Helens in 1980.

Some volcanoes discharge poisonous gases that can suffocate people and animals in nearby lowland areas. Eruptions can also release choking dust, and trigger landslides and mudflows that cause loss of life and property. In early 2010, an eruption of the volcano Eyjafjallajokull (EYE-a-fyat-la-jo-kutl) in Iceland spewed a mixture of water vapor and ash high into the atmosphere. Winds blew the ash cloud over northern Europe, disrupting air traffic for six days. No one was injured, but many people were inconvenienced.

Volcanoes sometimes show signs of an approaching eruption. The best way to protect yourself from their dangers is to move far enough away to avoid direct effects such as lava flows and gas clouds, and high enough to escape floodwaters.

Mass Movements

As tectonic forces build mountains, erosion powered by gravity wears them down. Nearly all erosion starts with earth materials at a high elevation and moves them to a lower elevation. Wherever the ground is too steep for friction to hold rock and soil in place, there is danger that it will move downslope. **Mass movement** is the motion of soil or rock down a slope without the influence of running water, wind, or glaciers.

Some soil movement happens slowly and is only noticeable over time. These slow movements are most common in clay-rich soils, particularly when they become saturated with water. If roads or buildings are constructed along these slopes, the structures can lean and break apart as downslope motion stresses and moves them.

Sometimes there is the rapid, downslope movement of rock and soil known as a **landslide**. In unpopulated areas, mass movement is of little concern. But each year downslope mass movement breaks up or covers roads, causes damage to property, and occasionally causes injury or loss of life.

Landslides are often triggered by water seeping into the ground. Clay minerals in soil can absorb many times their dry weight in water. Clay also offers little resistance to gravitational force. If the soil is composed mostly of moist clay, it can even slide down relatively gentle slopes. In the spring of 1993, a 200,000-m² (55-acre) area near the bottom of Tully Valley south of Syracuse, New York, slid downhill, covering a road and damaging three homes. The slope above the slide was steep, but held in place by bedrock.

FIGURE 16-6. In April of 1993, 55 acres of farmland slid into Tully Valley and across Tully Farms Road in New York. The road was covered to a depth of 3 m (10 ft). It destroyed three houses and caused the evacuation of four others.

However, the valley had been the site of a large post-glacial lake that left deposits of soft clay. When the clay was saturated by groundwater it moved downslope. (See Figure 16-6.)

An **avalanche** is a large amount of snow and rock that moves rapidly downhill over a steep slope. Avalanches can move at 150 km/h (100 mph). Uprooted trees in mountain valleys provide evidence of past avalanches. Figure 16-7 shows an avalanche shed built to protect a mountain road in Colorado.

The best way to protect yourself and your property is to be aware of where landslides and avalanches have occurred before.

FIGURE 16-7. Repeated avalanches across U.S. Highway 550 along Red Mountain in the San Juan Mountains of Colorado led to the construction of this snow shed to protect the road.

People should avoid building on or below steep or unstable land. Be aware of avalanche dangers when skiing or traveling in mountain areas in the winter or early spring.

Engineering Decisions and Geologic Hazards

Some hazards are a result of poor decisions by builders. When structures are built, design judgments are made about cost and safety. For example, the bridge on the New York State Thruway over Schoharie Creek washed out in a flood in 1987. Ten people were killed when their vehicles washed into the raging creek. An investigation found that the bridge supports were not sunk deep enough to resist flood erosion.

STUDENT ACTIVITY 16-4 —EVALUATING SCHOOL PROPERTY

1: ENGINEERING
DESIGN 1

Construct a map of your school grounds and identify places where ground failure could occur. Suggest preventive measures that could be taken. Determine the costs and benefits of your proposed preventive measures.

CHAPTER REVIEW QUESTIONS

Part A

1. Which location is most likely to experience a severe earthquake?

    (1) the east coast of North America
    (2) the east coast of Australia
    (3) the west coast of Africa
    (4) the west coast of South America

2. Sometimes one destructive natural event is caused by another event. Which of the following is most likely?

    (1) a thunderstorm caused by a landslide
    (2) a tsunami caused by an earthquake
    (3) a hurricane caused by a landslide
    (4) lightning and thunder caused by an earthquake

3. After a geological disaster, some homes were found to have severe roof damage with no other significant damage. What geological feature is most likely nearby?

   (1) an ocean
   (2) a volcano
   (3) a geological fault
   (4) a large-scale rock fold

4. Which kind of rock is most abundant in the Hawaiian Islands?

   (1) rhyolite   (3) basalt
   (2) granite   (4) gabbro

5. Experience shows that wooden homes are more resistant to earthquake damage than homes made from other materials. Why is this true?

   (1) Wood has a high hardness on Mohs' scale.
   (2) Wood is more dense than stone or brick.
   (3) Wood frames can bend and still hold together.
   (4) Wood does not conduct heat energy as well as stone or brick.

6. To avoid hazards related to volcanic eruptions, which is the best place to build a house?

   (1) along the Pacific Ocean
   (2) on New York's Long Island
   (3) in a location that has fine-grained igneous rock
   (4) where the bedrock is mostly basalt and rhyolite

7. If your area has many earthquakes, where is the safest place to build a home?

   (1) on the bank of a river
   (2) on a thick layer of sediment
   (3) at the base of a weathered cliff
   (4) on solid bedrock

8. Which location is most likely to have a volcanic eruption?

   (1) New York State
   (2) the east coast of South America
   (3) Central Australia
   (4) Iceland

9. A certain volcano is high enough that large glaciers and snowfields cover its top. Which of the following is likely to cause the most property loss, injury, and deaths in an eruption of this volcano?

(1) flooding
(2) thunder
(3) violent shaking of the ground
(4) people falling into cracks in the ground

10. Which geographic location is surrounded by a zone of mountain building, volcanoes, and frequent earthquakes?

(1) North America
(2) New York State
(3) the North Pole
(4) the Pacific Ocean

Part B

Base your answers to questions 11 through 14 on the map below, which shows the risk of seismic damage in the United States.

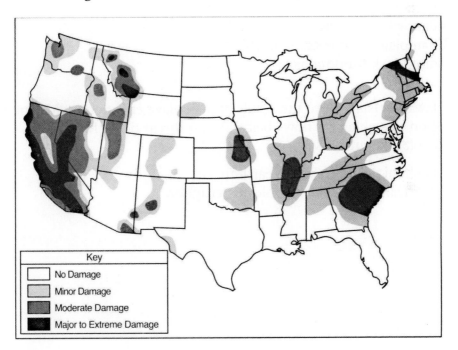

Key	
	No Damage
	Minor Damage
	Moderate Damage
	Major to Extreme Damage

11. The map shows that the greatest danger of future earthquakes is in states near

(1) divergent plate boundaries
(2) convergent plate boundaries
(3) transform plate boundaries
(4) inactive plate boundaries

12. Which city in New York State seems to have the greatest risk of earthquake damage?

    (1) Binghamton   (3) Plattsburgh
    (2) Buffalo   (4) Elmira

13. What is the most probable reason earthquakes are considered more likely in South Carolina than in Florida?

    (1) South Carolina is near a plate boundary
    (2) South Carolina has many active volcanoes
    (3) Florida has no history of large earthquakes
    (4) Florida has a cooler climate

14. Which New York landscape region is the *least* likely to have a major earthquake?

    (1) Allegheny Plateau
    (2) Atlantic Coastal Plain
    (3) Champlain Lowlands
    (4) Adirondack Mountains

15. The Hawaiian Islands are the result of many lava eruptions, with one lava flow resting on unstable lava flows below. What threat does this pose to California, 4000 km (3000 mi) away?

    (1) a tsunami   (3) an earthquake
    (2) a hurricane   (4) a volcanic eruption

Part C

16. The picture shows road damage found by a driver in California. The damage occurred during a dry season and was not related to weather events.

What sudden natural event probably caused this road damage?

17. Students read an article in their local newspaper stating that a major earthquake can be expected to affect their region within the next year. As a result, the students decide to help prepare their homes and families for this expected earthquake.

State three specific actions the students could take to reduce injury or damage from an earthquake in their home?

18. What part of the United States is in the greatest danger from volcanic eruptions?

19. Volcanoes damage homes and other structures in many ways. Some of those ways do not involve exposure to molten rock. Name one non-lava danger.

20. While on a hike, some students discovered a pile of rubble that included broken trees, but contained no large rocks. What geological hazard probably caused this destruction?

UNIT 6

Revealing Earth's History

Perhaps you have read some of the Sherlock Holmes mysteries written by Sir Arthur Conan Doyle. In each story, Holmes's careful observations and attention to detail help him solve mysteries that at first seem unsolvable. Geologists are often faced with mysteries.

To reveal Earth's long and complex history, geologists must learn to act and think like detectives. Written records go back only a tiny fraction of Earth's 4.6-billion-year history. People began to keep records only about 10,000 years ago. Therefore, historical records are of little help in telling Earth's story. To solve this mystery, a geologist must interpret the much older record written in the rocks.

Each layer in the Grand Canyon, pictured here, has a story to tell. Geologists have learned to read the clues in this rock record. The information contained in the rocks provides clues to the geology, the climate, and the animals and plants that lived in an area. The clues are not always easy to understand. Geologists who learn to read them can sometimes make discoveries that might astonish the legendary Holmes himself.

17

Sequencing Geological Events

WORDS TO KNOW

absolute age	isotopes	radiometric dating
decay product	original horizontality	relative age
decay-product ratio	outcrop	superposition
geologist	radioactive	unconformity
half-life period	radioisotope	uniformitarianism
inclusion		

This chapter will help you answer the following questions:

❶ Why do geologists use bedrock features to determine the order of geologic changes?

❷ How can we construct a geologic history based on observations of an outcrop?

❸ How can the decay of radioactive elements be used to determine the absolute age of rocks and other objects?

HOW CAN WE DETERMINE THE SEQUENCE OF EVENTS?

4: 1.2j

Scientists who study the origin, history, and structure of Earth are **geologists**. One fact geologists need to know is the age of rocks. There are two ways to communicate age. **Relative age** is used to express the order, or sequence, of events. It tells us that some

FIGURE 17-1. If we know when the tree was cut down or died, its rings can be used to establish numerical or absolute ages of events.

things are older or younger than others, or that certain events happened before or after other events. For example, you must get to school before you can go to class. This does not tell us when you got to school or how long you were there before class began, but it does put these two events in sequence.

Sometimes a sequence is not enough. You may want to know exactly when or how long ago an event took place. This is known as numerical, or absolute, time. If you say that Earth and the solar system formed about 4.6 billion years ago, you are expressing an absolute age. **Absolute age** always includes a number and a unit of time, such as years, days, or seconds. Scientists sometimes use tree rings (Figure 17-1) to establish numerical age. For example, if a large tree is cut down in the year 2010 and the fifth ring from the outside is very narrow, that usually indicates that 2005 was an unusually dry year.

STUDENT ACTIVITY 17-1 —RELATIVE AND ABSOLUTE TIME

1: MATHEMATICAL
ANALYSIS 1, 2
6: MODELS 2
6: MAGNITUDE AND
SCALE 3

For this activity, you will work in laboratory groups. Your teacher will set out a variety of objects. Divide your paper into four columns. List the name of each object in the first column. In the next column, estimate the absolute age of each object. The age of some objects will be easy to estimate, while the age of others will probably be uncertain. In the third column, indicate the reliability of your estimate by assigning a margin of error. (For example, if you are confident that the true age is within 10 years of your estimate, the margin of error will be + or − 10 years.) In the last column, record the method your group used to estimate the age of each object. After you have filled in the four columns, list the objects by their relative age, from youngest to oldest.

Geologists have several guiding principles to help them interpret geologic history. With these ground rules and the record preserved in Earth's rocks, they have been able to piece together a remarkable account of Earth's history.

Uniformitarianism

Geological features sometimes show the results of catastrophic events that happened in the past such as volcanic eruptions and continental glaciers. Geologists have adopted uniformitarianism as one of their guiding principles. **Uniformitarianism** is the idea that the geological processes that happened in the past are generally similar to processes we can observe today. This does not mean that all geological changes are slow and steady. For example, erosion caused by most streams is very slow except when the streams are in flood. Those limited flood events do most of the erosional work of streams and rivers. Most volcanic eruptions are sudden events that are not very frequent. Scientists can understand most of the features in the geological record by observing similar events in the present.

Superposition

Layers of rock are formed by a series of events. In most places, the lowest layer of sedimentary rock is the oldest. After all, the lowest layer must be in place before younger sediments can be deposited on top of it. The principle of **superposition** states that unless rock layers have been disturbed, each layer is older than the layer above it and younger than the one below it. Therefore, when geologists assign relative ages to rocks, the oldest rocks are on the bottom and the layers become progressively younger toward the top.

Figure 17-2, on page 384, is a profile of the rock layers at Niagara Falls. Each rock layer is named for the location where it can be observed and studied. For example, the hard cap rock at the top of the falls is Lockport Dolostone, which was deposited 420 million years ago. If you want to observe this layer in the field, there is a bedrock exposure near Lockport, New York. The first layer at Niagara Falls is the layer on the bottom, the Queenston Shale, which was deposited 460 million years ago. After this layer was deposited,

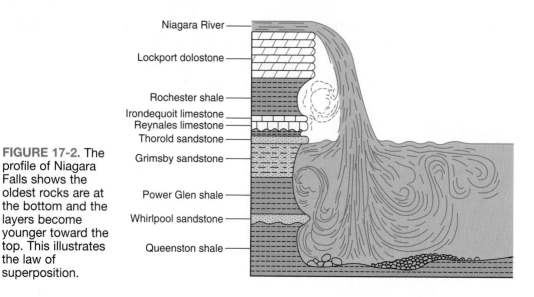

Niagara River

Lockport dolostone

Rochester shale

Irondequoit limestone

Reynales limestone

Thorold sandstone

Grimsby sandstone

Power Glen shale

Whirlpool sandstone

Queenston shale

FIGURE 17-2. The profile of Niagara Falls shows the oldest rocks are at the bottom and the layers become younger toward the top. This illustrates the law of superposition.

sand was deposited that became the Whirlpool Sandstone. One by one, each layer was deposited before the layer above it.

Superposition does not always apply. In some places, folding has overturned the rocks or faulting has pushed older layers on top of younger layers. These exceptions to the principle of superposition will be explained in the next section. However, if you are looking at rock layers or a diagram of rock layers, unless you see evidence for overturning or faulting you should assume that the principle of superposition can be applied. Visit the following Web site to see pictures of and read about the ages of the rock layers in the Grand Canyon: *http://www.nature.nps.gov/geology/parks/ grca/age*

Original Horizontality

Sedimentary rock is usually deposited in layers. In some places, the layers are tilted rather than horizontal. The principle of **original horizontality** states that no matter the present angle or orientation of sedimentary layers, it is almost certain that the layers were deposited horizontally and were tilted after deposition. The cause of the tilting could be folding or an uneven regional uplift. The tilted and folded layers in Figure 17-3 were originally horizontal. Forces within Earth pushed them into their present position.

FIGURE 17-3.
These layers were originally deposited in horizontal layers. Mountain-building forces tilted and bent them into their present shape.

Figure 17-4 shows the sequence of events that formed a bedrock outcrop. An **outcrop** is a place where bedrock is exposed at Earth's surface. In diagram 1, sandy sediments washed into a curved basin. Most sedimentary rocks begin as sediment deposits in water. The low spots in the basin are filled first with sediment. Therefore, no

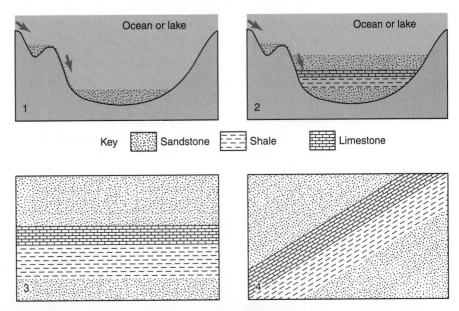

FIGURE 17-4. When sediments are deposited in a basin, they accumulate as flat, horizontal layers, as shown in diagrams 1 and 2. This occurs no matter the original shape of the basin. Diagrams 3 and 4 represent a later time and a bedrock outcrop that is from the center of diagram 2. When geologists see tilted sedimentary layers as in diagram 4, they usually infer that the layers were deposited flat and horizontal.

matter what the shape of the basin, the layers of sediment are flat and horizontal. Diagram 2 shows a sequence of layers that are horizontal in spite of the curvature of the basin. Diagram 3 represents an outcrop of rock layers from the depositional basin. In diagram 4, the layers have been pushed up on one side (or down on the other) to become tilted layers in the outcrop.

Inclusions

Geologists sometimes find that bedrock contains and surrounds pieces of a different kind of rock. An **inclusion** is a piece of one kind of rock enclosed in another rock. This can happen in at least two ways. First, an intrusion of basaltic magma following a zone of weakness through sedimentary rocks can pick up fragments of the sedimentary rock. When the magma solidifies it contains and surrounds pieces of the rock it invaded. Second, consider a region in which granite weathers to granite boulders sitting on granite bedrock. If layers of sediment are deposited, the sediments will surround the granite boulders. The result could be sedimentary rocks that contain granite boulders. Figure 17-5 shows an inclusion of sandstone rock surrounded by granite.

Whenever a body of rock contains inclusions of another rock, the inclusions are older than the surrounding rock. Some of the oldest masses of rocks found on Earth contain inclusions. This tells us that another rock unit existed before these very old rocks. Unfortunately, it may not be possible to determine how much older the inclusions are than the main body of rock, but the sequence can be inferred. Particles of sediment can be thought of as inclusions in sedimentary rock. Most sedimentary particles are the weathered

FIGURE 17-5.
It is clear that this fragment of sandstone was surrounded by magma when the granite was still a hot liquid. Therefore, the sandstone inclusion must be older than the granite that surrounds it.

remains of an older rock. Therefore, grains of sand in sandstone or pebbles in conglomerate are actually older than the rock they now occupy.

Cross-Cutting Relationships

When a fault appears in an outcrop, the fault must be younger than the rocks it cuts through. Some rocks also contain igneous intrusions. Cross-cutting means that something, such as the two white quartz veins in Figure 17-6, goes through previously existing rock. The cross-cutting feature, usually a fault or an intrusion, is always younger than the rock in which it is found. After all, the rocks must be there before they are faulted and before magma cuts through them.

FIGURE 17-6. The two white quartz veins were injected as molten material while the pre-existing darker rock was a solid. The thicker, vertical vein is offset by the horizontal vein. So the horizontal vein was probably injected later along a fault surface.

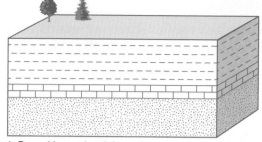

1. Deposition and rock formation

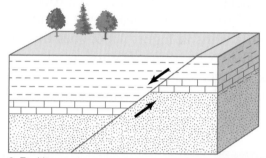

2. Faulting

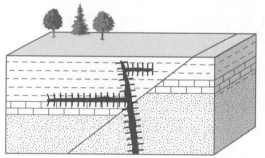

3. Intrusion

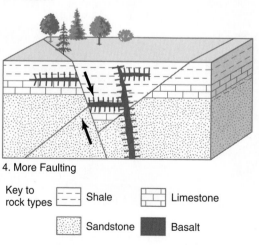

4. More Faulting

Key to rock types

- - - Shale
[bricks] Limestone
[dots] Sandstone
[black] Basalt
[cracks] Zone of contact metamorphism

FIGURE 17-7. The cross-cutting relationships in this series of diagrams show that faults and intrusions must be younger than the rocks in which they are found. The layers in diagram 1 must exist before they can be faulted as shown in diagram 2. The layers and the fault are crossed by the intrusion in 3, so both must be older than the intrusion. The second fault in diagram 4 is younger than everything it cuts through: sedimentary layers, the intrusion, and the original fault. Notice the zones of baked rock (contact metamorphism) that formed along the magma intrusions.

Figure 17-7 represents changes over time in the bedrock at a single location. This series of diagrams illustrates cross-cutting relationships. The layers shown in diagram 1 are offset by the fault in diagram 2. Therefore, the layers are older than the fault. Both rock layers and the fault cut by basaltic magma in diagram 3 are older than the magma. The fault shown in diagram 4 cuts through the sedimentary layers, the original fault, and a branch of the basalt intrusion. Therefore, all three of these features are older than the second fault.

Contact Metamorphism

Metamorphism includes the various changes that occur when a rock is subjected to extreme heat and/or pressure. Contact metamorphism occurs in narrow zones next to intrusions of molten magma. *Intrusion* is an internal process. Hot magma squeezes into cracks and zones of weakness; the magma cools as it passes its heat energy to the nearby rock. This creates a zone of contact metamorphism along the intrusion. Because the magma is surrounded by pre-existing rock, the zone of contact metamorphism extends from the intrusion in all directions.

Extrusion occurs at the surface; it is external. While rock below an extrusion is changed by contact metamorphism, there is no rock on top to be changed. Sometimes geologists find a layer of igneous rock that has not changed the rock layer immediately above it. This is evidence that the layer above was deposited after the magma cooled.

Figure 17-8, on page 390, shows layers of sedimentary rock that are intruded by magma that also flowed, as lava, onto the surface as an extrusion. Later, a new layer of sedimentary rock was deposited on top of the extrusion. From the final diagram, it is clear that the top layer is younger than the lava. It is also evident that the lava was extruded at the surface because the layer above the extrusion does not show contact metamorphism.

Unconformities

No location contains a continuous record of geological events throughout Earth's history. If an area was above sea level for part of its past, it is likely that sediments or rock were removed by

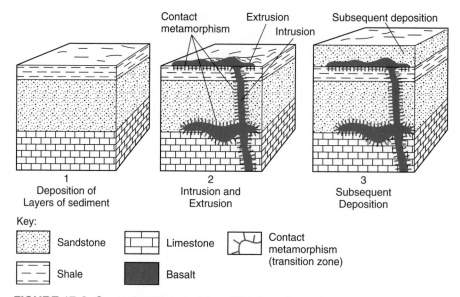

Key:

Sandstone		Limestone		Contact metamorphism (transition zone)	
Shale		Basalt			

FIGURE 17-8. Contact metamorphism. The first diagram shows sedimentary rocks that will be intruded by magma in diagram 2. The portion of the magma that comes to the surface is known as an extrusion. In diagram 3, new deposition has left a sandstone layer on top of the extrusion. By looking at diagram 3 alone, you can tell that the top sandstone layer is younger than the extrusion because there is no contact metamorphism above the extrusion.

weathering and erosion. Thus erosion causes gaps in the geologic record. When new layers are deposited on the erosion surface, the buried erosion surface is known as an **unconformity**. Figure 17-9 shows the formation of an unconformity. Deposition, which most often occurs in water, formed sedimentary rock. When the region was uplifted, the layers were exposed to weathering and erosion. After a period of erosion, the land sank (subsided) and flooding led to the deposition of a new layer on top of the erosion surface.

Unconformities are common throughout New York State. New York's sedimentary rocks show a fairly continuous record of the early development of fish and land animals. For much of this time, a shallow sea covered New York. Then, about 350 million years ago, the land was pushed up above sea level. This was probably the result of a collision between ancient North America and Africa. The disappearance of the sea not only stopped the deposition, but it also led to the erosion of some layers of previously deposited sediments. Wherever new layers were deposited on top of the erosion surface, a buried erosion surface, or

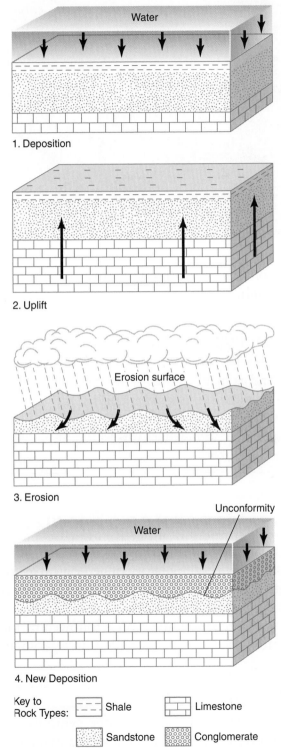

1. Deposition

2. Uplift

3. Erosion

Unconformity

4. New Deposition

FIGURE 17-9. Formation of an unconformity—layers of sedimentary rock form by deposition and compression. Uplift exposes the layers to weathering and erosion, which wear down the land to a new level. Deposition on top of the erosion surface forms an unconformity: a gap in the geologic record.

Key to Rock Types:

⎯⎯⎯ Shale

Limestone

Sandstone

Conglomerate

FIGURE 17-10. The cut-off layers along this cliff make it clear that the top layer was deposited over an erosion surface, forming this unconformity in Paria, Utah.

unconformity, formed. Figure 17-10 shows an unconformity in Utah.

STUDENT ACTIVITY 17-2 —LOCAL ROCK FEATURES

4: 1.2j
6: SYSTEMS THINKING 1

Find as many of the listed rock features (inclusions, cross-cutting relationships, contact metamorphism, and unconformities) as you can in local bedrock outcrops or in road cuts. For each feature, give the location where it is found. If possible, take photographs of these features.

HOW CAN WE INTERPRET GEOLOGIC PROFILES?

4: 1.2j
4: 2.1m, 2.1p, 2.1w

The profile of an outcrop shows a series of layers that help geologists interpret the history of geological events. Most profiles begin with deposition or solidification. (The rock needs to be formed before it can be changed.) If the bottom layer is metamorphic rock, deposition or solidification is followed by metamorphism. Erosion is often the last process evident in a geologic profile.

Keys to Rock Types

Do not expect to see the same symbols on every map or profile. However, there should be a key to make the meaning of the symbols clear. Figures 5-4, 5-8, and 5-16 from the *Earth Science Reference Tables* are charts to help you to identify igneous, sedimentary, and metamorphic rocks.

The charts for sedimentary and metamorphic rocks contain map symbols that are sometimes used to indicate various kinds of rock. Many of these map symbols illustrate characteristic textures of the rocks they represent. For example, sandstone, which is usually represented by a dotted pattern, is composed of gritty particles. Shale, which is usually shown by a pattern of short horizontal line segments, tends to break into thin layers. Limestone is shown by a symbol that looks like stacked bricks because limestone often breaks along joints and bedding surfaces into large, somewhat rectangular blocks.

STUDENT ACTIVITY 17-3 —SYMBOLS AND ROCKS

ESRT

Copy map symbols from the *Earth Science Reference Tables,* or profile diagrams and/or from geologic maps. Next to each symbol tell how the symbol resembles the texture of the rock it represents. Adding photographs or samples of the various kinds of rocks can make your work even better.

Interpreting Profiles

The profiles in Figure 17-11, on page 394, illustrate how a geologic history can be inferred for each profile. Profile *A* starts with deposition of sediments to form conglomerate, sandstone, and shale, in that order. These layers were faulted. This was followed by deposition (of limestone). The wavy top surface indicates that erosion has also taken place.

Profile *B* has granite, an igneous rock, at the bottom, so it started with crystallization of magma. This must have happened before the deposition of sediments because no metamorphism is shown in the siltstone layer. The shale, sandstone, and conglomerate above were deposited and then folded. Later, the top layers were eroded. Intrusion and extrusion at the surface are the final events, unless you wish to point out that erosion at the surface is an ongoing process.

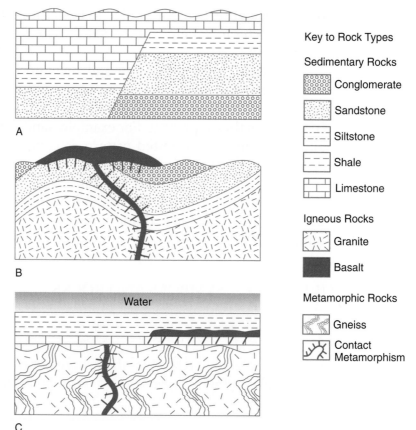

Key to Rock Types

Sedimentary Rocks

Conglomerate

Sandstone

Siltstone

Shale

Limestone

Igneous Rocks

Granite

Basalt

Metamorphic Rocks

Gneiss

Contact Metamorphism

FIGURE 17-11. Geologic profiles contain clues to the sequence of events that created them. The series of events that formed each of these three profiles is explained in the text.

Profile *C* has gneiss, a metamorphic rock, at the bottom. Gneiss can begin as a sedimentary or an igneous rock. Therefore, this history begins with either deposition and sedimentary rock formation followed by metamorphism, or solidification of an igneous rock followed by metamorphism. Intrusion of basalt occurred next. Erosion (note the unconformity) is followed by deposition of limestone. The second (horizontal) basalt body must have been an extrusion because there is contact metamorphism below it, but not above. Deposition of shale sediments is probably the last step in this profile because the water above (probably a lake or an ocean) would support ongoing deposition.

Do you know the square root of 4? Most people think only of 2, but negative 2 is also correct. This question has two equally valid answers. In the same way, geologic profiles are often subject to different interpretations. As long as the sequence of events supports the features shown in the profile, and all the features are explained, a sequence is considered correct. Multiple interpretations are common in science. This is especially true in an environment as complex as the geosciences.

HOW DO GEOLOGISTS ESTABLISH ABSOLUTE TIME?

4: 1.2j

In the first part of this chapter, you learned about the techniques used to determine a sequence of events in relative time. However, scientists want to know more. They want to know how old Earth is, how long mountain building and erosion have been taking place, and how long living creatures have inhabited Earth. Recent discoveries about the structure and stability of atoms, along with improved technology, have finally made it possible to answer these questions.

Radioactive Isotopes

Not all atoms are stable. The atoms of some elements are unstable. They break down spontaneously into atoms of different elements. These unstable elements are called **radioactive**. In the process of breaking down, they emit energy and subatomic particles. Radioactivity was discovered when a scientist working with uranium left it on a photographic plate. When the plate was developed, he realized that the film was exposed even though it had never been exposed to light. This discovery led to further investigations and the recognition of a group of substances that are now known as radioactive elements.

STRUCTURE OF ATOMS All matter is composed of extremely small atoms. Atoms, in turn, are made of protons, neutrons, and electrons. Protons and neutrons are found in the core of the atom, the atomic nucleus. The nucleus makes up 99.9 percent of the mass of the atom. Each element has a different number of protons in the nucleus of its atoms. That is, all hydrogen atoms have just 1 proton, atoms of oxygen have 8 protons; and iron atoms have 26 protons. To balance the positive electrical charge of the protons, neutral atoms have the same number of electrons in orbit around the nucleus. In your daily life, you will come across only a few of Earth's elements. Figure 17-12, on page 396, is a representation of an atom of carbon.

Unlike protons, the number of neutrons in any particular element can vary. It appears that the number of neutrons in the nucleus affects the stability of atoms. Carbon-12, which contains

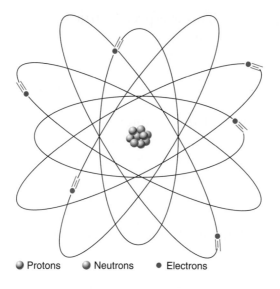

FIGURE 17-12. All carbon atoms have six protons in the nucleus. The most common form of carbon also has six neutrons in the nucleus. If the atom is neutral, six electrons orbit the nucleus. This diagram is not to scale. Any diagram of an atom is a model because the particles in an atom are too small to be visible at any magnification.

○ Protons ○ Neutrons ● Electrons

six neutrons in its nucleus, is stable. However, carbon-14, whose nucleus contains eight neutrons, is radioactive and unstable. Atoms of the same element that have different numbers of neutrons in the nucleus are **isotopes**.

Uranium has two common isotopes, both of which are radioactive. Uranium-238 is the more stable isotope, whereas uranium-235 is more unstable. Uranium-235 is the isotope used in nuclear power plants and atomic weapons.

RADIOACTIVE DECAY When a radioactive isotope, or **radioisotope**, breaks down, it often changes to a stable isotope of a different element. The stable end-material of radioactive decay is known as the **decay product**. For example, carbon-14 changes to nitrogen-14. Uranium-238 transforms through a series of more than a dozen steps until it reaches a stable decay product, lead-206. Potassium-40 can change to either argon-40 or calcium-40.

Radioactivity can be used to find absolute ages because the rate at which these nuclear changes take place is predictable. No matter how much carbon-14 you start with, after 5700 years just half of it will remain carbon-14 and half of it will have changed to nitrogen-14. The time it takes for half of the atoms in a sample of radioactive isotope to decay is its **half-life period**. Each radioactive isotope has its own measurable and consistent half-life period. Some radioisotopes have a half-life period of a fraction of a second. For others, the half-life period is billions of years.

Table 17-1. Radioactive Decay Data

Radioactive Isotope	Disintegration	Half-life (years)
Carbon-14	$^{14}C \rightarrow {}^{14}N$	5.7×10^3
Potassium-40	$^{40}K \rightarrow {}^{40}Ar$ or $^{40}K \rightarrow {}^{40}Ca$	1.3×10^9
Uranium-238	$^{238}U \rightarrow {}^{206}Pb$	4.5×10^9
Rubidium-87	$^{87}Rb \rightarrow {}^{87}Sr$	4.9×10^{10}

Table 17-1 is from the *Earth Science Reference Tables*. It shows the half-life periods of four substances commonly used by geologists to determine the absolute age of rocks or fossils. It is important to note that chemical combination or environmental factors such as heat or pressure do not affect the decay rate of these isotopes. This fact makes measurements of radioisotopes such a powerful tool in determining absolute time.

STUDENT ACTIVITY 17-4 —A CLASSROOM MODEL OF RADIOACTIVE DECAY

4: 1.2j
6: MODELS 2

You and your classmates are going to pretend that you are radioactive atoms of the same element. Stand by your seats and hold a coin. (Borrow a coin if you need to. Don't forget to give it back.) At regular intervals of about a minute, your teacher will tell you to flip your coin. Everyone whose coin lands "tails" sits down. Keep track of the number of coin tosses. Continue until only one student in the class is left standing.

What determines how the coins will land? Can you reliably predict how any coin will end up? What is special about the last student standing? (That is, why is it that particular person?) How many flips did it take until only one student was left standing? Is the number of flips fairly consistent? (Try it.) How is this like or unlike radioactive decay? (List as many similarities and differences as you can.)

There is no way to tell when an individual atom will decay. However, the atoms of each radioisotope have a characteristic half-life period that is predictable and unchanging. Visit the following Web sites to watch an animation that illustrate radioactive decay:

http://www.newcastle-schools.org.uk/nsn/chemistry/Radioactivity/ Halflife%20Page.htm or *http://einstein.byu.edu/~masong/HTM stuff/Radioactive2.html*

Working with Half-Life Periods

Scientists can tell the age of a sample of a radioactive isotope by measuring how much of it remains unchanged. Figure 17-13 shows a model of radioactive decay in which each section of the circle represents one half-life period. (That period of time depends on the radioisotope.) Therefore, every section of this circle, no matter how small, represents the passing of the same amount of time. However, as time goes on, the amount of the sample that remains unchanged becomes smaller. Therefore, the amount that can decay also becomes smaller. In theory, this circle can be divided into smaller and smaller sections without limit. The later sections at the top would eventually become too small to be seen, although each would represent the passing of the same amount of time. In fact, a sample of a radioactive substance does have a limited number of atoms. But that number is so large that, in practice, the number of atoms does not become an issue.

The way scientists determine the age of some radioactive samples is to compare the amount of the original radioisotope with the amount of its decay product. This comparison is known as the

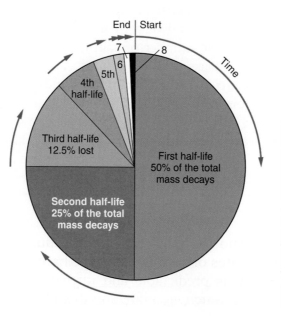

FIGURE 17-13. A model of radioactive decay—each arrow around the outside of this circle represents one half-life period. In that time, half of the atoms in the original radioactive sample change to the decay product. Through time, the amount of decay product increases, and the amount of original radioisotope decreases in a predictable way.

decay-product ratio. For example, if half the atoms in a sample are decay product, the age of the sample is one half-life period. For carbon-14 this would be 5700 years. (See Table 17-1 on page 397.)

One half-life period for potassium-40 is 1.3 billion years. Suppose that in a sample, three-fourths of the radioisotope has changed to the decay product. In this case, it took one half-life for half of the radioisotope to change and an additional half-life for half the remaining radioisotope to change. That is, the sample is 2.6 billion years old, two half-life periods.

Perhaps the best way to solve problems like this is to think through each half-life period until you get to the needed decay-product ratio. For example, suppose you want to know the number of half-life periods that have passed for a sample that is seven-eighths decay product. It must contain one-eighth of the original radioisotope. After one half-life period, half of the original radioisotope would remain. After a second half-life period, one-fourth would remain. After three half-life periods, one eighth would remain unchanged. The answer is three half-lives. Keep in mind that with the passage of each successive half-life period, one-half of the remaining mass of radioisotope decays. Figure 17-14 is

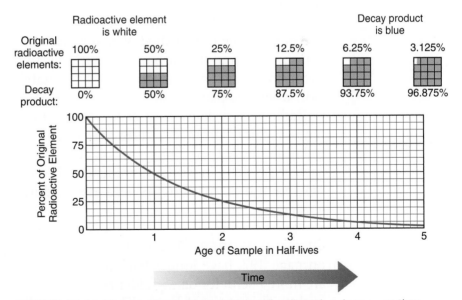

FIGURE 17-14. Two graphic representations of radioactive decay—as time progresses, the amount of decay slows because less and less of the original radioisotope remains. However, the half-life period does not change. The line on the graph shows the radioisotope remaining. The shading in the boxes above the graph shows the increase in the amount of the decay product through time.

a graphic model of radioactive decay showing that the radioactive element decreases as the decay product increases through time.

Selecting the Best Radioactive Isotope

A carpenter has many tools. Selecting the right tool for a job is the first step in any project. **Radiometric dating** (determining absolute age with radioactive isotopes) involves a similar choice. Different radioisotopes have different uses. The two critical issues are whether a particular radioisotope is present in the rocks being dated, and the estimated age of the rocks.

CARBON-14 Carbon-14 is often used to find the absolute age of organic material that is less than about 50,000 years old. Carbon-14 decays so quickly that there is little left in samples older than 50,000 years. This makes measurements difficult and precision poor beyond that age. Organic materials always contain carbon. Plants absorb carbon from the atmosphere. The air contains the more common isotope, carbon-12, and a smaller amount of carbon-14. The ratio of these two forms of carbon in the atmosphere is thought to be in equilibrium and relatively stable through time.

In living plants, carbon-14 decays to nitrogen-14. However, it is continuously replenished. When plants die, they no longer absorb carbon, and the carbon-14 continues to decay. Scientists use the ratio of carbon-12 to carbon-14 to establish with great precision when the plant died.

Radiocarbon dates have been verified by checking them with dates obtained from counting tree rings. The oldest trees alive today are more than 5000 years old. Radiocarbon dating has been especially useful in work on prehistoric human habitations, Egyptian mummies, mastodonts, and events during or since the last ice age.

OTHER ISOTOPES The remaining three isotopes in Table 17-1 on page 397, have longer half-life periods. This means that they are more useful for dating older rocks, from 100,000 years old and back to the origin of Earth. These three radioisotopes are found in selected minerals. Each can be measured with great precision. The age of an igneous rock is generally the age of its minerals. In some places, these isotopes have been used to find the age of layers of volcanic ash. Ash beds are useful horizons of absolute time because they often cover a large geographic area but are deposited in a very short time period. The absolute age of the ash provides useful time limits to rocks found above and below it.

CHAPTER REVIEW QUESTIONS

Part A

1. A whalebone that originally contained 200 g of carbon-14 now contains only 25 g of carbon-14. How many half-life periods of carbon-14 have passed since the whale was alive?

(1) 1　　　　　　　　　(3) 3
(2) 2　　　　　　　　　(4) 4

Base your answers to questions 2 and 3 on the graph below, which shows the decay of a 50-g radioactive sample.

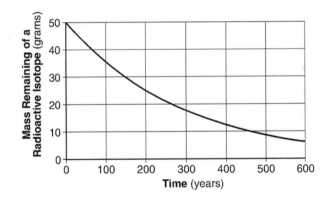

2. According to the graph above, what is the half-life of this substance?

(1) 100 years　　　　　(3) 200 years
(2) 150 years　　　　　(4) 300 years

3. What mass of this substance changed to its decay product after 100 years?

(1) 14 g　　　　　　　(3) 50 g
(2) 25 g　　　　　　　(4) 86 g

Base your answers to questions 4 and 5 on the figure below.

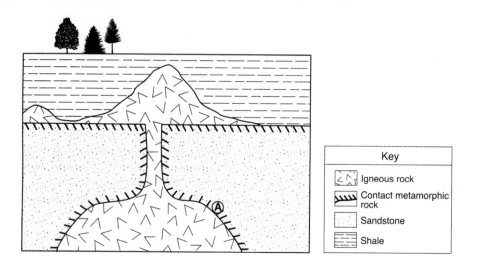

4. The metamorphic rock at location *A* in the geologic profile above is most likely

(1) marble   (3) phyllite
(2) quartzite   (4) slate

5. Which rock in the profile above is the youngest?

(1) shale
(2) sandstone
(3) igneous rock
(4) rock at location *A*

6. The models below represent the decay of radioactive atoms to stable atoms at 1 and 2 half-lives.

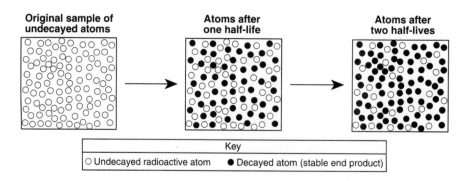

Which model below best represents the same sample after three half-lives?

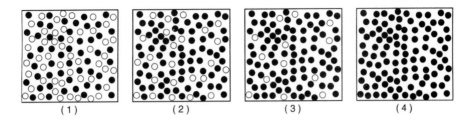

(1) (2) (3) (4)

7. What radioactive isotope has a half-life closest to 5000 years?

(1) carbon-14
(2) potassium-40
(3) uranium-238
(4) rubidium-87

Part B

Base your answers to questions 8 through 10 on the geologic cross section below in which overturning has not occurred. Letters *A* through *H* represent rock layers.

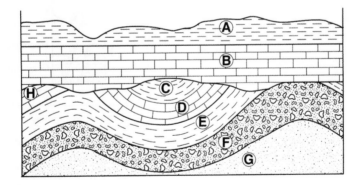

8. Which sequence of events most likely caused the unconformity at the bottom of rock layer *B* in the profile above?

(1) folding → uplift → erosion → deposition
(2) intrusion → erosion → folding → uplift
(3) erosion → folding → deposition → intrusion
(4) deposition → uplift → erosion → folding

9. The folding of rock layers *G* through *C* was most likely caused by

    (1) the erosion of overlying sediments
    (2) contact metamorphism
    (3) the collision of lithospheric plates
    (4) the extrusion of igneous rock

10. Which two letters represent bedrock of the same age?

    (1) *A* and *E*   (3) *F* and *G*
    (2) *B* and *D*   (4) *D* and *H*

Base your answers to questions 11 through 14 on the diagram below, which shows three rock outcrops, *A, B,* and *C.* Line *X-Y* is a fault. Overturning has not occurred in any of these locations.

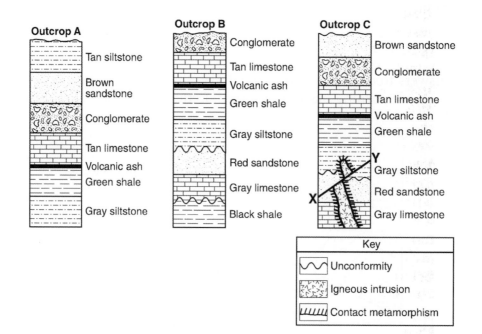

11. A volcanic ash layer is considered a good time marker for correlating rocks because it

    (1) has a dark color
    (2) can be dated using carbon-14
    (3) lacks fossils
    (4) was rapidly deposited over a wide area

12. Which sedimentary rock shown in these outcrops is the youngest?

(1) black shale   (3) tan siltstone
(2) conglomerate   (4) brown sandstone

13. What is the youngest geologic feature in the three bottom layers of outcrop *C*?

(1) fault   (3) unconformity
(2) igneous intrusion   (4) zone of contact metamorphism

14. What processes were primarily responsible for the formation of most of the rocks in outcrop *A*?

(1) melting and solidification
(2) heating and compression
(3) compaction and cementation
(4) weathering and metamorphism

Part C

Base your answers to questions 15 through 18 on page 406 on the geologic cross section below in which overturning has not occurred. Letters *A* through *I* represent rock layers.

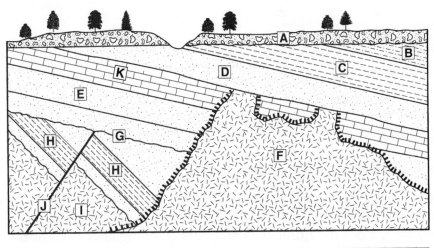

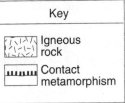

Key

Igneous rock

Contact metamorphism

15. Which letter identifies the oldest rock layer shown above?

16. What letter best shows where marble was formed?

17. State one form of evidence that rock layer *D* is younger than layer *F.*

18. Explain why rock layer *H* is not continuous.

Base your answers to questions 19 and 20 on the table below, which shows the result of a classroom activity to model radioactive decay. Each student stood by her or his desk and drew a random number from a container. Those who drew even numbers sat down. This process was repeated by those still standing until all students were seated.

DRAW	STUDENTS STANDING	STUDENTS SEATED
0 (Start)	30	0
1	16	14
2	7	23
3	4	26
4	2	28
5	1	29
6	1	29
7	0	30 (End)

19. Use a piece of graph paper or a grid to graph this information.

- Set up your graph with "Number of Draws" on the horizontal axis and "Number of Students Standing" on the vertical axis
- Mark with a dot the number of students standing after each draw
- Connect the dots with a solid line

20. How does the number of students in this experiment compare with the number of atoms in a 1-g sample of radioactive carbon-14?

CHAPTER

18

Fossils and Geologic Time

WORDS TO KNOW

correlation   extinction   paleontology

evolution   index fossil   species

This chapter will help you answer the following questions:

1 What are fossils? Where are they found?

2 How do scientists use fossils to interpret the past?

3 Have living organisms changed throughout Earth's history?

4 In what ways are fossils used to divide geologic time?

5 How do clues in the rocks help scientists match rock layers?

WHAT ARE FOSSILS?

4: 1.2i, 1.2j

Fossils are any remains or evidence left by prehistoric life. Historical records go back only about 10,000 years. Fossils must be older than this because they are not mentioned in these records. In fact, the earliest fossils are traces of primitive life-forms that lived on Earth very early in its history, billions of years ago. Fossils supply important information about the development of life and about past environments. The study of fossils is **paleontology**, and scientists who study fossils are paleontologists. Figure 18-1, on page 408, shows dinosaur fossils found at Dinosaur National Monument. Fossils are grouped into two broad categories: body fossils and trace fossils.

FIGURE 18-1. Dinosaur National Monument spans the Utah-Colorado border. Excavation at this site has been stopped to leave some bones in place. The remaining fossil dinosaur bones have been left in the rock where they were discovered.

Body Fossils

Among the most complete fossils are the mammoths, ice-age elephants. These creatures have been found in permanently frozen ground where they have been preserved since the last ice age. Mammoth fossils sometimes include soft parts such as hair and flesh.

Large mammals are not the only whole body fossils. Ancient insects have also been fossilized. Complete bodies of insects have been preserved in amber that is millions of years old. Amber is the hardened form of the resin that sometimes seeps from trees. When the insect landed on the sticky sap, or resin, of a pine tree, it was trapped and covered in the sap. The sap hardened and preserved the insect. However, fossil remains of soft tissue are relatively rare. Usually, scavengers eat the soft parts of an organism or it decays before it can be preserved.

Bones, teeth, and shells are more likely to be fossilized than soft tissue. For example, a homeowner in the Hudson Valley of New York discovered the bones of a mastodon in 1999. A mastodon, another ice-age elephant, is a relative of modern elephants. The

FIGURE 18-2. Complete skeletons of mammoths are discovered in the arctic tundra. Sometimes, flesh remains on the fossilized remains.

homeowner invited scientists and students from several universities to dig up the nearly complete skeleton. This is one of three recent finds of mastodon skeletons in New York State. Like mammoths, mastodons became extinct about 10,000 years ago. Figure 18-2 shows a complete mammoth skeleton found in arctic Russia.

How did the scientists know what animal left these bones? The bones of every animal are different. For this reason, an experienced paleontologist can often identify fossils from a single bone or tooth.

Replacement is another form of fossilization. In this form of fossilization, minerals in groundwater gradually replace organic substances. Petrified wood is a good example. The minerals take the shape and may even show the internal structures of the original living organism. However, these fossils are mineral material such as quartz rather than the wood or other organic substance that made up the original organism.

A fossil mold, or impression, forms when fine sediments surround an organism or part of an organism, preserving its shape. Leaf impressions are molds. Other sediments that fill in the hole left by the organism are known as casts. In fact, filling a hollow mold makes a cast. Fossils of seashells found in rocks that extend from the lower Hudson Valley through western New York State are often molds and casts. Figure 18-3, on page 410, is a sample of shale in which the shapes of shells of marine animals have been preserved.

FIGURE 18-3. You can see more than two dozen fossil molds of marine animal shells (*Brachiopods*) in this example of New York shale.

Trace Fossils

Trace fossils are signs that living organisms were present in an area. However, they do not include or represent the body parts of an organism. For example, the only dinosaur fossils found in New York State are footprints of a reptile about as tall as a large dog found in the lower Hudson Valley. Footprints can tell us the size and something about the structure of the animal as well as how it moved. It is likely that other dinosaurs existed in New York, but their fossils have not been found. Worm burrows are common in some sedimentary rocks in western New York State. Coprolites can be described as "the only material a fossil animal intended to leave behind." They are, in fact, fossilized dung, or feces, which can tell us what the animal ate.

STUDENT ACTIVITY 18-1 —NEARBY FOSSIL BEDS

4: 1.2i

Where are the nearest rocks that contain fossils? Neighbors, park naturalists, or college professors may be able to help you locate nearby fossils. You may even be able to visit if it is on public land or if you have the permission of the landowner. Road cuts sometimes reveal fossils, although there may be restrictions about stopping along busy highways. Be sure to get any needed permission before you visit a fossil bed or collect specimens. Collecting fossils can be fun and educational. There are many books and Internet resources that can tell you about the fossils you collect. As an alternative, some museums and universities have collections of fossils that you can observe.

STUDENT ACTIVITY 18-2 —INTERPRETING FOSSIL FOOTPRINTS

4: 1.2i

The pattern of footprints in Figure 18-4 was found in bedrock millions of years old. What do you think happened to produce this particular pattern of footprints? Propose a sequence of events that could explain what you see in the footprints. Give other explanations for this pattern. How can you determine which explanation is correct?

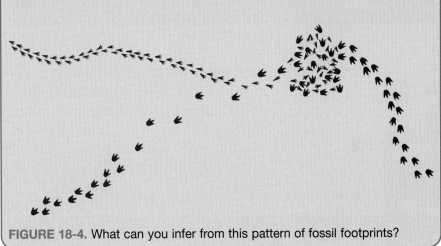

FIGURE 18-4. What can you infer from this pattern of fossil footprints?

Fossils tell us more than just what life-forms there were on Earth. They also provide clues to ancient environments. When paleontologists find fossil fish or the remains of other underwater creatures, they infer that they have found an ancient underwater environment. Although most sedimentary rocks form from material deposited in water, land-derived sedimentary rocks are also common. Fossil wood helps scientists identify land-based sediments.

Some fossils, such as chalk and diatomaceous earth, are important economically. Chalk, which is mostly calcite, is composed of millions of tiny fossil shells. It is used to neutralize acidic soil and as a part of many building materials such as cement. We also use chalk to draw on chalkboards or sidewalks. Diatomaceous earth (diatomite) is composed of millions of tiny quartz-rich fossil skeletons. It is used as a filtering agent, a mild abrasive, or a heat-insulating material.

HOW DID LIFE BEGIN ON EARTH?

4: 2.1i, 1.2j

Scientists are studying how life began on Earth. They know that chemical reactions in Earth's early environment could have made amino acids. Living organisms are made of amino acids, which are the building blocks of proteins. However, scientists do not understand how amino acids become organized to make even the simplest life-forms. Experiments to produce living material from nonliving processes have not been conclusive. Millions of years of chemical changes on the primitive Earth might have produced results that cannot be reproduced in short-term laboratory experiments.

Life could have started in places not usually considered good for living organisms. Recently, scientists discovered primitive forms of life in nearly solid rock many kilometers below Earth's surface. They have found living organisms near hot water vents in the deep ocean bottom beyond the reach of sunlight. Perhaps conditions in one of these places gave rise to the first life-forms.

It appears that the impact of objects from space, such as meteorites, blasted rock fragments from the surface of Mars. Some of these rocks fell to Earth. Scientists have found pieces of rock that came from Mars in Antarctica. These rocks provide another clue to the ability of life to exist in extreme environments. Microscopic examination of these rocks found tiny objects that some scientists think are evidence of life. Could very simple forms of life travel from planet to planet as spores in rock fragments like these? While this is an interesting possibility, scientists do not know for sure. They need more investigations to help decide how life began on Earth.

Scientists look at ancient rocks to find evidence of early life. However, there are very few rocks left from Earth's earliest history. Erosion and metamorphism have changed most of the oldest rocks that might have contained fossils. However, some scientists think that they have found evidence of carbon compounds made by organisms living nearly 4 billion years ago.

Stronger evidence comes from patterns of life-forms. The oldest fossils are clumps of primitive bacteria called stromatolites. There are fossil stromatolites north of Albany, New York, as you can see in Figure 18-5 (left). These fossils are about 3.5 billion

FIGURE 18-5. These circular patterns in bedrock north of Albany, New York, are fossilized colonies of algae (stromatolites) from 3.5 billion years ago (left). The location on the left must have looked something like this tropical bay in Western Australia, which is home to modern stromatolites (right).

years old. Stromatolites are alive today in shallow, tropical lagoons of Western Australia, as shown in Figure 18-5 (right).

WHAT IS ORGANIC EVOLUTION?

4: 1.2i, 1.2j

Fossils older than 2 billion years are remains of tiny single-celled organisms. These life-forms were so primitive that their cells did not contain a nucleus. They reproduced by splitting into two or more new cells. The first important advance in the development of more complex life-forms is found in rocks about 1.4 billion years old. Cells with a nucleus appeared. These cells could reproduce sexually and inherit characteristics from two parent cells.

Less than a billion years ago, colonies of single-celled organisms developed into simple multicellular organisms, such as jellyfish and worms. The first organisms with shells and skeletons appeared about 545 million years ago. These organisms eventually gave rise to the complex life-forms that exist today. The gradual change in living organisms from generation to generation is biological, or organic, **evolution**.

Biological evolution is sometimes called the "theory of evolution." Unfortunately, many people think that a scientific theory is a principle that has not been proved. They think that a theory is less certain than a law of science. However, this is not correct. Laws of science tell what happens. Theories help us explain why events happen. Theories and laws are so fundamental that they are the virtual certainties on which science is built.

The Work of Darwin

A **species** is a group of organisms so similar they can breed to produce fertile offspring. Humans are a good example of a species. Charles Darwin published *On the Origin of Species by Means of Natural Selection* in 1859. Soon after, most scientists accepted organic evolution as a fact. Darwin based his book on scientific observations he made while traveling around the world. He spent many years organizing his ideas about the development of life before he published his book. In fact, another biologist, Alfred Russel Wallace, independently arrived at similar conclusions at about the same time.

Four Principles of Evolution Organic evolution is based on four principles. First, within a species, there are variations, or differences, among individuals. There are differences in size and shape, which you can see in Figure 18-3, on page 410. In addition, individual organisms have different abilities to find food, resist disease, and reproduce effectively. Second, organisms usually produce more offspring than the environment can support. Think of the thousands of seeds a sunflower produces or the number of eggs one salmon can lay. If all these offspring survived, the world would be overrun with just sunflowers and salmon. The third principle is the effect of competition among individuals of a given generation. Organisms die for many reasons. However, those best suited to be successful in their environment are most likely to survive. Fourth, the individuals best suited to the environment will live long enough to reproduce and pass their traits to the next generation. This is what Darwin meant by natural selection.

STUDENT ACTIVITY 18-3 —VARIATIONS WITHIN A SPECIES

1: MATHEMATICAL
ANALYSIS 1, 2

Within a species, many characteristics, such as height or length, vary around an average value. Record the height in centimeters of all the members of your science class. Group the heights in 10-cm intervals (150 cm–159 cm, 160 cm–169 cm, etc.). Count the number of students whose height falls within each 10-cm interval. Graph your data as a histogram with height on the horizontal axis and the number of students on the vertical axis. Does the data show a pattern of distribution around an average value? What other characteristics that change from individual to individual might show a similar distribution?

Evolution and DNA

Darwin could not explain how organisms passed their characteristics to next generations. Chromosomes, genes, and DNA were not discovered and studied until about 100 years later. These chemical-messengers pass organic "building plans" from parents to offspring and guide the growth of individuals.

As this information passes from generation to generation, errors, or mutations, occur. Most mutations are harmful and are not passed on to offspring because they make the individual less able to survive and reproduce. However, the rare helpful changes in genetic plans that make an individual better adapted to the environment are passed to succeeding generations.

Many changes in the genetic makeup of species are slow and not easily seen. However, recent discoveries have shown that some changes occur relatively rapidly. In small interbreeding populations, genetic changes multiply. Eventually, the members of that group can no longer interbreed with other groups. This is how a new species begins. The gradual development of the great variety of life-forms on Earth is illustrated in "tree of life" diagrams. Visit the following Web site to learn about the Tree of Life Web project: *http://tolweb.org/tree*

Extinction

Competition occurs within a species as well as between species. Sometimes, one species is better adapted to an environment than another species. The better-adapted species may take food, shelter, or other natural resources needed by the less well-adapted species. Predation, or hunting, of one species by another can also eliminate whole species. The American bison, or buffalo, once roamed the prairies of North America in massive numbers. Hunting, especially by settlers with rifles, nearly eliminated the American bison. A much smaller number of bison survive today on farms and game preserves.

Extinction means that all the individuals of a particular species have died. A number of species including the marsupial Tasmanian tiger and the passenger pigeon have become extinct in the last century. Most of the life-forms found as fossils have become extinct.

STUDENT ACTIVITY 18-4 —AN EXTINCT SPECIES

2: INFORMATION
SYSTEMS 1

Choose a species that is in threatened with extinction or has become extinct in recent years. Report on why that species is threatened or became extinct.

The geologic record includes times when large numbers of species became extinct. Recent discoveries have shown that at least one of these mass extinctions occurred about the time that a giant asteroid collided with Earth. The impact of this object threw great amounts of dust into the atmosphere, which caused changes in global climate. The dust-laden atmosphere prevented much of the sunlight from reaching Earth's surface. The lack of sunlight killed plants and then the animals that depended on them. Most dinosaurs became extinct at that time.

Other scientists challenged the importance of a meteor impact. They suggested that similar atmospheric changes could be produced by volcanic eruptions. A third suggestion was that global changes in climate were related to plate tectonic motions and the formation of a supercontinent Pangaea. In March 2010, a panel of experts that had studied the competing theories concluded that it was indeed an asteroid that brought about the conditions that led to the extinction of the dinosaurs.

Just before the age of dinosaurs, about 225 million years ago, an even more catastrophic mass extinction occurred. The cause is still a matter of research and speculation by geologists. The trilobites, a very successful group of fossil organisms commonly found in New York State rocks, became extinct at this time. Scientists are now finding evidence for even more extinction events in the geologic past.

HOW HAS GEOLOGIC TIME BEEN DIVIDED?

4: 1.2h, 1.2i, 1.2j

In the late eighteenth and early nineteenth centuries, geologists working in Great Britain and Europe began to use fossils contained in rock layers to identify the layers. They also noticed that layers containing certain fossils were always located above or below layers that contained different fossils. In fact, they began to identify whole groups of fossils by where the first examples were found. For example, any rocks that have the same fossils found in Devon in the south of England are called Devonian. These rocks were always located above those with fossils found in certain parts of Wales and below the rocks of the coal-mining areas near Bristol. These observations were used to establish a geologic time scale of relative ages as shown in Figure 18-6 on pages 418–419. This chart is part of the *Earth Science Reference Tables.*

Some fossils are more useful than others in establishing the age of rocks. These are called index fossils. The best **index fossils** are easy to recognize and are found spread out over a large geographic area, but they existed for a relatively brief period of time. To geologists, trilobites are one of the most useful groups of index fossils.

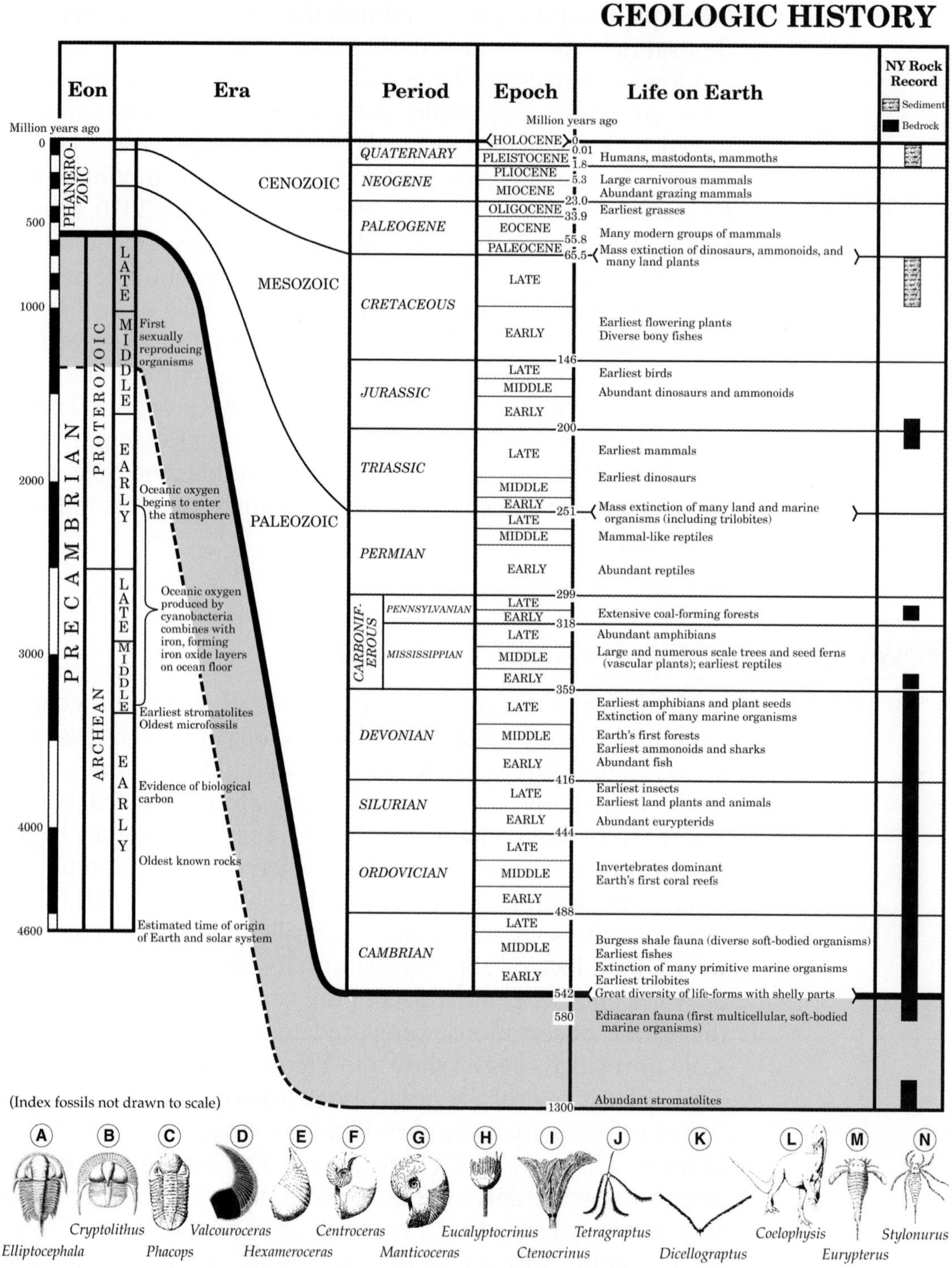

GEOLOGIC HISTORY

Eon	Era	Period	Epoch	Life on Earth	NY Rock Record

Million years ago (Eon) — 0, 500, 1000, 2000, 3000, 4000, 4600

Million years ago (Epoch)

NY Rock Record: Sediment / Bedrock

PHANEROZOIC

CENOZOIC
- QUATERNARY — HOLOCENE 0 / 0.01, PLEISTOCENE 1.8 — Humans, mastodonts, mammoths
- NEOGENE — PLIOCENE 5.3, MIOCENE 23.0 — Large carnivorous mammals / Abundant grazing mammals
- PALEOGENE — OLIGOCENE 33.9 — Earliest grasses; EOCENE 55.8 — Many modern groups of mammals; PALEOCENE 65.5 — Mass extinction of dinosaurs, ammonoids, and many land plants

MESOZOIC
- CRETACEOUS — LATE; EARLY — Earliest flowering plants / Diverse bony fishes — 146
- JURASSIC — LATE — Earliest birds; MIDDLE — Abundant dinosaurs and ammonoids; EARLY — 200
- TRIASSIC — LATE — Earliest mammals; MIDDLE — Earliest dinosaurs; EARLY — 251 — Mass extinction of many land and marine organisms (including trilobites)

PALEOZOIC
- PERMIAN — LATE; MIDDLE — Mammal-like reptiles; EARLY — Abundant reptiles — 299
- CARBONIFEROUS
  - PENNSYLVANIAN — LATE; EARLY — Extensive coal-forming forests — 318
  - MISSISSIPPIAN — LATE — Abundant amphibians; MIDDLE — Large and numerous scale trees and seed ferns (vascular plants); earliest reptiles; EARLY — 359
- DEVONIAN — LATE — Earliest amphibians and plant seeds / Extinction of many marine organisms; MIDDLE — Earth's first forests / Earliest ammonoids and sharks; EARLY — Abundant fish — 416
- SILURIAN — LATE — Earliest insects / Earliest land plants and animals; EARLY — Abundant eurypterids — 444
- ORDOVICIAN — LATE; MIDDLE — Invertebrates dominant / Earth's first coral reefs; EARLY — 488
- CAMBRIAN — LATE; MIDDLE — Burgess shale fauna (diverse soft-bodied organisms) / Earliest fishes / Extinction of many primitive marine organisms / Earliest trilobites; EARLY — 542 — Great diversity of life-forms with shelly parts

PRECAMBRIAN

PROTEROZOIC
- LATE / MIDDLE — First sexually reproducing organisms
- EARLY — Oceanic oxygen begins to enter the atmosphere

ARCHEAN
- LATE / MIDDLE — Oceanic oxygen produced by cyanobacteria combines with iron, forming iron oxide layers on ocean floor
- EARLY — Earliest stromatolites / Oldest microfossils; Evidence of biological carbon; Oldest known rocks; Estimated time of origin of Earth and solar system

580 — Ediacaran fauna (first multicellular, soft-bodied marine organisms)
1300 — Abundant stromatolites

(Index fossils not drawn to scale)

A — Elliptocephala
B — Cryptolithus
C — Phacops
D — Valcouroceras
E — Hexameroceras
F — Centroceras
G — Manticoceras
H — Eucalyptocrinus
I — Ctenocrinus
J — Tetragraptus
K — Dicellograptus
L — Coelophysis
M — Eurypterus
N — Stylonurus

FIGURE 18-6.

OF NEW YORK STATE

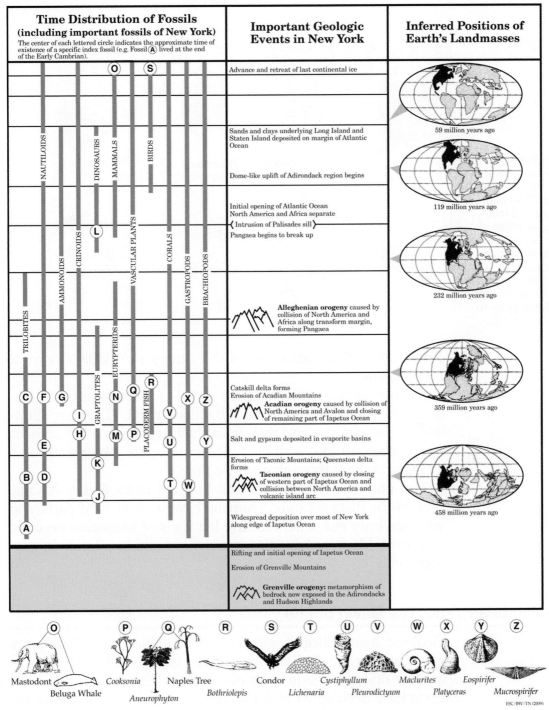

Time Distribution of Fossils (including important fossils of New York)
The center of each lettered circle indicates the approximate time of existence of a specific index fossil (e.g. Fossil Ⓐ lived at the end of the Early Cambrian).

Important Geologic Events in New York

Inferred Positions of Earth's Landmasses

59 million years ago

119 million years ago

232 million years ago

359 million years ago

458 million years ago

Advance and retreat of last continental ice

Sands and clays underlying Long Island and Staten Island deposited on margin of Atlantic Ocean

Dome-like uplift of Adirondack region begins

Initial opening of Atlantic Ocean North America and Africa separate
⟨ Intrusion of Palisades sill ⟩
Pangaea begins to break up

Alleghenian orogeny caused by collision of North America and Africa along transform margin, forming Pangaea

Catskill delta forms
Erosion of Acadian Mountains
Acadian orogeny caused by collision of North America and Avalon and closing of remaining part of Iapetus Ocean

Salt and gypsum deposited in evaporite basins

Erosion of Taconic Mountains; Queenston delta forms
Taconian orogeny caused by closing of western part of Iapetus Ocean and collision between North America and volcanic island arc

Widespread deposition over most of New York along edge of Iapetus Ocean

Rifting and initial opening of Iapetus Ocean

Erosion of Grenville Mountains

Grenville orogeny: metamorphism of bedrock now exposed in the Adirondacks and Hudson Highlands

(Fossil column labels: TRILOBITES, NAUTILOIDS, AMMONOIDS, CRINOIDS, DINOSAURS, MAMMALS, VASCULAR PLANTS, BIRDS, CORALS, GASTROPODS, BRACHIOPODS, EURYPTERIDS, GRAPTOLITES, PLACODERM FISH)

Ⓞ Mastodont / Beluga Whale
Ⓟ Cooksonia
Ⓠ Naples Tree / Aneurophyton
Ⓡ Bothriolepis
Ⓢ Condor
Ⓣ Lichenaria
Ⓤ Cystiphyllum / Pleurodictyum
Ⓥ Maclurites
Ⓦ Platyceras
Ⓧ Eospirifer
Ⓨ Mucrospirifer

ESC/BW/TN (2009)

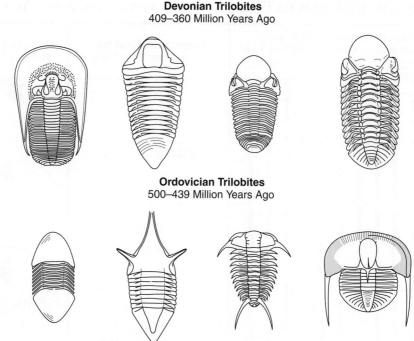

Devonian Trilobites
409–360 Million Years Ago

Ordovician Trilobites
500–439 Million Years Ago

FIGURE 18-7. The rapid evolution of trilobite species over 300 million years and their large number of fossils make them excellent index fossils. Adult trilobites ranged in size from 25 mm to 100 mm (1 in. to 4 in.).

Figure 18-7 illustrates some of the variety of trilobites used as index fossils.

Marine animals with shells are useful as index fossils. The ammonites existed for more than 300 million years. They were abundant in ancient seas and their hard shells were preserved or made impressions in the sediment. Paleontologists observe distinct changes in them through geologic time.

It is possible that humans will be good index fossils at some time in the distant future. Humans have distinct hard parts, and we often bury our dead. We have left signs of habitation on every continent. Humans have also experienced rapid evolution in geologic time.

Precambrian Time: The Dawn of Life

Geologic history begins with Earth's formation 4.6 billion years ago, which is the start of Precambrian time. Earth's oldest rocks are just under 4 billion years old. However, some mineral inclusions have been dated at 4.4 billion years. Scientific evidence indicates that Earth and the solar system formed at the same time.

However, our planet is too active to preserve its oldest surface materials. The processes of plate tectonics along with metamorphism, weathering, and erosion have destroyed Earth's original crust.

At one time, fossil collectors could not recognize fossils in rocks older than those from the Cambrian Period. Some concluded that older rocks (Precambrian) did not contain fossils. However, scientists now know that Precambrian rocks do contain fossils. Organisms alive during the Precambrian time did not have hard parts, like shells and skeletons. Therefore, they did not form easily identifiable fossils. Precambrian time makes up about 88 percent of Earth's history.

Living things evolved in the oceans. The ocean water provided them with food and protected them from harmful solar radiation. Earth's early atmosphere probably was mostly carbon dioxide and nitrogen. It was similar in composition to the present atmosphere of Venus and Mars. Single-celled Precambrian organisms developed photosynthesis. They used carbon dioxide to store energy, releasing oxygen as a waste product. The addition of oxygen to the atmosphere had two very important results. The ozone layer formed. Oxygen in the form of ozone absorbs harmful radiation, such as ultraviolet rays, from outside Earth. Oxygen also is necessary for air-breathing animals. These developments, roughly 2 billion years ago, allowed living things to move out of the oceans.

Events of the Precambrian time are not as well known as more recent events. This is because most of the Precambrian rocks have been covered by later rocks, or the rocks were recycled by weathering and erosion, or changed by metamorphism.

Paleozoic Era: The Origin of Complex Life-Forms

Paleozoic means "the time of early life." This era began a little more than half a billion years ago. The presence of many fossils marks the beginning of the Paleozoic Era. This abundance of life was a result of rapid evolution. Skeletons and shells allowed some organisms to move rapidly in search of food. These parts allowed other organisms to protect themselves from becoming food. Figure 18-8, on page 422, is a New York State fossil eurypterid, sometimes called a sea scorpion. Today's most complex life-forms have many specialized features, such as sense organs. Our sense organs can be traced back to structures that appeared early in the Paleozoic Era.

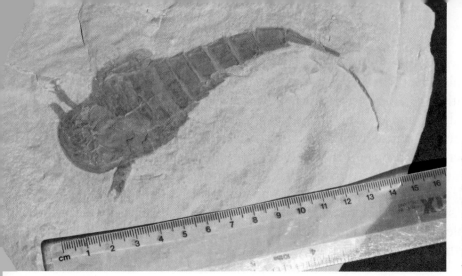

FIGURE 18-8. In 1984, the New York State legislature made the eurypterid the state's official fossil because of the quality of specimens found in New York's bedrock. Some individuals grew to as long as 2 m (6 ft).

In the oceans, trilobites and the first fish appeared early in the Paleozoic Era. Trilobites evolved many variations in shape before they became extinct at the end of the Paleozoic Era. Plants and amphibians that inhabited the land also appeared in the Paleozoic Era. Amphibians lay their eggs in water; but as adults, many move onto the land. Reptiles, which can lay eggs on land, followed the amphibians.

At the end of the Paleozoic Era, 95 percent of living species went extinct. Scientists do not know the cause of this extinction. They suspect some dramatic change in the world's climates. The change might have been related to the formation of the supercontinent Pangaea. It may have been the impact of a large meteorite. There may have been great volcanic eruptions. Whatever the cause, this catastrophic event led to the appearance of new lifeforms in the next era. Figure 18-9 shows several major mass extinctions that occurred in the geological past.

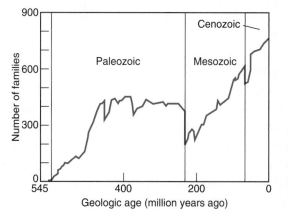

FIGURE 18-9. Mass extinctions marked the ends of the Paleozoic Era, 251 million years ago, and the Mesozoic Era, 65 million years ago.

Mesozoic Era: The Age of Dinosaurs

Mesozoic means "middle life." This era began about 251 million years ago, following the end of the Paleozoic Era. Some forms of fish, insects, and reptiles had survived the Paleozoic Era extinction. Mammals appeared in the Mesozoic Era. However, they remained small creatures. The first birds may have evolved from flying dinosaurs, such as *Archaeopteryx,* in the Mesozoic Era.

Dinosaur evolution produced a wide variety of animals that ruled the land during the Mesozoic Era. These animals inhabited nearly every terrestrial environment. Dinosaurs were very successful; they existed for more than 150 million years. Some may have been remarkably intelligent. Dinosaurs became extinct at the end of the Mesozoic Era. Recent evidence links that extinction to the impact of an asteroid and climatic change.

Cenozoic Era: The Age of Mammals

Cenozoic means "recent life." The Cenozoic Era began 65.5 million years ago and continues to the present. With the extinction of dinosaurs, mammals evolved as the most successful group of vertebrate animals. Mammals inhabit nearly every terrestrial environment. Whales, dolphins, and seals live in the seas, and bats fly through the air. The first humans evolved in the late Cenozoic Era. The oldest human fossils were found in Africa and are about 2 to 4 million years old.

In New York State, the most dramatic geological events of the Cenozoic Era were the repeated advances of continental glaciers. Glaciers nearly covered the state until about 15,000 years ago. The formation of the glaciers resulted in a drop in sea level. This exposed a land bridge between Asia and North America. The bridge allowed humans to migrate from Asia into North America. Unlike earlier eras, the Cenozoic Era is the only era with no specific end.

A few geologists have proposed that a new era be added. The Anthropocene Era: The Age of Humans would begin with the twentieth century, during which human technology caused dramatic changes worldwide. Extinctions of species have been fast and numerous. New life-forms have been created through genetic engineering. Climate changes, caused mostly by the use of fossil fuels, may lead to rapid changes in Earth's climates. The majority of geologists have not recognized the Anthropocene Era as a new era. This era is not listed on the *Earth Science Reference Tables*. However, most do agree that human influence on our planet is unequaled. Visit the following Web site, at the bottom left click full-window, go up and click on Evolution to watch Flash animations of Earth's evolution: *www.johnkyrk.com/index.html*

STUDENT ACTIVITY 18-5 —GEOLOGIC TIME LINE

6: MODELS 2
6: MAGNITUDE AND
SCALE 3

Use a strip of paper about 5 m (16 ft) long to make a time line of the events listed below. Develop a scale to tell how much time is represented by each unit of distance. According to your scale of time, plot each event where it belongs on the strip.

Origin of Earth and solar system
Oldest known rocks on Earth
Oldest microfossils
Fossil organisms become abundant
Earliest traces of terrestrial (land) organisms
First birds
Extinction of dinosaurs
Earliest humans in Africa
End of Wisconsin Glacial Episode
Columbus sails to the Americas
The first Americans land on the moon

When the events have been plotted, divide your time line into Precambrian, Paleozoic, Mesozoic, and Cenozoic eras. You may make each era a different color.

WHAT IS THE GEOLOGIC HISTORY OF NEW YORK STATE?

USING THE *ESRT*
4: 1.2j

Figure 18-6 on pages 418–419 is from pages 8 and 9 of the *Earth Science Reference Tables*. It is a chart of Earth's geologic history. Geologic events and fossils of New York State are highlighted. You should be familiar with the information in this chart and how to use it. Look at the figure and note the following:

- The left page shows the major divisions of geologic time and the absolute age of these divisions in millions of years before the present. For example, Precambrian time ended at the beginning of the Paleozoic Era, 542 million years ago.
- The column labeled "Life on Earth" identifies important changes in the evolution of life. This information expands on the information given in the area near the left side of the page.
- The next column has a thick black broken vertical line and two textured bars that indicate the age of bedrock exposures and sediment in New York State. The open parts of the thick line represent intervals for which there is no bedrock or sediment in New York. Notice that there are whole geological periods missing. For example, New York State has no bedrock exposures of the Permian Period and only a partial record of several other periods. These incomplete parts of the line represent major unconformities in New York bedrock.
- In the column labeled "Time Distribution of Fossils," thinner vertical lines indicate when certain fossil organisms lived. Notice the creatures labeled *A-Z* at the bottom of the page. These letters appear on the vertical lines to show when these organisms were alive. For example, the specific trilobite species labeled *A*, *B*, and *C* lived at different times during the Paleozoic Era.
- The maps at the right side of the right-hand page show plate tectonic motions of the continents. Specific positions of North America are indicated in black. The maps also show the position of North America with respect to other continents at five specific times in the past.

New York Bedrock

Figure 18-10 is the "Generalized Bedrock Geology of New York" map from the *Earth Science Reference Tables*. The oldest rocks in New York are Precambrian. These rocks are exposed in the Adirondack Mountains of northern New York State and the Hudson Highlands between New Jersey and Connecticut. Paleozoic sedimentary rocks extend westward from these regions to lakes Erie and Ontario. Long Island has the youngest "bedrock." It is composed of geologically recent glacial sediments. Note that you can determine the absolute (numerical) age of the bedrock by using

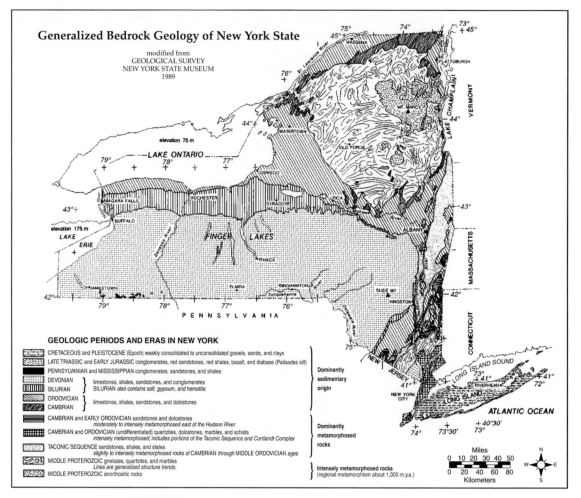

FIGURE 18-10. Paleozoic sedimentary rocks underlie the largest part of New York State, beginning in the western part of the state and ending near the Hudson River. The Adirondack region and Hudson Highlands are composed mostly of metamorphic rocks. Long Island is primarily composed of geologically recent sediments deposited by glaciers.

this map along with the geologic history chart, Figure 18-6 on pages 418–419.

HOW DO GEOLOGISTS CORRELATE ROCK LAYERS?

4: 1.2i, 1.2j

Geologists often compare rock layers in two or more locations. They do this to determine whether they are the same layers, or their relative ages. Matching bedrock layers by rock type or by age is **correlation**. Applying the principle of superposition (explained in Chapter 17) and using index fossils helps geologists make correlations. Correlation is an important part of drawing geologic maps such as Figure 18-10. Geologic maps tell the geologic history of an area.

Correlation is also useful in locating natural resources. Petroleum (crude oil), mineral ores, and groundwater are often found by drilling to a particular level in a rock sequence. Geologists can determine whether the drill is above or below the proper level by studying pieces of rock from the drill hole and knowing the local strata.

The first step in correlating rock layers in different locations is to find outcrops of bedrock. In most places, soil and sediment cover the rocks. An *outcrop* is a place where the solid lithosphere is exposed at Earth's surface. Erosion, such as on the steep bank along a stream or river, exposes bedrock. Fresh outcrops are also seen at road cuts. At these places, road crews had to drill and blast solid rock. Sometimes builders expose bedrock when they level the ground to construct homes or other structures.

Just because two places have the same kind of rock does not mean they are the same age. According to the principle of uniformitarianism, the processes and events that make different kinds of rock have occurred throughout geologic history. Furthermore, at any given time, different kinds of rock are formed in different environments all over the world. However, if geologists find the same sequence of rock types in two nearby locations, there is a good chance that rocks in both locations are the same age. The probability that the layers correlate is even greater if the geologists can follow the rock units from one outcrop to the next. Geologists try to do this when they draw geologic maps.

Comparing Fossil Types

Index fossils are very useful in correlation. Although many sedimentary rocks contain fossils, they are often difficult to see and of limited use in correlation. A geologist must look hard to find fossils that can be used to correlate rocks from one location to another.

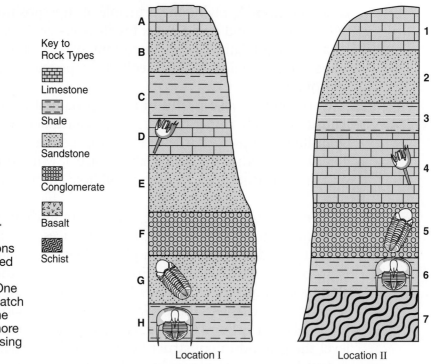

**Key to
Rock Types**

Limestone

Shale

Sandstone

Conglomerate

Basalt

Schist

FIGURE 18-11.
Rock layers in
different locations
can be correlated
(matched) in
several ways. One
method is to match
rock types. Time
correlation is more
reliable when using
fossils.

Location I   Location II

Figure 18-11 shows how rock types and fossils are used in correlation. If locations *I* and *II* are not too far apart, you can make the following inferences. Layers *A, B,* and *C* are probably the same age as layers *1, 2,* and *3*. The rock type and the sequence are the same. Layers *D* and *4* not only are the same rock type and occur in the same sequence, but they contain the same fossil. Fossils indicate that layers *G* and *H* correlate with layers *5* and *6*, respectively, although *G* and *5* are not the same kind of rock. (Fossil correlation is usually more dependable than correlation by rock types.) It appears that layers *E* and *F* are missing at location *II*. *E* and *F* may not have been deposited at location *II* or they may have been eroded. This would indicate an unconformity between layers *4* and *5* at location *II*. Furthermore, it is likely that the oldest rock layer in both locations is the schist at the bottom of location *II*.

In Figure 18-12, the dashed lines help to illustrate the following: (1) The mastodon, the condor, as well as the dinosaur layers at *B* are missing from location *A*. (2) The brachiopod fossils at *A* and *B* show that the limestone and sandstone are similar in age.

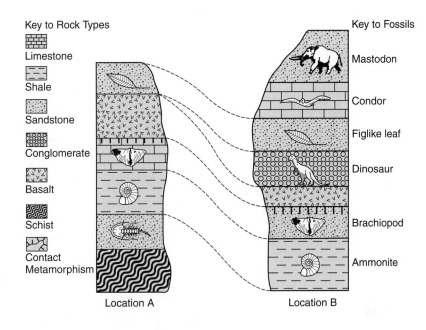

Key to Rock Types

Limestone

Shale

Sandstone

Conglomerate

Basalt

Schist

Contact
Metamorphism

Key to Fossils

Mastodon

Condor

Figlike leaf

Dinosaur

Brachiopod

Ammonite

Location A   Location B

FIGURE 18-12.
Fossils can be used to time-correlate layers in different outcrops even when some layers are missing in one location or the other.

(3) The two bottom layers at location *A* seem to be older than any exposed rock at *B*.

Figure 18-13, on page 430, is an exercise in correlation. In using these diagrams, assume that if a fossil species is not shown in its ocean or land environment, it did not exist at that time. How can the following be concluded from this diagram? (1) One outcrop is a marine environment, while the other is terrestrial. (2) Layer *A* can be only Cambrian or Ordovician in age. (3) Outcrop *I* (only) is overturned. (4) Layers *D* and *F* must both be Cretaceous in age.

Catastrophic Events

A large volcanic eruption sends great amounts of ash into the atmosphere. When Mount St. Helens exploded in 1980, it threw a cubic kilometer (35 billion ft³) of ash and rock debris into the air. The material went as high as 18 km (12 mi). In 1815, the Indonesian volcano Tambora ejected an estimated 30 times as much ash. Most of the debris from these explosions settled near the volcano. However, fine volcanic ash carried into the winds of the stratosphere was deposited over a large area in days or weeks, an instant of geological time.

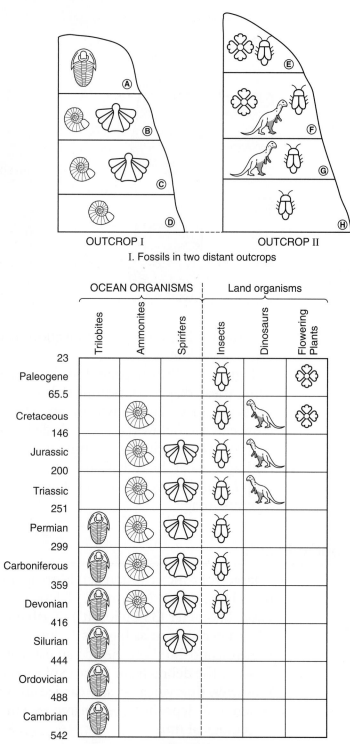

I. Fossils in two distant outcrops

II. Correlation chart

FIGURE 18-13.
What can you infer information about Outcrop I and Outcrop II at the top of this diagram by using the key below?

Catastrophic events leave time markers in rock layers. Scientists may find a layer of ash in the rock strata over a broad area. They use it is as a precise time marker, which can be found in almost any environment of deposition. These precise time markers are very useful in the regional correlation of rock layers.

The impact of large asteroids, meteorites, or comets creates similar time markers. The impact of an object about 10 km (6 mi) across ended the Mesozoic Era. The impact produced a crater more than 100 km (60 mi) in diameter on the Yucatan Peninsula of Mexico. Rock from the crater and the remains of the object formed a dense cloud of dust. The dust was so widespread it blocked the sun. This led to dramatic climatic changes over the whole planet. The event also deposited a distinctive layer of ash. This layer can be found in sedimentary rocks on all the continents and in places on the floor of the ocean. This layer of ash very precisely marks the upper boundary of rocks of Mesozoic age. Visit the following Web site to learn about New York State Geology field trip locations: *http://geology.about.com/od/geolog_y_ny/New_York_State_Geology.htm*

CHAPTER REVIEW QUESTIONS

Part A

1. What is the sequence of the geologic age of the bedrock as you travel from Ithaca, New York, to Watertown, New York?

   (1) Devonian, Silurian, Ordovician
   (2) Ordovician, Tertiary, Pleistocene
   (3) Devonian, Silurian, Cambrian
   (4) Ordovician, Taconic, Cambrian

2. What does the presence of sea-living eurypterid fossils in New York State bedrock indicate?

   (1) Eurypterids lived in land environments.
   (2) Parts of New York State were once covered by shallow seas.
   (3) Most of New York State was once a mountain region.
   (4) Eurypterids first appeared on Earth during the Devonian Period.

3. Which sequence of New York State index fossils shows the order in which the organisms appeared on Earth?

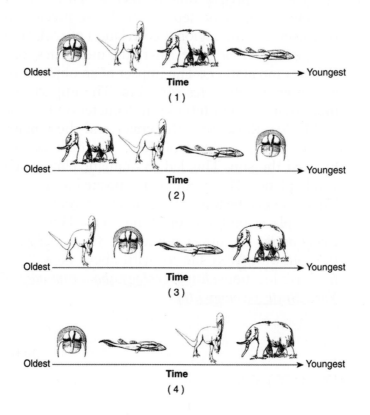

4. Based on fossil evidence, most scientists infer that

   (1) life-forms have not changed significantly throughout Earth's history
   (2) life has evolved from complex to simple forms
   (3) many organisms that lived on Earth have become extinct
   (4) mammals first appeared early in the Precambrian Era

5. *Cystiphyllum,* a solitary coral, *Baragwanathia,* a lycopod (an early land plant), and *Palaeophonus,* a scorpion (one of the first land animals) all lived at the same time. Which other life form reached its peak development during the same geologic period during which these three life-forms first appeared?

   (1) dinosaurs
   (2) stromatolites
   (3) eurypterids
   (4) mastodons

6. Which group of fossil organisms is inferred to have existed for the *least* amount of time in geologic history?

(1) placoderm fish   (3) dinosaurs
(2) trilobites   (4) eurypterids

7. Which pie graph below best represents the four major divisions of geologic time?

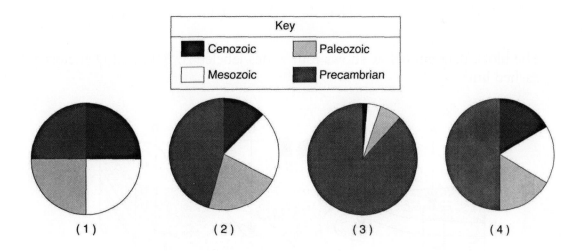

8. Which index fossil shows that forests existed in New York State approximately 400 million years ago?

(1) *Aneurophyton*   (3) *Centroceras*
(2) *Cystiphyllum*   (4) *Bothriolepis*

9. The geologic profile below represents a bedrock outcrop in New York State that has not been overturned. Certain index fossils are shown in the outcrop.

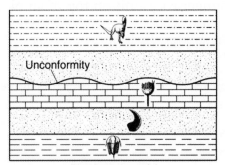

(Not drawn to scale)

Which New York State index fossil may have been destroyed by the erosion that produced the unconformity?

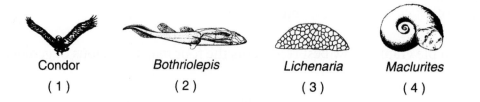

Condor	Bothriolepis	Lichenaria	Maclurites
(1)	(2)	(3)	(4)

10. The block diagram below shows four zones labeled *A, B, C,* and *D* enclosed by dashed lines.

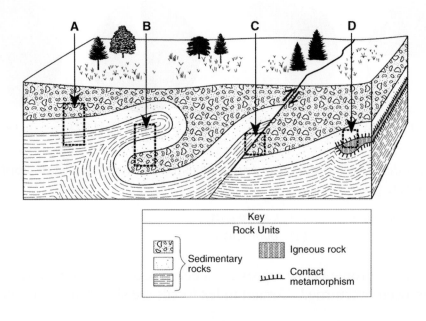

In which zone is a younger rock unit on top of an older rock unit?
(1) *A*   (3) *C*
(2) *B*   (4) *D*

11. Fossils of which absolute age existed in the Jurassic period?

(1) 50 million years
(2) 150 million years
(3) 300 million years
(4) 1 billion years

12. Which letter on the time line below best shows the first appearance of humans on Earth?

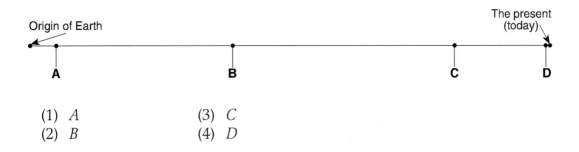

(1) *A*   (3) *C*
(2) *B*   (4) *D*

Part B

Base your answers to questions 13 through 15 on the map below, which shows where metamorphic bedrock of the Grenville Province is exposed at Earth's surface.

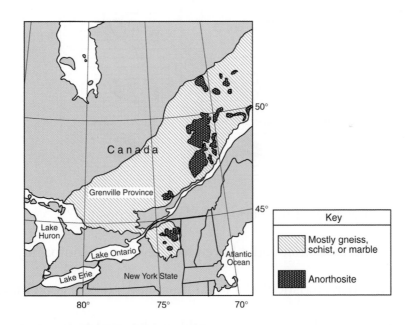

13. What is the approximate age of the bedrock of the Grenville Province?

(1) 250 million years
(2) 400 million years
(3) 560 million years
(4) 1,100 million years

14. Which New York State location has surface bedrock that is mainly anorthosite?

(1) Old Forge (3) Mt. Marcy
(2) Massena   (4) Utica

15. Which location has surface bedrock that consists mostly of gneiss, schist or marble?

(1) 43°N, 81°W (3) 47°N, 69°W
(2) 46°N, 78°W (4) 49°N, 71°W

Part C

Base your answers to questions 16 through 19 on the diagrams below, which represent two bedrock (*I* and *II*) outcrops several kilometers apart in New York State.

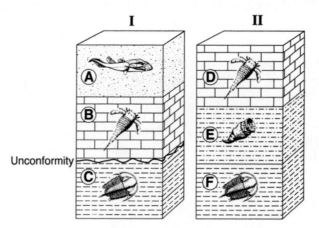

16. During which geologic time period was layer *C* deposited?

17. Identify *two* processes that could have produced the unconformity in outcrop *I*.

18. Describe one characteristic that a fossil must have to be a good index fossil.

19. Explain one reason that carbon-14 could not be used to find the ages of these fossils.

20. If you wanted to find evidence of dinosaurs in the bedrock of New York State, near which major city should you look?

The Dynamic Atmosphere

I t is easy to take air for granted. We know it is essential to keep us alive, but what larger role does Earth's atmosphere play? How would the physical characteristics of Earth's environments be different if Earth did not have an atmosphere?

The atmosphere is more than just the air we breathe. It is a blanket that keeps Earth warm. Our atmosphere helps distributes heat around the globe. The atmosphere is dynamic because it changes. It heats and cools, it absorbs moisture and releases moisture, and it transports moisture from one place to another. The atmosphere also blocks some of the harmful radiation from the sun.

Of the three nonliving parts of Earth (geosphere, hydrosphere, and atmosphere) the atmosphere is by far the smallest by mass, but it is also the most dynamic. Air can flow freely around the whole globe in a matter of days. At any given location, atmospheric conditions of temperature, humidity, and density (air pressure) can change very rapidly. A good example is the passage of a thunderstorm.

The principal role of the atmosphere is to distribute heat energy over Earth's surface. Our planet receives the most energy where the noon sun is always high in the sky in the tropics. The atmosphere, coupled with the oceans, moves some of this energy toward the poles to prevent extreme conditions of temperature in both the tropics and the polar areas.

19

Weather Variables and Heating of the Atmosphere

WORDS TO KNOW

angle of insolation	humidity	relative humidity
barometer	insolation	saturated
climate	meteorology	scatter
duration of insolation	nuclear fusion	temperature
fluid	precipitation	thermometer
fog	reflection	weather
greenhouse effect	refraction	

This chapter will help you answer the following questions:

1. What are the characteristics of the atmosphere that we call weather?
2. How do we measure weather variables?
3. How do cycles of sunlight affect Earth?
4. How do Earth systems distribute the energy Earth receives from the sun?

WHAT ARE THE ELEMENTS OF WEATHER?

4: 2.1c, 2.1d

Weather is the short-term condition of Earth's atmosphere at a specific time and place. The principal difference between weather and climate is the amount of time under consideration. **Climate**

is the average weather conditions of a specific region over a long period of time—at least 10 years, or even hundreds of years.

Scientists who study weather are *meteorologists*. The science of Earth's atmosphere and how it changes is **meteorology**. There are seven variables that meteorologists measure to describe the weather: temperature, humidity, air pressure, wind speed and direction, cloud cover, and precipitation. Visit the following Web site to see the current weather in the United States: *http://www.weather underground.com*

Temperature

Most people's first concern about weather is the temperature. This gives us important information about how we need to dress to stay comfortable and maintain our health. If we are too hot or too cold we change our clothing or we move to a more comfortable environment. Humans can thrive in a wide variety of temperature conditions.

Scientists define **temperature** as a measure of the average kinetic energy of the molecules in a substance. You learned that matter is made of extremely small units called atoms. Most atoms join in fixed ratios to form molecules. Molecules are the basic units of most elements such as oxygen (O_2) and hydrogen (H_2), as well as chemical compounds such as water (H_2O). At any temperature above absolute zero ($-273°C$ or $-460°F$), all atoms and molecules are in motion. This motion is a form of kinetic energy known as heat. In solids and liquids, most of the molecular motion is confined to vibrational motion. However, the molecules in a gas have more freedom of movement.

Matter exchanges kinetic energy with its surroundings. If matter is relatively cool, it will absorb more energy than it gives off and its temperature will increase. If matter is at a higher temperature than its surroundings, it will give off more energy than it absorbs and become cooler. Matter in temperature equilibrium (balance) with its environment continues to exchange energy, but the energy lost equals the energy gained.

MEASURING TEMPERATURE Temperature is usually measured with a **thermometer**. Common thermometers, such as the one in Figure 19-1, on page 440, work because matter usually expands

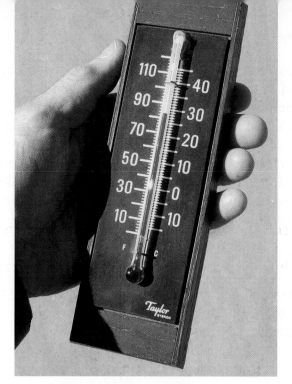

FIGURE 19-1. Some thermometers contain colored alcohol in a glass tube. The alcohol expands up the tube when it is heated, or contracts when it cools, registering changes in temperature.

when it absorbs heat energy. Many thermometers contain a liquid such as alcohol in a small bulb, or reservoir, at the bottom of a narrow tube. As the liquid absorbs energy, it expands and moves upward into the tube. This indicates an increase in temperature. When the liquid cools, it contracts and registers a lower temperature. A scale printed on the thermometer tube or on its base allows a person to read the temperature. A coloring agent is added to alcohol to make it easier to see.

STUDENT ACTIVITY 19-1 —MAKING A THERMOMETER

6: MODELS 2
6: MAGNITUDE AND SCALE 3

You can make a thermometer using a flask, a narrow glass tube, a one-hole stopper, and water as the liquid that expands and contracts. A ruler along the tube or attached to the tube can be the scale.

You can observe an interesting and unusual property of water if you start with water very close to the freezing temperature in the flask (thermometer bulb). As your thermometer warms to near room temperature, record the changing water level in the tube. What unusual property of water does this procedure show?

When measuring temperature, keep the thermometer in position for a minute or so. Heat energy must flow into or out of the thermometer. Only when the thermometer has reached equilibrium with the environment will it indicate the correct temperature.

TEMPERATURE SCALES   Figure 19-2 from the *Earth Science Reference Tables* shows three commonly used temperature scales. The Fahrenheit scale is generally used in the United States. On this scale, ice melts, or water freezes at 32°F and boils at 212°F. Since the difference between those values is 180°, each Fahrenheit degree is $\frac{1}{180}$ of the difference.

The Celsius scale is used in nearly every other country and in scientific investigations. On the Celsius scale, 0°C is the freezing

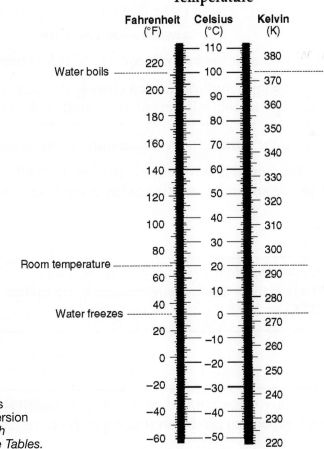

FIGURE 19-2. This temperature conversion chart is in the *Earth Science Reference Tables.*

point and 100°C is the boiling point of water. Since the difference between these values is 100°, each Celsius degree is $\frac{1}{100}$ of the difference. Therefore, a change of 1°C is a larger change than 1°F.

The third scale of temperature is the Kelvin scale. Although a change of 1 K (no degree sign is used) is the same as a change of 1°C, the Kelvin scale starts at absolute zero, the lowest possible temperature of matter. Molecules have no kinetic energy at this temperature. Like the speed of light (300,000,000 m/s), absolute zero is one of the few natural limits known to scientists. Therefore, there are no negative temperatures on the Kelvin scale.

You can use Figure 19-2, on page 441, to convert temperatures between the Fahrenheit, Celsius, and Kelvin scales. To convert from one scale to the other, simply look horizontally to the corresponding value on the proper scale.

SAMPLE PROBLEM

Problem 1 What Celsius temperature is equal to 0° Fahrenheit?

Solution On the temperature conversion chart in the *Earth Science Reference Tables,* 0° Fahrenheit is next to its equivalent of −18° Celsius.

Problem 2 What is the normal room temperature on the Celsius scale?

Solution On the temperature conversion chart in the *Earth Science Reference Tables,* room temperature is labeled at about 78°F or about 20°C.

Practice Problem
Which of the following temperatures is the coldest: −20°F, −20°C, or 220 K?

Humidity

The water vapor content of the air is **humidity**. The amount of water vapor that can be present depends on the temperature of the

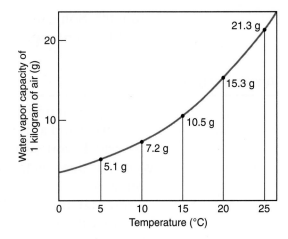

FIGURE 19-3. The amount of water vapor that can enter the atmosphere is a function of temperature. For every 10°C (18°F) rise in temperature, approximately twice as much water vapor can enter the atmosphere.

air. At higher temperatures, more water can evaporate and enter the air. Figure 19-3 shows the relationship between air temperature and the amount of water vapor that can enter the air. Sometimes people say that warm air can hold more water vapor than cooler air. This is not technically correct. The atmosphere is not like a sponge. Air does not "hold" water vapor. Water vapor could be present even if the atmosphere had no other gases.

We do not usually feel humidity as easily as we feel temperature. However, moist air may seem "clammy." In humid weather, sweat does not evaporate easily. This makes it more difficult to keep a comfortable body temperature. In the deserts of Arizona, the summer temperatures are often above 100°F (37°C); yet, low humidity allows sweat to evaporate and cool the body.

Meteorologists usually talk about **relative humidity**. This compares the actual water vapor in the air with the maximum amount of water vapor possible at a given temperature. For example, if the relative humidity is 50%, the air contains half as much moisture as it possibly could at the present temperature. At 100% relative humidity, the atmosphere contains all the water vapor it can at that temperature. The air is **saturated**. Most people find relative humidity more useful than humidity expressed in other ways.

Water vapor is the most important reservoir of energy in the atmosphere, and it plays a critical role in weather changes and violent weather. There are a variety of instruments to measure humidity. Some instruments use materials that expand or contract when they are exposed to moisture.

Air Pressure

You may be familiar with pressure in your daily life. For example, bicycle and automobile tires need a specific pressure to work properly. If you swim underwater, you may feel pressure when you dive deep below the water's surface. *Pressure* is force per unit area.

You may think air is light. However, Earth's atmosphere is so large that it has a mass of 5,000 million tons. Gravity acts on the atmosphere so it exerts a pressure of approximately 14.7 pounds on each square inch of Earth's surface. (See Figure 19-4.) We seldom notice changes in air pressure because it varies within a small range near Earth's surface.

MEASURING AIR PRESSURE Air pressure is measured with a **barometer.** There are two kinds of barometers, as shown in Figure 19-5. Air pressure pushing on the surface of a dish of mercury will push a column of mercury to a height of nearly 30 inches in a tube from which air has been removed. (Mercury is a dense, toxic metal that is a liquid at ordinary temperatures.) If the air pressure is high, the mercury is pushed a little higher in the tube. If it is low, the column of mercury does not rise as high. For this reason, atmospheric pressure is sometimes reported in inches of mercury.

The second kind of barometer, an aneroid barometer, contains a flexible airtight box. The box expands or contracts with changes in air pressure. This motion is transferred to an indicator dial that is calibrated in units of atmospheric pressure. Figure 19-6 is a scale to convert units of pressure from millibars (mb) to inches of mercury (in. Hg). For example, normal atmospheric pressure is 1013.2 mb, which is equal to about 29.92 in. of Hg.

Air pressure is something we seldom notice. Why, then, do meteorologists usually include it in their weather reports? In Chapter 22, you will learn how differences in air pressure cause winds. Atmospheric pressure (barometric pressure) can also be used to

FIGURE 19-4. Air pressure is caused by the weight of the atmosphere. Above each square inch of Earth's surface at sea level is a column of air that weighs 14.7 pounds.

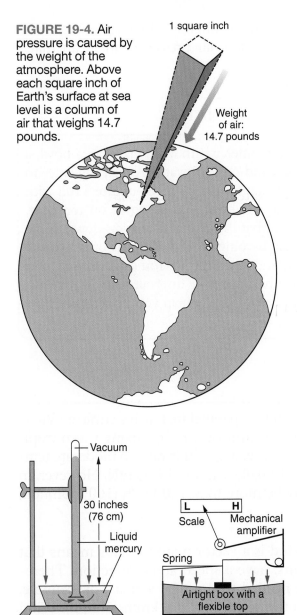

1 square inch

Weight of air: 14.7 pounds

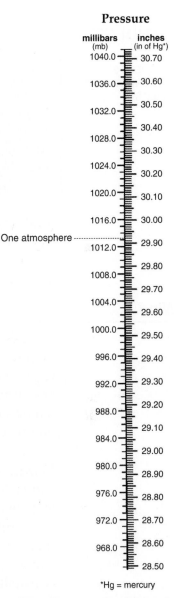

Pressure

millibars (mb)	inches (in of Hg*)
1040.0	30.70
1036.0	30.60
1032.0	30.50
	30.40
1028.0	30.30
1024.0	30.20
1020.0	30.10
1016.0	30.00
One atmosphere 1012.0	29.90
	29.80
1008.0	29.70
1004.0	29.60
1000.0	29.50
996.0	29.40
992.0	29.30
	29.20
988.0	29.10
984.0	29.00
980.0	28.90
976.0	28.80
972.0	28.70
968.0	28.60
	28.50

*Hg = mercury

FIGURE 19-5. Within a simple mercury barometer, (l) the force of atmospheric pressure balances the weight of a column of mercury. As air pressure changes, so does the height of the mercury in the tube. For this reason air pressure is sometimes reported in inches of mercury. An aneroid barometer (r) has an airtight box that expands and contracts as the air pressure changes. That motion moves a pointer on a scale.

FIGURE 19-6. Use this scale, found in the *Earth Science Reference Tables,* to convert between millibars (mb) and inches of mercury (in. Hg).

predict weather changes. Fair weather is often associated with rising barometric pressure. Falling air pressure can mean that a storm is approaching.

STUDENT ACTIVITY 19-2 —MAKING A BAROMETER

6: MODELS 2
6: MAGNITUDE AND SCALE 3

You can make a simple barometer with a tall glass jar, a bowl, a ruler, two rubber bands and colored water. Attach the ruler to the jar with rubber bands. Next, fill the jar three-quarters full with colored water. Securely hold the bowl, upside-down on top of it. Then flip the jar and bowl so that the jar stands upside-down in the bowl. Now you have a basic barometer. Changing air pressure will cause the water level to rise and fall in the jar. You can calibrate your barometer using a standard barometer or Internet sources for your location. Figure 19-5, on page 445, may help you understand how to make this barometer.

Wind

Wind is the flow of air along, or parallel to, Earth's surface. Winds help to transfer energy in the atmosphere by mixing warm tropical air with cold polar air. In this way, winds help to moderate temperatures on our planet. Usually, the wind is gentle. However, in major storms, winds may be more than 160 km/h (100 mph) and do great damage.

MEASURING WIND   Wind is a vector quantity. This means that winds, like ocean currents, have a speed and a direction. Therefore, two quantities are needed to describe winds. For example, the wind might be from the southwest at 8 km/h (5 mph). Wind speed is usually measured with an anemometer, which has a cup assembly or blades that rotate. The speed at which it rotates is usually transferred to an indicator on a calibrated dial. Anemometers are often combined with a wind vane, which points into the wind, indicating the wind direction. Wind is always labeled by the direction it comes from, not the direction it is moving toward. Figure 19-7 shows a combination anemometer and wind vane.

FIGURE 19-7. An anemometer measures wind speed. The turning blades send a signal to a device to indicate the wind speed. The tail automatically turns the instrument into the wind and shows the wind direction.

STUDENT ACTIVITY 19-3 —MAKING A WIND GAUGE AND WIND VANE

6: MODELS 2
6: MAGNITUDE AND
SCALE 3

You can make a simple wind gauge by wrapping one end of a piece of paper or thin cardboard around a pencil or coat hanger wire, letting the other end fall free as shown in Figure 19-8. Wind can lift the free end of the paper at an angle. If you attach a protractor, the angle can be related to the wind speed. To make the device work better in strong breezes, a paper clip or a coin can be added as a weight, as shown in Figure 19-8. A thread on the end of this device can be used with a magnetic compass to measure the wind direction.

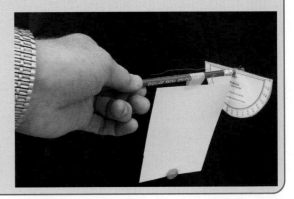

FIGURE 19-8. You can make a device to show wind speed and direction from very simple materials.

Cloud Cover

Clouds determine the amount of sunlight that reaches Earth's surface. The amount of cloud cover and types of clouds can help predict the likelihood of precipitation. Cloud cover is often expressed as the approximate percent of the sky that is covered at a given time. If clouds cover one-quarter of the sky, there is 25 percent cloud cover. Very low clouds that reach ground level are known as **fog**. Fog can reduce visibility and therefore affect transportation. Cloud formation and its role in weather development will be discussed in Chapter 20.

Precipitation

Clouds are made of tiny ice crystals or water droplets that are too small to fall through the air. The water droplets are called cloud droplets. Their behavior is similar to the smallest particles of sediment suspended in water. If these ice crystals or water droplets grow large enough, they can fall as **precipitation**. A raindrop can be a million times the size of a cloud droplet. Figure 19-9 shows a comparison of cloud drops and raindrops.

Most clouds are high enough in the atmosphere that the temperature is below freezing. Even when cloud droplets are below

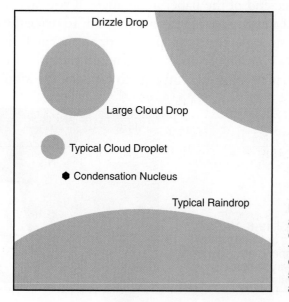

FIGURE 19-9. Raindrops are much larger than cloud droplets. That is why they fall as precipitation. These droplets are actually about 50 times smaller than they are shown in this diagram.

the freezing point, they can remain in the liquid state. However, if it is cold enough, separate snow crystals form and grow much faster than cloud droplets. The snow crystals also must become large enough to fall as precipitation. In warm weather, the snow melts as it falls into warmer air, changing to rain before it reaches the ground.

MEASURING PRECIPITATION Rain is measured with a rain gauge. Any cylindrical container with straight sides that is open at the top can be used as a simple rain gauge. The amount of rain is measured as a depth over the surface of the land. It can be reported in inches, centimeters, or any other convenient units of length. Snow can be measured as a depth of accumulation, or it can be melted to find the equivalent quantity of rain.

OTHER FORMS OF PRECIPITATION Rain and snow are the most common forms of precipitation. However, there are several others. *Drizzle* is very small raindrops that fall more slowly than normal rain. *Sleet* is a mixture of rain and snow that falls when the temperature is close to freezing. *Freezing rain* falls as a liquid but freezes, forming a coating of ice on exposed surfaces.

STUDENT ACTIVITY 19-4 —RECORDING WEATHER VARIABLES

4: 2.1e
6: PATTERNS OF
CHANGE 5

If you make or buy basic weather instruments (thermometer and barometer or hygrometer), you can record daily weather conditions. You may want to graph your observations through time. What patterns do you see in weather changes? How do the variables relate to one another? In Chapter 22, you will learn how meteorologists use weather maps and computers to make predictions of future weather. However, you can use your readings to make your own predictions. Future measurements will let you check your own predictions.

Hail is precipitation made of ice. It usually occurs in violent thunderstorms. Hailstones usually begin as snowflakes that start to melt. The flakes gather a coating of moisture as they fall. Strong updrafts in the storm blow the wet flakes back up into colder air where the coating of moisture freezes, forming small hailstones. Hailstones grow as they are water-coated and re-frozen many times, often forming rings that are visible in larger hailstones. Eventually they become heavy enough to fall through the updrafts and reach the ground. Hailstones are usually less than a centimeter in diameter. However, hail the size of baseballs has fallen during especially severe thunderstorms in the American Midwest.

Placement of Weather Instruments

When meteorologists place their weather instruments, they are very careful. For example, if a thermometer is in direct sunlight, its readings will not be accurate. The liquid in the thermometer will be heated by sunlight, not the air. For this reason, meteorologists place their thermometers and psychrometers in a weather shelter. (See Figure 19-10.) They measure precipitation where buildings will not block or add to the readings. Snow depth can be particularly difficult to measure. Wind moves snow from open areas. Drifts form where the wind is blocked. Meteorologists measure wind in an open area where it is not affected by buildings or other structures. Visit the University of Utah Web site to see state-by-state the current weather: *http://mesowest.utah.edu/index.html*

FIGURE 19-10.
Placement and protection of weather instruments is important in recording accurate weather data.

HOW DOES THE SUN WARM EARTH?

4: 2.1a, 2.1b
4: 2.2a, 2.2b

Sunlight is the primary source of atmospheric energy. For many years, scientists tried to understand how the sun could create so much energy and never appear to burn out. The discoveries that led to radiometric dating of rocks and fossils helped scientists understand solar energy.

The sun is mostly hydrogen. Under the extremely high heat and pressure within the sun, hydrogen nuclei join in **nuclear fusion**. When four hydrogen nuclei form a helium nucleus, a large amount of energy is released as solar mass is converted to energy.

The Sun's Energy

Solar energy moves to the surface of the sun mostly by convection currents. From the surface, energy radiates into space as electromagnetic waves in sunlight. *Radiation* is the transfer of energy in the form of electromagnetic waves. It is the only way that energy can travel through the vacuum of space. Therefore, radiation is the only way heat and light can travel from the sun to Earth. Radiation travels at the speed of light. In fact, it takes about 8 minutes for solar energy to reach Earth. Earth receives only a tiny portion of the energy given off by the sun. However, this energy powers hurricanes, tornadoes, and all other dynamics of the atmosphere.

Electromagnetic Waves

The electromagnetic spectrum includes a wide range of radiant energy. This energy travels as waves similar to the waves on a lake or an ocean. You may have been surprised to learn in Chapter 11 that waves on water do not carry the water with them. Like electromagnetic waves, waves on water carry energy. However, we are not able to see electromagnetic waves in the same way that we see waves on water. Still, they have similar measurable properties.

One property common to all energy waves is amplitude. In waves on water, amplitude is the height of the wave. For light, amplitude is a measure of strength or brightness. Another measure of energy waves is wavelength. This is the distance from the top of

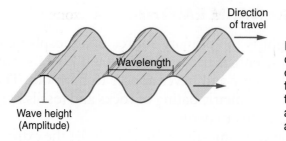

FIGURE 19-11. All waves carry energy. This diagram of waves on water illustrates two characteristics common to all waves: wavelength and wave height (or amplitude).

one wave to the top of the following wave. Figure 19-11 illustrates these properties for waves on water.

Figure 19-12, from the *Earth Science Reference Tables,* shows the range of wavelengths of electromagnetic energy. Only a narrow band of this energy is visible light. We cannot see other wavelengths of electromagnetic energy. However, they are important to us. Gamma rays and x-rays have the shortest wavelengths and are used in industry and medicine. Ultraviolet solar radiation damages skin and can cause skin cancer.

Most of the sun's energy reaches Earth's surface as visible light, which enables us to see. Beyond visible light is invisible long-wave radiation. Infrared rays and microwaves are longer than visible light. We experience infrared rays as heat. We use microwaves to heat our food. Radio waves are used in communications for radio and television transmission.

Ozone and the Sun's Energy

Earth's atmosphere stops high-energy gamma rays and x-rays from reaching Earth's surface. Some ultraviolet rays do reach Earth's

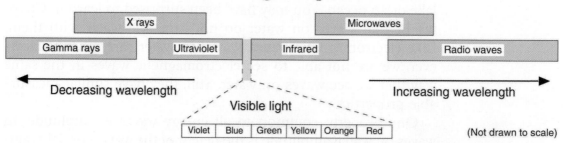

FIGURE 19-12. Sunlight contains a wide range of wavelengths. Most of the short-wave radiation is absorbed by the atmosphere. Insolation is about 10% ultraviolet, 40% visible light, and 50% infrared. The visible light is the only part of the electromagnetic spectrum that humans can see.

surface. However, the formation of ozone uses up most of them. In the upper atmosphere, ultraviolet radiation converts two-atom oxygen (O_2) to the three-atom molecule ozone (O_3).

There was no oxygen in Earth's early atmosphere. Photosynthesis did not begin until the mid-Precambrian time. This added oxygen to the atmosphere. The ozone layer formed in the upper atmosphere. Until the formation of the ozone layer, there was no life on the land. Life existed only in the oceans, where it was protected from harmful ultraviolet radiation. The formation of the ozone layer made the evolution of terrestrial life-forms possible. Even with the protection of the ozone layer, exposure to too much ultraviolet radiation is harmful and can be fatal.

In the late twentieth century, scientists discovered that the ozone layer was thinning rapidly. Scientists found that CFCs (chlorofluorocarbons), used as propellants in spray cans and in air conditioners as refrigerants, were drifting into the upper atmosphere. CFCs cause the conversion of ozone (O_3) to oxygen (O_2). Fortunately, compounds that do not destroy ozone have been substituted for CFCs. This is a good example of a human-caused threat to our environment that seems to be decreasing due to international efforts and new technology.

Insolation and Earth

Solar energy that reaches Earth is known as **insolation**. This word can be thought of as a shortened form of the phrase *incoming solar radiation*. Insolation is about 10% ultraviolet, 40% visible light, and 50% infrared. As sunlight travels deeper into the atmosphere, its speed decreases. This decrease causes the light to bend toward the ground. The bending of light as it moves from one substance to another of different density is known as **refraction**.

STUDENT ACTIVITY 19-5 —OBSERVING REFRACTION

1: SCIENTIFIC INQUIRY1

Place a long pencil or any straight object the same size into a see-through container filled with water. If you look along the pencil, you can see how it appears to bend at the surface of the water. Viewed from the side, the pencil may appear in two straight segments. Both observations are caused by the refraction of light rays.

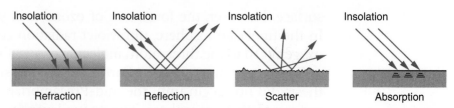

FIGURE 19-13. Sunlight reaching Earth is affected in four ways. It is refracted (bent) as it travels through the atmosphere. It is reflected by smooth surfaces or scattered by rough surfaces. It is mostly absorbed by dark-colored surfaces.

Solar energy that reaches the ground can be reflected or absorbed. Earth absorbs about half the insolation that reaches the ground. This raises the ground's temperature. Radiation that is not absorbed is reflected. **Reflection** is the process by which light bounces off a surface or a material. (Some objects such as clouds have no distinct surface, but they still reflect light.) Figure 19-13 shows four ways that Earth affects insolation.

Whether light is reflected or absorbed depends on the properties of the surface it strikes. Dark surfaces, such as soil and vegetation, absorb more insolation than light-colored surfaces, such as snow. After all, dark surfaces appear dark because relatively little light bounces off them to reach your eyes. In addition, rough surfaces absorb even more light. When light reflects off a rough surface, it is reflected in many different directions, as shown in Figure 19-13. This is called **scatter**. Some of the reflected light may strike the surface more than once. The result is more opportunities for the surface to absorb energy.

What portion of the energy that Earth receives from the sun is absorbed and what part is lost by reflection? Figure 19-14 shows that the ground absorbs about half of the insolation energy reaching Earth. Nearly one-third of it is reflected back into space. The rest is absorbed within the atmosphere.

Angle of Insolation

The strength of sunlight depends on the sun's position in the sky. The angle between Earth's surface and incoming rays of sunlight is the **angle of insolation**. The angle of insolation is also the angle of the sun above the horizon. If the sun is directly overhead, the angle of insolation is 90°. At sunrise and sunset, when the sun

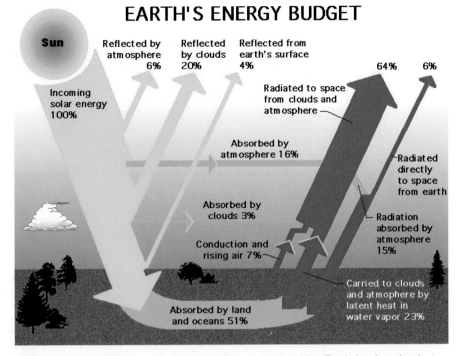

FIGURE 19-14. About half of the insolation received by Earth is absorbed at the surface. Eventually, all the incoming solar energy is reflected or radiated back into space. Since energy input and energy lost are equal, Earth's temperature remains roughly constant.

appears along the horizon, the angle of insolation is near 0°. The angle of the sun changes with time of day and with the seasons. When the sun appears highest in the sky, as it is around 12 noon, early in the summer, the strength of sunlight is greatest. There are several reasons for this.

First, when the sun appears high in the sky, light spreads over a smaller area as it reaches Earth's surface. A given amount of sunlight concentrated on a smaller area results in energy that is more intense. Figure 19-15, on page 456, shows that light is more intense when the source shines straight down on a surface. For most locations on Earth's surface, sunlight is most intense at noon on the first day of summer. You should note that for any location in New York State the sun is never directly overhead. However, the angle of insolation in New York is highest at noon on the first day of summer about, $73\frac{1}{2}°$. Only within tropical latitudes can the angle of insolation be 90°. Visit the following Web site to watch an

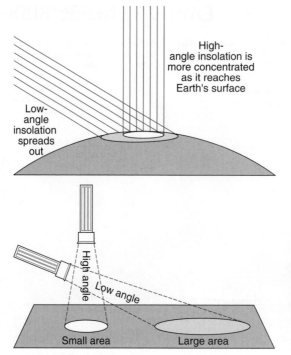

FIGURE 19-15. When the sun is directly overhead, its rays are concentrated in a small area. When the sun is low in the sky, the same energy covers more area, so the sunlight is not as intense.

applet that models the Sun's path in a geocentric view: *http:// engnet.anu.edu.au/DEpeople/Andres.Cuevas/Sun/SunPath/ SunPath.html*

The second reason sunlight is more intense when the sun appears high in the sky is related to absorption of energy by the atmosphere. Earth's atmosphere is not completely transparent, even to visible light. The lower the sun appears in the sky, the more atmosphere light must pass through to reach Earth's surface. Red light penetrates the atmosphere better than blue light. That is why the sun appears red when it is rising and setting, but not when it is high in the sky.

Furthermore, reflection increases when the sun appears low in the sky. You may have noticed that if you look into a still lake when the sun is nearly overhead, you can see deep into the water. If the sun appears low in the sky, reflected light bounces off the lake's surface and prevents you from seeing under the surface. Most surfaces absorb more light when the angle of insolation is greater.

In addition to the time of day and the season, the intensity of insolation at Earth's surface depends on latitude. Near the equator, the noon sun always appears high in the sky, so sunlight is

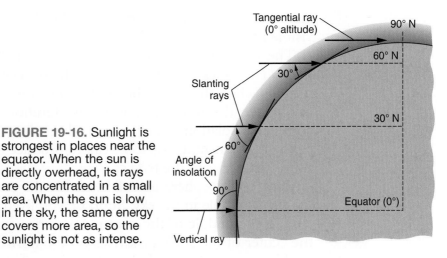

FIGURE 19-16. Sunlight is strongest in places near the equator. When the sun is directly overhead, its rays are concentrated in a small area. When the sun is low in the sky, the same energy covers more area, so the sunlight is not as intense.

relatively strong. However, near the poles the sun never appears far above the horizon. Polar regions receive less solar energy because the sun's rays spread over a larger area and the light of the sun has to pass through a greater thickness of the atmosphere. Figure 19-16 shows how the angle of the sun depends on latitude. The angles on this diagram represent noon on the first day of spring or the first day of autumn. Throughout the year, the tropical latitudes have the greatest angle of insolation.

Duration of Insolation

The longer you leave a container under an open water faucet, the more water flows into the container. The same principle is true for solar energy. Duration refers to a length of time. Therefore, the **duration of insolation** is the amount of time the sun is visible in the sky or the number of hours between sunrise and sunset. In 1 year, every place on Earth has a total of 6 months of daylight and 6 months when the sun is below the horizon.

Tropical regions have 12 hours of daylight every day throughout the year. For this reason, tropical climates are sometimes called the seasonless climates. Neither the hours of daylight nor the temperature changes very much throughout the year.

At higher latitudes, such as New York State, daylight is longer in summer and shorter in the winter. This is true both north and south of the equator. These areas receive more solar energy in the summer because the sun appears higher in the sky and the sun is visible in the sky for a longer time.

The greatest range in the length of daylight occurs near the poles. The poles experience 6 months of daylight followed by 6 months when the sun is below the horizon. Although the poles experience the longest continuous duration of insolation, the angle of insolation is always low. Therefore, although summer sunlight is constant, it is too weak to produce the warm temperatures that occur in the tropics.

Earth's atmosphere actually extends the amount of time that the sun is visible in the sky. Suppose you lived on a very small island in the middle of the ocean. At sunrise, you could see when the center of the sun's disk rises on the horizon in the east. You could measure accurately how long it took the sun to cross the sky and set when its center was level with the western horizon. On the equinox (the first day of spring or the first day of autumn) you might expect the sun's trip to take exactly 12 hours. However, it is actually a little longer than 12 hours. Refraction (bending) of sunlight in the atmosphere allows the sun to be visible after it has actually dipped below the horizon. Similarly, we can see the morning sun before it would be visible if Earth had no atmosphere.

HOW DOES SOLAR ENERGY CIRCULATE OVER EARTH?

4: 2.2a, 2.2b

Heat energy can move from one place to another in three ways. You already learned that the only way solar energy can reach Earth is by radiation through space. Although radiation is very fast (300 million, or 3×10^8, m/s) it can occur only through empty space or through transparent materials. Once solar energy reaches Earth, it is distributed over the planet's surface by conduction and convection.

Conduction

Earlier in this chapter you learned that when molecules absorb energy and their temperature rises, the molecules move more rapidly. When molecules touch, this energy is passed along from molecule to molecule. The transfer of energy by collisions of mol-

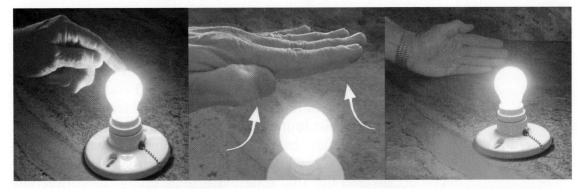

FIGURE 19-17. Energy travels in three ways. Conduction (*left*) occurs when one object touches another. Convection (*center*) carries energy with the air, water or other fluid that moves. Radiation (*right*) is the only form of heat (energy) flow through space. (You should not touch very hot objects such as light bulbs.)

ecules is *conduction*. Conduction occurs most efficiently in solids. Heat energy is conducted within Earth as well as between the solid Earth and the atmosphere.

Some substances such as metals are good conductors. We use metal containers such as frying pans to cook our food because heat is conducted easily through the metal from the heat source to the food. On the other hand, you do not want to burn yourself when you hold the frying pan. So, the handle of a frying pan is often made from a poor conductor, an insulator. Wood and plastic are good insulators.

Conduction (Figure 19-17, *left*) can be an effective way for heat to flow between two materials if they are touching. The atmosphere exchanges some heat energy with Earth's surface by conduction. However, conduction is slow, and the atmosphere is not a good conductor of heat energy.

Convection

The atmosphere distributes energy over Earth's surface mostly by convection. You have learned that all matter above a temperature of absolute zero (0 K) has energy. When matter moves, the energy it contains goes with it. Because the atmosphere is made up of gases, it flows freely from one place to another, carrying its energy with it. Liquids and gases are **fluids** because they can flow from place to place.

What makes air flow? When most substances are heated they expand and become less dense. Air is no exception. Under the influence of gravity, cooler air, which is more dense, flows under warmer air, replacing it. Therefore *convection* can be defined as a form of heat flow that moves both matter and energy as density currents under the influence of gravity. You will learn more about how convection distributes energy over Earth in Chapter 21.

Without heat flow by convection, differences in temperature over Earth's surface would be much greater than they are now. The oceans and the atmosphere distribute energy by convection currents. Energy is carried from tropical regions toward the poles. Cold fluids (water and air) flow from the polar regions to the tropics, where they can absorb solar energy. This cycle tends to even out the extremes in temperature on the planet.

Earth Radiates Energy

If Earth is always absorbing energy from the sun, why is the planet's temperature relatively constant? Constant insolation should cause Earth to become warmer, unless the planet loses as much energy as it absorbs. If the inflow of energy is equal to the outflow of energy, the energy of the object is in *equilibrium.*

Earth cannot lose energy by conduction because no other object in space is in contact with Earth. Heat loss by convection would require the flow of matter into space. However, very little matter escapes from Earth. We do not see radiation leaving the planet because Earth does not give off visible light like the sun. So how does Earth give off energy to maintain an energy balance?

You learned that most forms of electromagnetic energy are invisible. Experiments show that all objects above a temperature of absolute zero radiate electromagnetic energy. The form of energy radiated depends on the temperature of the object. The sun is hot enough to give off light energy. Earth is much cooler than the sun, so it must radiate energy in long wavelengths. In fact, the planet loses its energy as heat in the infrared portion of the spectrum. If our eyes were sensitive to infrared (like some military devices used to detect warm objects), we would see the planet glow with a constant flow of escaping heat.

There have been times in Earth's history when it was not in energy equilibrium. The ice ages occurred when the planet radiated more energy than it received from the sun. This imbalance led to colder temperatures and the growth of glaciers. Since the ice age glaciers melted, Earth's average temperature has risen.

The Greenhouse Effect

The atmosphere is only partly transparent to heat energy radiated by the ground. Carbon dioxide and water vapor in the atmosphere absorb infrared (heat) radiated by the ground. Eventually, the gases radiate this energy, some escapes, some is absorbed by other atmospheric gases, and some is again absorbed by earth. This keeps Earth's temperature warmer than it would be if carbon dioxide and water vapor were not in the atmosphere. (See Figure 19-14 on page 455.)

The process by which carbon dioxide and water vapor in the atmosphere absorb infrared radiation is known as the **greenhouse effect**. Perhaps you have seen a greenhouse used to grow flowers or vegetables in a cold location. It may seem strange that a transparent glass building can support plant growth. After all, the greenhouse has little or no heating and the ground outside may be covered with snow. To understand how a greenhouse works, you need to know more about light and heat. Short-wave visible sunlight passes through the glass and heats the inside of the greenhouse. The warm objects in the greenhouse give off long-wave infrared radiation. The glass is not transparent to infrared radiation. Thus these rays are reflected back into the greenhouse, as you can see in Figure 19-18 on page 462. By letting short-wave sunlight enter and not allowing the longer heat rays to escape, the greenhouse traps heat energy.

The atmosphere affects Earth in a similar way. However, Earth's atmosphere absorbs and reradiates energy rather than simply reflecting it. Although carbon dioxide and water vapor are the principal greenhouse gases, they are not the only greenhouse gases. Methane, nitrous oxide, and several other gases contribute to the greenhouse effect. Scientists estimate that the combined effect of these gases is to make Earth's surface about 35°C warmer than it would be without them. If it were not for the greenhouse effect, Earth would be too cold for life as we know it. Visit the following Web site to see digital images of weather phenomena: *http://www.weatherscapes.com/index.php*

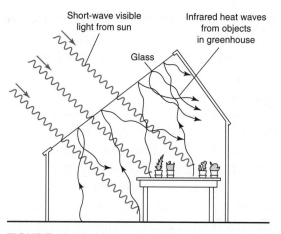

FIGURE 19-18. Visible light enters a greenhouse through the glass. The light warms objects inside the greenhouse. The heated objects give off infrared (heat) radiation, which cannot pass through the glass. The trapped energy warms the inside of the greenhouse.

A LESSON FROM VENUS The surface of Venus is hot enough to melt lead. Why is Venus so hot? It is not just because it is closer to the sun than Earth is. The surface of Venus is actually hotter than the hottest places on the surface of Mercury, even though Mercury is much closer to the sun than Venus. Venus is hotter because the atmosphere of Venus is 96 percent carbon dioxide and 100 times as thick as Earth's atmosphere. Some planetary scientists call Venus the planet with "the greenhouse effect gone wild."

CHAPTER REVIEW QUESTIONS

Part A

1. Which weather instrument is used to measure wind speed?

    (1) anemometer   (3) psychrometer
    (2) wind vane   (4) thermometer

2. Which surface below is most likely to absorb the greatest amount of insolation?

3. Which weather variable is measured by the instrument shown below?

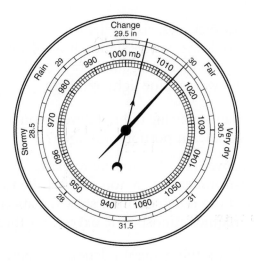

    (1) wind speed   (3) cloud cover
    (2) precipitation   (4) air pressure

4. Which process transfers energy primarily by electromagnetic waves?

    (1) radiation   (3) conduction
    (2) evaporation   (4) convection

5. When Earth cools, most of the energy transferred from Earth's surface to space is transferred by the process of

    (1) conduction   (3) refraction
    (2) reflection   (4) radiation

6. Great volcanic eruptions send gas and dust into the stratosphere. In the following weeks, air temperatures at Earth's surface are generally

    (1) cooler than normal because the atmosphere is less transparent
    (2) cooler than normal because the atmosphere is more transparent
    (3) warmer than normal because the atmosphere is less transparent
    (4) warmer than normal because the atmosphere is more transparent

7. Most of the solar energy absorbed by Earth is radiated back into space as which type of radiation?

    (1) x-ray   (3) infrared
    (2) ultraviolet   (4) radio waves

8. Which list contains electromagnetic energy arranged in order from longest to shortest wavelength?

    (1) gamma rays, ultraviolet rays, visible light, x-rays
    (2) radio waves, infrared rays, visible light, ultraviolet rays
    (3) x-rays, infrared rays, blue light, gamma rays
    (4) infrared rays, radio waves, blue light, red light

9. Which statement explains why life did not develop on land before Earth's atmosphere contained a significant portion of oxygen?

    (1) All land locations were too hot for life to exist.
    (2) All terrestrial environments were too cold for life to exist.
    (3) The land received too much harmful solar radiation.
    (4) All forms of life require atmospheric oxygen for their vital processes.

10. Which graph best shows the general relationship between the altitude of the noontime sun and the intensity of insolation received at a location?

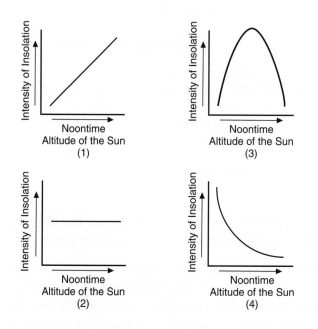

11. How do clouds affect insolation?

    (1) Clouds are transparent to all solar energy.
    (2) Clouds reflect all solar radiation.
    (3) Clouds absorb all solar radiation.
    (4) Clouds absorb and reflect insolation.

Part B

12. After conversion from inches of mercury to millibars (mb), which reading is indicated by the instrument shown below?

(1) 1009.0 mb   (3) 1015.5 mb

(2) 1012.5 mb   (4) 1029.9 mb

Base your answers to questions 13 through 15 on the graph below, which shows the amount of insolation at four different locations on Earth's surface.

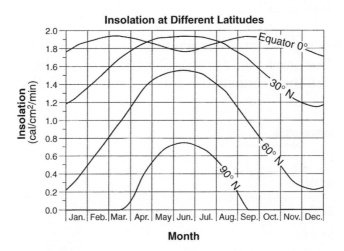

13. This graph shows that insolation varies with

(1) latitude and time of day

(2) latitude and time of year

(3) longitude and time of day

(4) longitude and time of year

14. Why is less insolation received at the equator in June than in March or September?

(1) The daylight period is longest at the equator in June.

(2) Winds blow insolation away from the equator in June.

(3) The sun's vertical rays are north of the equator in June.

(4) Thick clouds block the sun's vertical rays at the equator at noon.

15. Why is insolation 0 cal/cm²/min from October through February at 90°N?

    (1) Snowfield reflects sunlight at that time.
    (2) Dust in the atmosphere blocks sunlight during that time.
    (3) The sun is continually below the horizon during that time.
    (4) Intense cold prevents insolation from reaching Earth's surface at that time.

Part C

16. How does a greenhouse keep radiant energy from escaping?

17. Explain what a meteorologist means when he or she reports that the relative humidity is 50%.

18. What measure of the atmosphere tells us how much vibrational energy the air contains per unit mass?

19. The data below represent the temperatures at a mid-latitude location on a sunny, summer day. Use graph paper or a similar grid to graph this data. Show each data point with a dot, then connect the dots to form a graph line.

Time	Temperature
6 A.M.	17°C
8 A.M.	20°C
10 A.M.	22°C
Noon	23.5°C
2 P.M.	24°C
4 P.M.	23.5°C
6 P.M.	22°C

20. When you describe the wind, what two factors must be included?

20 Humidity, Clouds, and Atmospheric Energy

WORDS TO KNOW

absolute humidity	dry-bulb temperature	melting
boiling	freezing	phase of matter
cloud	frost	psychrometer
condensation	hygrometer	specific heat
condensation nuclei	joule	transpiration
dew	lake-effect storm	vaporization
evaporation	latent heat	wet-bulb temperature
dew-point temperature		

This chapter will help you answer the following questions:

❶ What is the role of water vapor in atmospheric energy?

❷ What factors affect evaporation of water?

❸ How is water vapor within the atmosphere measured?

❹ How do clouds form?

LET IT SNOW!

4: 2.1c
4: 2.2c

Most students like "snow days." The opportunity to sleep late, have some free time, and even enjoy the snow can be a welcome change from your usual school routine. The likelihood of winter storms

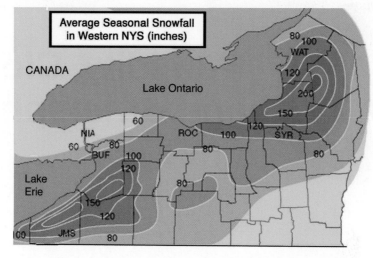

FIGURE 20-1. The two locations with the greatest average winter snowfall in New York State are located east of Lake Erie and Lake Ontario.

leaving a thick blanket of snow depends on where you live. People who live near the eastern shore of Lake Erie and Lake Ontario receive more snow than any other locations in New York State.

The Tug Hill Plateau, east of Lake Ontario, holds the state record for snowfall. Fortunately, the Tug Hill Plateau is one of the least populated areas of New York State and local residents have learned to cope with these monster storms. The Buffalo area, near the eastern end of Lake Erie, also receives more than its share of the winter snow. Unlike the Tug Hill Plateau region, the Buffalo metropolitan area has a population of more than 1 million people. Heavy snowfalls can cause many problems for the people of Buffalo. Figure 20-1 shows the average annual snowfall in inches for parts of western New York State. Notice the two areas of maximum snowfall just east of each of the two Great Lakes.

Lake-Effect Storms

Why do these areas receive so much snow? Both are in the path of **lake-effect storms**. These are mainly winter-weather events that are the result of a combination of atmospheric phenomena and geographic factors. The most common wind direction in New York State is from the west. Air traveling for more than 320 km (200 mi) across the Great Lakes picks up large amounts of moisture. In the winter, lakes do not cool as quickly as the air, and generally remain free of ice cover. Therefore, in early winter, air blowing over the warm water readily absorbs moisture.

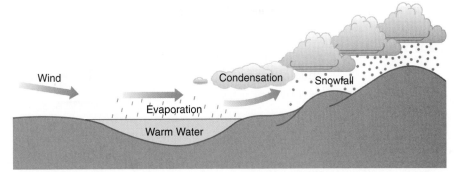

FIGURE 20-2. Lake-effect storms are most common in the early winter, when the water in the lakes is warmer than the surrounding land. Winds blowing over the Great Lakes absorb moisture. When the air cools over land, clouds and precipitation form. Additional cooling caused by the moist air rising over the hills increases precipitation.

However, as this moist air moves over land, it cools. When air cools, clouds and precipitation form. Precipitation increases when air rises over the hills at the ends of the lakes. If the westerly winds continue, the process is like a conveyor belt, picking up moisture from the lakes and dumping it on the land over a period of days. Figure 20-2 shows how lake-effect storms form.

STUDENT ACTIVITY 20-1 —RATE OF EVAPORATION

1: MATHEMATICAL
ANALYSIS 1
1: SCIENTIFIC
INQUIRY 1, 2
6: OPTIMIZATION 6

You can use a small sponge cube to investigate the variables that affect evaporation. Attach the cube to a string or thread by pulling the string through the cube.

After allowing the cube to absorb water, gently squeeze out excess water that might drip. Measure the mass of the wet cube with an electronic balance at the beginning and the end of a uniform evaporation time.

Each group should select a location in the classroom where the cube will experience evaporation. Make sure cubes are placed throughout the classroom. Consider limitations such as requiring that cubes be placed at least a minimum distance from electrical devices.

The percent of original mass lost indicates evaporation. Nothing other than air should touch the wet cube, which must be gently handled by its string.

Report on factors that you think were most important in maximizing evaporation.

HOW DOES THE ATMOSPHERE STORE ENERGY?

4: 2.1a, 2.1b
4: 2.2a, 2.2b

It is clear that large quantities of energy are stored in the atmosphere and released in weather events. That energy is stored in a special form of heat energy called latent heat. Latent means hidden or not apparent.

States of Matter

Water, like other forms of matter, exists on Earth in three states: solid, liquid, and gas. Solid water is ice. Liquid water is simply called "water." In the form of a gas it is known as water vapor. The temperatures that occur on our planet allow us to experience water in all three states in the natural surface environment. The expression **phases of matter** is sometimes used as a synonym for states of matter.

When water changes its state, a great deal of energy is involved. Heating that causes a change in temperature shows up as increased kinetic energy of molecules. However, energy that transforms solids to liquids and liquids to gases does not show up as a temperature change. In this sense, it is hidden, or latent energy. **Latent heat** is therefore energy that is absorbed or released when matter changes state.

A variety of units can be used to measure energy. Scientists use **joules** (J) to measure energy. Different substances absorb different amounts of energy as their temperatures change. The reason is that they have different specific heats. **Specific heat** is the energy needed to raise the temperature of 1 gram of a substance 1 Celsius degree, without changing its state. The higher the specific heat of a substance, the more energy needed to raise its temperature. For example, it takes about 33 times as much energy to heat 1 g of water (specific heat = 4.18 J/g ● °C) 1° C than it takes to heat 1 g of lead (specific heat = 0.13 J/g ● °C) the same amount. Figure 21-4 on page 502, from the *Earth Science Reference Tables*, lists the specific heats of some common materials.

Water is often used to carry away excess heat energy. For example, a car has a radiator and cooling system. A solution of water and antifreeze circulates through the engine and picks up energy to prevent the engine from overheating. That solution carries the excess heat to the radiator where it transfers the energy to the external environment. Then the cooler solution is sent back into the engine to do its work again. Water-cooling is used in many such systems. On Earth, water (and air) transfers energy from the tropics to the polar regions.

Changes in State

For now, you will deal only with the heating and cooling of water. Consider what happens when 1 g of water that starts as ice at a temperature of −20°C is heated. Figure 20-3 graphs the change and shows the amount of energy used in each part of the process. Ice has a lower specific heat than liquid water's, as you can see on page 1 of the *Earth Science Reference Tables*. So ice warms more easily than liquid water.

After the ice reaches 0°C, it stays the same temperature as it continues to absorb energy. In fact, it takes energy to melt ice with absolutely no change in temperature. That is why the term "latent" is applied to energy used for a change in state.

MELTING   The change from a solid to a liquid is called **melting**. Ice must absorb energy to change to liquid water. Only after the ice has melted can its temperature increase. The energy used to melt 1 g of ice at a constant temperature of 0°C is nearly as much energy

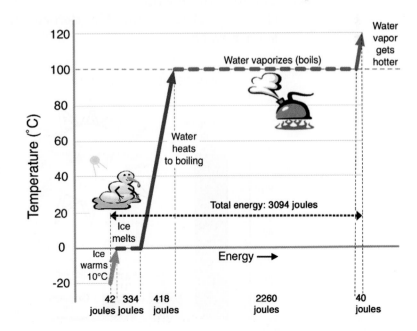

FIGURE 20-3. This graph shows what happens as water absorbs energy, changing from ice at −20°C to water vapor at 120°C. Notice that most of the energy is in the dotted part of the heating line. Of the total energy, most is used just to change liquid water into water vapor (with no change in temperature). The level parts of this graph indicate absorption of latent heat.

as it takes to heat the gram of water all the way from its melting temperature of 0°C to the boiling temperature of 100°C.

Vaporization   The change in state from liquid to gas (vapor) at any temperature is **vaporization**. Liquid water must absorb energy to change to water vapor. Water vapor is the phase of water with the highest energy because of the latent heat it contains. **Boiling**, a form of vaporization, is the change in state from a liquid to a gas at the boiling temperature of the substance. Of all water's changes in state, vaporization requires the most energy. Again, during boiling there is no change in temperature until all the liquid becomes vapor.

After all the water becomes vapor, it heats to a temperature of 120°C. Of the total energy involved in these five changes, vaporization alone took 73 percent. So you can now understand why the evaporation of water is the principal way that the atmosphere absorbs solar energy, as shown in Figure 20-4.

Condensation   You have read that vaporization is the most important way that insolation adds energy to the atmosphere. How does the atmosphere release that energy to make wind and weather? The energy absorbed by the atmosphere in phase changes is released when water returns to lower-energy phases. For example, the energy used to vaporize water is released when it changes back to a liquid. **Condensation** is the change from a gas, such as water vapor, to a liquid, such as water.

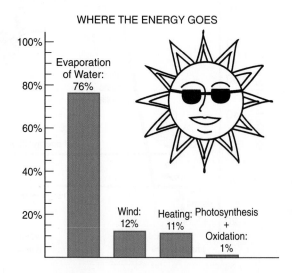

FIGURE 20-4. More than three-quarters (76%) of the solar energy absorbed by Earth evaporates water into the atmosphere. Latent heat is the major reservoir of atmospheric energy.

Have you ever noticed that grass is sometimes wet on a cool, clear summer morning? Water that collects on outdoor surfaces when moist air is cooled is called dew. Where did the moisture come from? **Dew** comes from water vapor in the atmosphere. Dew does not fall from the sky, so it is not considered a form of precipitation.

FREEZING   Ice is the lowest-energy phase of water. For example, snow is composed of delicate ice crystals. **Freezing** is the change in state from a liquid to a solid. The energy absorbed to melt ice is released when water freezes. Although freezing releases much less energy than condensation does, ice formation is still an important contribution to heating the atmosphere.

Frost is ice crystals that form directly from water vapor. It is similar in origin to dew. The change of a vapor directly to a solid, skipping the liquid phase, is known as deposition. This change releases both latent heat of vaporization and latent heat of fusion.

Figure 20-5, from the *Earth Science Reference Tables,* lists the physical properties of water. (The density of water is listed at 3.98°C because at this temperature water is most dense. Consequently, it is generally cold water rather than warm water that sinks to the bottom of the ocean.)

Water vapor can release energy to become liquid water and then ice. Figure 20-6, on page 474, shows the three phases of water relative to their latent energy. It also shows when energy is absorbed or released during changes in state.

What happens when energy is released by condensation or freezing? That energy warms the air, or at least prevents it from cooling as much as it would without the addition of latent heat. For example, as air rises, expansion causes it to cool. If a cloud forms as water vapor condenses into cloud droplets, the cooling is slowed. This can keep air in a cloud warmer than its surroundings

Properties of Water

Heat energy gained during melting . . . . 334 J/g
Heat energy released during freezing . . . . 334 J/g
Heat energy gained during vaporization . . . . . 2260 J/g
Heat energy released during condensation . . . 2260 J/g
Density at 3.98°C . . . . 1.0 g/mL

FIGURE 20-5. Properties of Water, from the *Earth Science Reference Tables.*

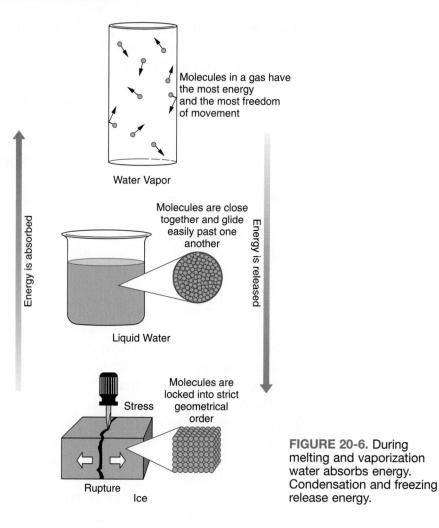

Molecules in a gas have the most energy and the most freedom of movement

Water Vapor

Molecules are close together and glide easily past one another

Liquid Water

Molecules are locked into strict geometrical order

Stress

Rupture

Ice

Energy is absorbed

Energy is released

FIGURE 20-6. During melting and vaporization water absorbs energy. Condensation and freezing release energy.

and keep the air rising. The release of latent heat causes rising air currents that can lead to violent weather, such as thunderstorms and hurricanes.

Have you ever seen water sprayed onto fruit trees to prevent frost damage? It works because the freezing of the water spray releases energy. This can prevent the fruit from freezing solid.

HOW DOES WATER VAPOR ENTER THE ATMOSPHERE?

4: 2.1e
4: 2.2c

Evaporation is the most important way that moisture enters the atmosphere. If you have ever seen puddles disappear after a rainstorm, or dew vanish in the morning sunlight, you have seen evaporation. **Evaporation** is defined as the vaporization of water when the temperature is below the boiling point.

Evaporation and Kinetic Energy

How can water change to water vapor below the boiling temperature? Within a substance some molecules have more energy than others. Within liquid water, only those molecules that absorb enough energy to become water vapor escape as a gas. The cooler molecules are left behind. It is like separating the tallest individuals from a crowd of people. With the tallest people gone, the average height of the crowd is less than it was when the tall people were included. Evaporation is a similar process. The escape of the most energetic molecules reduces the average energy of the molecules remaining.

Sources of Water

Evaporation from the oceans is the primary source of water vapor in the atmosphere. Covering almost three-quarters of our planet, the oceans provide a vast area for evaporation. On land, freshwater and groundwater also contribute to atmospheric humidity through evaporation.

Moisture also enters the atmosphere from plants. Plants draw water from the soil through their roots, and use water to transport nutrients from the soil throughout the plant and to carry on photosynthesis. Plants, especially through their leaves, add water vapor to the air by **transpiration**. In some places on land, more water enters the atmosphere by transpiration than by evaporation from surface water and from the ground.

Rate of Evaporation

The rate at which water evaporates depends on four factors:

- *Availability of water*—The greater the exposed surface area, the more water that can evaporate. This is why oceans are the primary source of atmospheric water vapor.
- *Energy to support evaporation*—Tropical regions receive more intense sunlight than polar areas do. Therefore, evaporation is most active in the tropics.

- *Ability of air to contain moisture*—Only a certain amount of water vapor can enter the atmosphere, depending upon the temperature. When it is warmer, more water vapor can enter the atmosphere. If that limit has, in fact, been reached, we say that the air is *saturated*.
- *Wind*—If the air over a body of water does not move, the air in contact with the water surface becomes saturated, and evaporation stops. You may have noticed that water evaporates faster on a windy day than when the air is calm. As wind speed increases, so does the supply of unsaturated air.

HOW DO WE MEASURE WATER IN THE ATMOSPHERE?

1: MATHEMATICAL
ANALYSIS 3
4: 2.1c, 2.1d, 2.1f

The *humidity* of air indicates its moisture content. Compared with dry air, humid air contains more water vapor. Humidity can be described in terms of absolute humidity and relative humidity. **Absolute humidity** is the mass of water vapor in each cubic unit of air. It is sometimes measured in grams per cubic meter. Figure 20-7 shows that for each 10°C rise in temperature, there can be roughly twice as much water vapor in the air. Knowing the absolute humidity does not indicate whether the air is close to saturation unless the air temperature is also known. Visit the following Web site to see current water vapor concentrations across the United States: *http://weather.unisys.com/satellite/sat_wv_us.html*

Dew Point

The dew-point temperature is an indication of absolute humidity. Figure 20-7 shows that as air is cooled, the potential moisture

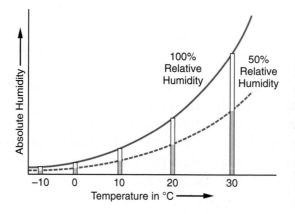

FIGURE 20-7. For every 10°C increase in temperature, about twice as much water vapor can be present in the atmosphere. The dotted line shows how much water vapor is present in the air at any given temperature when the relative humidity is 50 percent.

content decreases. If moisture does not leave the air, air can be cooled until it is saturated. The temperature at which saturation occurs is the **dew-point temperature**. This temperature is often called the dew point because dew (or frost) forms on exposed surfaces when the outside temperature falls below the dew point.

The dew-point temperature can be demonstrated with a glass of ice water or any cold object. If the object cools the nearby air below its dew-point temperature, a film of water will form on the cold outside surface of the glass. In theory, you can measure the dew point with the glass of water. The dew point is the temperature of the water in the glass when a film of water just starts to condense on the outside of the glass.

STUDENT ACTIVITY 20-2 —REMOVING MOISTURE FROM AIR

1: SCIENTIFIC
INQUIRY 1
1: ENGINEERING
DESIGN 1
6: MODELS 2

Place a mixture of ice and water in a glass or a metal container. Construct a torsion balance with a long arm, using a meterstick or a plastic tube. Attach a pencil to one end as an indicator. Tape a napkin to the bottom of the glass to catch any water that might drip off the outside of the glass. Then tape the glass of ice and water to the other end of the torsion arm. Balance the arm on a thin edge, as shown in Figure 20-8. Place a vertical scale at the pencil tip and mark its position, then wait a while.

FIGURE 20-8.

Did droplets of water appear on the outside of the glass? Where did the moisture come from? Did the indicator point move up or down? Why? What did the movement of the indicator show you? Where can you see a similar event in the natural environment? Will this procedure work better on some days than others? Why?

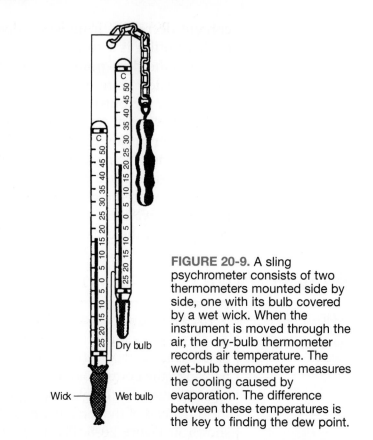

FIGURE 20-9. A sling psychrometer consists of two thermometers mounted side by side, one with its bulb covered by a wet wick. When the instrument is moved through the air, the dry-bulb thermometer records air temperature. The wet-bulb thermometer measures the cooling caused by evaporation. The difference between these temperatures is the key to finding the dew point.

A better way to find the dew point is to use a **psychrometer** (sigh-CRAH-met-er). A psychrometer is a form of **hygrometer**, an instrument used to measure humidity. Figure 20-9 illustrates a sling psychrometer. It is made from two thermometers mounted side by side on a narrow frame. The sling is the cord or chain that is used to swing the instrument through the air. One thermometer records air temperature. This is the **dry-bulb temperature**. A wet cloth called a wick covers the bulb of the second thermometer. This thermometer gives us the **wet-bulb temperature**. The wet bulb is usually mounted lower than the dry bulb to keep the water from splashing on the dry bulb.

When the psychrometer is swung through the air, evaporation from the wet wick causes the wet bulb to register a lower temperature. Depending on the design of the psychrometer, it may be necessary to swing the instrument for a minute or so to get accurate temperature readings. The dryer the air, the more evaporation from the wet wick and the greater the cooling effect. It is the amount of evaporation and cooling that allows us to determine the dew-point temperature. Unless they are equal, the wet-bulb temperature is always lower than the dry-bulb temperature.

Dew point can be confusing. Although it is expressed as a temperature in degrees Celsius (or Fahrenheit), it is not the actual temperature of the air. Instead, it tells you to what temperature the air can be cooled before condensation will begin. Dew point is a temperature, but it is used to indicate how much moisture is in the air. You can use a table to translate dew-point temperatures into absolute humidity, measured as the mass of water vapor per cubic meter. That is why dew point is a good indicator of absolute humidity.

Use the *Earth Science Reference Table,* reproduced in Figure 20-10, with dry-bulb and difference between wet- and dry-bulb temperature readings to determine the dew-point temperature. Follow the vertical column below the difference between wet- and dry-bulb temperatures until it meets the horizontal row to the right of the dry-bulb (air) temperature. Read the dew-point temperature where the row and column meet. A common mistake students make in using this table is to read the top scale of numbers as wet-bulb temperatures. The label clearly indicates that this scale is the *difference* between the thermometer readings. Remember that there is no wet-bulb temperature on this table.

Dewpoint (°C)

Dry-Bulb Temperature (°C)	Difference Between Wet-Bulb and Dry-Bulb Temperatures (C°)															
	0	1	2	3	4	5	6	7	8	9	10	11	12	13	14	15
−20	−20	−33														
−18	−18	−28														
−16	−16	−24														
−14	−14	−21	−36													
−12	−12	−18	−28													
−10	−10	−14	−22													
−8	−8	−12	−18	−29												
−6	−6	−10	−14	−22												
−4	−4	−7	−12	−17	−29											
−2	−2	−5	−8	−13	−20											
0	0	−3	−6	−9	−15	−24										
2	2	−1	−3	−6	−11	−17										
4	4	1	−1	−4	−7	−11	−19									
6	6	4	1	−1	−4	−7	−13	−21								
8	8	6	3	1	−2	−5	−9	−14								
10	10	8	6	4	1	−2	−5	−9	−14	−28						
12	12	10	8	6	4	1	−2	−5	−9	−16						
14	14	12	11	9	6	4	1	−2	−5	−10	−17					
16	16	14	13	11	9	7	4	1	−1	−6	−10	−17				
18	18	16	15	13	11	9	7	4	2	−2	−5	−10	−19			
20	20	19	17	15	14	12	10	7	4	2	−2	−5	−10	−19		
22	22	21	19	17	16	14	12	10	8	5	3	−1	−5	−10	−19	
24	24	23	21	20	18	16	14	12	10	8	6	2	−1	−5	−10	−18
26	26	25	23	22	20	18	17	15	13	11	9	6	3	0	−4	−9
28	28	27	25	24	22	21	19	17	16	14	11	9	7	4	1	−3
30	30	29	27	26	24	23	21	19	18	16	14	12	10	8	5	1

FIGURE 20-10.

SAMPLE PROBLEMS

Problem 1 After swinging a sling psychrometer, you find that the air temperature (dry bulb) is 18°C and the wet-bulb thermometer reads 12°C. What is the dew point?

Solution The difference between 18°C and 12°C is 6°C. Use the dew-point temperature table. (See Figure 20-12.) Follow the vertical column down from 6 until it meets the horizontal row from the dry-bulb (air) temperature of 18°C. They meet at 7, therefore the dew-point temperature is 7°C.

Problem 2 A student used a sling psychrometer to obtain readings of 8°C and 11°C. What was the dew point?

Solution This problem introduces two new issues. First, although it is not stated in the problem, the dry-bulb temperature must be the higher temperature: 11°C. (Evaporation always makes the wet-bulb temperature the lower of the two temperatures.) Second, 11°C is not listed on the dry-bulb scale. It will be necessary to estimate its position between 10°C and 12°C. The difference between 11°C and 8°C is 3°C. The vertical column below 3°C intercepts the two closest dry-bulb temperatures at 4°C and 6°C. Therefore, you estimate the dew point at an intermediate value of 5°C.

Problem 3 If the dry-bulb temperature is 24°C and the wet-bulb temperature is also 24°C, what is the dew point?

Solution This can be solved logically. If there is no difference between the two temperatures, it is because water could not evaporate from the wet bulb. When no evaporation takes place, the air must be saturated. If the air is saturated (relative humidity is 100%), the temperature is at the dew point. Consequently, the dew point is also 24°C.

Practice Problem 1

If the dry bulb on a psychrometer reads 14°C, while the reading on the wet bulb is 5°C, what is the dew point? What does this indicate about the humidity?

STUDENT ACTIVITY 20-3 —A STATIONARY HYGROMETER

6: MODELS 2

Materials: Two thermometers, ring stand or frame, 30 cm of string, 2- to 4-cm section of a hollow, cotton shoelace (the wick), small dish, fan made of cardboard or any other convenient material.

Hang two thermometers next to each other from a ring stand or frame. Place one thermometer bulb in the wet, hollow shoelace. The other end of the shoelace should rest in water in the dish as shown in Figure 20-11. Before you use the fan, record the two temperatures. Use your readings with the Dew-point Temperature Table to determine a dew point.

Repeat the procedure, but this time fan the thermometers for one minute before taking wet- and dry-bulb readings. Use this data to determine a second dew point.

Which determination of dew point is more accurate? Why?

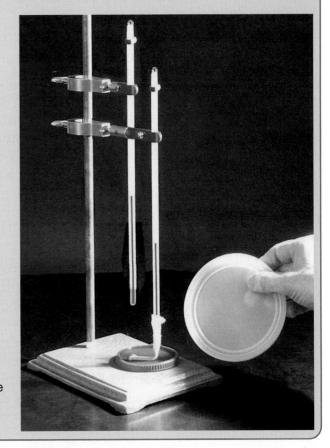

FIGURE 20-11.
A stationary hygrometer is not swung through the air. Fanning the apparatus ensures that a fresh supply of dry air surrounds the wet bulb. Evaporation from the wet thermometer bulb lowers the wet-bulb temperature.

Relative Humidity

Relative humidity compares the actual water-vapor content with the maximum amount of water vapor that could be present in the air at that temperature. In the summer, the absolute humidity can be high. No matter how much water vapor is in the air on a hot day, if the relative humidity is low, more water can evaporate and enter the air. Under these conditions, sweating cools us.

In the winter, the air usually contains much less water vapor. However, because so little water vapor can exist in cold air, the relative humidity might be 100%. Under those conditions of low absolute humidity, the air still feels "clammy." The bottom line is that the relative humidity tells us more about our level of comfort than does the absolute humidity.

Relative humidity is a measure of how close to saturation the air is. *Relative humidity* can be calculated by dividing the absolute humidity by the absolute humidity that would exist if the air were saturated. This ratio is expressed as a percent of saturation:

$$\text{Relative humidity} = \frac{\text{absolute humidity}}{\text{maximum absolute humidity at this temperature}} \times 100$$

For example, if the relative humidity is 20%, the air contains 20%, or one-fifth, of its moisture capacity. That is quite dry. Evaporation can easily add moisture to the air. However, if the relative humidity is 80%, the air is near its saturation point. Evaporation will be slow. The humidity is likely to make you feel uncomfortable. Sweat may remain on your skin, and clothing does not dry well. These conditions apply at a constant relative humidity regardless of air temperature. That is the advantage of reporting relative humidity.

Relative humidity changes when the air temperature changes. Consider air that is cooling. If there is no evaporation or condensation, the absolute humidity will not change. Remember that relative humidity is a ratio between absolute humidity and maximum absolute humidity. When the temperature decreases, the maximum absolute humidity decreases, and the air is closer to saturation. Therefore, the relative humidity increases. This often occurs at night when temperatures decrease, leading to the formation of dew or frost.

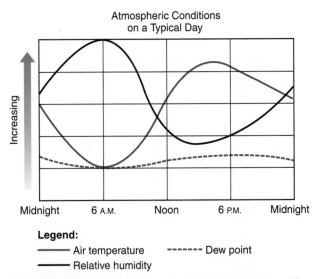

FIGURE 20-12. The daily cycle of temperature and humidity shows that if the dew point stays nearly constant, changes in air temperature will cause the opposite changes in relative humidity. Although the absolute humidity has changed very little, the relative humidity is affected by air temperature.

Figure 20-12 shows typical changes in air temperature, dew point, and relative humidity over a period of one day. The blue line indicates changes in air temperature. The changes in temperature are the result of overnight cooling and daytime heating by insolation. Over this period there is little change in the dew point, indicating that the absolute humidity is relatively constant, as shown by the dotted line. However, changes in relative humidity depend on air temperature and dew point. At night as the air temperature cools to the dew point, the black line indicates that the relative humidity increases to 100%. This situation changes when the sun warms the air in the morning. Relative humidity drops as the temperature rises, which allows more water vapor to enter the atmosphere.

To determine relative humidity, use the *Earth Science Reference Table,* reproduced in Figure 20-13 on page 484. You will also need the dry-bulb temperature to determine the relative humidity. Just as in Figure 20-10, on page 479, follow the vertical column below the difference between wet- and dry-bulb temperatures until it meets the horizontal row to the right of the dry-bulb (air) temperature. However, with this chart, where the columns meet you read the percent of relative humidity.

Relative Humidity (%)

Dry-Bulb Temperature (°C)	Difference Between Wet-Bulb and Dry-Bulb Temperatures (C°)															
	0	1	2	3	4	5	6	7	8	9	10	11	12	13	14	15
−20	100	28															
−18	100	40															
−16	100	48															
−14	100	55	11														
−12	100	61	23														
−10	100	66	33														
−8	100	71	41	13													
−6	100	73	48	20													
−4	100	77	54	32	11												
−2	100	79	58	37	20	1											
0	100	81	63	45	28	11											
2	100	83	67	51	36	20	6										
4	100	85	70	56	42	27	14										
6	100	86	72	59	46	35	22	10									
8	100	87	74	62	51	39	28	17	6								
10	100	88	76	65	54	43	33	24	13	4							
12	100	88	78	67	57	48	38	28	19	10	2						
14	100	89	79	69	60	50	41	33	25	16	8	1					
16	100	90	80	71	62	54	45	37	29	21	14	7	1				
18	100	91	81	72	64	56	48	40	33	26	19	12	6				
20	100	91	82	74	66	58	51	44	36	30	23	17	11	5			
22	100	92	83	75	68	60	53	46	40	33	27	21	15	10	4		
24	100	92	84	76	69	62	55	49	42	36	30	25	20	14	9	4	
26	100	92	85	77	70	64	57	51	45	39	34	28	23	18	13	9	
28	100	93	86	78	71	65	59	53	47	42	36	31	26	21	17	12	
30	100	93	86	79	72	66	61	55	49	44	39	34	29	25	20	16	

FIGURE 20-13.

SAMPLE PROBLEMS

Problem 4 If wet- and dry-bulb temperatures recorded with a sling psychrometer are 14°C and 24°C, respectively, what is the relative humidity?

Solution The difference between these temperatures is 10°C. The column leading down from 10°C meets the row extending to the right of the dry-bulb temperature, 24°C, at the number 30. Therefore, the relative humidity is 30%.

Problem 5 If both thermometers on a sling psychrometer record −4°C, what is the relative humidity?

Solution This is similar to Sample Problem 3. The table can be used but it is not needed. If there is no cooling of the wet-bulb thermometer, there was no evaporation, and the air is saturated: the relative humidity is 100%.

Practice Problem 2
What is the relative humidity when the wet-bulb temperature is 17°C and the dry-bulb temperature is 12°C?

Predicting Precipitation

Meteorologists measure air temperature and determine the dew point to predict precipitation. When the air temperature and dew point are close, the chance of precipitation increases. Figure 20-14 illustrates weather conditions over one day at a specific location. The top graph shows temperature dropped in the late afternoon until it reached the dew point. Under these surface conditions, precipitation is likely to occur. (Up in the clouds, the relationship between these readings is likely to be similar.) The bottom graph makes the prediction clearer by adding a line indicating relative

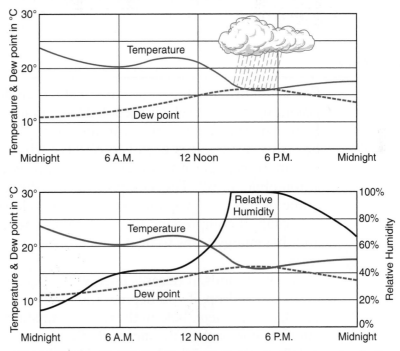

FIGURE 20-14. The top graph shows air temperature dropping to the dew-point temperature in the late afternoon. This is a sign that the chance of precipitation is greatest in late afternoon. The bottom graph has a third line showing the relative humidity during the same period. The chance of precipitation is greatest when relative humidity is high.

humidity. Rising relative humidity in the late afternoon shows that the air is coming closer to saturation. This is also a strong indicator of precipitation.

Predictions for future weather are based on past weather observations. Precipitation predictions are often given as a percent chance of rain (or snow, etc.). For example, an 80 percent chance of precipitation means that in the past, during 10 days with similar conditions, there was precipitation on eight of the 10 days. A 20 percent chance of precipitation means that precipitation is possible, but not likely. A 50 percent chance of rain indicates that the chance of having rain equals the chance of not having rain.

HOW DO CLOUDS FORM?

4: 2.1e, 2.1f

When droplets or ice crystals are too small to fall as precipitation, they remain suspended in the air. A **cloud** is a visible mass of tiny ice crystals or droplets of water. For clouds to form, several conditions must be met.

Cooling of Air

For condensation to occur, air must be cooled below its dew point. At ground level, air is cooled by contact with cold surfaces. How can air be cooled higher in the atmosphere? To understand how this occurs, it is useful to observe what happens when air is compressed. Have you ever felt a pump used to force air into automobile tires? The compressor on the pump gets warm. Some of this energy comes from friction. However, most of it is the result of compressing the air. It takes energy to compress air, and that energy turns to heat, which makes the air and the pump warm.

As air expands, it releases energy. This causes the temperature of the air to drop. If you depress the pin in a tire valve, air rushes out. A thermometer held in the rushing air shows that the expanding air is cooler than the air inside the tire.

When air rises into the atmosphere, it expands due to reduction of atmospheric pressure with altitude. Expansion of a gas causes it to become cooler. If the air cools below its dew point, deposition or condensation can create a cloud. The bottom of the cloud is at the level where the temperature equals the dew point (the relative humidity is 100%).

What causes air to rise? Sometimes air rises as it blows over mountains. Sometimes warm, moist air is pushed up by cooler, dryer air that is more dense. As it rises, it cools and cloud formation can begin. However, once the process of cloud formation gets started, it tends to keep going. Remember that condensation and deposition release energy that warms the surrounding air. This causes the air in the cloud to continue rising and results in more cloud formation. When does it stop? Eventually the cloud runs out of moist air. Without water vapor to feed cloud formation, the atmosphere stabilizes and cloud formation stops.

STUDENT ACTIVITY 20-4 —MODELING TEMPERATURE CHANGES

6: MODELS 2

You can use a table tennis ball, a paddle, and a very hard surface to model temperature changes caused by compression and expansion. The moving ball represents kinetic energy in air molecules. First, drop the ball vertically on the hard surface. While the ball is still bouncing, hold the flat side of the paddle about 20 cm (10 in.) above it. Note the change in speed of the bouncing ball as it meets the paddle. Where did its added energy come from? You can model the release of adiabatic energy by removing the paddle and allowing the bouncing ball to escape. How does the speed of the ball change?

Condensation Nuclei

On a cold night, for dew or frost to form on grass, the air must be below its dew-point temperature. Nevertheless, we do not usually see a cloud (fog) at this time. This is because water must condense on a surface. In the atmosphere, tiny particles suspended in the air provide those surfaces. These particles are called **condensation nuclei.** Dust storms, fires, and the exhaust from automobiles, homes, factories and power plants add condensation nuclei to the atmosphere. Volcanic eruptions and even organic sources, including plant pollen and microorganisms, also contribute condensation nuclei.

You have probably seen the white trails left by jets flying at high altitude. Actually, the trails are clouds that form on the exhaust particles from the jets' engines. At that altitude, air is usually below the dew point, but the lack of suspended particles prevents cloud formation. After a few minutes, the cloud is scattered, and the trail is no longer visible. These streaks are sometimes called "vapor trails." But if they were actually water vapor, we couldn't see them. They're actually thin clouds that quickly disperse.

You may have noticed how clear the atmosphere looks after a rainstorm. Precipitation not only takes moisture out of the atmosphere, but also removes condensation nuclei. In addition, the falling droplets pick up other suspended particles. In this way, precipitation cleans the atmosphere.

How do clouds disappear? Clouds lose some of their moisture by precipitation. However, precipitation stops before the cloud is gone. Even in a hard downpour, clouds seldom lose more than 20 percent of their water. Furthermore, most clouds do not produce precipitation. Clouds disappear when solar radiation vaporizes the cloud faster than cloud formation takes place. The tiny ice crystals and water droplets change back to water vapor and the cloud disappears. Visit the following Web site to see the Cloud Appreciation Society's clouds gallery: *http://www.cloudappreciationsociety.org/gallery/*

STUDENT ACTIVITY 20-5 —HOMEMADE CLOUDS

6: MODELS 2

Materials: large, transparent wide-mouthed, container (glass or plastic jar, or a very large beaker), 1-gallon plastic food storage bag, match, 300 mL of ice, hot and cold tap water.

Run a few centimeters of hot tap water into the bottom of the container. Place the crushed ice cubes and about 100 mL of cool water in the plastic bag. (Crushed ice or snow works better than big ice cubes.) Cover the mouth of the container with the bag of ice and water. (See Figure 20-15.) Do you see a cloud?

Try this again. Take out the bag of ice and water. Carefully hold a lit match in the container and blow it out. Allow some smoke from the extinguished match to circulate in the container. Place the bag

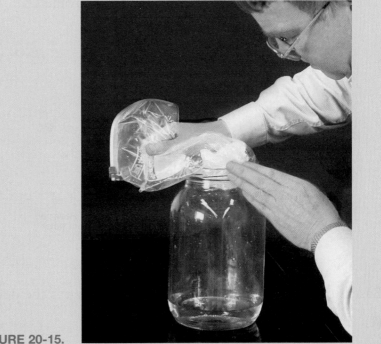

FIGURE 20-15.

of ice and water over the mouth of the container. This time you should see a cloud inside the container. What is the purpose of the hot water, the ice and water, and the smoke?

CHAPTER REVIEW QUESTIONS

Part A

1. What is the relative humidity when the dry-bulb temperature is 16°C and the wet-bulb temperature is 14°C?

   (1) 90%   (3) 14%
   (2) 80%   (4) 13%

2. The upward movement of air in the atmosphere generally causes the temperature of that air to

   (1) decrease and become closer to the dew point
   (2) decrease and become further from the dew point
   (3) increase and become closer to the dew point
   (4) increase and become further from the dew point

3. The instrument shown below is a sling psychrometer.

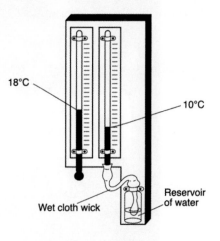

18°C

10°C

Wet cloth wick

Reservoir of water

Based upon the temperatures shown here, what is the dew point?

(1) 2°C   (3) 33°

(2) 19°C   (4) 40°C

4. Which statement best explains why an increase in the relative humidity of a parcel of air generally increases the chance of precipitation?

(1) The dew point is farther from the condensation point, causing rain.
(2) The air temperature is closer to the dew point, making cloud formation more likely.
(3) The amount of water vapor in the air is greater, making the air heavier.
(4) As the temperature of the atmosphere increases, it is less able to contain water vapor.

5. How many joules are required to evaporate 1 g of water?

(1) 4.18   (3) 2260

(2) 334   (4) 4180

6. Under what atmospheric conditions will water most likely evaporate the fastest?

(1) hot, humid, and calm
(2) hot, dry, and windy
(3) cold, humid, and windy
(4) cold, dry, and calm

7. As wind flows down the leeward side of the mountain range, the air becomes

   (1) cooler with lower relative humidity
   (2) cooler with higher relative humidity
   (3) warmer with lower relative humidity
   (4) warmer with higher relative humidity

8. What is the relative humidity if the air temperature is 29°C and the wet-bulb temperature is 23°C?

   (1) 6%   (3) 54%
   (2) 20%   (4) 60%

9. The diagram below represents a sling psychrometer.

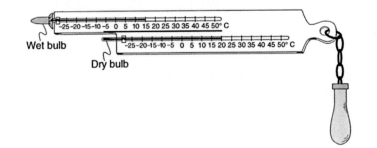

   Based on the temperatures shown, what is the relative humidity?

   (1) 66%   (3) 51%
   (2) 58%   (4) 12%

10. There is little snowfall at the South Pole because the air over the South Pole is usually

   (1) rising and moist   (3) sinking and moist
   (2) rising and dry   (4) sinking and dry

11. Which of the following changes releases the most energy?

   (1) the formation of dew from 1 g of water vapor
   (2) the formation of frost from 1 g of water vapor
   (3) 1 g of water heating 1C°
   (4) 1 g of water cooling 1C°

Part B

12. A student used a Fahrenheit instrument to find the dew point. If the Fahrenheit air temperature was 68°F and the dry-bulb temperature was 50°F, what was the dew point?

(1) 10°F   (2) 20°F   (3) 28°F   (4) 68°F

13. A student determined the dew point of the air using wet- and dry-bulb temperatures recorded on a sling psychrometer. Which of the following best represents those three temperatures?

(1) dry bulb temperature 10°C, wet-bulb temperature 8°C, dew point 6°C
(2) dry bulb temperature 6°C, wet-bulb temperature 8°C, dew point 10°C
(3) dry bulb temperature 8°C, wet-bulb temperature 6°C, dew point 10°C
(4) dry bulb temperature 10°C, wet-bulb temperature 6°C, dew point 8°C

Part C

Base your answers to questions 14 through 16 on the diagram below, which shows the changing temperature of a rising parcel of relatively warm air that forms a cloud. Location A is at the base of the cloud.

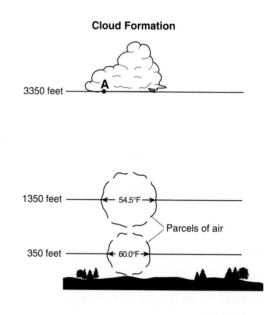

Cloud Formation

3350 feet

1350 feet — 54.5°F →

Parcels of air

350 feet — 60.0°F →

14. Explain why the relatvely warm air rises.

15. Assume that the cooling rate of this parcel of air is constant. Determine the temperature of the air at a height of 3350 ft. Express your answer to the *nearest tenth of a degree.*

16. State the relative humidity of the air at location A.

17. As a lake surface freezes in the winter, how many joules are released by each gram of water?

18. Explain why salt and dust particles are important in cloud formation.

19. State one natural process that allows large amounts of dust to enter the atmosphere.

20. How do we know that clouds are *not* composed of only water in the gaseous (water vapor) state?

21 Air Pressure and Winds

WORDS TO KNOW

convection cell   isobar   monsoon

convergence   jet stream   prevailing wind

divergence   land breeze   radar

Doppler radar   low-pressure system   sea breeze

high-pressure system

This chapter will help you answer the following questions:

1. Why do we have winds?
2. How can local winds and regional winds be different?
3. What winds occur in the upper atmosphere?
4. How do isobaric maps help us understand wind velocity and direction?

FAST AS THE WIND

4: 2.1c, 2.1d, 2.1e

Determining a world's record for wind speed is not a simple matter. On April 12, 1934, an anemometer on the summit of Mount Washington in New Hampshire registered a sustained wind speed of 373 km/h or 231 mph. Mount Washington, at 1910 m (6262 ft), is the highest peak in the northeastern United States. This record has stood for more than 75 years. Prevailing westerly winds are forced up the mountain. The mountaintop and the overlying layers of the atmosphere squeeze the winds, increasing their speed. The average wind speed at this location is 57 km/h (35 mph) making it the windiest surface location in the United States. Visit the following Web site to see the webcam view of the current conditions at the top of Mount Washington: *http://www.mount washington.org/weather/cam/*

On the island of Guam in 1997, scientists reported a wind speed of 380 km/h (236 mph). The measurement was made during a typhoon (hurricane). This report led to an investigation by the National Climate Extremes Committee. The committee found that the anemometer was not properly calibrated. It was inaccurate for winds greater than 274 km/hr (170 mph). The committee concluded that the combination of high wind and heavy rain caused the Guam instrument to malfunction. They said that the true wind speed was probably less than 322 km/h (200 mph).

Was the wind speed recorded on Mount Washington the fastest surface wind ever to occur? It is not likely. Meteorologists estimate that winds in the strongest tornadoes reach more than 483 km/h (300 mph). Scientists have tried to measure tornado winds with ground-based instruments. However, it is hard to place instruments in the narrow path of a tornado. In addition, flying debris and damage done by strong tornadoes make this nearly impossible. Teams of "storm chasers" have tried to put instrument packages where a tornado would run over them. However, none have succeeded in obtaining anemometer measurements of the strongest tornado winds.

There is a new tool available for measuring extreme winds. Meteorologists can now use radar to measure wind speed and direction from a distance. **Radar** works by bouncing long-wave radio signals off distant objects. It was developed during the Second World War primarily to observe enemy aircraft. The distance is determined by how long the signals take to return as reflected energy. The name radar comes from "*r*adio *d*etection *a*nd *r*anging." Using advanced technology, engineers have developed radar that can measure the speed of objects or winds. It also tells whether they are moving toward or away from the radar station. This is called **Doppler radar**. Doppler radar was used to record a wind gust of 512 km/h (318 mph) in a tornado in Oklahoma in 1999. However, this wind speed is not considered as reliable as the 1934 measurement on Mount Washington.

WHAT CAUSES WINDS?

1: MATHEMATICAL
ANALYSIS 1, 2, 3
4: 2.1b, 2.1 e
4: 2.2b

Surface winds blow in response to differences in air pressure. Winds always move from places of higher pressure to places of lower pressure. When you exhale, you do so by squeezing the air in your lungs, increasing the pressure. Air escapes from your body to equalize the pressure inside and outside your lungs. An air pump works in a similar way. By compressing the air inside the pump, air is forced out of the pump to where the pressure is lower. Visit the following Web site to watch a demonstration of air pressure: *http://www.youtube.com/watch?v=zl657tCCudw*

You learned earlier that atmospheric pressure is caused by the weight of the atmosphere. Earth's atmosphere is not confined the way air is in your lungs or in an air pump. The atmosphere has a uniform depth from whatever is defined as the "top of the atmosphere" to Earth's surface. However, differences in the density of air directly relate to changes in the weight of the air. Primarily, temperature and humidity determine the density of air. (As either temperature or humidity increases, air becomes less dense.) When air density increases, so does air pressure at Earth's surface, forcing the air to move to places with a lower surface pressure. Figure 21-1 is a map of the continental United States that shows a dramatic change in air pressure over the Northeast. Notice that the isolines (isobars) on the map are close together over New York State. This indicates a strong pressure

FIGURE 21-1.
The fastest wind speeds occur where the change in air pressure (pressure gradient) is greatest. On the day of this map, strong winds occurred in New York State. The lengths of the arrows indicate the speed of the winds.

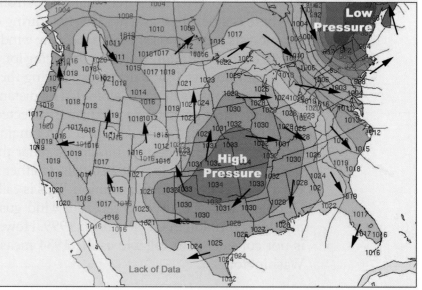

Atmospheric Pressure and Surface Wind Directions, Nov. 13, 2003

Low Pressure

High Pressure

Lack of Data

gradient. Therefore, strong winds occurred in this region where there is a large change in pressure over a short distance.

You may recall from Chapter 3 that to calculate the gradient, you need to use the formula from page 1 of the *Earth Science Reference Tables*. For example, the following sample problem shows you how to calculate the average pressure gradient.

SAMPLE PROBLEM

Problem 1 Calculate the pressure gradient between a location where the air pressure is 1020 mb and another location 100 km away where the air pressure is 1025 mb.

Solution

$$\text{Gradient} = \frac{\text{change in field value}}{\text{distance}}$$

$$= \frac{1025 \text{ mb} - 1020 \text{ mb}}{100 \text{ km}}$$

$$= \frac{5 \text{ mb}}{100 \text{ km}}$$

$$= 0.05 \text{ mb/km}$$

You can use a similar procedure to find the rate of change using another formula from page 1 of the *Earth Science Reference Tables*. For example, the next sample problem shows how to find the rate of change of air pressure with time.

SAMPLE PROBLEM

Problem 2 What is the rate of change when over a period of 10 h the air pressure goes from 965 mb to 985 mb?

Solution

$$\text{Rate of change} = \frac{\text{change in field value}}{\text{time}}$$

$$= \frac{985 \text{ mb} - 965 \text{ mb}}{10 \text{ h}}$$

$$= \frac{20 \text{ mb}}{10 \text{ h}}$$

$$= 2 \text{ mb/h}$$

STUDENT ACTIVITY 21-1 —THE FORCE OF AIR PRESSURE

1: SCIENTIFIC
INQUIRY 1
6: MODELS 2

Perform this activity over a sink or a large container to catch any spilled water. Fill a small glass completely to the top with water. Place an index card over the top of the glass. Carefully invert the glass while holding the index card to maintain an airtight seal. Remove your hand from the index card. Air pressure will hold the card and the water in place until the wet card loses its stiffness.

Air pressure depends on altitude. Mountain climbers know that as you climb higher, the atmosphere becomes thinner. At the altitude where most jets fly, they must be pressurized. If they were not pressurized, each breath would take in only about half the air that enters the lungs at sea level. A sealed bag of chips that was packaged near sea level and taken high into the mountains expands due to the lower air pressure high on the mountain.

STUDENT ACTIVITY 21-2 —PRESSURE AND DEPTH

1: SCIENTIFIC
INQUIRY 1, 2, 3
6: MODELS 2

You can model differences in air pressure at different depths with a 2-liter plastic soda bottle or a similar tall plastic container. Make three holes in the bottle at different heights. Cover the holes with plastic tape and fill the bottle with water. Hold the bottle over a sink or a container to catch the water. Remove the tape from the holes; notice that the water coming out of each hole travels a different distance. This illustrates that pressure increases with depth in a fluid. Each hole represents a different level in the atmosphere.

Temperature, Air Pressure, and Winds

Heating increases the motion of air molecules and pushes them apart. If you have seen air rising over a campfire, you have seen convection currents caused by density differences. The fire heats the air, causing it to expand. The low-density air floats higher into the atmosphere. It is replaced by cooler air that flows in from the surrounding area. This cooler air is then heated by the fire and expands. This keeps the air constantly flowing upward, carrying the heat of the fire into the atmosphere. Expansion by heating and contraction by cooling cause changes in atmospheric pressure. When air pressure is higher in one place and lower in another, winds blow to equalize the pressure differences.

Humidity, Air Pressure, and Winds

The role of humidity is not as obvious as that of temperature. Scientists have determined that under the same conditions of temperature and pressure, the same number of molecules of any gas occupies the same volume. Therefore, if lighter gas molecules are substituted for heavier molecules, there is no change in volume, but the density of the gas decreases. The mass of the individual molecules determines the density of any gas.

You usually think of water as a substance that is more dense than air. Although it is true that liquid water is far more dense than air, this changes when water becomes water vapor. Dry air is 78 percent nitrogen. If you look at the Periodic Table of Elements, you will see that each atom of nitrogen has a mass of 14 atomic mass units (amu). Like many other gases, nitrogen, hydrogen, and oxygen exists in molecules made up of two atoms. Therefore, the mass of a molecule of nitrogen (N_2) is 28 amu. Oxygen (O_2), which makes up most of the rest of dry air, has an atomic mass of 16 amu and a molecular mass of 32 amu. Oxygen is just a little more dense than nitrogen. Therefore air is composed mostly of molecules with a mass of about 28 amu, as shown in the left of Figure 21-2.

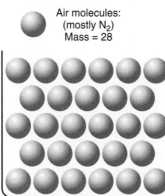
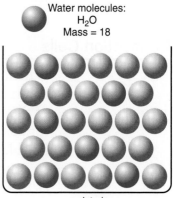

FIGURE 21-2. When water vapor is added to air, the air becomes less dense. Water molecules have less mass than molecules of nitrogen, which make up most of the atmosphere. Therefore, substituting water vapor for dry air makes the air less dense. (The units of mass below these diagrams are atomic mass units.)

Air molecules: (mostly N_2) Mass = 28

dry air

Total mass of air molecules = 30 × 28 = 840

Water molecules: H_2O Mass = 18

moist air

Air molecules = 24 × 28 = 672

Water molecules = 6 × 18 = 108

Total mass = 780

Water vapor is a compound made of two atoms of hydrogen and one atom of oxygen (H_2O). The water molecule has three atoms: one more than either nitrogen or oxygen. However, the hydrogen atoms are very light. They have an atomic mass of just 1 amu. You may recall that the oxygen atom has a mass of 16 amu. Therefore, the mass of the water molecule is 18 amu (1 + 1 + 16). This is much less than the molecular mass of nitrogen (28 amu) and oxygen (32 amu), which together make up 99 percent of dry air. Therefore, when water vapor molecules replace dry air molecules, the air becomes less dense. Figure 21-2 models dry air as 30 molecules of nitrogen with a total mass of 840 amu. On the right side of this diagram, six water molecules have taken the place of the same number of nitrogen molecules. The total mass decreases to 780 amu. Therefore, adding water vapor to the atmosphere makes air less dense.

The effects of temperature and humidity are confirmed when you use a barometer to measure air pressure in different weather conditions. As temperature and humidity increase, the barometric pressure decreases. Conversely, a change to cooler and dryer weather results in increasing barometric pressure.

WHY DO LOCAL WINDS OCCUR?

4: 2.1b, 2.1e, 2.1f, 2.1h
4: 2.2a, 2.2b, 2.2c

There are two categories of wind currents. Regional winds extend over a large area, such as several states of the United States. Local winds are those that extend only for a few miles before they die out.

Convection Cells

Whenever air is heated in one place and cooled in another, circulation tends to occur. Consider a room with a radiator on one side of the room and a cold window on the other. Air near the radiator absorbs energy. This causes the air to expand and rise. At the far side of the room, air is cooled as it loses its energy. Heat is lost by contact with the cold window and the wall (conduction) as well as by radiating heat toward these surfaces. Air near this end of the room contracts and sinks to the floor, where it flows toward the radiator. As long as the air is heated in one place and cooled in another, circulation will continue. This pattern, shown in Figure

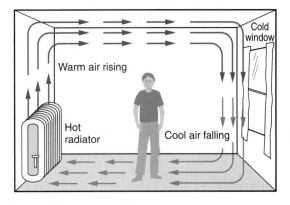

FIGURE 21-3. Air heated on one side of this room expands and rises. Cooling air on the opposite side contracts and sinks. This energy exchange maintains a flow of air and heat energy called a convection cell.

21-3, is called a **convection cell**. The air currents in the convection cell carry energy from the radiator to the cold side of the room and the window.

Winds on Earth are not confined to a closed space the way air is in this diagram. Convection cells do occur within the atmosphere. Rising air in some locations must be balanced by sinking air in other places. Winds that blow in one direction at Earth's surface must be balanced by a return flow somewhere else. The return flow usually happens in the upper atmosphere.

STUDENT ACTIVITY 21-3 —OBSERVING CONVECTION

4: 2.1f
6: MODELS 2

You can use smoke from an extinguished match or a stick of incense to show convection currents in a classroom. This works best in very cold weather when strong downdrafts overpower the heating effect of the match or incense. The smoke will identify places in the room where the air is moving in different directions. If people do not move around, you may be able to map complete convection cells with updrafts, downdrafts, and horizontal airflow. Can you identify the net flow of energy within the classroom?

Land and Sea Breezes

You can often see the wind-producing effects of temperature changes at the shore. During summer weather in coastal regions such as Long Island or along the Great Lakes, the wind direction often reverses on a daily cycle.

Specific Heats of Common Materials

MATERIAL	SPECIFIC HEAT (Joules/gram • °C)
Liquid water	4.18
Solid water (ice)	2.11
Water vapor	2.00
Dry air	1.01
Basalt	0.84
Granite	0.79
Iron	0.45
Copper	0.38
Lead	0.13

FIGURE 21-4.

Of the common Earth materials, water has the highest specific heat. This means that water can absorb large amounts of heat with only a small change in temperature. Water also yields the most energy when it cools. To understand why, it may help to look at Figure 21-4, from the *Earth Science Reference Tables*. This table shows the specific heat of seven common substances. The specific heat of ice and water vapor is only half that of liquid water. This means that a unit of heat energy absorbed by a given mass of ice or water vapor will cause twice the temperature increase it causes in liquid water.

Basalt and granite, which are very common rocks, have relatively low specific heats. Given the same energy input per gram, they would heat up about five times as much as liquid water. Most beach sand is similar in mineral composition to these two igneous rocks. Therefore, the sand on the beach heats relatively quickly. Metals, such as iron, copper, and lead, have even lower specific heats. Therefore, they heat up still faster when they absorb energy. The key point is, when water absorbs sunlight, it heats more slowly than most other materials.

During the day, the land heats up more than the water. Radiation and conduction from the land's surface heat the air over the land. This heated air expands and becomes less dense, causing it to rise. The result is a breeze that comes from the water to replace the rising air over the land. **Sea breezes** are light winds that blow from the water to the land. They usually develop in the late morning or afternoon when the land becomes warm. These breezes continue into the evening until the land cools. Figures 21-5 and

FIGURE 21-5.
During the day, the surface of the land heats up more than the water, and the warmer air over the land rises. This pulls in a cooling sea breeze off the ocean.

Warmer air rises over the land

FIGURE 21-6.
At night, the land cools more, so the ocean is warmer. The warmer air over the ocean then rises and pulls the air off the land. This is a land breeze.

Warmer air rises over the ocean

21-6 show sea and land breezes on New York's Long Island. Visit the following Web site to watch simple air pressure and wind animations: *http://serc.carleton.edu/NAGTWorkshops/visualization/ collections/atmospheric_pressure.html*

Sea breezes provide two benefits. First, a breeze off the ocean moves the cooler ocean air onto the beach. This brings relief from summer heat at the hottest time of day. Second, the breeze replaces humid air that builds around your body, allowing sweat to evaporate and cool you.

The wind reverses direction at night and through the early morning, becoming a **land breeze**. Land not only warms faster than the ocean, but it also loses its heat more quickly. The lower specific heat for rock materials means that at night the same amount of energy lost has a greater cooling effect on the land than on water. When the land cools at night, so does the air over it. The air over the water is now warmer than the air over the land. Instead of the air rising over land, air begins to rise over the water during the evening. This causes the wind to change direction, blowing from the land to the water. The conditions that lead to a land breeze also are shown in Figure 21-6, on page 503.

Land and sea breezes do not always occur along the shore. They require large areas of land and water. Therefore, these breezes do not occur at small lakes or ponds or on small islands. Nor do they develop when daily temperature changes are small, such as during cloudy weather. Strong regional weather events such as the passage of fronts can easily overpower land and sea breezes. However, when these breezes do occur, they can bring welcome relief from summer heat. People who live near the ocean sometimes talk about their "natural air conditioning" from these breezes.

WHAT CAUSES REGIONAL WINDS?

4: 2.1b, 2.1g, 2.1h, 2.1i
4: 2.2a, 2.2 b, 2.2c

The fastest winds develop in larger and more powerful atmospheric events than land and sea breezes. If you have watched a television weather report you have probably seen maps of the United States with large areas marked "H" and "L." These are regional high- and low-pressure systems.

Low- and High-Pressure Systems

Low-pressure systems are areas where warm, moist air rises. In the last chapter, you learned that when the air is humid, rising air produces clouds. Cloud formation, which occurs by condensation, releases latent energy and warms the air even more. This warming accelerates the rising air. Therefore, once a low-pressure system develops, it tends to strengthen and continue as long as it can draw in moist air. In fact, some low-pressure systems build into major storms that release great amounts of energy. Low-pressure centers

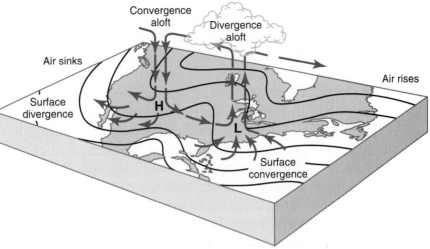

FIGURE 21-7.
Regional high- and low-pressure zones often last for days as they move from west to east across the country. The differences in air pressure cause winds to blow from areas of higher pressure to areas of lower pressure.

are also called zones of **convergence**. Convergence means coming together.

High-pressure systems are usually places where cool, dry air sinks lower into the atmosphere. Although the air warms as it sinks, it may still be cooler than the surrounding air. The air spreads out at the surface, making room for more descending air. High-pressure regions can also remain strong for many days. High-pressure systems are sometimes called zones of divergence. **Divergence** means moving apart.

Rising and falling air are part of convection cells. Vertical air movements generate surface winds that move from regions of high pressure to places where the pressure is lower. At Earth's surface, descending air spreads as it moves out of a high-pressure system. Conversely, winds come together as they blow into regions of low pressure. Rising air at the center of the low maintain these winds. Figure 21-7 is a diagram of North America that shows regional high- and low-pressure centers.

STUDENT ACTIVITY 21-4 —MOVEMENT OF PRESSURE SYSTEMS

6: PATTERNS OF
CHANGE 5

Use a daily weather map from a newspaper, televised weather report, or the Internet to locate centers of high- and low-pressure on a map of the United States. Over the next three days, plot the movements of these pressure systems across the country. Is there a general direction in which they usually move?

The Coriolis Effect

The *Coriolis effect* produces the curved path that objects, including winds and ocean currents, appear to follow as they travel over Earth's surface. It was named after the French scientist, Gaspard-Gustave Coriolis, who first described it. Look at the three people in Figure 21-8. To conduct an experiment, they are using a rotating platform similar to those often found in playgrounds. In part *A,* the boy on a rotating platform is about to throw a ball toward the two people opposite him. Part *B* shows the same people 1 or 2 seconds later. From the point of view of the boy on the ground, the ball travels straight toward him as the two people on the platform move. However, the people on the moving platform see the path of the ball curve to the right. If the platform in Figure 21-8 rotated in the opposite direction, the observers on the platform would see the ball curve left. Visit the following Web site to watch animations of the Coriolis effect: *http://www.youtube.com/watch?v=49JwbrXcPjc*

Ball

A

Apparent path as seen by observers on rotating platform

Actual path

B

FIGURE 21-8. Whether the ball appears to curve or travel in a straight line depends on whether you are on the rotating platform or standing still. The straight path of the ball, shown by the dotted line, is what the person on the ground sees. But to the moving person throwing the ball, the ball appears to curve, as shown by the dashed line.

STUDENT ACTIVITY 21-5 —THE CORIOLIS EFFECT

1: SCIENTIFIC
INQUIRY 1
6: MODELS 2

You can demonstrate the Coriolis effect. You will need a plastic milk bottle, a flexible tube, a wooden stick, string, and duct tape. Fill the bottle with water, place the tube in the bottle, and start the siphon. Spin the apparatus. As the water spurts out, notice that it seems to curve. It actually goes straight, but the moving spout makes the stream of water seem to curve.

The difference between a turn to the right and a turn to the left can be confusing. For example, if you stand facing another person, what you call the right side of the room will be the other person's left. If each of you steps to the right, you will be going in opposite directions. Obviously, we need some kind of rule to distinguish which way is "to the right." This is not so different from the way winds are labeled. Therefore, a wind is named according to the direction from which it comes, not for where it is going.

Whether a wind curves to the right or to the left is determined by its direction of movement, as shown in Figure 21-9. One way to think of this is to imagine that you are walking in the direction of the arrow. A right curve would be to your right only if you are looking forward in the direction of motion. Winds and ocean currents in the Northern Hemisphere appear to curve to the right as

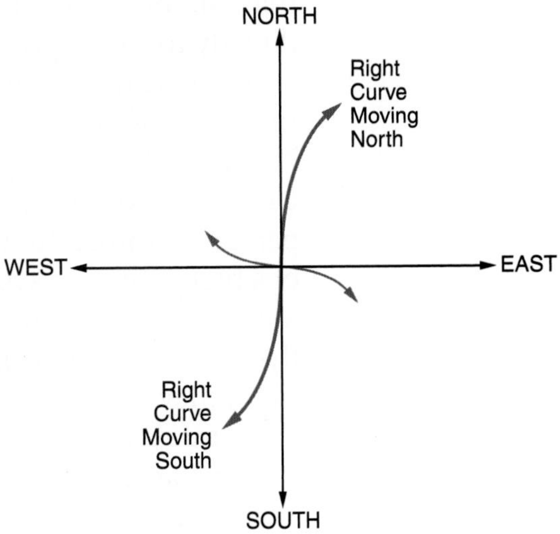

FIGURE 21-9. The straight lines show the four compass directions. The diagram also shows four paths of motion curving to the right. Whether the curve is to the left or right is defined according to movement in the direction of the arrow. Follow the arrows to see why each is a curve to the right.

they move forward. In the Southern Hemisphere, winds and currents curve to the left.

If Earth did not rotate, patterns of convection on our planet would be relatively simple. Air would descend in high-pressure regions and blow directly toward low-pressure centers. However, because Earth spins on its axis, wind patterns are more complicated. The winds follow a straight path, but as they blow over long distances, the planet turns under them. Note that the Coriolis effect is not noticeable over small distances such as those covered by land and sea breezes.

When you look at much larger regional wind patterns, the apparent change in wind direction is very easy to see. The apparent curvature of winds as they move along Earth's surface is the result of the Coriolis effect. Look back at Figure 21-1 on page 496. It is the map of high- and low-pressure systems over North America. The isolines, called **isobars**, connect locations with the same atmospheric pressure. These lines highlight the high- and low-pressure centers. The arrows show wind speed and direction. Longer arrows indicate higher speeds. The winds do blow out of the high-pressure systems and into the low-pressure systems. The apparent curvature caused by the Coriolis effect swings them to the right of their path in the Northern Hemisphere. In the Southern Hemisphere, the Coriolis effect shifts the winds to the left of their path.

In fact, over long distances, the Coriolis effect is so important that winds generally blow almost parallel to the isolines rather than simply following the pressure gradient from high pressure directly to lower pressure.

The ocean currents map in the *Earth Science Reference Tables* show that most of them circle clockwise (to the right of their path) in the Northern Hemisphere. South of the equator they turn counterclockwise (to the left of their path). The winds show a similar apparent curvature. In the Northern Hemisphere, winds that flow out of a high-pressure system turn clockwise. The winds flowing into a low-pressure system turn counterclockwise. Notice that although the winds leaving a high-pressure system in Figure 21-10 curve right, they bend in the opposite direction as they approach a low-pressure system. What could cause them to curve in the "wrong direction" as they blow into a low? The easiest way to explain this change is to point out that if the winds continued to circle to the right, they would move away from the center of the low-pressure

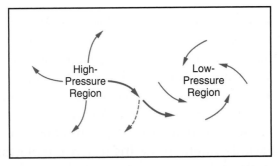

FIGURE 21-10. If Northern Hemisphere winds curve right due to the Coriolis effect, how can they converge into a low? The dotted line shows what would happen if winds only curved right. But, as the solid arrows show, the pressure gradient pulls the wind left into a low.

system. The dashed line in Figure 21-10 illustrates this. Therefore, to follow the pressure gradient, regional winds change their curvature as they converge into low-pressure systems.

In the Southern Hemisphere the situation is reversed. Winds that flow out of a high-pressure system turn counterclockwise. The winds flowing into a low-pressure system turn clockwise. Figure 21-11 clearly shows the opposite circulation of regional winds in the two hemispheres. These are two satellite images of mature hurricanes. The cloud patterns in these pictures show winds converging into the eye of each storm. However, the winds circulate counterclockwise in the northern hemisphere, while they circulate in the opposite direction, clockwise, in the southern hemisphere.

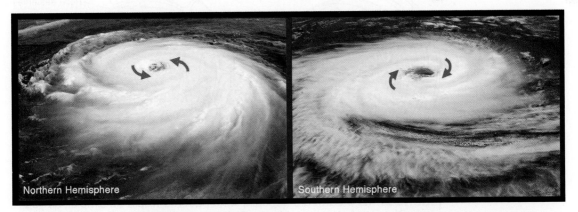

FIGURE 21-11. The arrows drawn on these two satellite images highlight converging winds in the Northern and Southern hemispheres. The left image is a Northern Hemisphere hurricane with the winds rotating counterclockwise. The right image is a Southern Hemisphere hurricane in which the winds rotate clockwise.

Prevailing Winds

In New York State, winds blow from the west or southwest more often than from any other direction. Remember that winds are labeled according to the direction you face when you look into the wind. **Prevailing winds** refer to the most common wind direction and speed at a particular location and time of year. Figure 21-12 shows two diagrams of Earth. Diagram *A* shows how terrestrial winds might blow if Earth were not spinning. Cold air would sink at the poles and travel along the surface toward the equator. Strong sunlight heating the air near the equator would cause the air to rise and move back toward the poles. Two large convection cells, as shown in the vertical profile, would dominate planetary winds.

However, Earth's rotation modifies this motion through the Coriolis effect. This is shown in Diagram *B*. Winds curving to the right in the Northern Hemisphere and to the left in the Southern Hemisphere form six convection cells. Within each cell, winds curving to the right in the Northern Hemisphere and to the left in the Southern Hemisphere change the North and South winds into East and West winds. Regional weather systems (highs and lows) complicate the pattern even more. Winds can come from any direction depending on changes in the pressure gradient.

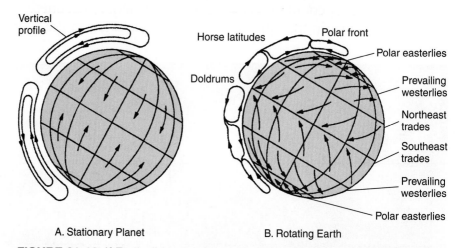

FIGURE 21-12. If Earth did not spin, wind patterns would be simpler, as you see in Diagram *A*. However, due to Earth's rotation and the Coriolis effect, winds appear to curve as you see in Diagram *B*. This results in prevailing winds from the west and the east and several convection cells in each hemisphere.

Monsoons

Large continents create seasonal changes in the direction of the prevailing winds. These seasonal wind directions are called **monsoons**. They are similar to land and sea breezes, but monsoon winds last for months and move over greater distances. The monsoons of India are a good example. During the winter, when the sun is always low in the sky, the continent of Asia cools. Cooling of the air creates a long-lasting high-pressure zone over central Asia. Sinking air spreads southward over India, bringing in dry air from central Asia. Rain is very scarce in India during this part of the year. In fact, the dry air becomes warmer as it descends from the high plateaus; so the relative humidity actually decreases.

The monsoon climate is very different in summer. Central Asia becomes warmer as the sun moves higher in the sky. By midsummer, rising air over the continent draws in moist winds from the Indian Ocean. This brings much needed rain to the Indian subcontinent. The rains allow farmers to grow crops. Some years the summer monsoon winds are very weak and the rains come late, or not at all. This causes crop failures and economic hardship for the millions of people who depend on the rain brought by the summer monsoon. Figure 21-13 is a simplified map of the seasonal changes in wind direction over India called monsoons.

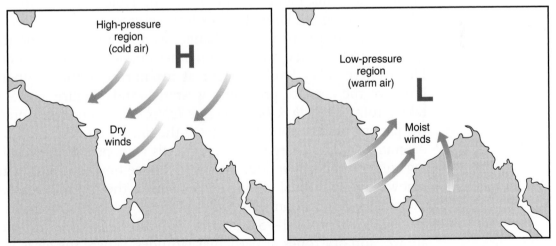

FIGURE 21-13. Seasonal changes in temperature and atmospheric pressure over the continent of Asia result in seasonal winds called monsoons. Rainfall in India depends on these seasonal changes.

New York State does not have dramatic seasonal changes in wind direction and precipitation. However, the southwestern desert of the United States does have monsoons. In spring and early summer, conditions are dry. At the end of the summer, these conditions are replaced by moist winds and occasional thunderstorms. At that time, winds bring moisture off the Pacific Ocean and into the deserts.

Jet Streams

Jet streams are wandering currents of air high above Earth's surface. They were discovered during the Second World War. At that time, pilots of high-altitude aircraft found they were traveling much slower or faster than their air speed indicated. Today, pilots will sometimes change their flight paths to catch fast tail winds, or to avoid fighting head winds.

Jet streams circle the globe, usually in the middle latitudes. Their wind speeds can be greater than 160 km/h (100 mph). Jet streams seldom follow surface winds. These high-altitude winds usually occur where cold polar air meets warmer air in the mid-latitudes. They circle the globe from west to east, usually in the upper part of the troposphere.

Meteorologists need to know the location and speed of the jet streams. These winds influence the development and movements of storm systems. Figure 21-14 shows the typical path of the jet stream crossing the United States from west to east.

The path of the jet stream changes as it meanders around the globe. In fact, two jet streams sometimes develop in the Northern Hemisphere. They tend to occur at the northern and southern limits of the zone of prevailing westerly winds. Figure 21-15 is taken from the *Earth Science Reference Tables*. It is a generalization of the pattern of winds on our planet at the time of an equinox. This diagram shows the large convection cells responsible for prevailing surface winds at various latitudes. Notice how the jet streams (shown by the symbol ⊗) generally occur in the regions between the circular convection cells.

Notice in Figure 21-15 that rising and sinking air currents create wet and dry zones at particular latitudes. Where the air is rising, such as along the equator, the cooling of warm, moist air creates clouds and precipitation. (Remember that air expands as it rises and air pressure is reduced. Expansion often cools air below the dew point.) Most of the world's deserts are located approxi-

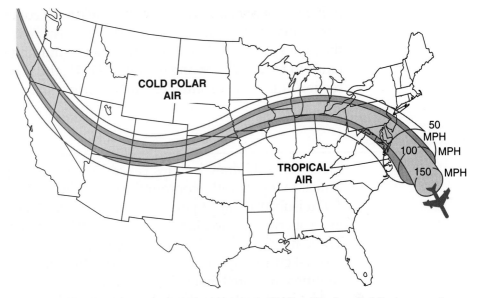

FIGURE 21-14. The jet stream is a band of high-altitude wind that separates cold polar air from warmer air to the south. The jet stream influences the development of weather systems and steers their movements.

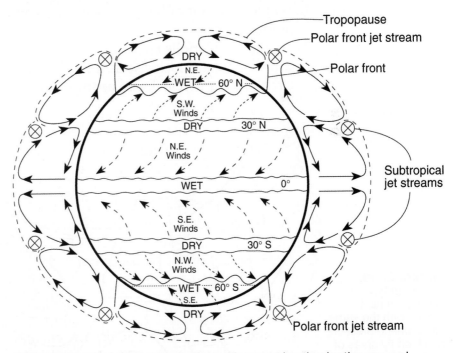

FIGURE 21-15. The combined effects of uneven heating by the sun and Earth's rotation (the Coriolis effect) set up patterns of atmospheric convection and prevailing surface winds. Zones of moist weather occur where rising air currents cause clouds and precipitation. Deserts are most common in the zones of sinking air. This diagram is in the *Earth Science Reference Tables.*

mately 30° north and south of the equator in zones of high pressure. This is where sinking air currents warm as they fall through the atmosphere. In these areas, the relative humidity at the surface tends to be low. According to Figure 21-15, the prevailing wind direction is from the southwest in New York State.

Isobaric Maps

Meteorologists draw isolines called isobars on maps. The idea is similar to topographic contour lines that you read about in Chapter 3, but isobars connect points that have the same air pressure. They use isobaric maps to identify weather patterns and predict weather. These maps are based on measurements of barometric (air) pressure taken throughout a large geographic region, such as the 48 continental United States. Figure 21-16 is a simplified isobaric map.

Meteorologists use the following principles to infer winds from an isobaric map.

1. Winds blow out of high-pressure areas and into low-pressure areas.
2. Due to the Coriolis effect, in the Northern Hemisphere, winds circulate clockwise as they diverge from highs. They circulate counterclockwise as they converge into the lows.

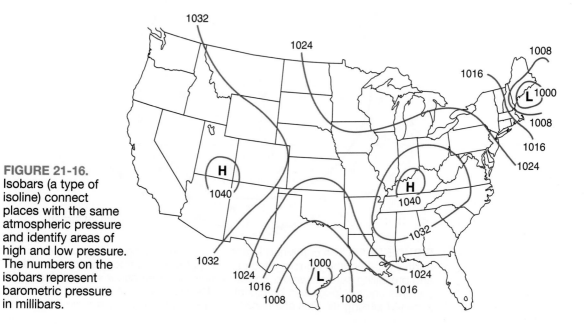

FIGURE 21-16.
Isobars (a type of isoline) connect places with the same atmospheric pressure and identify areas of high and low pressure. The numbers on the isobars represent barometric pressure in millibars.

3. Winds are the fastest where the pressure gradient is greatest. This is illustrated on the map in Figure 21-1 on page 496. The fastest winds are in New England, and the far West is relatively calm.

STUDENT ACTIVITY 21-6 —SURFACE WIND PATTERNS

6: MODELS 2
6: MAGNITUDE AND SCALE 3

Using a copy of Figure 21-16, draw arrows to represent the surface winds at the time this map was drawn. The arrows should show wind directions throughout the map region. Also indicate relative wind speeds by the length of the arrows. (Do not write in your book.)

CHAPTER REVIEW QUESTIONS

Part A

1. What causes the Coriolis effect?

   (1) Earth's tilt on its axis
   (2) the spin of Earth on its axis
   (3) the orbital motion of the moon around Earth
   (4) the orbital motion of Earth around the sun

2. Which diagram below best shows the correct pattern of surface winds around a Northern Hemisphere high-pressure system?

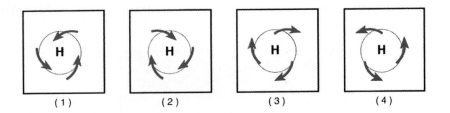

(1)   (2)   (3)   (4)

3. Surface winds blow from places of

   (1) higher temperature to lower temperature
   (2) higher air pressure to lower air pressure
   (3) higher latitude toward lower latitude
   (5) higher elevation toward lower elevation

4. Which weather condition most likely determines wind speeds at Earth's surface?

   (1) visibility changes
   (2) amount of cloud cover
   (3) dew point differences
   (4) air-pressure gradient

5. Which map below best represents the surface wind pattern around high- and low-pressure systems in the Northern Hemisphere?

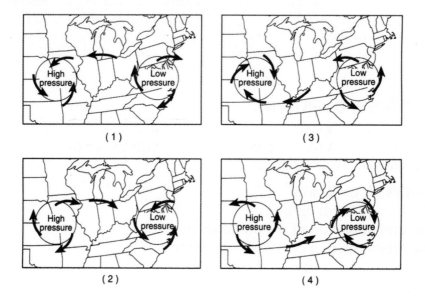

Base your answers to questions 6 and 7 on the weather map below, which shows isobars associated with a low-pressure center east of Poughkeepsie, New York.

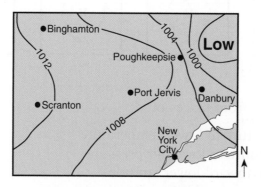

6. Which city is most likely experiencing winds of the highest wind velocity?

   (1) New York City
   (2) Binghamton
   (3) Poughkeepsie
   (4) Scranton

7. Surface winds are most likely blowing from

   (1) Binghamton toward Danbury
   (2) Poughkeepsie toward Scranton
   (3) Danbury toward New York City
   (4) Port Jervis toward Binghamton

8. Under what conditions would the wind blow from the water toward the shore of Lake Ontario?

   (1) at 2 A.M., when the air over land is 70°F and the air over the lake is 80°F
   (2) at 6 A.M., when the air over land is 70°F and the air over the lake is 70°F
   (3) at 2 P.M., when the air over land is 80°F and the air over the lake is 70°F
   (4) at 10 P.M., when the air over land is 70°F and the air over the lake is 72°F

9. On a windless day, the air temperature inside a house is 18°C, while the outside temperature is 10°C. When a window is opened, which diagram best represents the most likely air circulation pattern?

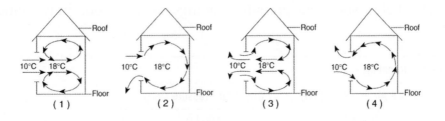

10. Which weather variable is a direct result of the force of gravity on Earth's atmosphere?

   (1) barometric pressure
   (2) cloud cover
   (3) relative humidity
   (4) atmospheric transparency

Part B

Base your answers to questions 11 through 13 on the map below, which shows Earth's planetary wind belts.

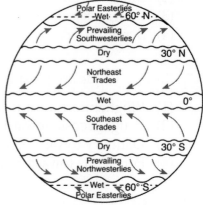

11. The curving of planetary winds is the result of

    (1) Earth's gravitation pulling on the moon
    (2) unequal heating of the atmosphere
    (3) unequal heating of Earth's surface
    (4) Earth's rotation on its axis

12. Which wind belt has the greatest effect on the climate of New York State?

    (1) prevailing northwesterlies
    (2) prevailing southwesterlies
    (3) northeast trades
    (4) southeast trades

13. Which climatic conditions exist near the equator where the trade winds occur?

    (1) cool and wet   (3) warm and wet
    (2) cool and dry   (4) warm and dry

14. The table below shows air-pressure readings taken at noon in two cities 100 mi apart.

Air Pressure Readings

Day	City A Air Pressure (mb)	City B Air Pressure (mb)
1	1004.0	1004.0
2	1000.1	1002.9
3	1002.2	1011.1
4	1010.4	1012.3

On which day was the wind speed between the two cities the greatest?

(1) 1   (2) 2   (3) 3   (4) 4

15. What is the most common wind direction 15° south of Earth's equator?

    (1) northwest   (3) southwest
    (2) northeast   (4) southeast

Part C

16. Describe the pattern of regional surface winds around the center of a low-pressure system in the Northern Hemisphere.

Base your answers to questions 17 and 18 on the barogram below, which shows the atmospheric pressure recorded in 1982 in Green Bay, Wisconsin.

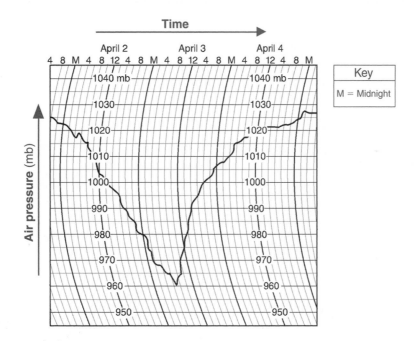

17. Calculate the rate of change in air pressure from 10 A.M. to 8 P.M. on April 3. Label your answer with the correct units.

18. What most likely caused the changes in air pressure for the time period shown on this graph?

19. The diagram below represents a view of part of the rotating earth as observed from directly above the North Pole. Use a copy of this map to draw a curved arrow starting at point B to show the general direction of surface winds in the zone from 30°N to 60°N.

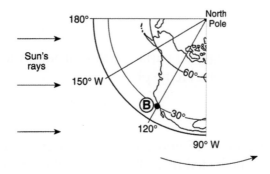

20. Why does atmospheric pressure usually decrease when humidity increases?

22 Weather Maps

WORDS TO KNOW

air mass	front	source region
anticyclone	maritime air mass	station model
arctic air mass	mid-latitude cyclone	stationary front
cold front	occluded front	tropical air mass
continental air mass	polar air mass	warm front
cyclone	polar front	

This chapter will help you answer the following questions:

1. How do we identify air masses?
2. In what ways do mid-latitude cyclones and anticyclones influence our weather?
3. What are the different types of weather fronts?
4. Why do cyclones form?
5. How can many forms of weather data be shown on the same maps?
6. How do weather maps help meteorologists?

WEATHER FORECASTING

4: 2.1c
4: 2.2c

"Everybody talks about the weather but no one does anything about it." This statement comes from the author Mark Twain. It means that while weather affects us, we cannot change the weather.

For many centuries, people depended on their limited understanding of weather to forecast weather events. "Red sky at night, sailors delight. Red sky in the morning, sailors take warning." These statements are based on the observation that a buildup of clouds in the morning is common before a storm. However, observations at one location may not give enough information to make forecasts accurate for more than a few hours into the future.

Weather forecasting improved near the end of the nineteenth century. The telegraph and radio increased communication between areas. People could learn about weather conditions in other areas. At the same time, the science of meteorology began to explain how regional weather systems develop.

In the second half of the twentieth century, meteorologists began to use computers, which quickly analyzed huge amounts of data. Meteorologists then had more data and mathematical models of the atmosphere. They hoped that these advances would greatly extend their forecasting ability. Some expected accurate weather predictions for weeks or even months into the future. Although forecasting has certainly become better, the level of accuracy they hoped for has not been reached. However, computer analysis has greatly improved short-term weather forecasts. In addition, weather satellites and regional radar allow meteorologists to see the broad scope of weather systems.

Scientists have learned that nature is not as predictable as they had hoped. The atmosphere is a complex, fluid system. Within that system, tiny changes grow to major importance through time. The atmosphere and its motions are too complex to permit accurate long-range forecasting. Visit the following Web site to see the current weather radar for the United States: *http://www.intellicast.com/Local/Default.aspx?redirectUrl=/Local/Weather.aspx&country=United%20States*

WHAT ARE AIR MASSES?

4: 2.1e, 2.1f, 2.1h

When a large body of air stays over a region, it takes on the temperature and humidity characteristics of the region. An **air mass** is a body of air that is relatively uniform in temperature and humidity. Meteorologists describe air masses by their temperature and humidity. These variables influence other weather variables.

For example, temperature and humidity affect air density. Where the air is cold and dry, barometric pressure is likely to be high. The air-pressure gradient affects winds. Where the gradient is steep, the winds are strong. Humidity affects cloud cover and precipitation. Therefore, when meteorologists describe an air mass by its temperature and humidity, they suggest other weather variables as well.

Source Regions

If you know the temperature and humidity of an air mass, you can infer the area over which it originated. This is the **source region**. For example, central Canada is most likely the source region of a cold, dry air mass moving from west to east across New York State. In central Canada, there is relatively little open water and the temperatures are cooler. A warm, moist air mass most likely originated over the South Atlantic Ocean or the Gulf of Mexico. This kind of air mass is likely to bring lower air pressure with clouds and precipitation. Figure 22-1 shows the source regions for air masses that move into the United States.

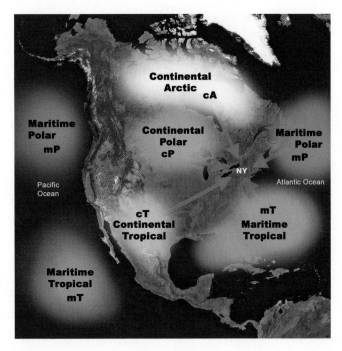

FIGURE 22-1. This map shows the principal source regions for air masses that affect New York State. Note that the characteristics of an air mass are determined by the prevailing weather conditions in the region over which it originates.

D‌ESCRIBING H‌UMIDITY   The terms *continental* and *maritime* describe the humidity of an air mass. **Continental air masses** have relatively low humidity. They originate over land. **Maritime air masses** have relatively high humidity. They originate over the ocean or another large body of water. (The term *maritime* refers to the oceans or human activities associated with oceans.)

D‌ESCRIBING T‌EMPERATURE   The terms *tropical, polar,* and *arctic* define the temperature characteristics of an air mass. **Tropical air masses** are warm. They originate close to the equator. **Polar air masses** are cool. They originate near one of Earth's poles. When an especially cold air mass that originated in the Arctic enters the United States from Canada, meteorologists call this an **arctic air mass**.

Air Mass Codes

Each type of air mass has a unique two-letter code that is made of a lowercase letter followed by a capital letter. These codes are listed in the *Earth Science Reference Tables*. Five air mass types are listed below along with their most common North American source regions.

mT   *maritime tropical:* A warm, moist air mass that probably came from the South Atlantic Ocean or the Gulf of Mexico.

cT   *continental tropical:* A warm, dry air mass that may have originated in the deserts of the American Southwest or the land regions of Mexico.

mP   *maritime polar:* A cold, moist air mass that most likely moved in from the northern parts of the Atlantic or Pacific Ocean.

cP   *continental polar:* A cool and dry air mass that most likely entered the United States from Canada.

cA   *continental arctic:* An unusually cold and dry air mass from Arctic Canada.

Note that the first letter, indicating humidity, is lowercase, while the second letter, which designates temperature, is capitalized. These codes are also shown in Figure 22-1 on page 523.

STUDENT ACTIVITY 22-1 —IDENTIFYING AIR MASSES

6: MODELS 2
6: PATTERNS OF
CHANGE 5

Use weather maps from a newspaper or from the Internet to identify different air masses entering the United States and infer their source regions. You may decide to use temperature isoline (isotherm) maps to help you locate regions with relatively uniform temperatures.

WHAT ARE MID-LATITUDE CYCLONES AND ANTICYCLONES?

4: 2.1c, 2.1e, 2.1h
4: 2.2b

The term *cyclone* has many meanings, even in meteorology. In general, a **cyclone** is a region of relatively low atmospheric pressure. Hurricanes in the Indian Ocean are often called cyclones. It also is used as a synonym for tornado. Hurricanes and tornadoes are intense low-pressure weather events. In this chapter you will learn about **mid-latitude cyclones**. These are areas of low-pressure or storm systems, such as those that usually move west to east across the continental United States.

When air is heated or when moisture is added, the air becomes less dense. This results in lower barometric pressure and causes the air to rise. Rising air is part of heat flow by convection. Rising air pulls air from all sides into the low-pressure center.

A low-pressure system is a cyclone. A high-pressure system is the opposite of a low-pressure system. Therefore a high-pressure system is an **anticyclone**. Anticyclones usually bring clear weather. Within an anticyclone, cool, dry air descends and spreads over the surface. Anticyclones do not pull in air masses. A high-pressure system is a single, spreading air mass that generally covers the surface with cool, dry, and stable fair weather.

Remember, while winds generally blow from high-pressure areas to low-pressure areas, the Coriolis effect makes them appear to curve along Earth's moving surface. Consequently, winds seem to curve to the right moving out of a high-pressure zone. Then they turn left into a cyclone. You will see this wind pattern around all mid-latitude cyclones in the Northern Hemisphere. (In the Southern Hemisphere, winds curve left when moving out of a high-pressure zone and turn right into a cyclone.)

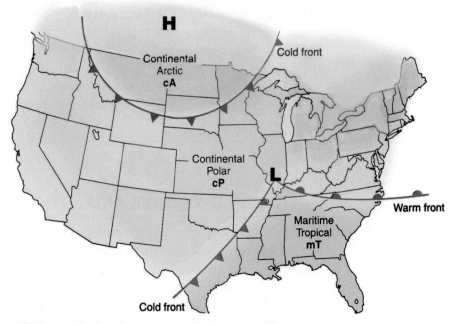

FIGURE 22-2. The three air masses on this weather map are separated by interfaces called fronts. The symbols on the frontal lines show the direction of wind movement, spreading outward from the high and converging in a counterclockwise circulation around the cyclone (low). The arctic air mass following the polar mass identifies this as a winter weather map.

Figure 22-2 is a weather map of the United States that shows the location of three air masses in the winter. The boundaries, or interfaces, where air masses meet are weather **fronts**. Fronts separate one air mass from another. Weather changes occur rapidly along fronts as one air mass replaces another. You will learn about the symbols on the fronts later. Remember that winds diverge from a high-pressure center and circulate counterclockwise as they converge into the center of the low. The triangles and half circles along the front lines indicate the direction in which the front is moving.

HOW ARE WEATHER FRONTS IDENTIFIED?

4: 2.1c, 2.1g, 2.1h

Like other interfaces in nature, energy is exchanged along fronts. Weather changes generally occur along fronts. For example, stormy weather is usually associated with the passage of fronts and strong cyclonic systems. Meteorologists keep track of weather fronts as

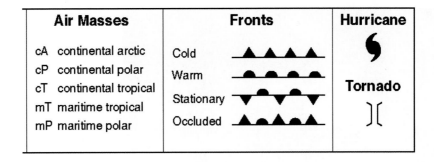

Air Masses	Fronts	Hurricane
cA continental arctic cP continental polar cT continental tropical mT maritime tropical mP maritime polar	Cold Warm Stationary Occluded	↻ **Tornado**)(

FIGURE 22-3. Weather map symbols from the *Earth Science Reference Tables.*

they observe and forecast weather over broad regions. Meteorologists identify four kinds of weather fronts: cold fronts, warm fronts, stationary fronts, and occluded fronts.

Figure 22-3 from the *Earth Science Reference Tables* shows the standard symbols used to label fronts on weather maps. For example, triangular points on the south side of the front line represent a cold front moving south. The "swirl" symbol on the right side of Figure 22-3 is used to mark the center of a hurricane.

Cold Fronts

Cold fronts are so named because they bring cooler temperatures. Figure 22-4 is a three-dimensional view of a cold front. Because cold air cannot absorb as much moisture as warm air, it is also likely to be less humid. Cold and dry air is more dense than

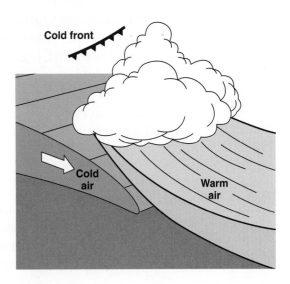

FIGURE 22-4. When a cold front passes, cold, dense air wedges under warmer air, pushing it upward. The rising air often causes cloud formation and precipitation along the front. Following the passage of a cold front, the weather is cooler and usually dryer. On weather maps, cold fronts are labeled with triangles pointed to show the direction in which the front is moving.

FIGURE 22-5. A bank of puffy cumulus clouds often announces an approaching cold front.

air that is warmer and more moist. Therefore, the denser cold air will wedge under the warmer air it is replacing and force the warm air mass to rise. Rising air expands and becomes cooler. If the warmer air mass being pushed up is humid, there will probably be cloud formation and precipitation along the cold front. Dark, puffy cumulus clouds such as those in Figure 22-5 are typical along cold fronts. Visit the following Web site to learn more about the different kinds of clouds: *http://asd-www.larc.nasa.gov/SCOOL/cldchart.html*

Cold fronts pass quickly, usually in an hour or two. Precipitation tends to be brief. It is followed by clearing skies and falling temperatures. Cold air is more dense than warm air. Therefore, the passage of a cold front usually produces higher barometric pressure, as you can see in Figure 22-6. In the summer, cold fronts often bring thunderstorms and relief from hot, humid weather. In winter, the passage of a cold front usually results in crisp, clear conditions.

Warm Fronts

Just as cold fronts cause the temperature to fall, **warm fronts** bring warmer temperatures. Figure 22-7 is a three-dimensional representation of a warm front. Unlike the quick passage of cold

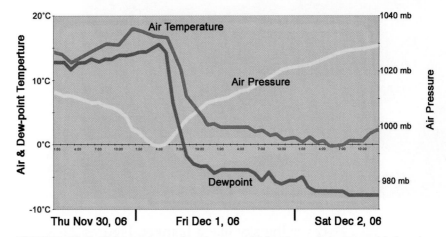

FIGURE 22-6. The passage of a strong cold front can bring a rapid drop in both temperature and dew point. Air pressure tends to decrease as the low-pressure center approaches and then increase quickly with the arrival of the cold air mass.

fronts, warm fronts may take a day or two to pass through. Warm air is lighter and less dense than cold air, so warm air does not have as much ability to push aside cold air. As a result, the warm air drags the cold air mass. Therefore, the warm front moves slower than the cold front. A wedge of warm air rides up and over the air mass it is replacing. You can predict the approach of a warm front when you see high, thin cirrus clouds that gradually become thicker and lower. As the warm front approaches, stratus clouds

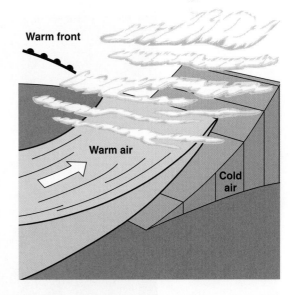

FIGURE 22-7. A wedge of warm air forced up over cooler air exerts the gentle push of a warm front. Meteorologists label warm fronts with half circles drawn on the advancing side of the front.

cover the sky like a blanket. Figures 22-8 and 22-9 show clouds that may appear as a warm front approaches.

Rising air along the warm front often produces thick layered clouds and gentle but long-lasting precipitation. As the front passes, the temperature gradually increases. More humid conditions commonly follow the frontal interface, as well. Warm air is less dense than cool air. Therefore, atmospheric pressure generally decreases with the passage of a warm front. On weather maps, half circles along the front line indicate advancing warm air.

Look again at the weather map in Figure 22-2 on page 526. It shows two cold fronts and a warm front. The cold front that extends southwest from the low-pressure center is the advancing edge of a cold, polar air mass. The second cold front moving south

FIGURE 22-9. As you learned in Chapter 5, strata are layers. Thickening stratus clouds like these often indicate an approaching warm front.

from Canada defines an outbreak of especially cold arctic air. Meanwhile, circulation about the low is drawing in a warm air mass and slowly pushing a warm front to the north over the eastern seaboard.

Occluded Fronts

Because cold fronts move faster than warm fronts, a cold front sometimes overtakes a warm front. When this happens, the interface between warm and cold air masses is pushed off the ground and up into the atmosphere. The body of warm air is held up by two cooler air masses that merge beneath it. This is called an **occluded front**, as shown in Figure 22-10.

Passage of an occluded front produces cloudy weather. It may or may not bring precipitation. The warm air-cold air interface is isolated above the surface. Therefore, as an occluded front passes, people at ground level might not notice a change in temperature. On a weather map, alternating triangles and half circles show the direction in which an occluded front is moving.

Stationary Fronts

The word *stationary* means not moving. **Stationary fronts** can remain in one location, keeping skies cloudy for hours or even days. The eventual direction of movement of the frontal interface

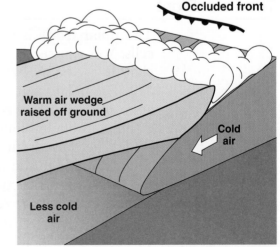

FIGURE 22-10. When a cold front overtakes a warm front, the cooler air masses push the whole warm mass high above Earth's surface. This is an occluded front.

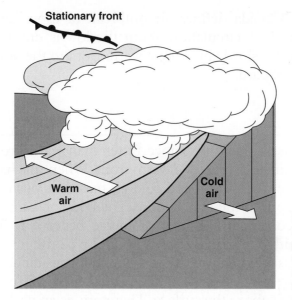

FIGURE 22-11. Winds blowing in opposite directions on opposite sides of an interface between air masses often characterize a stationary front.

is difficult to predict. Therefore, temperature changes along the front are not as predictable as they are when a cold or a warm front passes through. Winds blow in opposite directions on either side of the front, as shown in Figure 22-11.

On a weather map, triangles and half circles are drawn on opposite sides of the front. A stationary front can become a warm front if it advances toward the side of the half circles. If the stationary front advances toward the triangles, it becomes a cold front.

The Polar Front

The **polar front** is the boundary between two great convection cells. You can see it on the Planetary Wind and Moisture Belts chart in the *Earth Science Reference Table* and Figure 21-15 on page 513. The polar front is also the most common path of the upper atmosphere polar jet stream. The jet stream begins the formation of mid-latitude cyclones and steers them across Earth's surface. When the polar front moves south, it advances as a cold front. When moving to the north, it becomes a warm front. In theory, the polar front is a continuous boundary. It circles Earth at the northern edge of the zone of prevailing westerly winds. However, the polar front is not very stable and it is often broken apart by cyclonic circulation.

In the summer, the polar front is usually located north of New York State. In warm weather, New York is more often affected by weather systems that develop along the subtropical jet streams at lower latitudes. However, in the winter, the global wind belts shift to the south. This brings the polar front closer to New York. Outbreaks of arctic air push the polar front southward through New York, bringing winter storms. In spring, the polar front returns to arctic Canada as summer weather patterns settle in.

HOW DO MID-LATITUDE CYCLONES EVOLVE?

4: 2.1e, 2.1f, 2.1g, 2.1h   Low-pressure systems, or mid-latitude cyclones, often develop as swirls along the polar front. Figure 22-12 shows the evolution of a typical mid-latitude cyclone. This is a cycle in which the last stage returns to the beginning pattern. Not all lows develop in this clear sequence. However, the model is useful in understanding the structure and formation of mid-latitude cyclones. As they develop,

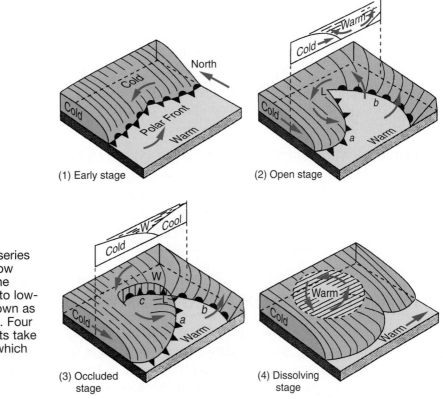

FIGURE 22-12. This series of diagrams shows how swirls, or eddies, in the polar front develop into low-pressure systems known as mid-latitude cyclones. Four types of weather fronts take part in this process, which eventually returns to stationary polar front conditions.

(1) Early stage

(2) Open stage

(3) Occluded stage

(4) Dissolving stage

low-pressure systems seldom stay in one place. They generally move with the prevailing winds. In the United States, most low-pressure systems move across the country toward the east or northeast. For this reason, people sometimes say that if you want to know tomorrow's weather in New York, look at today's weather in the Midwest.

The early stage (*1*) shows the polar front as a stationary front. Winds blowing in opposite directions along the front create friction at the interface. A swirl starts to develop. In the open stage (*2*), winds moving toward the front change the stationary front to a cold front at *a* and to a warm front at *b*. The symbols for these two kinds of fronts match the key to front symbols in the *Earth Science Reference Tables*. The profile above diagram (*2*) shows cold air starting to push up the warm air mass. The third stage (*3*) is the occluded stage. The cold front is overtaking the slower-moving warm front. The profile above shows the warm air now pushed completely up above the front. In the dissolving stage (*4*), the warm air eventually loses its identity as it mixes with the air around it. Although this is the final stage of the mid-latitude cyclone, a stationary front is forming, which returns to first-stage conditions.

These diagrams do not show cloud formation and precipitation. Wherever warm, moist air is pushed upward, expansion and cooling cause clouds and precipitation. This occurs primarily along the fronts.

STUDENT ACTIVITY 22-2 —STAGES OF CYCLONIC DEVELOPMENT

4: 2.1g
6: MODELS 2
6: PATTERNS OF
CHANGE 5

Find weather maps from newspapers or other media sources to illustrate each of the four stages of cyclonic development shown in Figure 22-12. Label each stage with the correct name.

HOW ARE WEATHER DATA SHOWN ON WEATHER MAPS?

4: 2.1c, 2.1g

To draw regional weather maps, meteorologists begin by collecting weather data. These data come from a wide network of land-based weather stations as well as from observations made on ships at sea. On the map, meteorologists group the information around a circle that indicates where the measurements were made. They use a standard system to display abbreviated weather data. This is called the **station model**. They indicate the time of these simultaneous measurements somewhere on the weather map. It is important to know how to interpret the information from the station models. Therefore, meteorologists follow certain rules.

Interpreting Weather Data

The first rule is that each weather variable has a unique position around a circle that indicates the location of the weather station. For example, barometric pressure in millibars is always indicated as a three-digit number above and to the right of the circle. The dew-point temperature in degrees Fahrenheit is always found at the lower left. Figure 22-13 is the weather station model printed in the *Earth Science Reference Tables*. It is common for weather stations to leave out some information, often because it is not important. However, the information shown will always occupy the standard position with respect to the station circle.

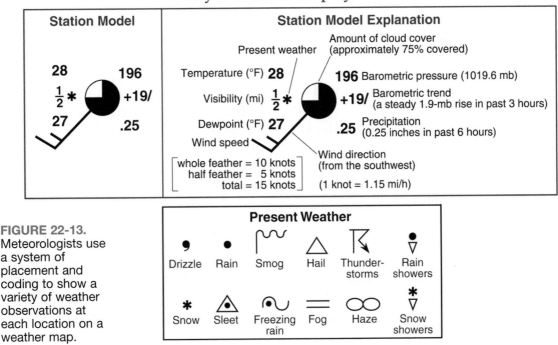

FIGURE 22-13. Meteorologists use a system of placement and coding to show a variety of weather observations at each location on a weather map.

Air Temperature   Also known as dry-bulb temperature, air temperature on a station model is always recorded in degrees Fahrenheit. You can convert this temperature to Celsius using the temperature conversion scale on the same page of the *Earth Science Reference Tables* as the station model. (For example, the 28°F shown here is approximately −2°C.)

Present Weather   This indicates conditions such as precipitation or limited visibility. Symbols used to indicate the present weather are shown in the box below the station model. The United States Weather Service uses many more symbols than you will find in the *Earth Science Reference Tables*. The station model in Figure 22-13 shows that it was snowing at the time of the observations.

Visibility   The visibility is the greatest distance, measured in miles, at which objects can be identified. Precipitation, fog, or haze reduce visibility. In Figure 22-13, the half-mile limit of visibility is probably caused by snowfall.

Dew Point   You may recall that the dew point is the temperature at which the air would be saturated. Although dew point is a temperature, it used here as an indicator of the absolute humidity. In this station model, the dew point of 27°F is very close to the air temperature of 28°F. This indicates a high relative humidity, which is not surprising because it is snowing.

Wind Direction   A line connected to the circle indicates wind direction. Just as north is at the top of most maps, these symbols follow the standard compass directions. Think of the circle as the head of an arrow. In this case, the wind is blowing out of the southwest. (Wind, you will recall, is always named according to the direction from which it comes.) A line extending directly above the circle would indicate a north wind.

Wind Speed   The small lines at the end of the wind direction indicator are called feathers. Wind speeds are rounded to the nearest 5 knots. (As printed on the chart, 1 knot is 1.15 mph.) A half feather indented a short distance from the end of the wind direction line indicates a 5-knot wind. A 10-knot wind is shown by a single feather at the end of the wind direction line. Two feather lines of equal length would indicate a 20-knot wind. This example shows a 15-knot wind. A circle surrounding the weather station dot indicates a calm area on the weather map. In this way, a calm area is not mistaken for a station in which the data is missing.

CLOUD COVER   The amount of the circle that is dark indicates the portion of the sky covered by clouds. The example shows that the sky is mostly cloudy (75% covered). A circle that is all white in the center represents a clear sky.

AIR PRESSURE   Air pressure is reported by a three-digit code number. Normal barometric pressure at Earth's surface is generally about 1000 millibars (mb). The first digit or the first two digits (9 or 10) of the air pressure reading are not shown. Furthermore, barometric readings are recorded to the nearest tenth of a millibar. Follow the rules below to interpret air pressure codes.

Rule 1: If the three-digit code is less than 500, add a "10" at the beginning and a decimal point before the last number. For example, if the barometric pressure were listed as 162, this would translate to 1016.2 mb.

Rule 2: If the three-digit code is more than 500 add a "9" before the number and a decimal before the last digit. For example 884 would become 988.4 mb. Do whichever procedure brings you closer to 1000 mb. If you do not get a value close to 1000 mb you have probably decoded the number incorrectly. The barometric pressure indicated in Figure 22-13, on page 535, is 1019.6 mb.

PRECIPITATION   The amount of rain or snow in the past 6 h is sometimes shown. In Figure 22-13, the weather station has received an amount of snow equivalent to one-quarter of an inch (0.25) of rainfall. (Precipitation on a station model is always shown as its liquid water equivalent.)

BAROMETRIC TREND   Rising air pressure usually brings clearing weather. Stormy weather is more likely if the atmospheric pressure is falling. The symbol "+" indicates rising air pressure and the symbol "–" indicates falling air pressure over the past 3 h.

The number is in millibars and tenths of millibars. (You need to insert a decimal between the two numbers.) Next to the two-digit number is a line indicating the pattern of change. The data shown in Figure 22-13 indicates a steady rise of 1.9 mb in the last 3 h.

Note: There is a guide to the placement on this information, as well as how to interpret weather station data, in the *Earth Science Reference Tables.*

SAMPLE PROBLEMS

Problem 1 What weather conditions does the station model shown in Figure 22-14 indicate?

FIGURE 22-14.

Solution This station model indicates a temperature of 68°F and a dew point of 65°F. Hail is falling. The barometric pressure is 996.4 mb. Winds are from the east at 5 knots. There has been an inch of rain in the past 6 h. The sky is half (50%) covered with clouds.

Problem 2 Draw a station model to represent the following conditions: Temperature and dew point are 62°F, and there is fog. Visibility is one-tenth of a mile. Winds are from the west at 10 knots. The air pressure is 1002.0 mb and it has gone down steadily by 0.9 mb in the past 3 h. Cloud cover is 25 percent.

Solution Figure 22-15 shows these conditions. The fog symbol and visibility fraction can be shown above or below the wind direction line. The feather indicating wind speed can slant up or down. Otherwise the numbers and symbols must be in the positions shown in the figure.

FIGURE 22-15.

Problem 3 Draw a station model to indicate a temperature of 20°C, a dew point of 11°C, and clear skies.

Solution Temperatures on the station model are always shown in degrees Fahrenheit. The temperatures in Sample Problem 3 must be converted from Celsius to Fahrenheit before they are placed in their proper positions next to the station circle, as shown in Figure 22-16.

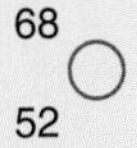

68
○
52

FIGURE 22-16.

STUDENT ACTIVITY 22-3 —CURRENT STATION MODELS

4: 2.1g

Create several station models to describe your local weather conditions as recorded at a particular time of day over a period of several days. (Your teacher will tell you how long.) You can gather weather data from online sources or from your own weather instruments.

HOW ARE WEATHER MAPS DRAWN AND USED?

4: 2.1c, 2.1g
6: PATTERNS OF
CHANGE 5

You have already seen several weather maps used to illustrate wind patterns and air masses. In this section, you will learn how to draw a weather map. In addition, you will learn how weather maps are used to predict the weather. Weather includes many variables (temperature, air pressure, humidity, wind, sky conditions, and precipitation). Meteorologists find it useful to show many of these factors over a wide geographic area. A map that combines several weather variables is called a synoptic map because it shows a summary, or synopsis, of weather conditions.

Gathering Data

Remote sensing has become a standard observation technique available to meteorologists. Remote sensing is the use of instruments to gather information at a distance from the instrument.

Radar images and photographs from satellites are good examples. You can access current satellite photographs, radar images, and other regional data on the Internet, through a local office of the National Weather Service, or the National Oceanic and Atmospheric Administration (NOAA) of the United States Government. Other organizations such as *The Weather Channel, Weather Underground,* and similar Web sites provide current online weather maps and images.

STUDENT ACTIVITY 22-4 —DRAWING WEATHER FRONTS

4: 2.1c, 2.1g
6: PATTERNS OF
CHANGE 5

From the Internet, download and print a current satellite weather image of a part of the United States. On that image, label the centers of high- and low-pressure systems. Draw appropriate weather fronts using weather maps or the pattern of development you see. Label air masses by their two-letter codes. Use your drawing to predict tomorrow's weather at your location. One source of satellite images is the *National Oceanic and Atmospheric Administration:* *http://www.weather.gov/sat_tab.php?image5ir*

Using Weather Data

Professional meteorologists often have years of experience and access to more information than can be shown on a single weather map. For example, observations of the jet stream help them predict how weather systems will move. They can also use computers to access how weather changed in the past when similar conditions were in effect. Some of the most powerful computers ever built have been used to work with complex equations that represent the atmosphere. Meteorologists sometimes construct mathematical models of the dynamics (changing nature) of the atmosphere to find how far into the future accurate predictions can be made. Visit the following Web site to learn how to understand weather maps: *http://ww2010.atmos.uiuc.edu/(Gh)/guides/maps/home.rxml*

STUDENT ACTIVITY 22-5 —RELIABILITY OF WEATHER FORECASTS

2: INFORMATION
SYSTEMS 1
6: PATTERNS OF
CHANGE 5

Make a table to record the expected changes in weather conditions over a period of a week. See the sample shown below. You may add more weather factors if you wish. Each day, make your own ("My") predictions of how you expect the weather to change the following day. Also record how these changes are predicted in a particular news media source.

Name _____

Week of _____

	Monday		Tuesday		Wednesday		Etc. . . .
Changes In	My	Media	My	Media	My	Media	
Temperature							
Cloud cover							
Precipitation							
Winds							

The next day, circle any predictions that were incorrect and record your own predictions as well as the media predictions for the following day. Continue this procedure for at least a week.

How often do you differ from the media predictions?

What is the percentage of correct predictions for you and for the news source?

Drawing Weather Maps

Figures 22-17 *A* through *D,* on page 542, illustrate how weather data is transformed into a weather map. Diagram *A* shows selected weather station data. Diagram *B* shows isobars of air pressure. Diagram *C* shows isotherms of temperature. Diagram *D* is a weather system map. Note that isolines are drawn as if the numbers were located at the center of the circles. The pattern of a cyclone

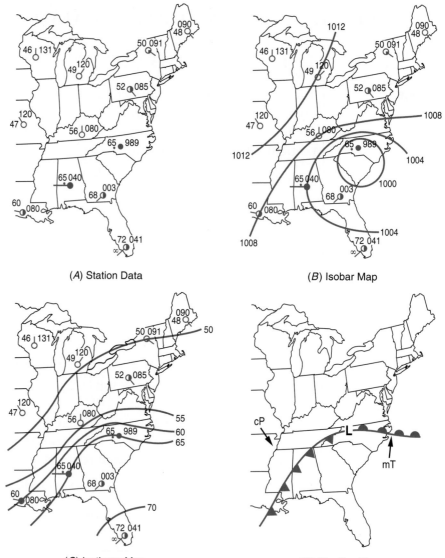

(A) Station Data

(B) Isobar Map

(C) Isotherm Map

(D) Weather Map

FIGURE 22-17. Drawing synoptic weather maps is a step-by-step process. (A) Weather data is collected from weather observers over a geographic region. The data can be used to draw a weather map. Next (B), the isolines of barometric pressure (isobars) are drawn based on the air-pressure data displayed at each station. (C) The temperature field is shown with isotherms. (D) The position of the low and the two weather fronts are based on wind direction and other data from the previous maps. Precipitation usually occurs along weather fronts.

(low-pressure system) is apparent from the data and from the circulation of winds around the North Carolina station. The changes in temperature and air pressure along with the winds and precipitation guide the placement of the cold and warm fronts.

Using Weather Maps to Make Predictions

Weather maps are an important tool used by meteorologists in forecasting weather. Figure 22-12 on page 533 shows how low-pressure systems generally develop. They mature and move across the United States from west to east, or toward the northeast. You can use current weather maps to make predictions of your local weather for the next day or two. Figure 22-18 is a weather map of the United States. Notice how this map can be used to make the following forecasts for these cities:

- *New York City* Rain will probably end as the warm front advancing northward passes through the city. Warmer weather and partial clearing will occur as a warm, moist air mass slides over New York. The density of warm, moist air is low; therefore, the barometric pressure will remain low.

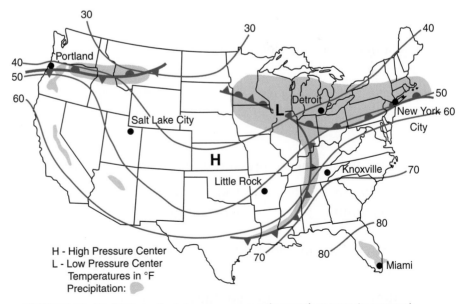

FIGURE 22-18. Meteorologists use a synoptic weather map to present information about weather conditions over a large region. The map may be used to make local weather forecasts for locations throughout the area.

- *Knoxville* The cold front approaching from the west will bring rain, followed by cooler and clear weather. The temperature and humidity will then drop, and barometric pressure will rise as a cool, dry air mass moves in with the passage of the cold front.
- *Little Rock* Cool, dry weather is likely to continue. It will probably be a few days before the anticyclone (high-pressure system) that is building into the area no longer dominates local weather.
- *Detroit* The spell of rainy weather is likely to continue for another day or so until the low-pressure, storm system moves away to the east or northeast. Visit the following Web site to see animated weather maps that predict national weather for the next four days: *http://www.wunderground.com/modelmaps/ maps.asp*

CHAPTER REVIEW QUESTIONS

Part A

1. Which type of air mass most likely formed over the deserts of the American Southwest?

   (1) cT   (3) mT
   (2) cP   (4) mP

2. A weather station model is shown below.

   029

   What is the barometric pressure indicated by this station model?

   (1) 0.029 mb   (3) 1002.9 mb
   (2) 902.9 mb   (4) 1029.0 mb

3. From what direction do most weather systems move as they begin to affect New York State?

   (1) northeast   (3) southeast
   (2) northwest   (4) southwest

4. The weather map below shows a strong mid-latitude cyclone moving across eastern North America.

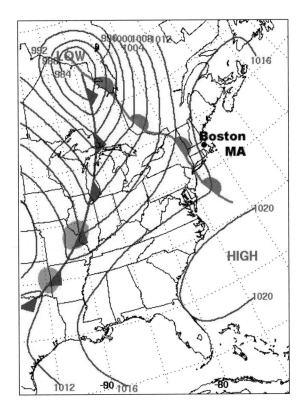

What are the most likely weather changes in Boston, Massachusetts over the next few hours?
(1) increasing temperature and increasing air pressure
(2) increasing temperature and decreasing air pressure
(3) decreasing temperature and increasing air pressure
(4) decreasing temperature and decreasing air pressure

5. Which description below best describes high- and low-pressure systems?

(1) A high-pressure system is a cyclone and a low-pressure system is an anti-cyclone.
(2) A low-pressure system is a cyclone and a high-pressure system is an anti-cyclone.
(3) A cyclone includes regions of both low and high pressure.
(4) An anticyclone includes regions of both low and high pressure.

6. Why do cold fronts usually travel faster than warm fronts?

   (1) cold air is more dense than warm air
   (2) warm air is more dense than cold air
   (3) cold fronts bring higher temperatures
   (4) warm fronts bring lower temperatures

Part B

Base your answers to questions 7 through 9 on the graph below, which shows changes in air temperature and dew point at Dallas, Texas. Symbols below the graph indicate weather conditions during the time of this graph.

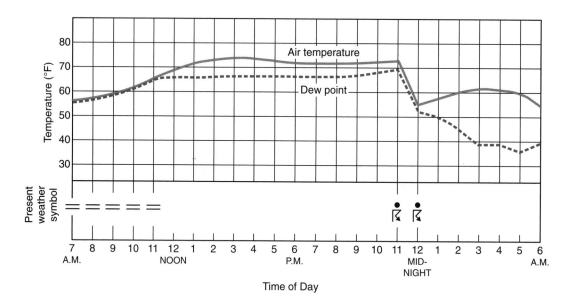

7. Which choice below best describes the weather event that was most likely occurring in Dallas at the time indicated?

   (1) At 6 P.M., a warm front was passing through.
   (2) At 11 P.M., a warm front was passing through.
   (3) At 6 P.M., a cold front was passing through.
   (4) At 11 P.M., a cold front was passing through.

8. What were the weather conditions in the late morning hours of the first day?

   (1) thunderstorms   (3) haze
   (2) fog   (4) heavy rain

9. When was the relative humidity the highest at Dallas?

    (1) 10 A.M. on the first day
    (2) 12 noon on the first day
    (3) 10 P.M. on the first day
    (4) 4 A.M. on the second day

10. The map below shows precipitation patterns on May 27, 2008. The darkest areas are experiencing the most precipitation.

What weather event is most likely shown by this map?

    (1) a warm front moving to the southeast
    (2) a warm front moving to the northeast
    (3) a cold front moving to the southeast
    (4) a cold front moving to the northeast

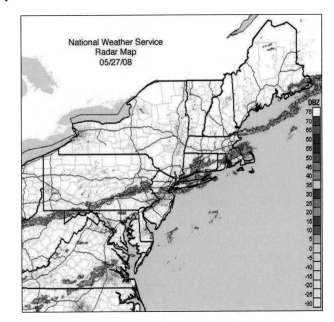

National Weather Service
Radar Map
05/27/08

Part C

Base your answers to questions 11 and 12 on diagram below, which shows weather conditions over New York State on a specific day in July.

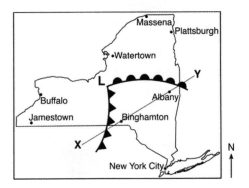

11. Make a sketch of the diagram below and on it show the proper abbreviations for the air masses at 1 and 2.

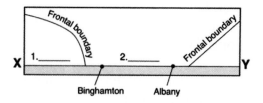

12. For which city is the following weather forecast most appropriate? "In the next hour heavy rain is expected with possible lightning and thunder. Temperatures will become much cooler."

13. Draw a box about 2 cm on each side to represent a regional weather map. To the right of the box, draw an arrow indicating that north is toward the top of the box. Draw a warm front moving to the northeast and indicate the positions of continental polar and maritime tropical fronts in their proper positions.

14. Draw a weather station model to properly represent the weather conditions shown in the diagram of a coastal lighthouse below. Include the present weather, temperature, dew point, atmospheric pressure, and wind speed and direction. The temperature is 36°F, the relative humidity is 100%, and the air pressure is 995.1 mb. The sky above the lighthouse is overcast.

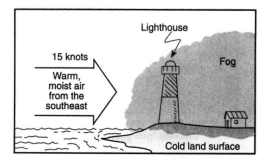

Base your answers to questions 15 through 17 on the weather station symbols below, which indicate weather conditions at the following cities in New York State: Niagara Falls, Syracuse, Utica, and New York City.

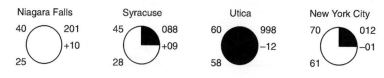

15. What was the air pressure at Niagara Falls? Be sure to give the correct units of measure.

16. Explain why the symbols indicate that Utica has the greatest chance of precipitation.

17. Make a copy of the weather station model for New York City. Add the following information in the proper locations and format: winds from the south at 15 knots and hazy conditions with three-quarters of a mile visibility.

Base your answers to questions 18 through 20 on the weather map below. Isolines on the weather map show air pressures in millibars.

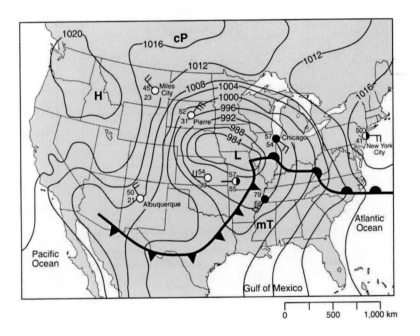

18. Over what geographic feature is the mT air mass likely to have originated?

19. Write the names of these cities in the sequence from lowest relative humidity to the highest relative humidity: Albuquerque, Chicago, and New York City.

20. What is the present Celsius temperature in New York City?

CHAPTER

23 Weather Hazards and the Changing Atmosphere

WORDS TO KNOW

acid precipitation	hail	rain shower	thunder
blizzard	hurricane	sleet	thunderstorm
drought	lightning	smog	tornado
freezing rain	outgassing	snow shower	weather hazard
global warming	rain		

This chapter will help you answer the following questions:

1 How can weather events place lives and property in danger?
2 What can we do to stay safe during severe weather events?
3 Why are Earth's atmosphere and climate changing?

THE COST OF NATURAL DISASTERS

4: 2.1c, 2.1d, 2.1h

Scientists measure natural disasters in several ways. In terms of money, Hurricane Katrina in 2005 was the most expensive weather event to strike the United States so far this century. Economists estimate the damage at 81 billion dollars. However, hurricanes that devastated Miami, Florida, in 1926 and Galveston, Texas, in 1900 are estimated to have cost even more in today's dollars.

Death is the most feared outcome of any natural disaster. In the United States, the greatest loss of life caused by a natural disaster

FIGURE 23-1.
Damage caused by the 1900 Galveston hurricane.

occurred in the Galveston hurricane of 1900. At that time, Galveston was the wealthiest city in Texas and its major shipping port. Galveston's geographic setting and the lack of advanced warning were the most important factors in this disaster. The city was built on a low-lying barrier island at the entrance to Galveston Bay. The highest point in the city was only 3 m (10 ft) above sea level. People did not know that a storm was approaching and that they should evacuate the island. Figure 23-1 is a photograph taken in Galveston just after the 1900 hurricane.

Galveston's location will always make it vulnerable to hurricanes. Shortly after the 1900 hurricane, a 5-m (17-ft) seawall was built to protect the city from hurricanes. Now, the data from weather radar and satellite images let forecasters know when a major hurricane is approaching. In 2008, Hurricane Ike slammed ashore as a category 2 storm. Planned evacuations limited deaths to about 35 people, primarily people who ignored the warnings.

WHAT WEATHER EVENTS POSE HAZARDS?

4: 2.1c, 2.1d, 2.1h

You know that water vapor in the atmosphere stores large amounts of solar energy. Some weather events concentrate that energy and release it, generating strong winds, excessive precipitation, and other hazards. These events are called **weather hazards**. Understanding storm systems and predicting their movements can save lives and property.

Thunderstorms

Many of you have been caught in a **rain shower**, a short period of **rain** (liquid precipitation). In addition, you have probably seen the buildup of clouds leading to a thunderstorm. A **thunderstorm** produces rain, lightning, thunder, strong winds, and sometimes, hail. Although thunderstorms can occur at any time of year, they are especially common during humid summer weather. Rapid updrafts caused by an approaching cold front may trigger the formation of massive storm clouds. Thunderstorms also are common in humid, tropical air masses. Some thunderstorm clouds extend to the top of the troposphere. Figure 23-2 shows the vertical development and flared top characteristic of clouds that produce thunderstorms.

Scientists do not yet fully understand why positive and negative electrical charges separate in thunderstorm clouds. However, the separation of charges leads to sudden electrical discharges we see as **lightning**. These electrical discharges heat the air to temperatures higher than the surface of the sun. We see the electromagnetic energy they give off as a flash of light. The sudden expansion of the hot air causes the sound of **thunder**.

Lightning occurs within clouds, between clouds, and between clouds and the ground. Lightning strikes cause about 100 deaths

FIGURE 23-2. The birth of a thunderstorm is evident when we see a massive thunderhead that builds to the top of the troposphere where stratospheric winds shear off the top of the cloud. This image was taken from a commercial aircraft over the central United States.

each year in the United States. Most deaths caused by lightning are isolated events so they are seldom reported in the news outside the local area. Lightning often is the leading cause of weather-related human deaths in the United States. Visit the following Web site to see what happened to a man who was struck by lightning: *http://www.youtube.com/watch?v=1Ed3aV0NcLo&feature=user*

STUDENT ACTIVITY 23-1 —LIGHTNING DISTANCE

1: MATHEMATICAL ANALYSIS 1, 2, 3
1: SCIENTIFIC INQUIRY 1

(*Note:* Thunderstorms can be dangerous. Perform this activity only if you are in a safe place, indoors would be best, and are under the supervision of a responsible adult.)

Light travels so fast that you see lightning just about the moment it occurs. However, the sound of thunder travels much more slowly. Here is a way to estimate the delay in seconds between the time you see a flash of lightning and hear the sound of the thunder. When you see the flash of lightning, begin counting slowly: one thousand one, one thousand two, one thousand three, etc., until you hear the clap of thunder. The distance to the lightning strike is 1 km for each 3-s delay or 1 mi for a 5-s delay (about 1000 ft per second of separation).

Flooding, strong winds, and damage from hail are other costly effects of violent thunderstorms. The pellets of ice that fall during thunderstorms are called **hail**. Hail begins as small drops of frozen rain high in a thunderstorm cloud. As the ice is blown around within the cloud, it is repeatedly coated by liquid water that freezes and forms a new layer of ice. Eventually, the hail falls to the ground. Hailstones the size of softballs (12 cm or 4.75 in.) have been recorded in the midwestern United States. Although rare, such large hailstones can break windows and dent cars. Figure 23-3, on page 554, shows the parts of the United States that have the most frequent thunderstorms.

Hail should not be confused with sleet. **Sleet** forms when drops of rain fall through a layer of cold air near Earth's surface and freeze completely. Unlike hail, the ice particles in sleet are always the size of a raindrop or smaller (less than 5 mm), and sleet pellets

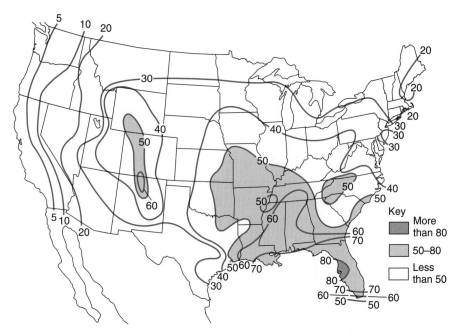

FIGURE 23-3. Thunderstorms are most common in the southeastern states where warm, humid conditions occur and cool continental air masses clash with tropical air masses

are transparent. Sleet shows no evidence of multiple cycles of coating by water followed by uplift into cold air aloft where each coating freezes. Therefore, the formation of sleet does not require the violent updrafts that produce hail. **Freezing rain** forms when liquid rain freezes on contact with Earth's surface.

Blizzards

A **snow shower** is a short period of snow. A **blizzard** is a winter snowstorm that produces heavy snow and winds of 56 km/h (35 mph) or greater. Blowing snow, limited visibility, and cold temperatures are typical of these storms. People can be stranded by blizzard conditions. Deep snow on country roads or city streets can shut down businesses and services. The wind and snow can break electric and telephone lines. Fire trucks and ambulances may not be able to reach places where they are needed. Some residents, especially the elderly, may be in danger because they run out of food or fuel before roads are cleared of snow and debris. People caught outdoors in the storm or those who perform vital emergency services may suffer from frostbite and loss of body heat (hypothermia).

Major lake-effect winter storms were discussed in Chapter 20. Away from major lakes, cold continental air masses usually pro-

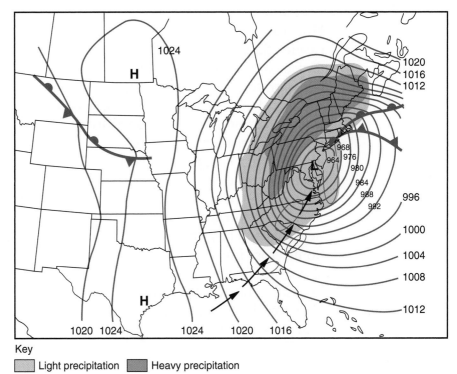

FIGURE 23-4.
In 1993, a March blizzard moved up the East Coast dumping as much as 4 ft of snow in some areas. Notice the tightly spaced isobars, indicating an intense pressure gradient and unusually strong winds.

Key

Light precipitation ░ Heavy precipitation

duce limited snowfall amounts. Greater snowfall is often associated with blizzards that draw moist air off the ocean as they move north through the Atlantic coastal states. In February 2010, two unusually heavy blizzards left several feet of snow in Washington, D.C., paralyzing the city and much of the Atlantic Coast.

Figure 23-4 is a weather map of a 1993 late-winter blizzard. This unusual blizzard included snowfalls as high as 125 cm (4 ft) in some inland areas. Notice the extreme pressure gradient shown by the closely spaced isobars. This pressure gradient caused winds of hurricane force during this storm.

STUDENT ACTIVITY 23-2 —STORM SURVIVAL

1: ENGINEERING
DESIGN 1
7: STRATEGIES 2

Make a list of supplies that can be stored at home to help your family get through a major storm. When your list is finished, number the items from most important to least important to keep in stock. Compare your list with classmates' lists.

Hurricanes

Hurricanes are the most destructive storms on Earth. Most hurricanes develop in the late summer and early autumn when the surface of the tropical ocean's water is warmest. Solar energy and the warmth of the ocean support evaporation. This carries water vapor and energy into the atmosphere. Hurricanes begin as tropical depressions (mild low-pressure regions) that become tropical storms when their winds exceed 63 km/h (39 mph). Vertical air currents and energy released by cloud formation (condensation) strengthen the storm. When sustained winds in a tropical storm exceed 120 km/h (74 mph), the storm is classified as a **hurricane**.

Atlantic hurricanes often begin as small disturbances over or near Africa. They drift to the west with the northeast trade winds. Some cross into the Gulf of Mexico where they may again gather strength. Others turn north before they reach North America and lose strength over the cooler water of the North Atlantic Ocean. However, hurricanes also come ashore along the East Coast or move parallel to the coast, causing destruction for hundreds of miles.

Hurricanes also begin over the Caribbean Sea or the Gulf of Mexico. When they move over North America, they weaken because they are deprived of water vapor from the warm ocean surface. Friction with land features is also an important reason that hurricanes usually weaken after they come ashore. Wind speed quickly decreases as the weakened hurricanes break up into a disorganized group of rainstorms. These storms may cause flooding in inland areas. After turning northward, the remains of the storm are picked up by the eastward flowing jet stream or the zone of prevailing southwest winds. Figure 23-5 shows the most common paths taken by these tropical storms. Visit the following Web site to learn about record breaking hurricanes by selecting the question you want answered: *http://www.aoml.noaa.gov/hrd/tcfaq/tcfaqE.html*

Hurricanes are very large. Some are more than 600 km (400 mi) in diameter. As in other Northern Hemisphere cyclones, winds circulate counterclockwise as they converge toward the center. From the outer bands of clouds to the center of the storm, wind speed generally increases, as does the intensity of precipitation. At the center of the strongest hurricanes, there is a small round area of relative calm known as the eye of the storm. In this region, the winds are light and the skies may be mostly clear. However, the calm of the

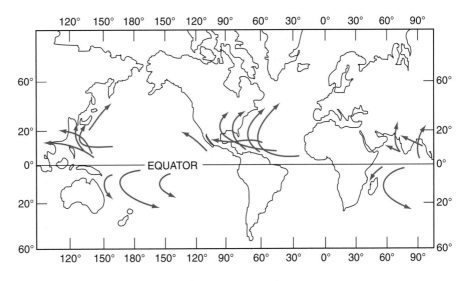

FIGURE 23-5.
Hurricanes in the Atlantic Ocean usually drift westward toward North America. Most turn north until they enter the zone of prevailing west winds and finally drift to the east as they die out over land or cooler ocean water.

eye is surrounded by the most violent part of the hurricane called the "eye wall," where the fastest winds and most intense rainfall occur. Some people think the storm is suddenly over when the eye of the storm passes over them. The approaching eye wall quickly brings them back to the reality of the second half of the storm. Visit the following Web site to create your own virtual hurricane and watch what happens: *http://www.nationalgeographic.com/forcesof nature/interactive/index.html?section=h*

Meteorologists measure the strength of hurricanes using the Saffir-Simpson scale of hurricane intensity. This scale is shown in Table 23-1. Category 1 hurricanes are the weakest. They are likely to do little damage to homes and other well-constructed buildings

TABLE 23-1. Saffir-Simpson Scale of Hurricane Intensity

Category	Sustained Winds		Storm Surge		Potential Damage
	km/h	mi/h	m	ft	
1	119–154	74–95	1–2	4–5	Minimal
2	155–178	96–110	2–3	6–8	Moderate
3	179–210	111–130	3–4	9–12	Extensive
4	211–250	131–155	4–6	13–18	Extreme
5	>250	>155	>6	>18	Catastrophic

that are not directly along the coast. A category 5 hurricane is the strongest. It will cause widespread damage and pose danger to nearly anyone caught in the strongest part of the storm.

Hurricanes present the greatest danger to people in low-lying coastal cities, including parts of New York City and Long Island. The storm surge is the greatest cause of damage and fatalities. A storm surge is the result of the combination of very low atmospheric pressure and strong onshore winds. These conditions can raise the level of the ocean so high that waves smash homes and other structures along the oceanfront. People who have not evacuated the area may find themselves with no escape route. They may be stranded for hours by violent seawater in a collapsing building. If the storm comes ashore at high tide, the storm surge may be even higher and more dangerous. Visit the following Web site to see pictures of the Galveston/Bolivar Peninsula before and after Hurricane Ike: *http://coastal.er.usgs.gov/hurricanes/ike/photo-comparisons/bolivar.html*

Figure 23-6 shows how the cost of hurricane damage and deaths due to hurricanes has changed over the past century. The death toll has dropped for two reasons. When evacuation routes have been established, people can move safely to higher ground.

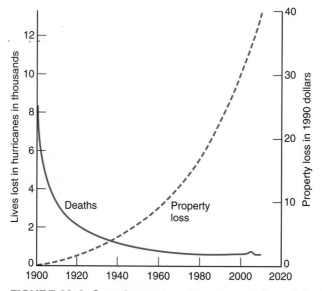

FIGURE 23-6. Over the past century, the number of deaths and injuries from hurricanes has decreased due to advanced storm warnings and emergency planning. However, property losses have increased due to the development of beachfront areas.

In addition, radio and television provide updated warnings. At the same time, the cost of hurricane damage in the United States has increased. The principal cause of the increase is the development of shoreline areas. Although people can escape to higher ground, the homes and resorts along the ocean remain vulnerable. If global warming causes sea level to rise, how will this affect damage along coastlines?

STUDENT ACTIVITY 23-3 —HURRICANE TRACKING

2: INFORMATION
SYSTEMS 1

For this activity you will need an outline map of the United States that includes the Atlantic Ocean, the Caribbean Sea, and the Gulf of Mexico. The map must clearly show latitude and longitude. Use the Internet or printed references to find tracking coordinates of a major Atlantic or Gulf of Mexico hurricane. Plot its daily path over the period of time that it was a strong storm system. Explain why the storm formed where it did and how it died out.

Tornadoes

In terms of area, **tornadoes** (Figure 23-7) are very small storms, usually less than 0.5 km (0.3 mi) in diameter. Most touch down for fewer than 10 minutes. However, some can last for hours and skip from place to place over Earth's surface. Tornadoes have the fastest winds on Earth. Their winds sometimes are in excess of 500 km/h (300 mph). This is twice as fast as winds in the most violent hurricanes.

FIGURE 23-7.
Tornadoes are small weather disturbances, but they have the highest wind speeds of any storm. Damage may be limited in area, but severe in destructive power.

Tornadoes can occur anywhere in the United States and at any time of year. Nevertheless, they are most common in the Midwest and southern states from March through May. At this time, strong cold fronts form in the region where polar and tropical air masses meet. Most tornadoes are born as swirling winds within a thunderstorm. These winds intensify and extend upward into the cloud as well as down to the ground. Most tornadoes, like the larger cyclonic storms, rotate counterclockwise in the Northern Hemisphere. However, because tornadoes are very small, the Coriolis influence is slight, so about 1 percent of tornadoes in the United States rotate "in the wrong direction."

The Enhanced Fujita scale (EF-scale) classifies tornadoes by their wind speed and damage. The scale goes from an EF-0, which causes light damage, to EF-5, which causes total destruction along its path. Table 23-2 is the Enhanced Fujita scale of tornado intensity. Meteorologists speculate that global climate change may extend the American tornado season, exposing people to increased risk of tornadoes.

Tornadoes can occur in swarms such as the outbreak of April 3–4, 1974. On those dates there were 127 tornadoes, some of them EF-5. These tornadoes killed 315 people and injured more than 6000. The path of destruction ranged from Mississippi to western New York State. New York usually gets one or two confirmed tornadoes each year. Blizzards, hurricanes, and tornadoes are all seasonal storms. Figure 23-8 illustrates when these weather events tend to occur throughout the calendar year.

TABLE 23-2. Enhanced Fujita Scale of Tornado Intensity

Rank	Wind Speed km/h	mph	Damage
EF-0	105–137	65–85	Light
EF-1	138–177	85–110	Moderate
EF-2	178–217	111–135	Considerable
EF-3	218–266	136–165	Severe
EF-4	267–322	167–200	Devastating
EF-5	>322	>200	Incredible

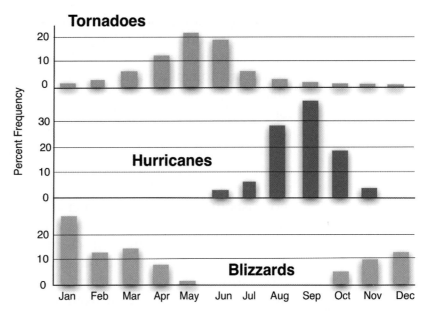

FIGURE 23-8. Different kinds of weather hazards in the United States generally occur at different times of the year. Tornadoes are most common in late spring and early summer. Most hurricanes occur in late summer and early autumn. Blizzards are usually winter events.

Research indicates that tornadoes may form within strong hurricanes. These tornadoes cause some of the worst damage done by major hurricanes. Tornadoes are difficult to notice in the fury of the hurricane. However, investigators have found evidence of tightly swirling winds in the debris left by some major hurricanes. In 2005, Hurricane Katrina produced tornadoes in Georgia and Mississippi.

Floods

Of weather-related events in the United States, floods have taken the greatest toll in lives and property. Flash floods are sudden events. On a summer Saturday in 1976, a strong late afternoon thunderstorm hovered over the Rocky Mountains of Colorado. The storm dropped an estimated 25 to 30 cm (10 to 12 in.) of rain. Over a period of 3h, the discharge of the Big Thompson River increased by a factor of 200. The water swept into a narrow canyon. This canyon was popular for recreational and camping activities. Along the river, 400 houses were destroyed and 145 lives were lost. Many of the victims had taken refuge from the rain and rising floodwaters in their cars. The flood swept the cars into the river, carrying away the cars and their passengers. Others, who abandoned their cars and scrambled up the rocks at the side of the canyon, survived.

Flash floods are a hazard throughout the United States. They are a special hazard in the desert southwest, where local roads often cross dry streambeds without bridges. Summer monsoon thunderstorms are common in this region. These storms may wash away small bridges. However, the floodwaters block these crossings only for a few hours at a time. The majority of deaths due to floods in this country are people who are swept away in their cars when they try to drive along roads that cross flooded streams. Figure 23-9 shows two images of the same location in Arizona where a local road crosses a streambed. When floods occur, as in the right image, motorists must wait for the water to fall before they drive across the stream.

Other floods are long-term events associated with periods of above average precipitation or spring snowmelt. The deadliest flood in American history occurred in 1889, when a poorly maintained dam burst upstream from Johnstown, Pennsylvania. The city was destroyed in minutes and more than 2000 people lost their lives. Destructive floods along the Susquehanna and Delaware rivers in upstate New York have occurred when the remains of hurricanes brought unusually heavy rain into these watersheds.

The Mississippi River and its tributaries have flooded throughout recorded history. Heavy regional rains and rapid snowmelt lead to spring floods, especially when the ground in the watershed

FIGURE 23-9. This stream crossing in Arizona is dry about 99 percent of the time. But when it floods, motorists must wait for several hours until the water level is low enough to cross safely and get to their homes.

is saturated. In 1993, floodwater covered an area the size of the state of Indiana and set new high-water marks at St. Louis.

Levees are banks along a river. They can be of natural or human origin. Natural levees are broad areas that may be just a few feet above a river. Artificial levees as much as 6 m (20 ft) high have been built along parts of the Mississippi River and its tributaries. The levees confine the river and prevent flooding of valuable farmland and homes when the river runs high. However, artificial levees confine deposition to the river bottom, which can also make the river run higher when it floods. This increases the danger if the levees break or overflow at flood stage. The levees have allowed people to build homes in low areas near the river. Poor maintenance or overtopping, which leads to erosion of the levees have caused extensive property losses. Visit the following Web site to see the flooding of New Orleans in Hurricane Katrina in 2005: *http://www.nola.com/katrina/graphics/flashflood.swf*

Droughts

Many of the same areas that experience sudden flooding also suffer from drought. A **drought** is a long period of dry weather. Rainfall is especially unpredictable in the Great Plains region of the United States. The climate has cycles of wet and dry years. Encouraged by several years of good precipitation, farmers depend on these conditions. However, the droughts return, and without irrigation, farming becomes impossible. Droughts do not lead to sudden loss of life as do storm events. However, their long-term economic effects and loss of life due to famine can be even greater.

The Northeast can be severely affected by a drought that would be the normal conditions in an arid environment. Any area of the United States suffers when precipitation is below normal for many months. This may cause wells to run dry and cities to run out of water. It may also have harmful effects on farming and businesses.

HOW CAN WE PROTECT OURSELVES FROM WEATHER HAZARDS?

4: 2.1h
7: STRATEGIES 2

Major storms cannot be predicted more than a few days in advance. However, people can make long-range plans to deal with storms when they do occur. Advanced planning can prevent property losses. Educational programs about weather hazards help people learn what to do before the storm and how to survive if they are caught in potentially hazardous situations. Rapid evacuation from coastal areas may be the best survival strategy.

Know Your Area's Weather History

It is good to know the kinds of destructive weather events that have happened in your area in the past. These are the most likely weather hazards that will occur in the future. For example, hurricanes occasionally affect New York State. Hurricanes cause damage related to storm surges, wave action, and winds in coastal areas such as Long Island. The greater threat from hurricanes in upstate New York is flooding of low-lying areas along streams and rivers. Blizzards and thunderstorms cause power and telephone outages and make travel hazardous. Visit the following Web site to look at the hurricane history of various Eastern Seaboard and Gulf states: *http://www.csc.noaa.gov/hez_tool/history.html*

Local Building Codes and Zoning

One of the government's most important functions is to protect people from harm. For example, building codes require that buildings be strong enough to withstand most local weather events and protect the people inside. Most communities have zoning laws. These laws prevent the construction of homes in areas at risk from weather hazards such as floods. Building on steep and unstable land may require special permits and inspections. This is to ensure that the property is not subject to landslides and other forms of mass wasting.

Emergency Planning

You can learn to protect yourself in case of weather-related emergencies. When a thunderstorm occurs, you should avoid exposed areas such as hilltops and open fields. You should stay away from tall trees that may be hit or knocked down by a lightning strike. Inside buildings, avoid anything that could provide a path for electricity such as electrical wires and water pipes. Unplug computers and other sensitive electrical devices. Automobiles, as long as they are not located in a flood zone, are usually a safe place to stay in a thunderstorm. The metal shell of the car conducts electricity around the passengers and into the ground. Commercial airplanes are often hit by lightning. This can frighten the passengers, but it rarely causes a serious problem.

Cities should have plans in effect to notify citizens and evacuate people from hazardous areas during weather emergencies. Planning helps to coordinate police and emergency services. Roads and bridges in shore areas can be changed to one-way routes that lead people to safety and prevent others from entering a danger zone. Evacuation can save lives. However, it can also lead to massive traffic jams when roads become clogged and cars stall. These problems become worse when service stations are not open along the evacuation route. Visit the following Web site to learn about hurricane hazards in the New York City metropolitan area: _http://www. nytimes.com/2008/08/31/nyregion/nyregionspecial2/31R hurricane.html?_r=2&oref=slogin&oref=slogin_

STUDENT ACTIVITY 23-4 —COMPREHENSIVE EMERGENCY PLANNING

6: OPTIMIZATION 6
7: CONNECTIONS 1

Your teacher will divide the class into five groups. Each group will select a different weather hazard: thunderstorm, blizzard, flood, tornado, or hurricane. The group will prepare a community action plan to deal with its weather hazard. Each group will present its plan to the class. Plans should include a section on the greatest dangers, how to prepare for the event, and what people should do (and not do) if caught in a severe weather event.

Early Warning Systems

The National Weather Service can generally predict the arrival of hurricanes, blizzards, and floods several days in advance. Doppler radar enables the National Weather Service to identify swirling winds in storm systems many miles away. Real-time satellite coverage and Doppler radar are able to find and track more localized storms that would have escaped notice in the past. If a storm is approaching, radio or TV broadcasts can keep you informed. Some areas use sirens to warn people of approaching tornadoes.

The National Weather Service issues three levels of alerts. An advisory indicates conditions that could produce a storm are expected in the near future. A storm watch means that conditions already exist for storm development. A storm warning indicates that a storm has actually been observed in the area, or that a storm is likely to appear at any moment.

In November 2002, an early tornado warning in Van Wert, Ohio, saved many lives. After being warned, people in a movie theater, including a large school group, moved from the auditorium into bathrooms and interior hallways. A tornado passed through 28 minutes after the warning. The bathrooms and hallways were the only parts of the building left standing. Miraculously, no one in the theater was injured.

Government agencies have programs to educate the public about the dangers of thunderstorms and flash floods. Lives have been saved when people have followed the advice they heard in broadcasts and read in pamphlets. These publications help people avoid the most common fatal mistakes made during storm events.

STUDENT ACTIVITY 23-5 —COMMUNITY PLANNING MAP

7: STRATEGIES 2

Use a map of your community to identify places where natural hazards are most likely to cause loss of lives and property. Make recommendations about building codes and how residents in those areas should prepare for natural hazards.

HOW IS EARTH'S ATMOSPHERE CHANGING?

4: 1.2e, 1.2f, 1.2g, 1.2h
4: 2.2d

Earth's atmosphere is nearly as old at the planet itself. When magma comes near the surface, pressure is reduced and bubbles of hot gas escape into the atmosphere. This is **outgassing**. Gases from erupting volcanoes are about 80 percent water vapor, 10 percent carbon dioxide, and 10 percent other gases including nitrogen.

The world's oceans also may have come from water vapor vented during volcanic eruptions. The water vapor condensed and fell as precipitation. Carbon dioxide and other gases remained as Earth's atmosphere. Objects in space that are much smaller than Earth, such as our moon, do not have enough gravity to prevent gases from escaping into space. However, Earth has been able to hold onto its atmosphere and oceans.

A second theory proposes that the oceans came primarily from outer space. Comets are sometimes called "dirty snowballs." It is likely that there were more comets when the solar system was very young. Many comets hit Earth and left their dust and water on Earth's surface. Determining whether the oceans are primarily the result of outgassing from magma or collisions with space objects is an area of active scientific research.

Atmosphere and Life

If carbon dioxide is the most plentiful gas in magma after water vapor, why is the atmosphere mostly nitrogen and oxygen? Only a very small fraction of 1 percent of the present atmosphere is carbon dioxide. Both Mars and Venus have atmospheres that are mostly carbon dioxide.

Unlike our neighboring planets, conditions on Earth are favorable for the development of abundant life on the surface. Earth is at just the right distance from the sun so that it has moderate temperatures. The history of life and the evolution of the atmosphere go hand-in-hand. Earth's earliest life-forms were primitive bacteria that thrived in an atmosphere with no oxygen. Chemical evidence of these earliest life-forms has been found in rocks 3.5 billion years old; that is only 1 billion years after Earth formed.

Ocean water protected early life-forms from destructive solar radiation. Early plantlike life-forms in the oceans began to use the energy of sunlight to make food through *photosynthesis*. During photosynthesis, oxygen is produced as a waste product.

In fact, as oxygen built up in the atmosphere, the original life-forms could no longer live at the surface. Their descendants now live in environments where there is little free oxygen. As the atmosphere changed, organisms evolved that thrived in an oxygen-rich environment. Oxygen also maintains the ozone layer in the upper atmosphere. The ozone layer protects land dwellers from short-wave ultraviolet radiation. Today's atmosphere is in a dynamic balance maintained by plants that produce oxygen and animals that use it. Figure 23-10, on page 568, represents the changing composition of Earth's atmosphere throughout geologic history. The composition of Earth's atmosphere is still changing.

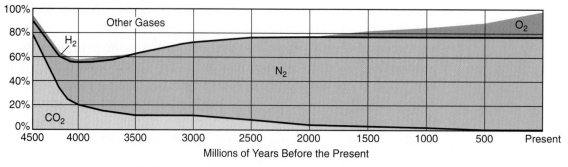

FIGURE 23-10. The geologic record and atmospheric samples from polar ice cores give us clues to the changing composition of Earth's atmosphere. The appearance of oxygen about 3 billion years ago has been especially important to the emergence of life on land.

People Affect the Atmosphere: Global Warming

The carbon dioxide content of Earth's atmosphere is small. However, of all Earth's greenhouse gases, carbon dioxide worries scientists most. Carbon dioxide alone is responsible for more than half of Earth's greenhouse effect. Furthermore, measurements of the concentration of carbon dioxide in Earth's atmosphere show that it is increasing at a rate of 1 percent every 2 years, as you can see in Figure 23-11.

The primary cause for the increase in carbon dioxide is the burning of carbon-based fossil fuels, such as gasoline, oil, and coal. At the same time, Earth's forests are being destroyed to supply wood and land for agriculture. As a result, the amount of carbon dioxide that can be absorbed by forests is reduced. At the current rate of change, the carbon dioxide content of the atmosphere could double by the end of the next century.

The addition of greenhouse gases to the atmosphere could result in an increase of several degrees Celsius in Earth's average temperature. Although this may not sound like a dramatic change, this is equal to the average global temperature change that has taken place since the end of the last ice age. This rise in Earth's average temperature is called **global warming**. Furthermore, a small amount of warming may cause the oceans to release carbon dioxide and accelerate global climate change.

How would global warming affect humans? Atmospheric circulation is a very complex system. The specific effects for any par-

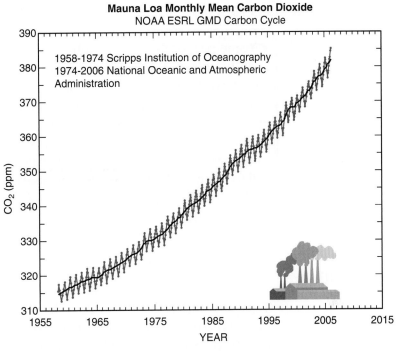

FIGURE 23-11. Over the past half century, scientists have measured a long-term increase in the carbon dioxide content of Earth's atmosphere. The primary cause is the burning of fossil fuels. The smaller fluctuations show cyclic, seasonal changes in plant growth that affect short-term carbon dioxide levels in the atmosphere.

ticular location are difficult to predict. Some locations that are now too cold for agriculture might eventually become important food producers. However, other regions that are now productive could become too hot or too dry for efficient crop growth at some point. Scientists have recorded rapid melting of glaciers and polar ice. This could raise sea level enough to cause flooding and increased storm damage in coastal locations, including many of the world's largest cities.

The United Nations has held international conferences to discuss global warming. Most industrialized nations have agreed to stabilize or even reduce their production of greenhouse gases. Some people in the United States have been unwilling to place limitations on economic growth. However, the majority of Americans have now accepted that the risk of global warming is too dangerous to ignore. Will the nations of the world unite in time to avoid a major disaster? Visit the following Web site to find the answers to frequently asked questions about global warming provided by the Intergovernmental Panel on Climate Change: *http://ipcc–wg1.ucar.edu/wg1/FAQ/wg1_faqIndex.html*

STUDENT ACTIVITY 23-6 —CARBON CAPTURE

1: SCIENTIFIC INQUIRY 1, 2, 3

Some proposals to ease global warming involve planting trees or other plants to absorb carbon dioxide and help the atmosphere pass terrestrial radiation into space. Place very dry but fertile soil into a planter. Carefully measure the mass of the soil and planter. Find the mass of seeds or cuttings of a fast growing houseplant such as coleus. Plant the seeds or cuttings and water the plant according to its needs. When the plant has grown large, allow the plant and soil to dry out like the dry soil you started with. Then find the mass of the dry soil and the dry plant. Where did the extra weight come from? What is carbon capture?

Air Pollution

The fuels that add carbon dioxide to the atmosphere also add oxides of nitrogen and sulfur. These compounds act as condensation nuclei to form clouds. They also make the cloud droplets acidic. Nitric acid and sulfuric acid are strong acids. Although some of the acidity in the atmosphere comes from volcanoes venting the same substances, the human-generated part of the problem is much more troubling. Figure 23-12 shows smog caused by air pollution over the city of Los Angeles, California. **Smog** is a mixture of fog and smoke.

Acid clouds form **acid precipitation** when the moisture falls as rain or snow. Some kinds of bedrock, such as limestone and marble that contain calcite, can react chemically with these acids to neutralize them. However, most rocks do not.

FIGURE 23-12. The clouds obscuring downtown Los Angeles and the San Gabriel Mountains are smog caused mostly by emissions from burning fossil fuels. Fossil fuels are our most important energy sources in transportation and other energy producing uses.

FIGURE 23-13. Trees in the highest parts of New York's Adirondack Mountains live near their climatic limits. This is a difficult environment for these trees. Acid-producing air pollution has added more stress, killing these trees.

The Adirondack Mountains have a three-part problem with acid damage. First, the prevailing southwesterly winds carry air pollution from the Midwest to the Adirondacks. Figure 23-13 shows damage to trees from acid clouds, especially in the higher parts of the mountains that are subjected to a harsh climate. Second, there is little calcite in most of the rocks of the Adirondacks so the acid runs into streams and lakes. Normal plants and fish cannot live in water that is too acidic. Third, environmental damage is most severe when spring snowmelt brings a sudden rush of acidic water just when fish and plants are most vulnerable.

A historical look at Earth's atmosphere reveals three important facts. First, the composition of Earth's atmosphere has changed through geologic history, primarily in response to living organisms. Second, in the past century, human activities have become the most important cause of change in Earth's atmosphere. Those changes may have even more dramatic results over the next few centuries. Third, people need to understand what can occur when they change any part of the environment and how the changes are likely to affect them. People may be facing a choice between reducing their effect on the atmosphere and adapting their lives to a wide range of unknown effects of global warming. Watch video insights into global warming: *http://www.manpollo.org*

CHAPTER REVIEW QUESTIONS

Part A

1. In the United States, most tornadoes are classified as intense

    (1) low-pressure funnel clouds that draw in air
    (2) low-pressure funnel clouds that spin out air
    (3) high-pressure funnel clouds draw in air
    (4) high-pressure funnel clouds that spin out air

2. Which graph below best shows the relationship between the concentration of carbon dioxide in Earth's atmosphere and the amount of infrared (heat) radiation absorbed by the atmosphere?

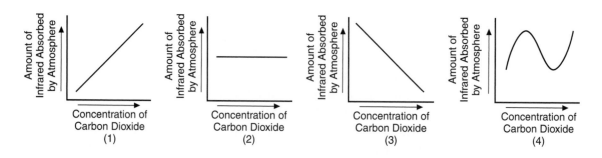

3. Earth's atmosphere was first formed during the Early Archean Era. Which gas was generally absent in the atmosphere at that time?

    (1) water vapor   (3) nitrogen
    (2) carbon dioxide   (4) oxygen

4. What kind of front sometimes spawns tornadoes between an advancing cold and dry air mass and a warm, moist air mass?

    (1) cold   (3) stationary
    (2) warm   (4) occluded

5. What component of Earth's atmosphere is most likely to cause climate change over the next few decades?

    (1) water vapor   (3) nitrogen
    (2) carbon dioxide   (4) oxygen

Part B

Base your answers to questions 6 through 8, on page 574, on the graph and table below.

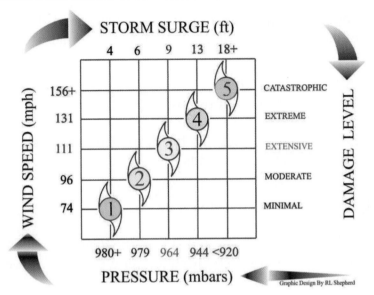

Characteristic Hurricane Damage by Category on the Saffir-Simpson Scale

Category 1—No real damage to sturdy buildings. Damage to poorly constructed older homes or those with corrugated metal. Some tree damage.

Category 2—Some damage to building roofs, doors, and windows. Considerable damage to poorly constructed homes. Some secondary power lines downed. Some shallow-rooted trees blown down.

Category 3—Some structural damage to well built small residences and utility buildings. Non-reinforced cinderblock walls blown down. Large trees blown down.

Category 4—Failure of many roof structures, window frames, and doors. Many well-built wooden and metal structures severely damaged or destroyed. Most shrubs defoliated. Many large trees blown down.

Category 5—Extensive or total destruction to wood-framed residences. Extensive glass failure due to impact of flying debris and explosive pressure forces during extreme gusts. Shrubs and trees lose up to 100 percent of their leaves.

6. If a hurricane has maximum winds of 100 mph, what is its category on the Saffir-Simpson scale?

   (1) category 1   (3) category 3
   (2) category 2   (4) category 4

7. How high above normal ocean height would we expect the sea to rise as the storm comes ashore as a hurricane that has an internal pressure of less than 920 mb?

   (1) less than 4 ft   (3) 9–13 ft
   (2) 6–9 ft   (4) 18 ft or more

8. If you observed damage to trees in a hurricane, but no trees blown down, what were the most likely maximum winds?

   (1) 74–96 mph   (3) 111–131 mph
   (2) 96–111 mph   (4) 131–156 mph

Base your answers to questions 9 through 11 on the diagram below, which shows the number of major natural disasters by state over a period of 29 years. Letters X and Y identify particular states.

9. What is the most likely cause of major destruction in the largest cities in the state labeled *X*?

    (1) hurricanes
    (2) blizzards
    (3) ice storms
    (4) earthquakes

10. Why does state *Y* have relatively few major storm events?

    (1) State *Y* has a warm, dry climate.
    (2) State *Y* is located along a boundary of Earth's tectonic plates.
    (3) State *Y* is far from a tropical ocean area.
    (4) State *Y* is the smallest of the 48 states shown on this map.

11. Look at the states that have the highest number of events. Which event accounts for most of that number?

    (1) blizzards
    (2) hurricanes
    (3) thunderstorms
    (4) earthquakes

Part C

Base your answers to questions 12 through 14 on the information below.

Ozone in Earth's Atmosphere

Ozone is a form of oxygen. The oxygen we breathe is a molecule composed of two atoms of oxygen (O_2). Ozone molecules have three atoms of oxygen (O_3). A layer of ozone between 10 and 30 mi above Earth's surface absorbs some of the harmful ultraviolet radiation from the sun. The amount of ultraviolet light reaching Earth's surface is directly related to the angle of the incoming solar radiation. The greater the sun's angle of insolation, the greater the amount of ultraviolet light that reaches Earth's surface. If the ozone layer were completely destroyed, the ultraviolet light reaching Earth's surface would most likely increase human health problems, such as skin cancer and eye damage.

12. In what temperature zone of Earth's atmosphere does most of the ozone layer exist?

13. Explain how the ozone layer benefits humans.

14. Assuming clear atmospheric conditions, on what day of the year do people out-doors in New York State receive the most ultraviolet radiation?

15. State one action that people should take to protect themselves from lightning.

Base your answers to questions 16 and 17 on the data below, which shows the average number of days with thunderstorms over land areas at various latitudes.

Data Table

Latitude	Average Number of Days a Thunderstorm Occurs Over Land
60°N	5
45°N	14
30°N	19
15°N	30
0° (equator)	56
15°S	44
30°S	21
45°S	8
60°S	0

16. Make a graph of the above data by labeling the bottom axis of a grid with lati-tude numbers from the table starting at 60°N and ending at 60°S. Label the ver-tical axis "Thunderstorm Days Per Year" from 0 to 60. Plot, with an *X*, the average number of days per year a thunderstorm occurs over a land area for each latitude shown on the data table. Finally, connect the *X*'s with a line.

17. State the relationship between latitude and the average number of thunder-storms that occur on Earth over land areas.

Base your answers to questions 18 through 20 on the map below, which shows the amount of snowfall in a December snowstorm in central New York and surrounding states. The units are inches of depth.

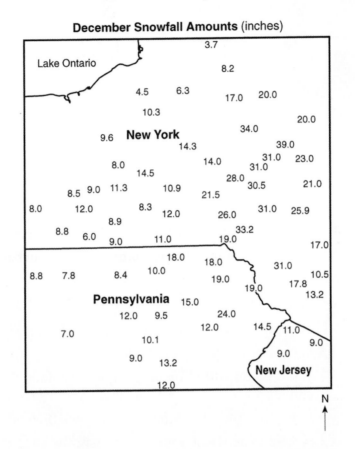

December Snowfall Amounts (inches)

18. On a copy of the map above, draw the 30.0-in.-snowfall isoline. Assume that the decimal point at each location represents the place where the depth was measured.

19. Most residents were aware several days in advance that the snowstorm was coming. State one action local residents could have taken to prepare for the snow emergency.

20. If the storm followed the usual path for mid-latitude cyclones in this area, toward what direction did the snowstorm move as it left New York State?

24

Patterns of Climate

WORDS TO KNOW

arid climate

climate

continental climate

deforestation

maritime climate

temperate climate

urbanization

This chapter will help you answer the following questions:

1. What do we mean by climate?
2. What natural and human geographic factors influence the climate?
3. How does the geography of New York State affect its climate?
4. How can we use graphs to compare the climates of different locations?
5. Why is Earth getting warmer?
6. How can humans speed or slow climate change?

ARE CLIMATES CHANGING?

4: 2.2d

A century ago, 150 glaciers covered the mountains of Glacier National Park in Montana. Today, only 27 glaciers remain. Meanwhile, the remaining glaciers are melting back so quickly that scientists estimate all could be gone in 20 years. The shrinkage in glacial ice is a worldwide event. However, a few glaciers are holding their own.

Measuring global climate change is not easy. The nature of global climate change is subtle. Scientists must take into account the natural cycles of temperature change. Nevertheless, analysis of ocean surface temperatures clearly shows a worldwide warming trend. This trend goes back to the beginning of the Industrial Revolution 130 years ago, as you see in Figure 24-1.

Within the ice of Earth's major glaciers is a record of past climates that goes back nearly half a million years. The thickness of annual layers and the crystal structure of the ice tell scientists about conditions of precipitation and temperature in the distant past. Air was trapped in the snow before it was buried within glaciers. Just as rock layers preserve a record of Earth's history, glaciers preserve a record of Earth's atmosphere. Visit the following Web site to view global warming experiments and watch and play games from NASA: *http://earthobservatory.nasa.gov/Laboratory/ PlanetEarthScience/GlobalWarming/GW.html*

Polar glaciers, especially in Antarctica, hold much of Earth's surface water. As a consequence of severe global warming, these glaciers would melt. The coastal regions of Earth would be flooded by

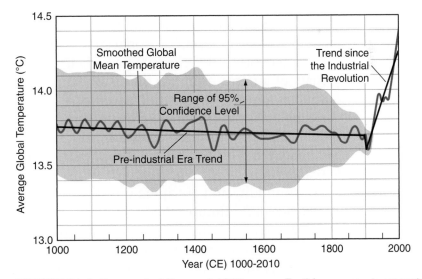

FIGURE 24-1. For most of the past 1000 years, Earth's average temperature has been very slowly decreasing. However, since the Industrial Revolution and our large-scale use of fossil fuels, the pattern has changed dramatically. In the past century we have seen an alarming spike in global temperatures.

rising sea levels. Antarctica holds enough ice to cause sea level to rise about 800 m (240 ft) worldwide. This would flood most of the world's largest cities, including Tokyo, Japan; Jakarta, Indonesia; and New York City. Visit the following Web site to work with interactive flood maps. These maps ask you to input a level of sea surface rise just from 1 to 14 m (3 to 50 ft) to see coastal flooding. *http://flood.firetree.net*

STUDENT ACTIVITY 24-1 —EFFECTS OF RISING SEA LEVELS

1: MATHEMATICAL
ANALYSIS 1
6: MAGNITUDE AND
SCALE 3

Estimate the total number of people who would need to move due to rising sea levels if the glaciers of Antarctica were to melt. How much do the estimates of different students differ and why are they different?

WHAT IS CLIMATE?

4: 2.1h, 2.1i

Weather describes the atmospheric conditions at a particular place and time, perhaps over a period of hours or days. **Climate** is the average weather based on measurements made over many years. Temperature and precipitation are the main elements of climate. However, humidity, winds, and the frequency of storms are also important. The normal seasonal changes in these factors are a part of climate as well. Scientists base their understanding of the climate of an area primarily on historical records. The more observations and the longer they have been kept, the more accurately scientists can describe the climate.

Scientists classify climates according to temperature. Tropical climates are usually warm. **Temperate climates** include large seasonal changes. Polar climates are usually cold. Scientists combine humidity and precipitation when describing a humid climate or **arid** (dry) **climate**. Visit the following Web site to see extreme records temperatures and ranges in the USA: *http://ggweather.com/climate/extremes_us.htm*

The plants found in an area are an indication of the climate. Rain forests, deserts, grasslands, and tundra are terms that describe both vegetation and climate. As you can see in the two images in

FIGURE 24-2. The kinds of vegetation in these images indicate two very different climates. How is the vegetation in each of these locations adapted to the local climatic conditions?

Figure 24-2, the leafy green trees indicate that this is a temperate climate. The cactus tells you that you are looking at a desert. Places where the natural vegetation changes are probably affected by changes in climate. Figure 24-3, on page 582, is a map of North American climates characterized by temperature, precipitation, and seasonal changes.

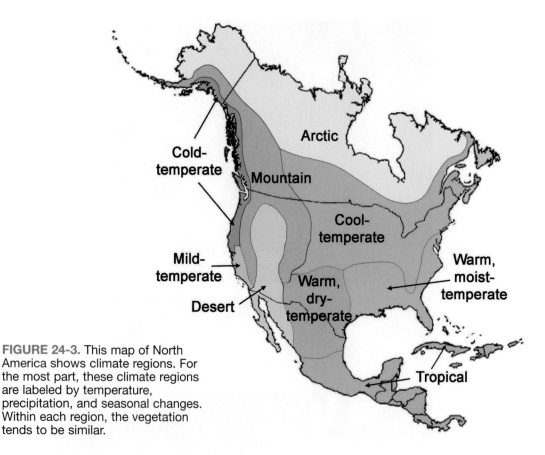

FIGURE 24-3. This map of North America shows climate regions. For the most part, these climate regions are labeled by temperature, precipitation, and seasonal changes. Within each region, the vegetation tends to be similar.

HOW DOES LATITUDE AFFECT CLIMATE?

4: 2.2c

Climate zones circle Earth at specific latitudes. Latitude determines the angle of insolation. The angle of insolation affects the temperature, which influences the climate. Latitude also has a role in precipitation.

Temperature

Variation in the intensity of insolation (sunlight) is the major cause of temperature differences over Earth's surface. In the tropics, where the noon sun is always high in the sky, solar energy is most intense. At the poles, where the sun is never high in the sky, solar energy is least intense.

TROPICS The tropics are sometimes called the latitudes of seasonless climate. Although the noon sun is a little higher in the sky

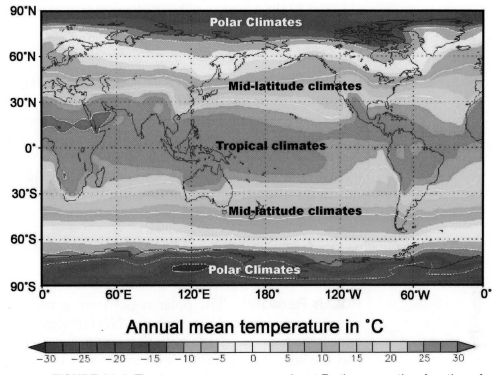

FIGURE 24-4. The temperature zones on planet Earth are mostly a function of latitude. Places near the equator receive the most direct sunlight while the poles receive far less solar energy.

in some parts of the year than in others, the change is small. The seasonal change in the length of daylight is also very small—in fact, hardly noticeable at all. Therefore the intensity of solar energy changes very little throughout the year. Except for high mountain locations, the weather is always warm.

The tropics extend from the Tropic of Cancer 23.5° north of the equator to the Tropic of Capricorn 23.5° south of the equator. (See Figure 24-4.) Here sunlight passes through the minimum thickness of Earth's atmosphere; so relatively little heat energy is lost within the atmosphere.

MID-LATITUDE Locations such as New York State have seasonal climates due to the annual cycle of changes in insolation. These are called temperate climates because the average temperature is neither hot nor cold. The largest seasonal changes actually occur in the mid-latitudes. The seasons in the Northern Hemisphere are the reverse of those in the Southern Hemisphere. When it is summer in the Northern Hemisphere, it is winter in the Southern Hemisphere.

1: MATHEMATICAL
ANALYSIS 1, 2, 3

Construct a single graph to show the average monthly temperature at three North American cities for a period of 1 year. Select cities at or near sea level. One location should be in Central America, a second in the main part of the United States, and the third in Canada or Alaska. What does your graph show about latitude and climate?

POLAR REGIONS The polar regions are generally cooler than other regions throughout the year. Nevertheless, they do experience large seasonal changes. In the winter, the days are very short and the sun, if it is visible, is always low in the sky. Insolation is extremely weak and temperatures may stay below freezing for months at a time. Even the summer sun is not very high in the sky, but the number of hours of daylight in the summer is very long. Because there is a large difference in the strength of insolation between winter and summer, most high-latitude locations are much warmer in the summer than they are in the winter. The interior of Alaska has bitter cold winters. However, even in central Alaska, summer temperatures can reach 30°C (about 90°F) at the height of summer. Visit the following Web site to see NOAA On-line Weather and Climate Data: *http://www.weather.gov/climate/*

Precipitation

Rainfall is generally plentiful in the tropics. Most of Earth's desert regions occur roughly 30° north and south of Earth's equator. Another zone of abundant precipitation can be found another 20 degrees or so from the equator. But why do these parallel zones of deserts and plentiful precipitation circle Earth?

These patterns are a result of Earth's rotation acting on terrestrial winds. You learned earlier that instead of one big convection

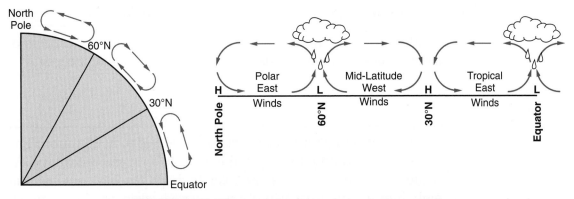

FIGURE 24-5. Places where rising air dominates experience more cloud formation and precipitation. Zones of descending air create high-pressure zones at the surface that have little precipitation.

cell in each hemisphere, the Coriolis effect forms three convection cells in each hemisphere.

The rotation of Earth and the position of the continents break convection in the Northern Hemisphere into three cells. The three convection cells are shown on the Planetary Wind and Moisture Belts diagram in the *Earth Science Reference Tables.* Figure 24-5 is a representation of part of that diagram. The left side of the diagram is a profile of Earth showing convection cells in the Northern Hemisphere. The diagram on the right shows Earth's surface as flat, the way it looks as you stand on it. Along the equator and at about 60° north latitude, air rises, forming low-pressure regions that circle Earth. The rising air causes cloud formation and generous precipitation at these latitudes. However, at latitude 30°N and at the North Pole (90°N) are regions of high pressure where sinking air warms and gets drier as it is compressed by atmospheric pressure. These latitudes have low relative humidity and relatively little precipitation. The 90° segment shown in Figure 24-5 is one of four similar profiles that circle Earth.

These high- and low-pressure belts are not still. Seasonal changes cause them to shift toward the equator in the winter and toward the poles in the summer. Furthermore, the wandering jet streams move these regions of high and low pressure and interrupt them with the passage of storm systems. Other geographic features you will soon read about also influence patterns of precipitation.

WHAT OTHER GEOGRAPHIC FACTORS AFFECT CLIMATE?

4: 2.2c, 2.2d

In addition to latitude other factors, such as elevation, nearness of large bodies of water, winds, and ocean currents, affect climate.

Elevation

The average temperature of a location is related to its elevation. The higher the elevation, the lower the average temperature will be. As rising air expands, it becomes cooler. Perhaps you have noticed that high mountains are often snow covered, even in the summer. Mount Chimborazo, shown in Figure 24-6, and Mount Kilimanjaro in Africa are near the equator. However, these mountains have permanent snow cover near their summits. Nearby locations at lower elevation have a tropical climate where it never snows. Rising dry air cools at a rate of about 1°C/100 m (4°F/1000 ft). Dry air warms at the same rate as it descends.

Mountain Barriers

Mountain ranges affect patterns of precipitation and temperature. Moist winds off the Pacific Ocean blow across California and rise up the western side of the Sierra Nevada Mountains. As the air rises into the mountains, it expands and cools below its dew point. This leads to cloud formation and precipitation. On the western side of the Sierra Nevada Mountains, the climate is cool with abundant precipitation. Seattle and the coast of northern California have a temperate, moist climate.

The air descends on the eastern side of the mountains. The air is compressed by increasing barometric pressure and becomes

FIGURE 24-6. Mount Chimborazo in South America is nearly on the equator. Yet, it is always covered with snow and ice. This occurs because air cools when it expands as it rises up the mountain.

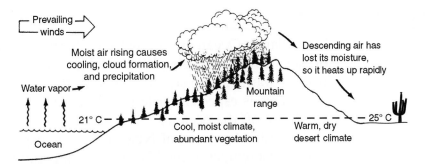

FIGURE 24-7. Cloud formation and precipitation usually occur on the windward side of a mountain range where moist air rises and cools. On the downwind side, descending air is warmed by compression, so the relative humidity quickly drops, generating an arid climate.

warmer. As the descending air warms without picking up moisture, its relative humidity decreases. Inland cities such as Spokane, Las Vegas, and Phoenix are located in the desert climate zone. Climate on the downwind, or leeward, side of mountains is sometimes called a "rain shadow" climate. Figure 24-7 illustrates the difference between climates on the opposite sides of a mountain range.

As moist air rises into the mountains, condensation (cloud formation) releases energy. This slows the rate of cooling. When clouds form, the air cools at a lower rate of about 0.6 C°/100 m (2°F/1000 ft). The descending air cannot pick up moisture. Therefore, the air heats at the greater, dry air rate of 1°C/100 m. Figure 24-8 illustrates this difference.

FIGURE 24-8. (1) As air rises, it expands and cools quickly. (2) When clouds form in the rising air, condensation releases the energy stored in water vapor, slowing the rate of cooling. (3) When the air sinks to a lower elevation and becomes warmer, there is no change in state to slow the rapid warming rate.

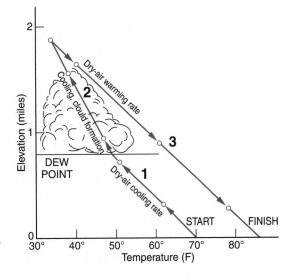

Large Bodies of Water

The Atlantic Ocean and Long Island Sound moderate the climate of New York's Long Island. Therefore, on Long Island, winters are usually warmer and summers cooler than in other parts of the state. This is especially true for places along the coast when the winds are off the ocean.

The inland regions of New York State experience the highest and lowest temperatures. The lowest temperature ever recorded in New York State was −47°C (−57°F) at Old Forge in the Adirondack Mountains. The record high temperature in New York State was measured in the capitol district: 42°C (108°F) at Troy. Both places are far from the moderating influence of the Atlantic Ocean and the Great Lakes.

STUDENT ACTIVITY 24-3 —MODELING TEMPERATURE CHANGE

6: MODELING 2

You can model the temperature changes over land and water using two containers. Fill one about a third full of sand and the other a third full of water. Place thermometers in both containers just above the sand or water surface. Be sure both thermometers stay completely dry! Position a lamp so it shines equally on both containers. For best results shield the thermometer bulbs from direct light. Compare the temperature changes in both containers.

Why do the oceans have such a great effect on climate? As you learned in Chapter 21, *specific heat* is a measure of the ability of a substance to warm as it absorbs energy or cool as it gives off energy. In general, metals and rock have low specific heats. They heat up and cool down rapidly. However, water has a very high specific heat. It heats and cools slowly.

Winds off the oceans or the Great Lakes, such as Lakes Erie and Ontario, control the temperature of nearby land areas. Even the winds off the Finger Lakes have some moderating affect. All these areas are cooler in summer and warmer in winter than places farther from the water.

There are three other reasons that land areas experience greater changes in temperature than the oceans. First, because water is

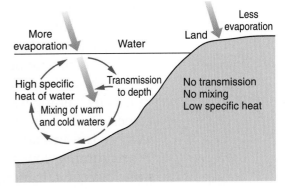

FIGURE 24-9. The ability of the oceans to absorb more heat energy than land absorbs is influenced by three factors: the relatively high specific heat of water, the depth of light penetration, and convective circulation of the water.

relatively transparent, sunlight penetrates deeper into water than it does on land. Rock and soil are opaque, so insolation energy is concentrated at the surface. Second, water is a fluid, so convection currents can distribute energy to the interior. Solids have no ability to mix. Finally, evaporation from the oceans uses some of the solar energy that would otherwise heat the oceans. Although there is some evaporation of water from soil, it is far less than evaporation from the oceans. Figure 24-9 summarizes these factors.

Scientists classify most terrestrial climates as maritime or continental. **Continental climate** is typical of inland areas. It is characterized by large seasonal changes in temperature. Inland areas generally do not experience the moderating influence of large bodies of water. Areas with a continental climate can be arid or moist, depending on the source region of the air masses that move into the area.

Maritime climate is sometimes known as the marine climate. It occurs over the oceans and along coasts, where water moderates the extremes in temperature. Areas that have a maritime climate experience moderate to high humidity.

Prevailing Winds

New York State has greater extremes of climate than many other coastal states. California, for example, is known for its mild climate. While inland areas of California experience greater ranges of temperature than the coastal locations, these extremes are not as great as those in New York State. The reason for this difference is the wind direction. Both states are in the global belt of prevailing west and southwest winds. However, in California those winds come off the Pacific Ocean. In most of New York State, the winds come from inland areas where temperatures are highly changeable. As a result, the nearby Atlantic Ocean has relatively little effect on the climate of most of New York State.

Monsoon climates include an annual cycle of weather patterns caused by shifting wind directions. In winter and spring, the wind comes from high-pressure centers over the continents. Spring weather is warm and dry with large changes in daily temperature. When summer low pressure builds over the continents, the wind shifts direction. It brings moist air from the ocean. Summer monsoon weather is more humid, with cooler days and warmer nights. The summer monsoons also bring clouds and precipitation, which reduce the temperature as well as the daily range of temperature.

Ocean Currents

Many tourists are surprised to see palm trees growing in the some parts of England and Ireland. Palm trees are not native to these countries. However, these imported plants can survive the mild weather conditions found in some parts of the British Isles. The Gulf Stream and the North Atlantic Current transport warm ocean water from the South Atlantic Ocean to the area surrounding Great Britain. These islands experience more moderate temperatures than does New York State. Most of the British Isles have damp and mild winters in which hard frosts are not common. This is true in spite of the fact that Great Britain is roughly 10° farther north than New York State. Along the East Coast of North America at the same latitude as Great Britain is the Labrador province of Canada, where the winters are even colder than in New York State.

Other locations are cooled by nearby cold ocean currents. The California Current keeps the coastal city of San Francisco in "sweater weather" throughout the summer. However, just a few miles inland people often experience desert heat. Even in the summer, local residents who visit the ocean may just wade in the surf. The water is too cold to swim in without a wetsuit. As shown in Figure 24-10, New York City's climate is warmed by the Gulf Stream. The Surface Ocean Currents map on page 58 and in the *Earth Science Reference Tables* provides a useful way to tell where warm and cold ocean currents affect the climate of coastal locations.

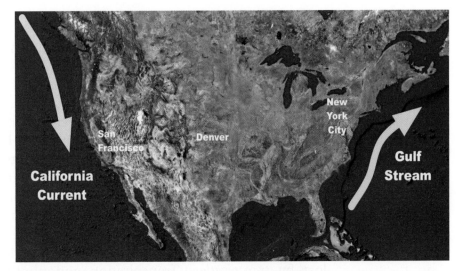

FIGURE 24-10. The California Current keeps San Francisco cool, even in the summer. The Gulf Stream tends to make New York City warmer than it would be otherwise. Denver, far from the moderating influence of the oceans, has large seasonal swings in temperature.

STUDENT ACTIVITY 24-4 —CLIMATES AND OCEAN CURRENTS

6: SYSTEMS
THINKING 1

Using a political map of the world and the Surface Ocean Currents map from the *Earth Science Reference Tables*, make a list of countries or regions that are affected by warm ocean currents. Make another list of places affected by cold currents. Alphabetizing your list will help you compare your locations with the lists of other students.

Ocean currents also affect patterns of precipitation. Cold air can evaporate far less water than warm air. In addition, cool air blowing over warmer land surfaces causes the relative humidity to decrease. Decreasing relative humidity makes precipitation unlikely. Therefore, coastal regions affected by cold ocean currents are usually places where rainfall is scarce. The Atacama Desert lies along the west coast of South America. A weather station there has been in place for decades without experiencing any measurable precipitation. On the other hand, the relatively warm Alaska current makes coastal Alaska one of the rainiest places in the United States.

Vegetation

The local climate and soil determine natural vegetation. Therefore, vegetation is a good indicator of the climate. For example, the temperate rain forests along the Pacific coast of the United States and Canada grow only in a cool, moist climate. However, vegetation also contributes to climate. Thick vegetation, such as the trees and plants in a forest, moderate temperature by holding in cool air during the day. At night, the vegetation prevents the rapid escape of warm air. Vegetation slows surface winds. In addition, plants contribute moisture to the air. During precipitation, the plants slow runoff. This gives water at the surface time to soak into the ground. Groundwater is then absorbed by the roots of plants and rises into the leaves, where over an extended period, water is slowly lost by *transpiration*. Transpiration and photosynthesis absorb solar energy, which would otherwise heat the land and air during daylight hours. So forest conditions are generally more moderate and consistently more humid than open land in the same area.

Human activities such as cutting wood, plowing fields, mining, or construction remove native plants. We replace plants with open ground, paved surfaces, or buildings. **Deforestation** is cutting forests to clear land. **Urbanization** is the development of areas with large populations. These activities have replaced natural vegetation with farmlands and cities at an ever-increasing rate. Bare ground and paved surfaces do not allow evaporation of groundwater. These surfaces heat up quickly during the day and cool quickly at night. As a result of urbanization, the local climate becomes drier and warmer with an increased daily range of temperatures.

Urban Heat Islands

Heating and air conditioning release heat to the outdoor environment. In addition, cars, trucks, buses, and other forms of transportation burn fuel and release heat. Businesses and industries, which are concentrated around cities, produce heat. In a city, the high concentration of human activities produces an urban heat island. In general, urban areas, such as New York City, warm more quickly and stay warmer than rural locations.

The effects of urban heat islands are easy to see. Have you ever noticed how much longer winter snow lasts in the country than it does in nearby urban areas? Even undisturbed parks in cities will be clear of snow before similar rural land is snow-free. On summer nights, city dwellers often need air conditioning all night, while in neighboring rural areas, residents can cool off by opening their windows to the cool evening air.

WHAT GEOGRAPHIC FEATURES OF NEW YORK STATE AFFECT THE LOCAL CLIMATE?

4: 2.2c, 2.2d

Throughout New York State differences in climate are relatively small. Many climatologists would classify the whole state as having a humid, continental, temperate climate with large seasonal variations in temperature. However, local geographic features do cause some significant differences in climate at various locations in the state.

As noted earlier, the Atlantic Ocean and Long Island Sound make the climate on Long Island more moderate than inland areas of New York State. Winter precipitation that falls as snow upstate is more likely to be rain on Long Island. Breezes off the ocean keep the humidity higher than in other parts of the state. Long Island is also more vulnerable to hurricanes and coastal storms.

Winter snow lasts longer in higher parts of the Adirondack Mountains and the Catskills for two reasons. First, the mountains, due to their elevation, are a little cooler than other areas of New York State. Second, mountains also influence patterns of precipitation. On the windward side, air rising into these mountain areas expands and cools. This causes increased precipitation throughout the year. Mountains also influence the climate on their downwind side. The land around Lake Champlain and the central Hudson Valley are in the rain shadow of mountains and may have as little as half the annual precipitation of the nearby mountains.

You read earlier that areas of New York State at the eastern end of Lakes Erie and Ontario are subject to "lake-effect" storms. This is especially true in late autumn and early winter, when the lake water is warmer than surrounding land areas. The lakes also moderate temperatures in nearby land areas. The first hard frost of autumn occurs later in these areas. The extended growing season makes land near the lakes valuable for agriculture.

HOW IS CLIMATE SHOWN ON GRAPHS?

4: 2.1g

Climate graphs are a way to illustrate different kinds of climates. On the following graphs, the dark line shows the average monthly temperature. The monthly bar graphs indicate the average monthly precipitation.

Figure 24-11 is a climate graph for Syracuse, New York. Notice the large seasonal changes in temperature and plentiful precipitation throughout the year. Remember that these are average conditions over many years. Therefore, unusual events such as droughts do not appear on these graphs.

Figure 24-12 is a climate graph for a desert location in the southwestern United States. It shows major seasonal changes in

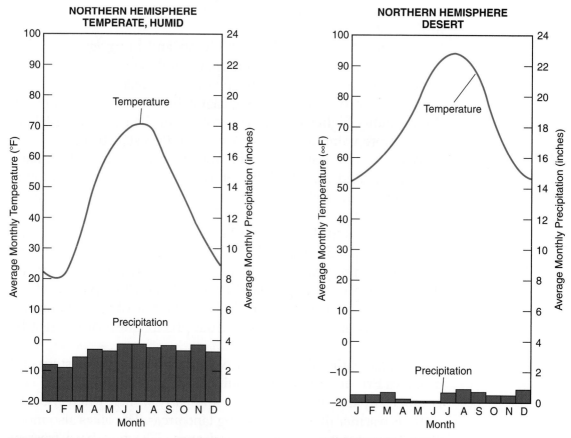

FIGURE 24-11. Temperate humid climate graph for Syracuse, New York.

FIGURE 24-12. Desert climate graph for Phoenix, Arizona.

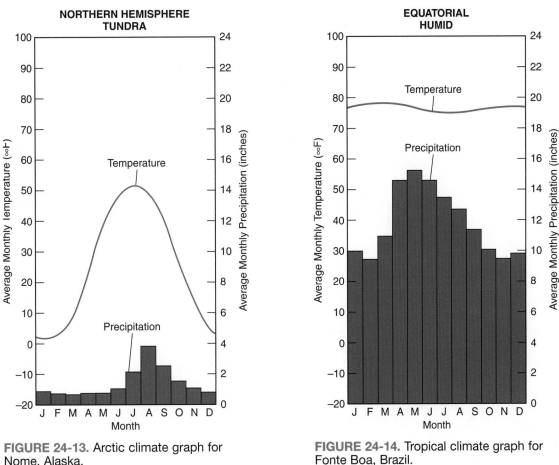

FIGURE 24-13. Arctic climate graph for Nome, Alaska.

FIGURE 24-14. Tropical climate graph for Fonte Boa, Brazil.

temperature like the Syracuse graph (Figure 24-11). However, this location is warmer in the winter and the summer. The precipitation is limited throughout the year, especially in the spring before the summer monsoon season.

Figure 24-13 illustrates a tundra climate in arctic Alaska. Temperatures are significantly lower than Syracuse, New York, throughout the year. Although precipitation is low, so is evaporation in this cold climate.

The rain forest of tropical Brazil provides the data for Figure 24-14. The average temperature changes very little throughout the year and precipitation is usually plentiful.

Figure 24-15 is the climate graph for Sydney, Australia. Notice that the highest and lowest temperatures are off by 6 months from those of the Northern Hemisphere locations. In the Southern Hemisphere, winter begins in June and summer begins in December. Sydney is a coastal city so the annual temperature range is not as great as it is at the previous temperate locations.

The last climate graph is a monsoon location in India. (See Figure 24-16.) Precipitation is very seasonal. Also notice how the temperatures fall off relatively early when the summer monsoons arrive in July. Compare factors that affect climate by observing climate graphs: *http://people.cas.sc.edu/carbone/modules/mods4car/ccontrol/index.html*

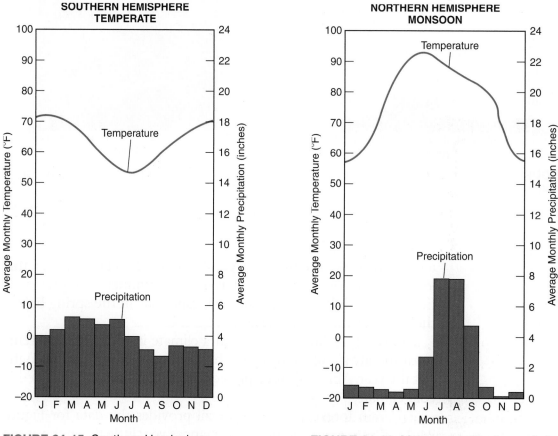

FIGURE 24-15. Southern Hemisphere climate graph for Sydney, Australia.

FIGURE 24-16. Monsoonal climate graph for New Delhi, India.

CHAPTER REVIEW QUESTIONS

Part A

Base your answers to questions 1 and 2 on the following diagram, which represents prevailing winds blowing over a high mountain range.

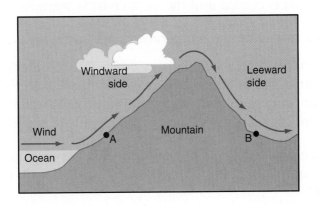

1. Why do clouds often form only on the windward side of the mountains?

   (1) Air rises and cools on the windward side.
   (2) Air rises and becomes warmer on the windward side.
   (3) Air sinks and cools on the windward side.
   (4) Air sinks and becomes warmer on the windward side.

2. Points *A* and *B* are at the same latitude and the same elevation above sea level. How does the climate at *A* probably differ from the climate at *B*?

   (1) Location *A* has less precipitation.
   (2) Location *A* has summer when location *B* has winter.
   (3) Location *A* has a greater annual temperature range.
   (4) Location *A* is usually cooler in the summer.

3. Why are temperatures in the tropic low-latitudes always higher than temperatures at the South Pole?

   (1) The South Pole receives more intense sunlight.
   (2) The South Pole receives less solar radiation.
   (3) The South Pole has more cloud cover.
   (4) The South Pole has more vegetation.

4. Which statement best summarizes the general effects of ocean currents at 20° south latitude on coastal regions of South America?

(1) The east coast and the west coast are both warmed by ocean currents.
(2) The east coast and the west coast are both made cooler by ocean currents.
(3) The east coast is warmed and the west coast is cooled by ocean currents.
(4) The east coast is cooled and the west coast is warmed by ocean currents.

5. Why do oceans moderate the temperatures of coastal areas?

(1) Temperature of ocean water changes slowly due to water's low specific heat.
(2) Temperature of ocean water changes slowly due to water's high specific heat.
(3) Temperature of ocean water changes rapidly due to water's low specific heat.
(4) Temperature of ocean water changes rapidly due to water's high specific heat.

6. The diagram below represents prevailing winds blowing from New York State across Lake Champlain and into Vermont. Compared with the climate at A, the climate at B is

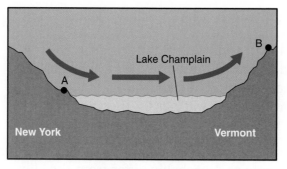

(Not drawn to scale)

(1) warmer and wetter
(2) warmer and drier
(3) cooler and wetter
(4) cooler and drier

7. London, England is approximately 51°N latitude, 0° longitude. Binghamton, New York, is about 42°N latitude, 76°W longitude. Why does Binghamton have colder winters?

(1) Binghamton is farther from the equator.
(2) Binghamton is closer to sea level.
(3) London is affected by the North Atlantic Drift.
(4) London has a longer duration of insolation in the winter.

Part B

Base your answers to questions 8 through 10 on the map below, which represents an imaginary continent.

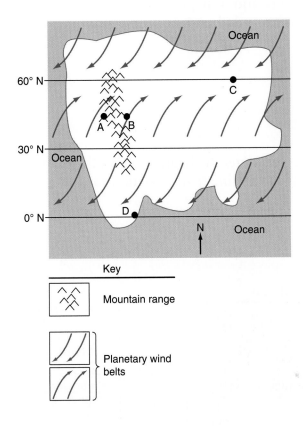

8. Locations *A* and *B* are at the same elevation below a mountain range. Compared with the climate at *A,* the climate at *B* is most likely

   (1) warmer and more humid
   (2) warmer and less humid
   (3) cooler and more humid
   (4) cooler and less humid

9. How can we best describe the wind pattern at point *D*?

   (1) converging winds and rising air
   (2) converging winds and falling air
   (3) diverging winds and rising air
   (4) diverging winds and falling air

10. Location *C* most likely experiences

   (1) low air pressure and little precipitation
   (2) low air pressure and abundant precipitation
   (3) high air pressure and little precipitation
   (4) high air pressure and abundant precipitation

Part C

Base your answers to questions 11 through 13 on the map and data tables below, which show geographic and climate information for two towns in Australia.

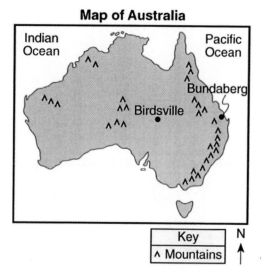

Map of Australia

Key

^ Mountains

N

Data Table 1
Average Monthly High Temperatures for Birdsville, Australia

Month	Temperature (°C)
January	39
February	38
March	35
April	30.5
May	25
June	22
July	21
August	23.5
September	28
October	32.5
November	36
December	38

Data Table 2
Information about Two Australian Cities

City	Latitude (° S)	Longitude (° E)	Elevation (m)	Average January Rainfall (mm)
Birdsville	25.9	139.4	47	25
Bundaberg	24.9	152.4	14	105

11. Why does Bundaberg have its lowest temperatures at the same time New York State is having its highest temperatures?

12. Make a copy of the graph below, and then plot an **X** to show the average temperature each month in Birdsville. Finally, connect the X's with a line and label the line "Birdsville."

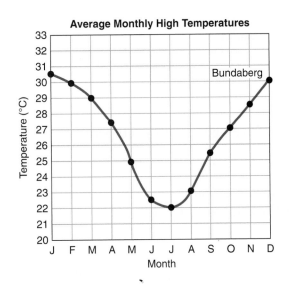

13. State one reason that Birdsville has less rainfall in January than Bundaberg.

Base your answers to questions 14 through 17 on the graph below, which shows the annual cycle of temperature changes at two cities, X and Y. Both X and Y are at the same latitude.

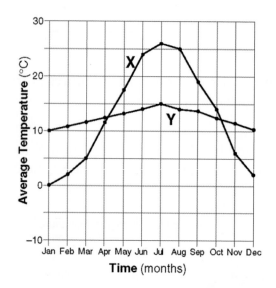

14. What is the range of average monthly temperatures at city Y?

15. What geographic factor explains why city X most likely has a greater temperature change over the course of a year.

16. What evidence indicates that both city X and city Y are in the Northern Hemisphere?

17. How would the climate of Britain and Western Europe change if Atlantic Ocean currents stopped moving?

Base your answers to questions 18 through 20 on the topographic profile and bar graph below, which represent rainfall and prevailing winds blowing over the mountains of California.

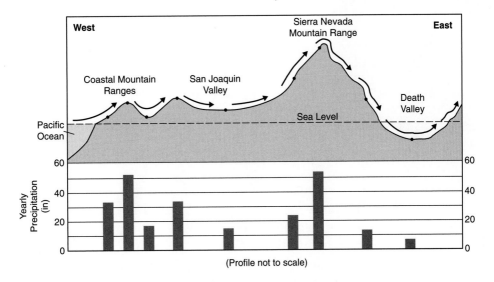

(Profile not to scale)

18. Explain why Death Valley has less rainfall than the top of the Sierra Nevada Mountains.

19. What is the total precipitation combined for the four points in the Coastal Mountain Ranges?

20. Why are temperatures colder in the Sierra Nevada Mountains than in the Coastal Mountain Ranges?

UNIT 8
Earth and Beyond

Why do we explore space? Since ancient times, wise and curious people have looked to the night skies for entertainment and inspiration. What they saw was mysterious, changing, and yet remarkably predictable. They wondered what they were looking at. Was this a magical palette of nature or some unknown world? They were curious.

In modern times, we are still curious. However, our investigations of the heavens have become far more sophisticated. Telescopes have extended our observations. Humans have visited the moon. We have sent robot explorers to Mars. Meanwhile, spin-offs of the space program such as enhanced global communications, GPS systems, and medical research have clearly changed all our lives. There are also artistic discoveries such as the view of the Eagle Nebula seen here. The beauty and the mysteries of space are clear in this Hubble Space Telescope image of the Eagle Nebula.

Will space colonization increase our possibilities for survival? Can we find resources of economic importance in space? Will we find evidence of new life-forms or even another civilization? The answers may come very soon, or they may take centuries to answer. However, these could be among the most important discoveries in human history. The sense of curiosity is the constant that drives inquiry through science.

25

Earth, Sun, and Seasons

altitude

Antarctic Circle

Arctic Circle

celestial object

equinox

geocentric

heliocentric

revolution

rotation

solar noon

solar time

summer solstice

Tropic of Cancer

Tropic of Capricorn

vertical ray

winter solstice

zenith

This chapter will help you answer the following questions:

1 What is astronomy?

2 How can we locate positions in the sky?

3 How does the daily path of the sun change with the seasons?

4 How does the path of the sun change with our location on Earth?

5 Do the sun and stars actually move in daily and annual cycles?

6 What are the patterns of the apparent motions of celestial objects?

WHAT IS ASTRONOMY?

4: 1.1a

The study of objects beyond Earth and its atmosphere is *astronomy*. Most of our knowledge of the heavens comes through the light (electromagnetic radiation) these objects give off or reflect. In some ways astronomy is one of the oldest sciences. Historians have found records of observations of the heavens that go back thousands of years. This branch of science also involves some of the most advanced technology. It has consistently yielded new insights into where we came from as well as the fundamental nature of matter and energy.

In this chapter, it will be important to distinguish between what is apparent and what is real. For example, we observe the daily path of the sun through the sky. We know that this cycle is caused by Earth's rotation on its axis. However, we conduct our lives as if the sun actually moves around Earth. We talk about the sun rising and setting as if these were real motions of the sun. We call the motion of the sun through the sky an *apparent motion* because this movement of the sun looks real. Later in this chapter, you will learn how scientists proved that Earth moves. However, for the sake of describing celestial events, it is convenient to stay with the Earth-centered point of view. Visit the following Web site to access a free, web-based planetarium simulator: *http://www.skyviewcafe.com*

STUDENT ACTIVITY 25-1 —ASTROLOGY ON TRIAL

1: SCIENTIFIC
INQUIRY 1
2: INFORMATION
SYSTEMS 1, 2, 3

Some people confuse astronomy with astrology. Astrology is a traditional art of using celestial observations to explain personality traits based on when you were born. It also uses similar information to make predictions of the future. Astrologers divide the year into 12 zodiacal periods with their associated birth signs. Is astrology science? Does it work? The following is a way to test the accuracy of astrology.

Students should work in pairs. Your teacher will read a brief description of the personality traits of people born during each of the 12 zodiacal time periods. Choose the one that best describes your partner. Then compare the dates related to each description with your partner's birthday. If astrology is reliable, a large portion of the students' birthdays should fall within the time periods of the chosen personality trait. However, a random distribution would find only about 10 percent of the students have chosen the correct description. What does your analysis reveal about astrology?

HOW CAN WE DESCRIBE THE POSITION OF CELESTIAL OBJECTS?

4: 1.1a, 1.1c

To any observer on Earth, the sky looks like a giant dome that stretches across the sky from horizon to horizon. **Celestial objects** are the things seen in the sky that are outside of Earth's atmosphere. These objects include the sun, moon, planets, and stars. You may know that some of these objects are farther away than others. Nevertheless, many appear as if they are points of light on the surface of the sky.

In locating celestial objects, it is handy to treat the sky as a two-dimensional surface. Two coordinates are all you need to specify a location on a two-dimensional surface, for example, positions based on the x and y axes of a graph. Angles of latitude and longitude are the coordinates used to locate places on Earth's two-dimensional surface. Like Earth, the sky appears as a curved surface. Therefore, scientists also use angles to locate positions in the sky.

In the sky, you can specify a location using *azimuth* (compass direction) and angular altitude. You learned about compass directions such as north or southwest when you read about maps in Chapter 3. For locations in the sky, the compass direction refers to the direction along the horizon directly below the position in the sky. Azimuth usually is expressed as an angle measured from north (0°), clockwise to east (90°), south (180°), west (270°), and around the horizon back to 360° (360° is also 0°, due north).

Using this coordinate system, altitude is the second coordinate. **Altitude** is the angle above the horizon. Altitude angles start at a level horizon (0°) and increase to 90° at the point directly overhead, sometimes called the **zenith.** The coordinates of compass direction and angular altitude allow us to locate any point in the sky. Figure 25-1, on page 608, shows the angle of direction and altitude for a star that is high in the southeastern sky.

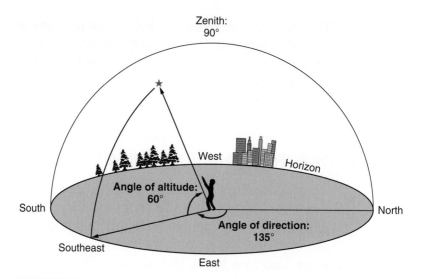

FIGURE 25-1. Any point in the sky can be located by two angles. The azimuth angle of direction is the clockwise angle from due north to a place on the horizon directly below the point in the sky. The angle of altitude is the angle from the horizon up to that point.

OUR INTERNAL CLOCK

4: 1.1d

Like most other animals on Earth, humans have an internal clock that works on a 24-h cycle. This cycle is called the circadian rhythm. Scientists have conducted experiments in which people were voluntarily isolated in special rooms. The people received no natural light and no clues to tell day from night. After a while, the subjects established their own periods of sleep and activity. They usually settled into a cycle that was about 24 h in length. This cycle is a natural part of our biochemical identity.

Travelers who journey east or west across a few time zones may notice the functioning of their internal clock. They experience jet lag. This may include difficulty sleeping when the people around them are going to bed. Their ability to concentrate can be harmed at the time they would have been sleeping at home. However, after a few days or a week, most people adjust to the new time routine, and the effects of jet lag go away.

How Is Clock Time Established?

The daily cycle of night and day is so much a part of our lives that our system of clock time is based on the length of the day.

The daily and yearly cycles of the sun have an enormous effect on us.

Most nations use the International System of Units (metric system). In this system, most measures are related to larger and smaller units by a factor of 10. However, the worldwide system of time is still a 24-h day with 60 minutes in an hour and 60 seconds in a minute. Calendar time is based on the cycle of seasons and Earth's revolution around the sun. A year is approximately $365\frac{1}{4}$ days.

Two regions on Earth do not have a 24-h cycle of night and day. The North and South Poles are on a yearly cycle of daylight and darkness. At both poles sunrise occurs about the first day of spring in that hemisphere. The sun remains continuously visible in the sky for 6 months, until the first day of autumn in that hemisphere. The dark of night begins a few weeks after that time, and continues for nearly the next 6 months. Humans are unable to adapt their activities to a 12-month cycle of day and night.

At these two locations there is no natural way to assign clock time. There is no settlement at the North Pole. However, the United States does maintain a permanent research station at the South Pole. Clocks at the South Pole are set to the time in New Zealand, the closest inhabited area, and the place from which most people fly to the South Pole research station.

WHAT IS THE SUN'S APPARENT PATH ACROSS THE SKY?

4: 1.1d

For most observers on Earth, the sun rises in the eastern part of the sky. The sun reaches its greatest angular altitude at solar noon. It then moves down to dip below the western horizon. This apparent motion of the sun is the basis for the earliest clocks: sundials.

Local **solar noon** occurs when the sun reaches its highest point in the sky. For mid-latitude locations in the Northern Hemisphere, such as New York State, at solar noon the sun reaches its highest point when it is due south. The noon sun is never directly overhead (at zenith) in New York State. In fact, the sun is never at the zenith except for observers located between the Tropic of Capricorn (23.5°S) and the Tropic of Cancer (23.5°N).

STUDENT ACTIVITY 25-2 —THE LENGTH OF A SHADOW

1: SCIENTIFIC INQUIRY 1, 2, 3

Select a place where buildings or trees do not block the sun for an extended period. During the course of 1 day, take measurements of the changes in length of the shadow of a vertical object 1 m tall (1 meter = 39.37 inches). Use these measurements to draw a graph of the change in length of the shadow from sunrise to sunset. Measurements must be made on a level surface. Think about the best times to take measurements and which values of shadow length you can determine by logical thinking. Your teacher will collect your data table and your graph.

According to your graph, what was the clock time of solar noon? As an extended activity, make the changing direction of the shadow part of your data and your graph.

STUDENT ACTIVITY 25-3 —CONSTRUCTING A SUNDIAL

1: ENGINEERING DESIGN 1

Internet sources (see below) as well as books have plans for making a simple paper or cardboard sundial. All individuals or groups of students can make the same style, or each can select a different design to construct.

You must position your sundial carefully in an open area on a level surface with a north–south alignment. (Be sure to determine true north, not magnetic north.) You can check your sundial by comparing its reading with a clock or watch. Solar time and clock time will differ depending on your longitude even if you account for daylight saving time. Visit one of the following Web sites to learn how to make your own sundial: *http://www.mysundial.ca/sdu/sdu_cd_sundial.html* or *http://www.scribd.com/doc/47530/Building-a-sundial*

HOW DOES THE SUN'S PATH CHANGE WITH THE SEASONS?

4: 1.1d, 1.1f, 1.1g, 1.1h

The previous section did not specify the exact direction of sunrise and sunset, nor did you learn how high the noon sun reaches in New York State. These observations depend on the time of year.

In fact, the seasons can be defined based on observations of the path of the sun through the sky.

For observers with a level horizon, such as anyone at sea, the sun rises due (exactly) east and sets due west on just two days of the year. People who live on land would also observe this if hills, trees, and buildings did not block their view of the horizon. These two days of the year are the **equinoxes**. Equinox is made of two parts, *equi-* means equal and *nox* means night. On these days, daylight and night are approximately equal in length: about 12 h of daylight and 12 h of night. Figure 25-2 is a series of four photographs taken

FIGURE 25-2. These four photographs were taken from the same location at the equinoxes and solstices. During winter and spring the point of sunrise moves to the north. Then in summer and autumn the sunrise moves steadily to its southern-most position.

from the same position. Note how the direction of sunrise changes in an annual cycle.

It is useful to remember that the complete apparent path of the sun is a circle. Part of this circle is above the horizon and part is below the horizon. The total angular distance around a circle is 360°. Furthermore, the apparent motion of the sun along that circle is constant. The rate at which the sun appears to move is determined by dividing 360° by 24 h, which equals 15°/h. No matter when or where you observe the sun, it always appears to move at a rate of 15°/h. However, the length of daylight changes. This is because the portion of the circle that is above or below the horizon changes on a yearly cycle.

Spring

Consider the changing path of the sun for an observer facing south at 42° north latitude in New York State. March 21 is our *spring equinox,* as shown in Figure 25-3. (This is also called the vernal equinox.) The sun rises due east at 6 A.M. It moves to the observer's right as it climbs to a noon altitude angle of 48°. (That is just a little more than half way from the horizon to the zenith, straight overhead.) In the afternoon, the sun continues to the right, setting due west at 6 P.M. Of the sun's circular path of 360°, half is in the sky and half below the horizon, so there are 12 h of daylight.

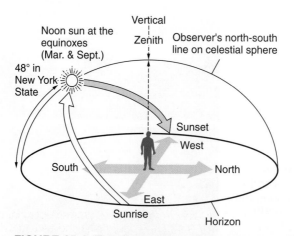

FIGURE 25-3. This figure shows the sun's path at 42° north latitude on the equinoxes. You are looking down on the dome of the sky, unlike the view of the observer under the sky, seen in Figure 25-2.

Summer

The longest period of daylight in New York State occurs 3 months after the spring equinox. It usually occurs on June 21. It is called our **summer solstice**. On this date, the sun rises in the northeast at about 4:30 A.M. standard time (5:30 A.M. daylight saving time) and sets at about 7:30 P.M. standard time (8:30 P.M. daylight saving time). (Figure 25-4.)

New York has about 15 hours of daylight at the summer solstice because the sun rises well into the northeast and sets in the northwest. Therefore, more than 60 percent of the sun's circular path is above the horizon. This is also when the noon sun is highest in the sky in New York State about 71.5° above the horizon (but still 18.5° below the zenith). Insolation is strongest at this time of the year. The season of summer begins on the summer solstice.

Autumn

The *autumnal equinox* occurs about September 22. On that day sunrise occurs at 6 A.M. due east, and sunset at 6 P.M. due west. The sun follows the same path it took 6 months earlier at the spring equinox. Figure 25-5, on page 614, is appropriate for both equinoxes. Because there are not exactly 365 days in a year, periodic

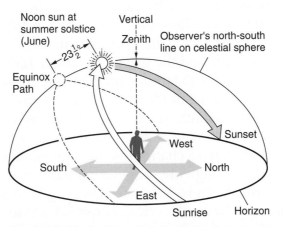

FIGURE 25-4. In New York State on the summer solstice, the sun rises in the northeast at about 4:30 A.M. (standard time) and sets in the northwest about 7:30 P.M., providing 15 h of daylight. The sun reaches its highest point at noon when it is nearly 72° above the southern horizon. The sun's path in March is shown by a dotted line.

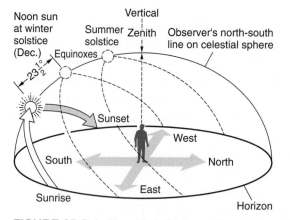

FIGURE 25-5. In New York State on the winter solstice, the sun rises in the southeast at about 7:30 A.M., and sets in the southwest about 4:30 P.M., giving just 9 h of daylight. The sun reaches its highest point at noon when it is 24.5° above the southern horizon. Dotted lines show the sun's path at the equinoxes and the summer solstice.

adjustments must be made in our calendars. Therefore, the dates of the beginning of the seasons vary slightly from year to year.

Winter

The first day of winter in the Northern Hemisphere occurs each year between December 21 and December 23, which is called the **winter solstice**. On that day, the sun rises in the southeast, as shown in Figure 25-5. Its path through the sky is relatively short and the sun sets in the southwest. Less than 40 percent of the sun's circular path is above the horizon. Sunrise does not occur until about 7:30 A.M., and the sun sets at about 4:30 P.M. The length of daylight at 42° north latitude on the winter solstice is about 9 h.

The maximum altitude of the sun, at solar noon, is only about 24.5° above the southern horizon. The sun is never very high in the sky, even at noon. This is why solar energy is weakest at this time of year. This pattern of change is similar for all mid-latitude locations on Earth, although the angular altitude of the noon sun does change with latitude. Visit the following Web site to learn about one of many Native American constructions to follow the seasons by the changing path of the sun: *http://solar-center.stanford.edu/AO/bighorn.html*

For mid-latitude locations such as New York State, the seasonal changes can be summarized as follows:

1. The sun generally rises in the east and sets in the west. However, the sun rises north of east when the days are longest in spring and summer and south of east during the short days of fall and winter. The sun always sets in the west, but the pattern of change is similar. The sun sets in the northwest during the long days of spring and summer. When the days are shorter in fall and winter, the sun sets in the southwest.

2. The direction of both sunrise and sunset moves northward along the horizon as the days get longer in winter and spring. During the summer and autumn, when the days get shorter, the sunrise and sunset position moves southward from day to day.

3. The noon sun is highest at the time of the summer solstice and lowest at the time of the winter solstice. At these times, the noon sun is 23.5° higher or lower than its equinox position.

4. When the noon sun is highest in the sky, the period of daylight (direct sunlight) is longest. Although our noon sun is highest in the summer, unlike in the tropics, the sun is never directly overhead for observers anywhere in the United States. Figure 25-6 summarizes these observations.

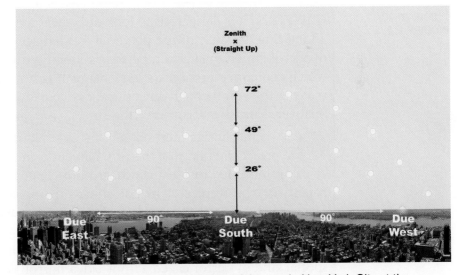

FIGURE 25-6. This figure shows paths of the sun in New York City at the summer solstice (top path), spring and autumn equinoxes (middle path), and the winter solstice (bottom path). Note the following: (1) The sun's path is higher and longer in the summer, and lower and shorter in the winter; (2) the sun rises and sets exactly east and west on only the equinoxes; (3) the sun is never overhead (marked zenith in this figure) for any observer in New York State.

STUDENT ACTIVITY 25-4 —TRACING THE DAILY PATH OF THE SUN

1: SCIENTIFIC
INQUIRY 1, 2, 3

For this activity, you will need a transparent hemisphere about 30 cm (12 in.) in diameter. You can get these hemispheres from educational materials suppliers. However, you may also be able to find a cheaper alternative such as a cake dome or a plastic container. Use a water-soluble marker to make a dot on the center of the dome. Get a square of cardboard about 45 cm (18 in.) on a side. Make a mark in the center of a piece of cardboard. Mount the dome on the cardboard base so that the dot on the cardboard is directly under the dot on the plastic dome. Place the mounted dome on a flat surface such as a sidewalk that has a broad view of the sky. You will need to mark this placement carefully so you can bring the dome back and place it in exactly the same position. As you repeat this procedure, you will make a line of dots that show a sun path on your hemisphere that looks similar to one of the paths shown in Figure 25-6, on page 615.

STUDENT ACTIVITY 25-5 —OBSERVING THE SUN

1: SCIENTIFIC
INQUIRY 1, 2, 3

CAUTION: It is dangerous to look directly at the sun, especially for a long time because of the potential for damage to your eyes. This danger is much greater if a person is looking through a telescope, which can concentrate the sun's energy. It is best to never look directly at the sun through any optical device. However, there are two relatively safe ways to observe the sun.

One relatively safe method is to use a small reflecting telescope to project an image of the sun. Direct the sunlight onto a sheet of paper mounted inside a cardboard box. Focus the image on the paper with the eyepiece. The box shields the paper screen from direct sunlight. **To repeat: *Never look directly at the sun through any optical device.***

A second method to project an image of the sun makes use of a special kind of mirror called a front-sided mirror. Most mirrors have the silver reflective coating behind the glass to protect the coating from scratches. A front-sided mirror has the coating on the front of the glass. Front-sided mirrors can be purchased from scientific supply companies. This activity works best with a tiny piece of mirror roughly 0.5 cm (0.25 in.) square. (The shape is not important.)

> Mount the small piece of front-sided mirror on a tripod so it reflects a beam of sunlight through a window. Project the reflection onto a light-colored interior wall of a darkened room. Although it can be difficult to aim this reflection, the image projected inside can be used for two purposes. This procedure creates a dim image of the sun, revealing details on the solar surface. It can also be used to observe and even measure the sun's apparent motion through the sky.

DOES THE SUN'S PATH DEPEND ON THE OBSERVER'S LOCATION?

4: 1.1d, 1.1f, 1.1h

At any particular time, half of Earth is lighted and half is in shadow. Whether it is day or night depends on where you are located. This is a consequence of living on a spherical planet.

Changes with Longitude

If you telephone a friend in California, you need to remember that the sun rises in New York 3 h before it rises in California. This is why it is better to telephone friends on the West Coast late in the evening, rather than early in the morning. For example, if it is noon in New York State, it is only 9 A.M. in California. This is due to Earth's rotation at 15°/h. Local time generally changes by 1 h for each 15° change of longitude east or west. This is why Earth is divided into time zones that generally are 1 h apart for each 15° change in longitude.

Changes with Latitude

Although the local **solar time** does not change with latitude, the path of the sun through the sky does change. You have already read that observers at the North or South Pole experience 6 months of daylight followed by nearly 6 months of night. (Of course dawn and dusk extend light for a while before sunrise and after sunset.) In fact, every location on Earth has a total of 50 percent of the year when the sun is in the sky and 50 percent when it is below the horizon. However, as a result of Earth's curvature, the cycle of day and night is most changeable at the poles and most constant near the equator.

People at the equator experience approximately 12 h of daylight every day. Changes from season to season are hardly noticeable.

The position of sunrise along the horizon changes, as it does in the mid-latitudes such as New York. However, the noon sun is always high in the sky. When it is spring and summer in the Northern Hemisphere, the sun rises to the north of east at the equator. The noon sun is a little north of the zenith (the point straight overhead). From late September through most of March the sun rises in the southeast and reaches its highest point a little south of the zenith. At the equinoxes the sun rises due east, sets due west, and the noon sun is directly overhead at the equator.

Moving out of the tropics, you would notice that the noon sun is never directly overhead. You would also notice distinct seasons and changes in the length of daylight. South of the equator, the seasons are the opposite of those in New York. When it is summer here and the days are long, it is winter south of the equator and daylight is short. When it is spring in New York, days are getting longer and the noon sun moves higher in the sky each day. At the same time, it is autumn south of the equator where the opposite changes are happening: the days are getting shorter and the sun moves lower in the sky each day. Visit the following Web site to calculate times of sunrise and sunset for any date and world location: *http://www.arachnoid.com/lutusp/ sunrise*

WHAT IS REALLY MOVING—SUN OR EARTH?

1: SCIENTIFIC
INQUIRY 1
4: 1.1d, 1.1e, 1.1f,
1.1g, 1.1h

This was an easy question for our ancestors to answer. They could not feel the ground moving. Their observations fit the idea that the sun circles Earth in a daily path that changes through the year. Self-evident (apparent) as this is, you have learned that this interpretation is not correct. Sometimes it is hard to know what is moving and what is not. You may have learned that Earth rotates at more than 1000 km/h at the latitude of New York State. (That is about 700 mph.) How can we be sure that we, standing on planet Earth, are spinning at this speed?

The Geocentric Model

For many centuries, people thought of the sky as a great dome over a flat Earth. For most people, the idea of a stationary Earth was logical and consistent with all their experiences. This is called the **geocentric** model. (The prefix *geo-* refers to Earth, as in geology, the study of the solid Earth.) Geocentric means Earth-centered. The geocentric model assumed Earth was stationary and the real motion of celestial objects caused the motions seen in the sky. For example, consider the photographs and diagrams of the sky that you have seen so far in this chapter. Every one of them seems to indicate that Earth is not moving. What would someone find if he could travel to the edge of a flat Earth?

As early as 2000 years ago, some scholars made observations that caused them to question this traditional view of Earth. These observations led some scholars to consider the idea that Earth is not a flat surface, but a huge sphere.

Heliocentric Model

Copernicus published his ideas about the sun-centered model in Poland in 1543. It was controversial. In the early 1600s, Italian mathematician and inventor Galileo published his own observations supporting the sun-centered model. Important clues came from observations of the motions of planets in the night sky. German astronomer Johannes Kepler developed the mathematical tools needed to predict the positions of the planets among the stars. However, his theories made sense only if he assumed that the planets, including Earth, orbit the sun. This is the central idea of the **heliocentric** model of the universe. (The prefix *helio-* refers to the sun.) The heliocentric model places the sun at the center of planetary motion. Figure 25-7 compares the geocentric and heliocentric models.

FIGURE 25-7. Arrows in the geocentric model show the direction of motion in orbits. The complicated paths of the planets among the stars required astronomers to add circles to the orbits, as shown by the dashed figures. Astronomers grew to prefer the heliocentric (sun-centered) model because it is simpler and easier to explain without the extra circles.

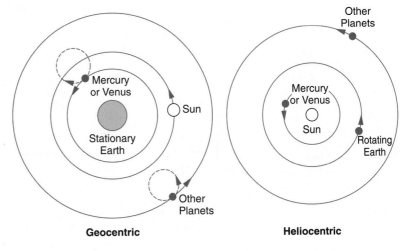

Geocentric **Heliocentric**

According to the heliocentric model, it is mainly the motions of Earth that cause the apparent motions of celestial objects. The motion of the planets in their orbits around the sun is known as **revolution**. The heliocentric model also includes the **rotation**, or spin, of Earth on its axis.

An easy way to remember the difference between rotation and revolution is to remember that a top is a spinning toy. Rotation has two *T*s, the first letter of *top*. The American Revolution started in 1776, which is a year. A revolution is the yearly cycle of Earth's motion in its orbit around the sun.

Experimental proof of Earth's motion was not discovered until 1851. By then, most astronomers already supported the heliocentric model. Motions of the planets in the geocentric model had become too complicated to explain by the laws of physics.

Proof of Earth's Rotation

Jean Foucault invented a pendulum that is free to rotate as it swings back and forth. Figure 25-8 is a drawing of a Foucault pendulum. If a Foucault pendulum were located at the North or South Pole, it would seem to rotate in a complete circle every 24 h (15°/h) as Earth spins underneath it. At the equator, a Foucault

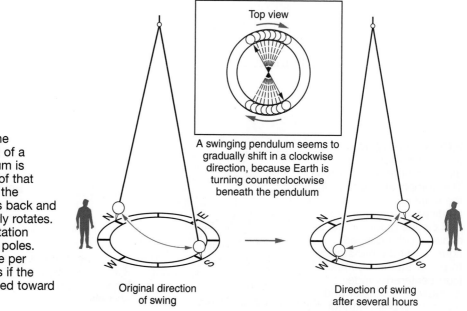

FIGURE 25-8. The apparent rotation of a Foucault pendulum is experimental proof that Earth rotates. As the pendulum swings back and forth, it also slowly rotates. One complete rotation takes 24 h at the poles. However, the time per rotation increases if the pendulum is moved toward the equator.

Top view

A swinging pendulum seems to gradually shift in a clockwise direction, because Earth is turning counterclockwise beneath the pendulum

Original direction of swing

Direction of swing after several hours

pendulum does not seem to rotate at all because of its orientation with Earth's axis. It is important to remember that the Foucault pendulum swings in the same plane as the Earth rotates beneath it. A Foucault pendulum in New York State takes about 36 h to complete one apparent rotation. Visit the following Web site to learn about the history and operation of a Foucault pendulum: *http://www.calacademy.org/products/pendulum*

The Coriolis effect is the second proof of Earth's rotation. Due to the Coriolis effect, winds in the Northern Hemisphere appear to curve to their right as they move out of a high-pressure system. Ocean currents in this hemisphere also curve to their right. The apparent curve to the right is actually a result of the winds and ocean currents trying to follow a straight path on a moving planet. Remember that this apparent curve is to the left in the Southern Hemisphere.

Earth's Motions as Viewed from Space

If you could view Earth from a stationary position in space beyond Earth's orbit, you might see the planet at one of the positions shown in Figure 25-9. This diagram shows Earth's two most important

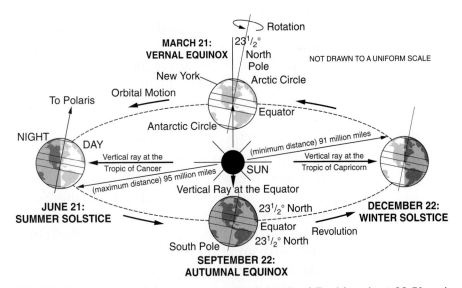

FIGURE 25-9. The seasons are the result of the tilt of Earth's axis at 23.5° and its annual revolution around the sun. Notice that Earth is slightly closer to the sun when it is winter in the Northern Hemisphere. So our seasons are *not* a result of the changing distance from the sun. The seasons labeled on this diagram apply only to the Northern Hemisphere.

motions: rotation and revolution. Note that this diagram is not drawn to a uniform scale. From outside Earth's orbit, the sun would appear as a tiny circle of light. Except near its closest position, Earth would be so small you couldn't see it at all. In addition, at any given time Earth is at only one position in its orbit. Earth revolves around the sun in an annual cycle. With 365 days in a year and 360° in a circle, the planet revolves a little less than 1° in its orbit each day.

Earth's Motions and the Seasons

The seasons are caused by a combination of two Earth motions. In addition to its revolution, Earth rotates on an axis that is tilted approximately 23.5° from a direction perpendicular to the plane of its orbit. The direction Earth's axis points is constant throughout the year. At this time in history, Polaris appears very close to the location in space directly above the North Pole. However, for half of the year, our spring and summer, the Northern Hemisphere is tilted toward the sun. For the other half of the year, our autumn and winter, the Northern Hemisphere is tilted away from the sun.

The latitude at which the vertical ray of sunlight strikes Earth determines the beginning of each season. The **vertical ray** is sunlight that strikes Earth's surface at an angle of 90°. It is sometimes called the direct ray of sunlight. This ray strikes Earth at a position where the sun is directly overhead, that is at the zenith. This vertical ray of sunlight always hits Earth within the tropics. However, its position changes in an annual cycle.

SPRING   The spring, or vernal, equinox occurs near the end of March. It is shown by the top Earth position in Figure 25-9. At that time, the vertical ray is at the equator, and spring begins in the Northern Hemisphere (autumn in the Southern Hemisphere). Earth's axis is pointed neither toward nor away from the sun, and most of the planet receives 12 h of daylight. The exceptions are the North and South Poles, where the sun circles along the horizon, resulting in twilight for 24 h.

SUMMER   Over the next 3 months, Earth moves toward the June solstice, the beginning of summer in the Northern Hemisphere. At this time, the North Pole is tilted toward the sun. The vertical ray reaches its most northerly latitude (23.5°N), the **Tropic of Cancer**. In the Northern Hemisphere, this is when the noon sun is highest in the sky, the sun's path through the sky is longest, and daylight is longest.

It is just after this time that Earth reaches its greatest distance from the sun. However, the change in the Earth-sun distance is very small. In fact, it is too small to affect the seasons. If Earth's distance from the sun caused the seasons, you would observe two major differences from the present seasons. First, summer and winter would be reversed. Our warmest weather would be in January when the Earth is very slightly closer to the sun. Second, summer in the Northern Hemisphere would be summer in the Southern Hemisphere instead of the present 6-month difference.

AUTUMN AND WINTER Over the next 3 months as Earth moves to its September position, the vertical ray moves back to the Equator. This is the autumnal equinox. Our winter begins after another 3 months, near the end of December, when the North Pole is tilted away from the sun. At this time, the vertical ray reaches its most southerly latitude (23.5°S), the **Tropic of Capricorn**.

Two other latitude positions are important. The **Arctic Circle** (66.5°N) is the latitude north of which the sun does not rise on our December solstice. North of this latitude, the sun is in the sky for 24 h at the June solstice. The **Antarctic Circle** (66.5°S) is the corresponding latitude in the Southern Hemisphere. On the December solstice (summer solstice in the Southern Hemisphere), the sun is in the sky for 24 h. On the June solstice (winter solstice in the Southern Hemisphere), the sun does not rise. Figure 25-9, on page 621, should help you understand the seasonal changes in the path of the sun through the sky and why the seasons are reversed north and south of the equator.

Figure 25-10 shows Earth at the solstice and equinox positions. Notice how the latitude at which the vertical ray of sunlight strikes Earth changes as does the length of daylight in New York State.

If Earth's axis were not tilted, but were perpendicular to the plane of its orbit, there would be no seasons. In places like New York State, there would be no annual cycle of temperature and changes in the sun's path. On the other hand, if Earth's axis were tilted more than 23.5°, New York State would experience greater changes in temperature from winter to summer. You would observe more

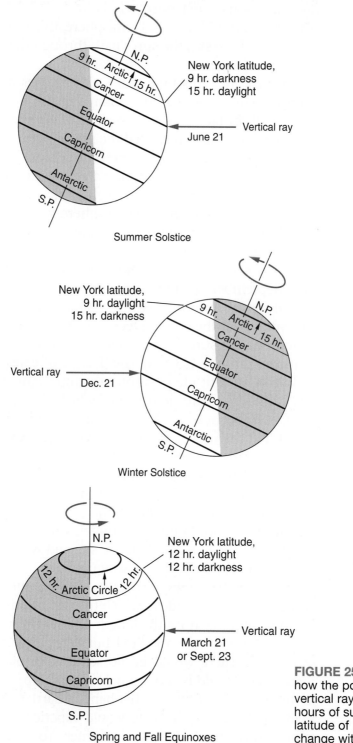

FIGURE 25-10. Notice how the position of the vertical ray and the daily hours of sunlight at the latitude of New York State change with the seasons.

extremes in the path of the sun in its annual cycle. The sun's path would be longer and higher in summer, but lower and shorter in winter. Visit the following Web site to watch animations and read explanations of the analemma ("equation of time"): *http://www. analemma.com/Pages/framesPage.html*

STUDENT ACTIVITY 25-6 —MODELING EARTH MOTIONS

6: MODELS 2

Place a single bright light, representing the sun, at the center of a darkened classroom. With a globe, illustrate Earth's motions of rotation on its axis tilted 23.5° and revolution around the sun. Notice how the height of the sun in New York State and the length of daylight change as Earth orbits the sun. In what ways is this model unlike the real Earth and sun?

HOW DO EARTH'S MOTIONS AFFECT THE APPEARANCE OF OTHER CELESTIAL OBJECTS?

4: 1.1a, 1.1g

The night sky was familiar to our ancestors. They had no electric lights, television, or computers, so they watched the sky. Although the stars are randomly distributed throughout the sky, ancient people imagined patterns in the stars.

Constellations

Certain patterns of stars are called constellations. Orion the hunter is a constellation prominent in New York's winter sky. Near Orion is a group of stars that represent his dog and Lepus the rabbit. Other patterns represent objects such as Lyra the harp.

Constellations were often associated with traditions and legends that were a part of the cultural heritage. The constellation called Cassiopeia was named for an Ethiopian queen who proclaimed that she was more beautiful than other women. Two bright stars, Castor and Pollux, dominate the constellation Gemini. In mythology, Castor and Pollux were twins.

Different cultures imagined different things about the same group of stars. For example, the group of stars we call The Big Dipper is known as The Plow in Britain. To the ancient Greeks, this

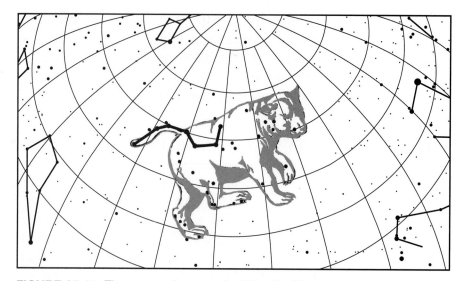

FIGURE 25-11. The group of stars called The Big Dipper is a part of the constellation Ursa Major (The Great Bear). For observers in New York State, this group of stars is always visible in the northern sky as it rotates once a day around Polaris.

was a part of the Great Bear, which we call by its Latin name, Ursa Major. Figure 25-11 shows this constellation with a drawing of a bear superimposed on the star pattern. For many constellations, it takes a good deal of imagination to see the objects they represent in the pattern of stars.

STUDENT ACTIVITY 25-7 —THE BIG DIPPER AND POLARIS

4: 1.1g

On a clear night, find the group of stars called the Big Dipper. Follow the pointer stars to Polaris. Sketch the Big Dipper and Polaris showing their orientation with respect to the Northern Horizon. You may wish to repeat this activity several hours later or in a few months at the same hour to see if their orientation changes. What name did slaves give this group of stars, which they followed, as they escaped from slavery before and during the American Civil War? Visit the following Web site to learn to recognize a few of the most well-known constellations visible in North America: _http://www.quietbay.net/Science/astronomy/nightsky_

Star Maps

Modern astronomers use the constellations to designate 88 regions of the night sky. This is a convenient way to establish a map of the stars. The stars always occupy the same position with respect to other stars. When a star is part of a particular constellation, that association helps observers locate the star in the night sky. Figure 25-12 is a map of the evening sky in the month of April as seen in New York.

To use the map, hold a copy of it upside down so that the compass directions printed on the map line up with the true compass directions. If you find the Big Dipper high in the sky, you can use it to locate other stars and constellations as shown by the arrows on the chart. The two stars at the end of the bowl of the Big Dipper point to Polaris. Binoculars or a small telescope will help you observe objects labeled OCl (open clusters), Dbl (double stars), and

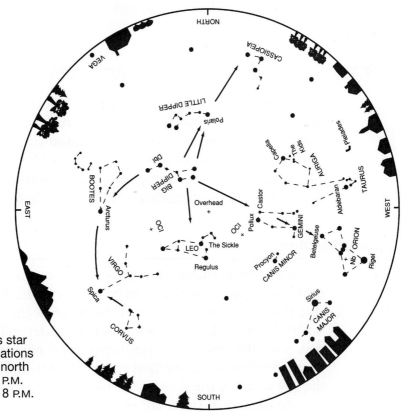

FIGURE 25-12. This star map shows constellations visible at about 40° north latitude at about 10 P.M. in March and about 8 P.M. to 9 P.M. in April.

Nb (nebulae). The planets wander among the stars, so they are not shown on this chart. Current information from a newspaper or the Internet can help you locate the visible planets. Visit the following Web site to print a copy of the star map for the current month: *http://www.kidsastronomy.com/astroskymap/constellations.htm*

STUDENT ACTIVITY 25-8 —LOCATING MAJOR CONSTELLATIONS

1: SCIENTIFIC INQUIRY 1

Use a copy of Figure 25-12 or a similar star map to locate constellations in the night sky. Try to observe on a clear night from a location that is far from artificial lights. The sky should not be obstructed by nearby buildings or trees. Make a list of the constellations you are able to find. Number them according to how easy they were for you to locate in the night sky. Visit the Astronomical Society of the Pacific's Web site, scroll down to the Activity Corner: Three-Dimensional Orion: *http://www.astrosociety.org/education/ publications/tnl/21/constellations2.html*

HOW DO EARTH'S MOTIONS SEEM TO AFFECT THE STARS?

4: 1.1a, 1.1d

We do not usually see stars shift their positions relative to each other. Because of that, they are sometimes called the fixed field of stars. However, the cycles of daily rotation and yearly revolution of Earth influence our observations of stars. Visit the following Web site to find 2300 nicely organized astronomy links to help your Web explorations: *http://www.linktospace.com/*

Daily Apparent Motions

Earth's rotation makes stars appear to move through the sky just as the sun appears to move. In New York State, most stars, like the sun, rise in the east, travel to the right across the southern sky, and set in the west. Like the sun, their path depends on where they rise. Stars that rise in the northeast travel high in the sky and set in the northwest. Those that rise in the southeast move lower in the sky to set in the southwest. Figure 25-13 shows the apparent motion of stars in the night sky.

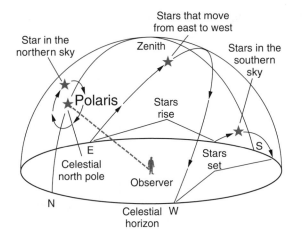

FIGURE 25-13. Due to Earth's rotation, the stars appear to rotate counterclockwise around Polaris.

You will observe a different kind of motion in the northern part of the sky, where the sun is never found. In earlier chapters, you read about Polaris, the North Star. Polaris is not one of the very brightest stars, but it appears to be located almost directly above Earth's North Pole near a point in the sky called the celestial north pole. Consequently, Polaris is the only star that does not seem to move. Actually, Polaris is not exactly above the North Pole; therefore, it does make a very small counterclockwise circle in 24 h.

Other stars that are not as close to the celestial pole make larger circles, all at a rate of 15°/h, Earth's rate of rotation. Due to Earth's rotation, the stars appear to rotate counterclockwise around Polaris. This motion includes stars that rise in the east and set in the west. In fact, all star paths can be thought of as circles around the north and south celestial poles. The planets and comets also show this apparent motion, although they slowly shift their positions among the stars from night to night.

Figure 25-14 illustrates the trails of stars that would be observed in each of the four compass directions by an observer in New York State. These diagrams show the stars rising in the east, moving through the southern sky, and setting in the west. Only stars in the northern sky do not seem to rise and set; they appear to circle counterclockwise around Polaris. Visit the following Web site to see animations of the apparent motions of the night sky: *http://astro.unl.edu/naap/motion2/animations/ce_hc.swf*

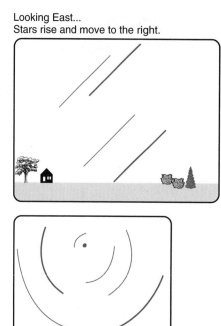

Looking East...
Stars rise and move to the right.

Looking South...
Stars circle to the right.

Looking West..
Stars set as they move to the right.

Looking North...
Stars circle counterclockwise around Polaris.

FIGURE 25-14. These drawings represent time exposure photographs of the night sky. True photographs would show light star trails moving through a dark sky.

STUDENT ACTIVITY 25-9 —PHOTOGRAPHING STAR TRAILS

1: SCIENTIFIC
INQUIRY 1

To take photographs of star trails, use a film camera or a good digital camera that allows you to take very long exposures. (Most digital cameras do not allow you to take very long exposures.) Mount the camera on a tripod or other object(s) to hold it steady while the shutter is open. A clear, dark sky away from artificial lights is essential. Exposures 15–30 s in length show star patterns in constellations. Exposures of 5 minutes to an hour will yield the best star trails. After viewing your first attempts at these photographs, you can plan changes to improve your pictures in later attempts. If color photography is used, these trails may be of different colors, indicating the surface temperature of the stars.

Yearly Apparent Motions

In addition to the apparent daily cycle of motion of stars through the sky, the stars in the evening sky also change in a yearly cycle. In the summer, Scorpio, the scorpion, with its red star, Antares, is prominent in the southern sky. At this time, the stars of Scorpio are located on the opposite side of Earth from the sun. This is why they are visible in the night sky at that time of year.

By winter, 6 months later, Earth has revolved halfway around its orbit. Orion, the hunter, is now visible in the southern sky in the evening. Scorpio is still in the same place, but it is now located in the direction of the sun. Ancient people suggested that Scorpio and Orion were enemies. Therefore, they were never seen in the sky together. During the day, the sky is too bright for most stars to be visible. Figure 25-15 shows Earth's orbit and some of the brighter constellations visible in the evening at different times of the year. Because Earth moves about 1° in its orbit each day, changes in the evening sky are not noticeable from night to night. However, over a period of months this change is easy to see.

Not all the constellations are seasonal. Stars and constellations in the northern sky, such as Polaris in Ursa Minor (the Little Bear), Ursa Major (including the Big Dipper), and Cassiopeia, are visible throughout the year. However, the polar constellations that are below Polaris at one time of year are high in the sky, above Polaris, 6 months later.

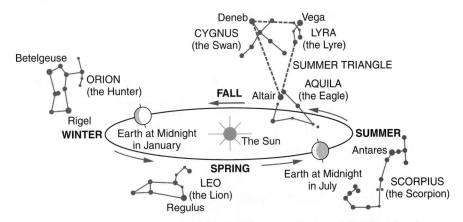

FIGURE 25-15. As Earth revolves around the sun, the night (dark) side of Earth faces different directions in space. Only constellations opposite the sun are visible because the light of the sun hides stars during the day.

STUDENT ACTIVITY 25-10 —CELESTIAL OBSERVATIONS

1: SCIENTIFIC
INQUIRY 1, 2, 3

In this chapter, you have learned about a number of changes that you can observe and/or measure quite easily. For example, you can measure the rate at which the sun moves through the sky or the apparent motions of stars.

The topics in this chapter relate to a number of long-range projects that are not difficult to perform. You might use a digital camera to photograph the horizon, recording changes in the position of sunset or sunrise over a period of weeks. The changing angular altitude of the noon sun can also be documented. Changes in the moon and the stars can also be observed and drawn or photographed.

CHAPTER REVIEW QUESTIONS

Part A

1. Earth's spin on its axis causes the

   (1) cycle of seasons
   (2) direction in which a wind vane points
   (3) movement of the Pacific tectonic plate
   (4) motion of a Foucault pendulum

2. The diagram below was drawn by a student who lives in a relatively flat part of New York State. The diagram shows the positions of sunrise at approximately the 20th of three consecutive months of the year. The horizontal line represents a horizon.

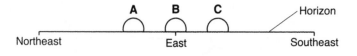

What months are best represented by positions A, B, and C in their proper order?

   (1) February, March, April
   (2) May, June, July
   (3) June, July, August
   (4) August, September, October

3. Approximately how many degrees does Earth move each day in its orbit of the sun?

    (1) 1°
    (2) 13°
    (3) 15°
    (4) 23.5°

4. The diagram below shows a demonstration of a pendulum mounted on a swiveling piano stool.

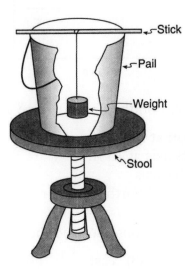

    What does the piano stool represent?

    (1) the Coriolis effect
    (2) the revolving Earth
    (3) the rotating Earth
    (4) the orbit of the moon

5. If the tilt of Earth's axis were to become less than the present 23.5°, which change would we probably experience in New York State?

    (1) Spring and fall would become cooler.
    (2) Spring and fall would become warmer.
    (3) Winter would become colder.
    (4) Summer would become cooler.

Part B

Base your answers to questions 6 and 7 on the diagrams below, which are drawings made by a student looking at a window at three different times during a day.

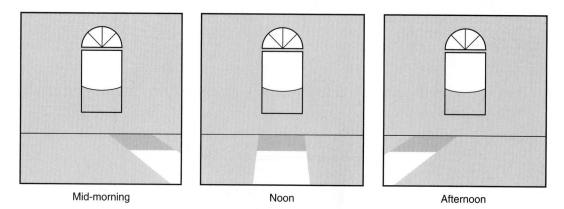

Mid-morning   Noon   Afternoon

6. What causes the sunlit area on the floor to move across the room?

   (1) Earth orbiting the sun at 1°/day
   (2) Earth spinning on its axis at 15°/h
   (3) the 23.5° tilt of Earth's axis
   (4) the 360° angle all around Earth's equator

7. In what direction are both the observer and the window facing?

   (1) north   (2) south   (3) east   (4) west

Base your answers to questions 8 through 10 on the following diagram, which represents the sky and the changing positions of the sun. Letters *A* through *D* represent positions in the sky.

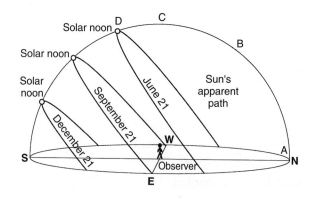

8. In which direction does sunrise occur on June 21?

(1) north of due west
(2) north of due east
(3) south of due west
(4) south of due east

9. How many hours are there between sunrise and solar noon on September 21?

(1) 6   (3) 12
(2) 8   (4) 24

10. Which letter best shows the position of Polaris in the night sky?

(1) *A*   (3) *C*
(2) *B*   (4) *D*

Base your answers to questions 11 through 13 on the map below, which shows the extent of daylight (white) and night (shaded) on Earth at a particular day and time. The dashed lines show the Arctic Circle and the Antarctic Circle. Point *A* is a particular location.

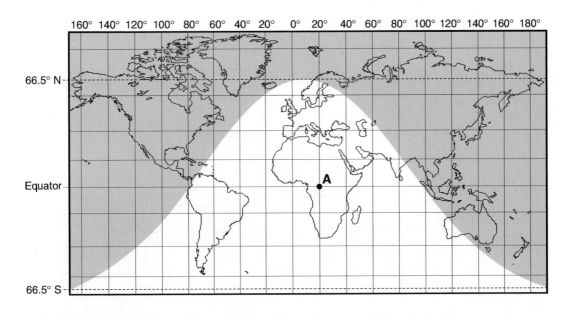

11. Approximately how many hours of daylight occur at position A on this day?

(1) 6   (2) 9   (3) 12   (4) 15

12. Which diagram below best shows how Earth is aligned with incoming sunlight on this day?

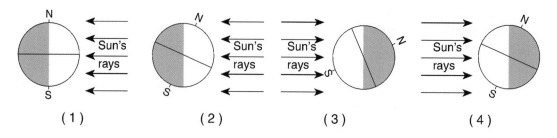

(1) (2) (3) (4)

13. On this day of the year, the duration of daylight from the equator to the Arctic Circle

(1) decreases, only
(3) decreases, then increases
(2) increases, only
(4) increases, then decreases

14. The constellation Virgo is visible high in the evening sky only in the spring. But Pegasus is visible high only in the autumn evening sky. Why does this occur?

(1) Earth rotates on its axis.
(2) Earth revolves around the sun.
(3) Earth's axis is tilted 23.5°.
(4) Earth's distance from the sun changes.

Part C

15. The diagram below is based on observations made at Ithaca, New York. What does this diagram show?

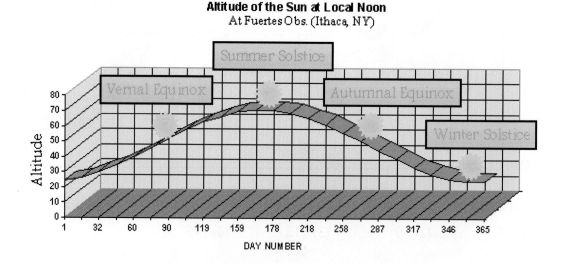

Altitude of the Sun at Local Noon
At Fuertes Obs. (Ithaca, NY)

16. Describe the path of the sun through the sky at the North Pole on June 21.

The data below was recorded over the course of a year by a student at Buffalo, New York.

Data Table

Date	Hours of Daylight	Altitude of the Sun at Noon (°)	Date	Hours of Daylight	Altitude of the Sun at Noon (°)
January 21	9.5	32.3	July 21	14.8	63.3
February 21	10.8	40.1	August 21	13.7	55.5
March 21	12.0	47.3	September 21	12.1	47.7
April 21	13.7	55.1	October 21	10.8	39.9
May 21	14.8	62.5	November 21	9.5	32.1
June 21	15.3	70.4	December 21	9.0	24.4

17. Sketch a simple graph of this data. You do not need to number the axes or plot exact graph points. Label the vertical axis "Hours of Daylight" and the horizontal axis "Altitude of the Noon Sun." Then draw a line that shows the relationship between these two variables.

Base your answers to questions 18 and 19 on the diagram below, which shows the apparent path of the sun on March 21 at Buffalo, New York. Copy this diagram so you can record your answers to the next two questions without writing in your book.

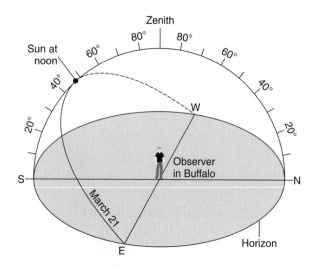

18. Draw a line that represents the path of the sun at the same location on May 21. Begin the line at sunrise and end it at sunset, showing the correct altitude of the noon sun.

19. Draw an asterisk (*) to show the position of the North Star on your copy of the diagram above. (*Reminder:* Do not write in your book.)

20. Throughout the year, the direction in which the setting sun reaches a level horizon changes. What can you say about how high the noon sun is in the sky as the compass direction of sunset moves southward?

26 Earth, Its Moon, and Orbits

WORDS TO KNOW

eccentricity	focus	lunar eclipse	phase	solar eclipse
eclipse	gravity	major axis	satellite	weight
ellipse	inertia			

This chapter will help you answer the following questions:

1 What have we learned about the moon from Earth-based observations and the Apollo space program?

2 What is the shape of orbits?

3 How can we explain the changing phases of the moon?

4 What causes lunar and solar eclipses?

APOLLO PROGRAM

1: SCIENTIFIC INQUIRY 1, 2, 3

American politicians and scientists were embarrassed by the early triumphs of Russia in space exploration. This led to the Apollo program. In July of 1969, American astronaut Neil Armstrong became the first human to set foot on the moon. That mission also returned the first samples of moon rock to Earth. In the next six Apollo missions, a total of 384 kg (845 pounds) of moon rocks were brought back to Earth for scientific study. Figure 26-1 is a photograph from the Apollo 16 moon landing. Since the last Apollo mission in 1972, no other humans have visited the moon.

FIGURE 26-1. Astronaut John W. Young stands on the powdery lunar soil at the Apollo 16 landing site. The Lunar Roving Vehicle is parked beside the Lunar Module "Orion." Stone Mountain dominates the background in this lunar scene.

From the Apollo program, scientists learned that the surface of moon contains minerals similar to those found in parts of Earth's mantle and deepcrust. Plagioclase feldspar, pyroxene, and olivine are the most common minerals in the lunar samples. Visit the following Web site to see images and animations by JPL of and about the exploration of space: *http://www.jpl.nasa.gov/multimedia*

WHAT IS THE HISTORY OF EARTH'S MOON?

4: 1.1a, 1.1b

A **satellite** is an object in space that revolves around another object under the influence of gravity. A moon is a natural satellite of a planet. The moon is Earth's only natural satellite. (Earth, like the other planets, is a satellite of the sun.) Earth is the only planet in our solar system with just one moon. Of the eight planets in our solar system, only Mercury and Venus do not have moons. Even dwarf planets, all of which are smaller than Earth's moon, now are known to have their own moons.

A curious feature of our moon is that its period of rotation and revolution are the same, about $27\frac{1}{3}$ days. As a result, the same side of the moon always faces Earth. This is why features of the far side of the moon were unknown until the Russians sent a satellite around the moon to take photographs early in the "space race."

The Origin of Earth's Moon

There have been several theories about the origin of the moon. Some astronomers have suggested that the moon and Earth formed as a double planet orbiting the sun. Others have proposed that the moon was an object in space that came close enough to Earth to be captured by Earth's gravity.

Most astronomers now agree that a collision between Earth and a smaller object probably created the moon shortly after Earth's formation. That impact destroyed the smaller body and created a debris field in space. The debris came together under the influence of gravity, forming our moon.

Lunar Surface

The surface of the moon has two landscape types, the lunar highlands and the maria (from the Latin word for "seas"). The rocks of the lunar highlands are mostly anorthosite, a rock type found in New York's Adirondack Mountains. However, it is not common in most places on Earth. The lunar maria are relatively flat and darker in color than the highlands. They are composed of basalt, a relatively common dark-colored, fine-grained, igneous rock also found on Earth.

Because the moon has no atmosphere or surface water, it has none of the common sedimentary rocks found on Earth. Chemical weathering does not occur on the moon. However, the surface is covered by material from meteorites and moon rocks broken by meteorite impacts. Breccia, a rock formed from the breakage and welding of rock fragments, is found in the lunar highlands. Radiometric dating has shown the oldest moon rocks are about the same age as Earth, 4.6 billion years. Visit the following Web site to enjoy moon images with full 3-D terrain taken by the Clementine satellite: *http://worldwind.arc.nasa.gov/moon.html*

The cratered surface of the moon contrasts sharply with Earth's surface. The highlands of the moon have been solid for more than 4 billion years. Without a doubt, many large objects have hit the moon and Earth. The highlands of the moon are so covered with impact craters that the surface seems to be made of crater upon crater.

The moon is small enough and its gravity weak enough that the moon cannot hold an atmosphere. As a result, there is no shell of gases surrounding the moon. Furthermore, unlike Earth, the surface of the moon is not composed of active tectonic plates that create and recycle surface material. Therefore, features on the moon last much longer than they do on Earth. The only active process on the moon is impact by meteoroids and other objects from space. Early in the history of the solar system, there were many more of these objects and impacts were much more common.

The dark lava flows that became the maria show that the moon once had a molten interior. Most of the moon's volcanic eruptions occurred between 3.8 and 3.1 billion years ago. The maria show less cratering than the highlands. This is true because most of the impacts that created the largest craters occurred before the surface of the maria had formed. By 3 billion years ago, the surface of the moon probably looked pretty much as it does today. Whether the moon still has molten rock in its interior is a question that scientists have not yet answered.

HOW CAN WE DESCRIBE ORBITS?

1: MATHEMATICAL
ANALYSIS 1, 2, 3
4: 1.1b

Astronomers noticed the wandering motion of the planets through the sky. This led to the idea that planets and the moon move in orbits independent of the fixed stars. These orbits were originally thought to be circles. However, careful observations of the planets showed that their motions under the influence of gravity could be explained best if their orbits were not circles, but ellipses.

An **ellipse** is a closed curve formed around two fixed points such that the total distance between any point on the curve and the fixed points is constant. The fixed points are the foci of the ellipse. (The singular of *foci* is *focus*.) If you use pins and a loop of string to draw an ellipse, as in Figure 26-2, the position of each pin is a **focus** of the ellipse. The string keeps the total distance to the two foci constant. Visit the following Web site to learn how to draw an ellipse: *http://www.youtube.com/watch?v=CaokHrXP8HM&feature =related*

In all orbits, the object the satellite revolves around, known as the primary, is located at one focus. For example, Earth is at one focus of the moon's orbit. The sun is also located at one focus of Earth's orbit. The other focus is a point in space. You should also note that Earth is not at the center of the moon's orbit and the sun is not at

FIGURE 26-2. You can draw an ellipse by using the materials shown in this image. Use the pencil to hold the string tight, and then move the pencil to make the ellipse.

the center of Earth's orbit. If the two foci of an orbit are far apart, the orbit is noticeably out-of-round. In that case, we say the orbit has a high eccentricity. The orbits of the moon and Earth have the two foci so close that the orbits are very close to perfect circles.

STUDENT ACTIVITY 26-1 —ORBIT OF THE MOON

4: 1.1b
6: MODELS 2

Get the following materials: one soft board approximately 30 cm (12 in.) square (fiberboard, ceiling tile, or soft pine work well), two straight pins, adhesive tape, one piece of light string about 30–40 cm (12–20 in.) long, one sharp pencil, one clean sheet of standard size paper ($8\frac{1}{2}$ in. by 11 in.), one metric ruler.

Find the center of the sheet of paper by drawing two lines that connect opposite corners of the paper.

1. Tie the string into a loop that is 10 cm long when fully stretched (± 0.5 cm).
2. Tape or pin the paper near the corners of the soft board.
3. Stick one pin through the center of the sheet of paper and into the soft board. Place the loop of string around it. Then stretch the loop to its greatest distance with the tip of the pencil. Keeping the loop taut and the pencil perpendicular to the paper, make a circle around the pin. Label it "circle" along its circumference.
4. Place the two pins 4.5 cm from the center of the paper along the paper's long axis. Stretch the string around both pins and then pull tightly on the string to draw an ellipse.
5. Make another ellipse by placing the two pins 1 cm apart (0.5 cm each side of the center). Draw this ellipse and label it "Orbit of the Moon."

Is the "Orbit of the Moon" noticeably elongated?

Calculating Eccentricity

Eccentricity describes the shape of an orbit, or how elongated the ellipse is. You will find the following equation, used to calculate the eccentricity of an ellipse, in the *Earth Science Reference Tables:*

$$\text{Eccentricity} = \frac{\text{distance between foci}}{\text{length of major axis}}$$

When asked to calculate eccentricity, you will be given either the values you need to solve the equation or a diagram that clearly shows the ellipse and the position of the two foci. If you are given a diagram, you will need to use a centimeter ruler to measure the distance between the foci and the length of the major axis. The **major axis** is the distance across the ellipse measured at its widest point. (The smaller axis, the width of the ellipse, is known as the minor axis.) These two features of an ellipse are shown in Figure 26-3. (Remember that a centimeter scale is printed on the front cover of the *Earth Science Reference Tables.*)

SAMPLE PROBLEMS

Problem 1 Calculate the eccentricity of the ellipse in Figure 26-3.

Solution Measuring the ellipse, you will find distance between the foci of this ellipse is 3 cm. The distance across the ellipse (the major axis) is 4 cm. The calculation is shown below.

$$\text{Eccentricity} = \frac{\text{distance between foci}}{\text{length of major axis}}$$

$$= \frac{3\,\text{cm}}{4\,\text{cm}}$$

$$= 0.75$$

distance between foci

length of major axis

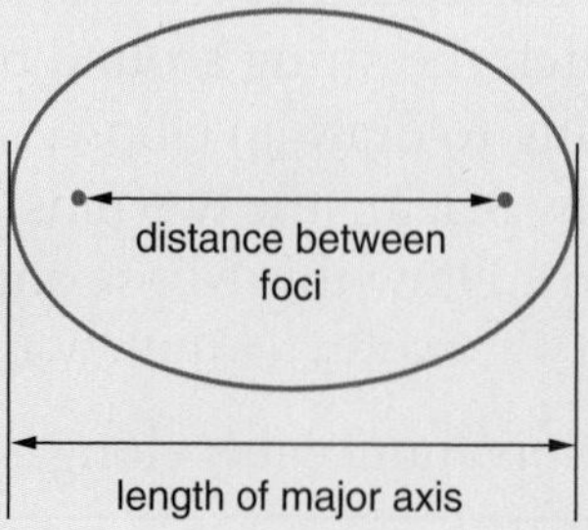

FIGURE 26-3. To calculate eccentricity, divide the distance between the foci of the ellipse by the length of the major axis. Eccentricity is a ratio without units.

Notice that there are no units of eccentricity because in the calculation the units cancel each other out. Eccentricity is therefore a ratio between two measured values.

Problem 2 The greatest distance across Earth's orbit is 299,200,000 km. The distance from the sun to the location in space that is the other focus of Earth's orbit is 5,086,400 km. Calculate the eccentricity of Earth's orbit.

Solution

$$\text{Eccentricity} = \frac{\text{distance between foci}}{\text{length of major axis}}$$

$$= \frac{5,086,400 \text{ km}}{299,200,000 \text{ km}}$$

$$= 0.017$$

Notice that the result is a very small number. Earth's orbit is nearly a perfect circle. That is why the changing distance between Earth and the sun is not a significant factor in seasonal changes in temperature on Earth.

Problem 3 The greatest distance across Mars' orbit is 455,800,000 km. The eccentricity of its orbit is 0.093. What is the approximate distance between the sun and the second focus of Mars' orbit?

Solution The first step is to rearrange the formula to isolate the distance between the foci. Multiplying both sides of the equation by *length of major axis* does this.

$$\text{Eccentricity} = \frac{\text{distance between foci}}{\text{length of major axis}}$$

The next step is to substitute the values and solve for the distance between the foci:

$$\text{Length of major axis} \times \text{eccentricity} = \text{distance between foci}$$

$$455,800,000 \text{ km} \times 0.093 = \text{distance between foci}$$

$$42,389,400 \text{ km} = \text{distance between foci}$$

This value is expressed to a much greater accuracy than the length of the major axis of the ellipse. So it should be rounded off to 42 million km. (Also note that the problem asked for the "approximate distance.")

Practice Problem 1

Calculate the eccentricity of an ellipse in which the distance between the foci is 20 cm and the length of the major axis is 60 cm.

Practice Problem 2

The greatest distance across the moon's orbit is 772,000 km; the eccentricity of its orbit is 0.055. How far apart are the foci of the moon's orbit?

Ellipses show a range of shapes. If the foci are at the ends of the ellipse, the ellipse will be a line segment with an eccentricity of 1. If the two foci are located near the ends of the ellipse, the ellipse looks flattened. A circle is the special case of ellipse in which the foci are at a single point. The eccentricity of a circle is 0. When considering the eccentricity of an ellipse, it may help to remember that just as the number "1" can be written as a line, an ellipse with an eccentricity of one is a straight line. Furthermore, just as the number "0" can be written as a circle, an ellipse with an eccentricity of zero is a circle.

The orbits of Earth and the moon look like circles. Only by carefully measuring the length and width of the orbits, or by locating the two foci, can scientists observe that the orbits of the moon and Earth are not quite circles.

WHAT DETERMINES A SATELLITE'S ORBIT?

4: 1.1a, 1.1b
6: EQUILIBRIUM AND
STABILITY 4

The path that Earth takes around the sun is determined by two factors: inertia and gravity. **Inertia** is the tendency of an object at rest to remain at rest or an object in motion to move at a constant speed in a straight line unless acted on by an unbalanced force. If you roll a heavy ball across a flat, hard floor, the ball continues in a straight line until some force causes it to change its speed or its direction. That force could be friction with the floor causing the ball to slow. It could be a force applied by an object or a wall as the ball collides with it. Or, it could be someone pushing the ball to one side, causing the ball to change direction. Similarly, an object moving through space will move in a straight line unless some force causes it change its speed or its direction.

Gravity

The moon, planets, and other satellites follow curved paths. This tells you that a force must be acting on them. That force is gravity. **Gravity** is the force of attraction between all objects. If the objects are relatively small, such as your body and familiar objects around you, the force between them is so small you cannot feel it. Although your body's mass is small, the mass of Earth is very large. Your **weight** is the force of attraction between your body and Earth.

When the mass of one or both objects increases, the gravitational force between them also increases. Therefore, you would weigh more on a more massive planet. Correspondingly, you would weigh less on Mars or the moon than you do on Earth because these smaller bodies have less mass than Earth.

Gravity also depends on the distance between the centers of the two objects. People who climb to the top of high mountains experience a very small but measurable decrease in their weight. This is because they are moving away from Earth's center. In fact, the higher a person goes above sea level, the less the person weighs.

The strength of gravity between two objects (such as your body and Earth) does not decrease with just the distance between their centers. The change actually follows the inverse square law. Suppose you were to rise from Earth's surface a distance from the center of Earth equal to twice Earth's 6000-km radius. Your weight would not decrease by half, it would actually be only one-quarter of your weight on Earth's surface ($2^2 = 4$). At three times Earth's radius from the center of the planet, your weight would be just one-ninth of your weight at Earth's surface ($3^2 = 9$).

The motion of Earth or any celestial object in its elliptical orbit can be thought of as a combination of two kinds of motion. The first is straight-line motion under the influence of inertia. The second is falling motion toward the primary of the satellite under the influence of gravity. The result is a curved path, as shown in Figure 26-4, on page 648.

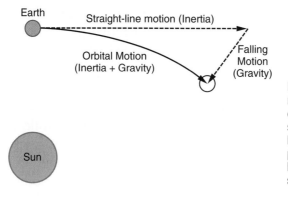

FIGURE 26-4. The curved motion of the moon in its orbit is the result of straight-line motion caused by inertia, and its falling path toward Earth caused by gravity. Earth and the sun are not drawn to scale.

Orbital Energy

The orbital energy that Earth has is a combination of kinetic and potential energy. Potential energy is a result of the distance between Earth and the sun. Kinetic energy is the speed of Earth in its orbit. The combined energy remains constant unless it is reduced by friction. However, there is no friction in space. The balance between the two components of orbital energy, speed and orbital distance, does change. As a satellite moves farther from its primary, its orbital speed decreases. As it moves closer, its speed increases.

Moving in an ellipse, Earth changes its distance from the sun by a small amount. Therefore, Earth's speed in its orbit changes. In fact, for all planets orbiting the sun, orbital speed is a function of the distance between the satellite and the primary. Earth moves a little faster when it is closer to the sun and slower when it is farther from the sun. Earth is closest to the sun in early January, which is when its orbital speed is greatest (although the change is relatively small). The moon also changes speed in its orbit as it moves slightly closer and farther from Earth. Visit the following Web site to see how the apparent size of the moon changes from apogee to perigee: *http://antwrp.gsfc.nasa.gov/apod/ap071025.html*

WHY DOES THE MOON SHOW PHASES?

4: 1.1a
6: PATTERNS OF
CHANGE 5

You have probably noticed that the moon seems to change its shape in a monthly cycle. The apparent shape of the moon, which is determined by the pattern of light and shadow, is called the **phase** of the moon. Figure 26-5 shows the range of moon phases.

FIGURE 26-5.
These 24 images show the monthly changes of light and shadow caused by the moon's revolution around Earth are called the phases of the moon. The new moon phase (bottom right) is dark and it is always seen near the sun. So the new moon may not be familiar to you.

For example, when the whole side of the moon facing Earth is lighted, we see the full moon. The gibbous phase occurs when most of the moon appears lighted. When half of the moon appears lighted, it is the quarter-moon phase. The crescent phase is a narrow, curved sliver of light. When the whole part of the moon's surface seen from Earth is in shadow, it is known as the new moon. The phase that you see depends on the relative position of the moon with respect to Earth and the sun.

Why is there a monthly cycle of moon phases? This cycle occurs because it takes roughly a month for the moon to orbit Earth. However, the daily rotation of Earth makes the monthly orbiting of the moon more difficult to follow. What you actually observe is the moon rising and setting nearly an hour later each day. Each day for half its cycle the moon moves about 13°away from the sun's position in the sky. For the other half of the cycle, the moon moves toward the sun's position in the sky. For this reason, each day or night you see progressively more or less of the moon's lighted surface (the side of the moon facing the sun). The cycle of the moon's phases takes about 29½ days. To fit exactly 12 months into a year, the average calendar month was made a little longer than the full cycle of moon phases.

In a large, darkened room, a friend can help you demonstrate these motions. Due to the dangers posed by a bare light bulb, this should be done under adult supervision. Use a bare light bulb at a distance from you of at least 2–4 m (7–13 ft) to represent the sun. Your head represents Earth; your eyes represent the position of the observer. A ball can represent the moon. The ball should be much closer to you than the light bulb (the sun). In the simplest representation, stand in one place (although you can turn your head) and observe the lighted part of the ball as your friend moves the ball around (orbits) you. You may want to name each phase as you see it. The phases are labeled in Figure 26-6.

Earth, Moon, and Sun

Figure 26-6 includes two ways of looking at the moon in a single diagram. The central part of this diagram shows the moon orbiting Earth as viewed from a point in space above the North Pole. Notice the arrows showing the orbital motion of the moon as it revolves around Earth. The straight arrows on the left represent rays of light from the sun, which is well outside the diagram to the left.

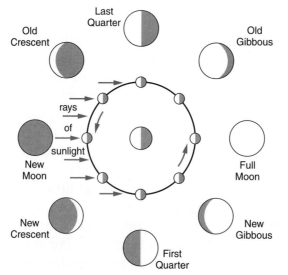

FIGURE 26-6. The center of this diagram shows the moon's orbit around Earth as viewed from a position high above the North Pole. The eight outer circles show how a person standing on Earth's surface would see each phase. This drawing is not to scale.

The eight outer circles show how the moon appears at each position as viewed from Earth. Notice that the first quarter phase at the bottom of the diagram is lit on the left as viewed from above the orbit (inner circle), but lit on the right as viewed from Earth at night. To understand why the bottom diagram is reversed, turn your book around and look across Earth to the first-quarter phase. Then the figure of the moon in its orbit will be lit on the right. Remember that the portion of the moon that is lighted, as well as the side that is lit, depend on the position of the observer. Visit the following Web site to watch a simple YouTube animation that shows the moon revolving around Earth and, at the top right, what the moon looks like from Earth: *http://www.youtube.com/watch?v= o1Ubjmnm06g*

Earth and Moon

The moon takes about $27\frac{1}{3}$ days to complete one orbit around Earth. However, the cycle of moon phases takes about $29\frac{1}{2}$ days. The reason it takes longer than one revolution of the moon to observe a full cycle of phases is shown in Figure 26-7. Consider Earth and the moon to start at position *A*. They will be at position *B* $27\frac{1}{3}$ days later. The moon has moved through one complete

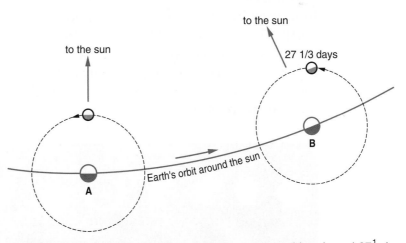

FIGURE 26-7. Consider Earth and the moon at position *A*, and $27\frac{1}{3}$ days later at position *B*. The moon has made one complete revolution of 360° from *A* to *B*, but the apparent position of the sun among the stars has changed. Therefore, the moon must travel another 2 days to get back to the new moon phase. (This is a view from above the North Pole.)

revolution of 360°. However, to get back to the new moon phase, it must move a little further in its orbit. The reason the cycle of phases takes longer than one revolution of the moon is that Earth is moving in its orbit around the sun.

WHAT IS AN ECLIPSE?

4: 1.1a, 1.1f
6: PATTERNS OF
CHANGE 5

The moon and Earth are not transparent. They cast shadows. Sometimes, the moon's shadow falls on Earth. This may affect the way we see the sun. At other times, Earth's shadow falls on the moon. This may affect the way we see the moon.

Solar Eclipses

As the moon orbits Earth, it sometimes comes between the sun and Earth, casting a shadow on Earth. When one celestial object blocks the light of another, an **eclipse** occurs. If the shadow is cast on Earth's surface, sunlight is blocked. To a person on Earth, the moon is observed to move in front of the sun and then move away, as in Figure 26-8. This is called an eclipse of the sun, or **solar eclipse**. If part of the sun is visible throughout the eclipse, it is called a partial eclipse.

A total solar eclipse occurs when the sun is completely blocked from view by the moon and the moon's full shadow passes over Earth. This is shown in Figure 26-9. In a total solar eclipse, the sky becomes much darker than usual and some stars may be visible briefly. A solar eclipse happens only in the new moon phase when the dark half of the moon faces Earth. **(CAUTION: Never look directly at the sun.)** Visit the following Web site to watch a flash

FIGURE 26-8. This series of images represents a total solar eclipse. The time from the beginning of the event to totality and then back to full sunlight can last nearly 2 h. Only the thin atmosphere of the sun is visible in the central image. The next total solar eclipse visible in the continental United States will be in 2017.

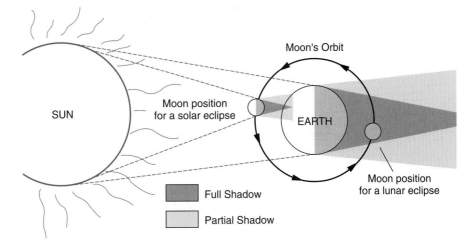

FIGURE 26-9. A solar eclipse occurs when the moon casts a shadow on Earth's surface. A lunar eclipse happens when the moon passes through Earth's shadow. If this drawing were to scale, the sun would be very, very small, while Earth and moon would not be visible at all.

movie of the March 2006 solar eclipse shadow crossing Earth's surface: *http://www.stormcenter.com/media/060331/*

Lunar Eclipse

There are also eclipses of the moon, **lunar eclipses**. Figure 26-9 shows the relative position of the moon, Earth, and sun for both solar and lunar eclipses. An eclipse of the moon occurs when the moon orbits to a position exactly opposite the sun. Viewed from Earth, the moon moves into Earth's curved shadow. The moon becomes relatively dark and takes on a coppery-red color from a small amount of light that is refracted through Earth's atmosphere. The eclipse may continue for an hour or more until the moon moves out of Earth's shadow. A lunar eclipse can occur only at the full moon phase because the moon must be on the side of Earth opposite the sun. Visit the following Web site or a similar YouTube film clip to watch the video of a lunar eclipse: *http://www.youtube.com/watch?v=IVkkCVh5t0E*

Predicting Eclipses

By making precise observations of the moon and sun over hundreds of years, astronomers are able to predict eclipses with great

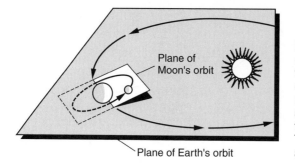

FIGURE 26-10. The moon's orbit is inclined about 5° from the plane of Earth's orbit. Therefore, the new moon and the full moon are usually above or below the plane of the sun. This explains why eclipses are rare.

accuracy. Lunar and solar eclipses occur about once or twice a year. Lunar eclipses are easier to see because they are visible to everyone on the night side of Earth. In addition, they may last more than an hour.

Solar eclipses, however, are usually visible only along a narrow path across Earth's surface. During a total eclipse of the sun, totality usually lasts only a few minutes. If the moon orbits Earth each month, why do you not see eclipses of the sun and moon every month? The reason is that the moon's orbit is tilted at an angle of about 5° from the plane of Earth's orbit around the sun. This is shown in Figure 26-10. Eclipses occur only when the moon is in the part of its path where the two orbits meet. Visit the following Web site to watch an animation of the inclination of the moon's orbit: *http://www.csulb.edu/~htahsiri/animate/Moon%20 tilt%20orbit.html*

STUDENT ACTIVITY 26-2 —THE NEXT LUNAR ECLIPSES

6: PATTERNS OF CHANGE 5

Use the Internet to find when the next partial and total lunar eclipses will occur. The time is often given in terms of Greenwich Mean Time. Convert the time of the next lunar eclipse into your local time. Where will the next solar eclipse be visible? When will the next solar eclipse be visible in your area?

Apparent Size of the Sun and Moon

It is a coincidence that as viewed from Earth the moon and sun are each about $\frac{1}{2}°$ in angular diameter. That is, two lines drawn from your eye to each side of the sun or moon are about $\frac{1}{2}°$ of angle apart. In fact, the sun's true diameter is about 400 times the diameter of the moon. However, the moon is much closer to Earth. Consequently, the moon and sun look about the same size to us. The distance from Earth to each of them changes slightly. Sometimes the moon looks slightly larger and sometimes the sun looks slightly larger. Usually this change is not noticeable. However, if a solar eclipse occurs when the sun is relatively close and the moon is relatively far away, the sun can be seen as a ring of light surrounding the dark moon. This is called an annular solar eclipse.

CHAPTER REVIEW QUESTIONS

Part A

1. Which factor determines the strength of gravity between Earth and the sun?

    (1) the degree of tilt of Earth's axis
    (2) the distance between Earth and the sun
    (3) Earth's period of rotation on its axis
    (4) the strength of insolation at Earth's surface

2. The diagram below represents Earth's revolution around the sun. Note positions *A, B, C,* and *D.*

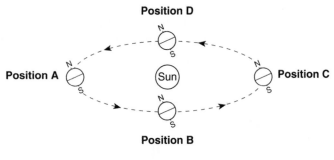

(Not drawn to scale)

Which position best represents Earth on the first day of summer in New York State?

(1) *A*   (3) *C*
(2) *B*   (4) *D*

3. Scientists can make plans to photograph a solar eclipse because most astronomical events are

   (1) cyclic and predictable
   (2) cyclic and unpredictable
   (3) random and predictable
   (4) random and unpredictable

4. Which sequence of moon phases could be observed from Earth over a period of 2 weeks?

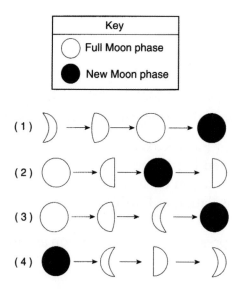

Base your answers to questions 5 and 6 on the diagram below, which shows the orbit of a planet around a star. F_2 is one focus of the ellipse. Four orbital positions of the planet are labeled A, B, C, and D.

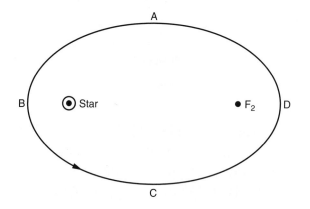

5. What is the approximate eccentricity of this ellipse?

   (1) 0.22   (3) 0.68
   (2) 0.47   (4) 1.47

6. In the diagram above, at which position of the planet is the gravitational attraction between the planet and the sun the greatest?

   (1) A   (3) C
   (2) B   (4) D

Base your answers to questions 7 and 8 on the diagram below, which shows Earth's orbit around the sun and the moon's orbit around Earth. Four Earth positions are labeled A, B, C, and D.

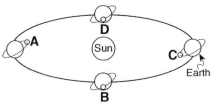

(Not drawn to scale)

7. An eclipse of the moon is most likely to occur at position

   (1) A   (3) C
   (2) B   (4) D

8. As observed from Earth, what phase of the moon occurs at position D on the diagram above?

   (1) new moon (dark)   (3) full moon
   (2) first quarter   (4) last quarter

9. Astronauts have observed the igneous rock basalt that covers large areas of the moon. What does this tell us about the history of the moon?

   (1) Earth and the moon are the same age.
   (2) Weathering and erosion are active on the moon.
   (3) Parts of the moon were once hot and molten.
   (4) Gravity on the surface of the moon is greater than on Earth's surface.

Part B

Base your answers to questions 10 through 12 on the diagram below. The arrows on the diagram indicate the motions of Earth, the sun, and the moon. The key labels each motion, *A, B, C,* and *D.*

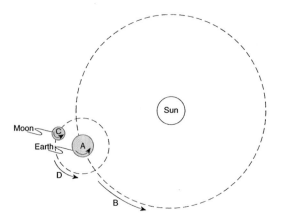

(Not drawn to scale)

Key	
Arrow	**Motion**
A	Earth's rotation on its axis
B	Earth's revolution around the sun
C	The moon's rotation on its axis
D	The moon's revolution around Earth

10. What is the time required for one complete circle of motion *A*?

    (1) 1 day   (3) 1 month

    (2) 1 week   (4) 1 year

11. Which two motions complete a full cycle in the same amount of time?

    (1) *A* and *B*

    (2) *B* and *C*

    (3) *C* and *D*

    (4) *D* and *A*

12. Which motion is responsible for the cycle of moon phases?

   (1) *A*
   (2) *B*
   (3) *C*
   (4) *D*

Part C

Base your answers to questions 13 through 16 on the diagram below, which shows Earth and the moon in its orbit from a position above the North Pole. Each of the eight positions labels the phase of the moon at that position. Light and shadow are indicated on Earth, but not on the moon.

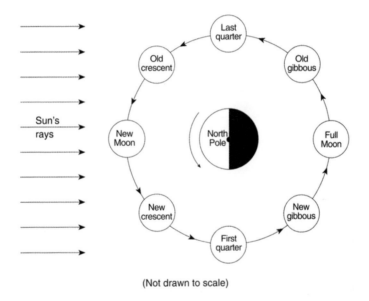

(Not drawn to scale)

13. Draw a circle approximately 2 cm in diameter. Use shading to show how the moon looks in the "last-quarter" position as viewed from New York State. *Note:* The shading should be used to show the unlit portion of the moon.

14. Explain what causes the moon phases as viewed from Earth.

15. Which moon phase occurs approximately one week after the new moon?

16. Explain why the same side of the moon always faces Earth.

Base your answers to questions 17 through 20 on the table below, which records the percentage of the lighted side of the moon visible from Earth on the first 14 days of July.

Date	Percentage of Lighted Side of the Moon Visible From Earth (%)
July 1	1
July 2	5
July 3	10
July 4	17
July 5	26
July 6	37
July 7	48
July 8	59
July 9	70
July 10	80
July 11	89
July 12	95
July 13	98
July 14	100

17. On what date of this month did the moon appear as shown below?

18. What motion of the moon causes the changed indicated in the table above?

19. Make a sketch of Earth, sun, and moon showing the positions of these bodies on July 14.

20. Why are the phases of the moon considered a cycle?

27

The Solar System

WORDS TO KNOW

asteroid

comet

Jovian planet

meteor

meteorite

meteoroid

terrestrial planet

This chapter will help you answer the following questions:

1 Why is Earth so important to us?

2 How did the solar system form?

3 What do the planets have in common and how is each unique?

4 What other objects orbit the sun?

COLONIZING SPACE

6: OPTIMIZATION 6
7: CONNECTIONS 1
7: STRATEGIES 2

The idea of sending people to live away from Earth is a familiar subject in science fiction. Astronauts have spent extended time in orbit around Earth and have spent several days on the moon. We may wish to take advantage of resources, such as precious metals, found on the moon or on other planets.

Living Away from Earth

People living in nonterrestrial, or non-Earthlike, environments need an enormous amount of materials for life support. In a space environment near a star, there would be a steady supply of electromagnetic energy. However, things we take for granted, such as oxygen and liquid water, are not available in space. In addition, they are rare on other planets in our solar system. The cost of transporting these necessities would be huge. Recycling these materials seems more practical.

Beyond the Solar System

In the past few years, astronomers have discovered hundreds of planets in orbit around other stars. If other planets like Earth exist, it seems logical that life and civilizations might have developed on some of them. However, attempts to detect radio signals from other civilizations in space have failed. Therefore, there is no direct evidence that humans could exist anywhere outside Earth without complex artificial support systems.

These discoveries have helped us appreciate how unique and how fragile our environment on Earth is. In spite of fears that a global disaster could make Earth uninhabitable, we have nowhere else to go. Clearly, the best alternative is to preserve our current terrestrial environment in a way that allows all living things to prosper.

WHAT IS THE ORIGIN OF THE SOLAR SYSTEM?

4: 1.2a, 1.2b, 1.2c

Evidence from space tells us that originally the universe probably was nearly all hydrogen and helium. The early universe most likely had a large number of very massive stars that used up their hydrogen fuel relatively quickly. These stars became unstable and exploded, creating clouds of debris full of heavy elements. In our solar system, there is a large quantity of heavy elements, such as carbon, iron, silicon, and oxygen. This indicates that our solar system formed from the remains of a stellar explosion known as a supernova. Visit the following Web site to see the relative sizes of Earth, the other planets, the sun, and our galaxy: *http://newsizeofourworld.ytmnd.com*

Nebular Contraction

The cloud of debris left over from the supernova explosion may have come together under the influence of its own gravity. This theory is sometimes called the nebular contraction theory. It suggests that the solar system began as a cloud of dust and gas in space. This cloud is commonly called a nebula. More than 99 percent of the mass condensed to form the sun, which is at the center of our solar system. Debris that remained formed the planets, dwarf planets, moons, and other objects orbiting the sun.

Just as an ice-skater spins faster when she brings in her arms, the collapse of this material produced the revolution of the planets in their orbits. It also caused the rotational motion of the sun and planets. All the planets revolve in the same direction that the sun rotates. Most of their moons also revolve in this direction. The spacing of the planets is remarkably ordered. Most planets are about twice as far from the sun as is their neighbor closer to the sun. In addition, the planets lie in nearly the same plane, orbiting the sun in a thin disk. This degree of order suggests that a single event formed the solar system.

Scientists have studied radioactive elements within material that has fallen to Earth and rocks recovered from the older parts of the moon. They estimate that these bodies are about 4.6 billion years old. Rocks from Earth are not as helpful because they have been recycled through the rock cycle and plate tectonics. Therefore, the oldest known terrestrial rocks are less than 4.6 billion years old. The patterns of change in stars indicate that the sun is about 5 billion years old. It therefore seems likely that the sun and the solar system formed in a single event that took place a little less than 5 billion years ago. This is only about one-third of the estimated age of the universe.

WHAT PROPERTIES DO THE PLANETS SHARE?

4: 1.1a, 1.1b,
4: 1.2c

The definition of a planet has evolved. Early observers of the night skies thought of the planets as special stars that wander among the other stars. Later, astronomers using telescopes were able to see differences between planets and stars. They began to think of planets as large objects that orbit the sun, reflecting the sun's light.

Like all satellites, the planets orbit the sun in ellipses with the primary, the sun, located at one focus. Their orbits have low eccentricities. The change in gravitational force between a planet and the sun causes the orbital velocity of the planet to change in a yearly cycle. All the planets move a little faster when they are closest to the sun because that is when the gravitational attraction of the sun is greatest. Visit the following Web site to use the Solar System Viewer, which shows motions of the planets with stop action and speed controls: *http://janus.astro.umd.edu/javadir/orbits/ssv.html*

Each planet in the solar system moves along its orbit at its own speed. As distance from the sun increases, the pull of the sun's gravity decreases. Therefore, orbital velocity is indirectly related to a planet's distance from the sun. That is, the farther the planet is from the sun, the slower it travels in its orbit. The innermost planet, Mercury, moves nearly 10 times as fast along its orbit as the outermost planet, Neptune. Mercury, because it is closer to the sun, also has a shorter orbit than any other planet. Therefore, Mercury revolves around the sun in only 88 Earth-days. Earth takes 1 year. Neptune takes nearly 165 Earth-years to revolve around the sun. The difference in orbital period is so large because of the combined effects of the longer orbits and the slower speeds of the outer planets.

All planets rotate on their axis. However, unlike the orbital properties of the planets, their rotation does not follow a regular pattern. Neither their size nor their distance from the sun is related directly to their period of rotation. The largest planet, Jupiter, takes only about 10 Earth-hours to rotate 360°. Therefore, the length of a day on Jupiter is 10 Earth-hours. Venus has the longest period of rotation, about 8 Earth-months. Furthermore, two of the planets (Venus and Uranus) do not rotate in the same direction as the other planets. Perhaps the rotation of these planets was affected by collisions with other objects after the solar system formed nearly 5 billion years ago.

The number of moons orbiting the planets shows a general pattern with distance from the sun. The two innermost planets, Mercury and Venus, have no moons. Earth has one moon. Then the number of moons generally increases as we move to the outer planets. However, the number of moons changes as more are discovered. It is difficult to distinguish between moons and other objects that orbit the outer planets. The question of how small an object can be and still be called a moon has no clear answer. For example, the rings of Saturn and less visible rings of other gas giants are made of millions of objects that orbit the planet in the same plane, but these objects are considered too small to be called moons. At this time, there is no known limit to the number of

Solar System Data

Celestial Object	Mean Distance from Sun (million km)	Period of Revolution (d=days) (y=years)	Period of Rotation at Equator	Eccentricity of Orbit	Equatorial Diameter (km)	Mass (Earth = 1)	Density (g/cm³)
SUN	—	—	27 d	—	1,392,000	333,000.00	1.4
MERCURY	57.9	88 d	59 d	0.206	4,879	0.06	5.4
VENUS	108.2	224.7 d	243 d	0.007	12,104	0.82	5.2
EARTH	149.6	365.26 d	23 h 56 min 4 s	0.017	12,756	1.00	5.5
MARS	227.9	687 d	24 h 37 min 23 s	0.093	6,794	0.11	3.9
JUPITER	778.4	11.9 y	9 h 50 min 30 s	0.048	142,984	317.83	1.3
SATURN	1,426.7	29.5 y	10 h 14 min	0.054	120,536	95.16	0.7
URANUS	2,871.0	84.0 y	17 h 14 min	0.047	51,118	14.54	1.3
NEPTUNE	4,498.3	164.8 y	16 h	0.009	49,528	17.15	1.8
EARTH'S MOON	149.6 (0.386 from Earth)	27.3 d	27.3 d	0.055	3,476	0.01	3.3

FIGURE 27-1. Solar System Data.

moons in the solar system. Figure 27-1 shows properties of the members of our solar system. (*Note:* Time is expressed in units of Earth-time. Thus, "days" are approximately 24 Earth-hours long.)

STUDENT ACTIVITY 27-1 —GRAPHING SOLAR SYSTEM DATA

6: MODELS 2
6: MAGNITUDE AND SCALE 3

Your teacher will divide the class into groups and give each group a piece of graph paper. On the piece of graph paper, write "Sun" followed by the names of the eight planets equally spaced along the horizontal axis. The vertical axis can be any one of the eight variables listed in Figure 27-1. Each group should make a graph displaying different data. Your teacher may share the graphs to the class. What relationship does each graph show between the position of the planets and the characteristic graphed?

HOW ARE THE PLANETS GROUPED?

4: 1.1b
4: 1.2c

Astronomers divide the planets into two groups. The four planets closest to the sun are the **terrestrial planets** because they are similar to Earth in their rocky composition. The four outer planets

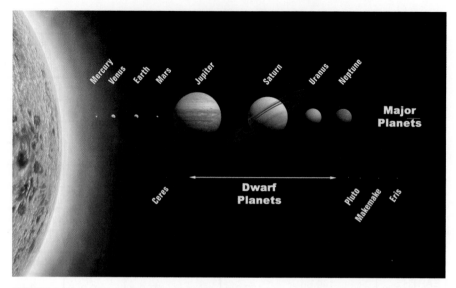

FIGURE 27-2. This diagram shows the sizes of the planets and dwarf planets to scale, but their distances from the sun are not to scale. (If distance and size were to the same scale, most planets would be too small to see.) The new category of dwarf planets will be discussed in page 672.

are called the **Jovian planets** because they are similar to Jupiter. Jupiter is a large planet that has a relatively low density. Most of Jupiter's volume is a thick shell of gas. Figure 27-2 compares the size of the sun and the planets.

Terrestrial Planets

Mercury, Venus, Earth, and Mars are the closest planets to the sun. These four planets are solid objects surrounded by a relatively thin atmosphere or no atmosphere at all. Mars, with the largest orbit in the group, is less than one-third of the distance to the next planet, Jupiter.

MERCURY The closest planet to the sun, Mercury, is a small planet. It therefore has weaker gravity than most of the other planets and it does not have a magnetic field. Electromagnetic radiation and charged particles from the sun (solar wind) have stripped Mercury of its atmosphere. Impact craters cover the surface of Mercury. Most craters probably formed from collisions with debris early in the history of the solar system. At that time there was more debris scattered through the solar system. With no atmosphere and no weather, these features are still evident after billions of years. The slow rotation of Mercury and the lack of an atmosphere cause an extreme range in temperature on the surface. Temperatures range from 400°C during the day to −200°C at night. Because it is relatively close to the sun, Mercury is visible from Earth only near sunset or sunrise. Visit the following Web site to see the current position of the planets in our solar system by selecting your time and your location: *http://www.fourmilab.ch/cgi-bin/Solar*

VENUS   Like Mercury, Venus is closer to the sun than is Earth; therefore, Venus is visible from Earth only near sunset or sunrise. Venus is sometimes called Earth's twin because its diameter and mass are similar to Earth's. Venus therefore has about the same gravity as Earth. For many years, astronomers wondered if Venus might have surface conditions similar to those on Earth. Thick clouds prevented direct observations of the solid surface. Radar images and mapping by artificial satellites revealed that Venus has volcanic features like Earth.

The surface conditions on Venus are very different from Earth's. Atmospheric pressure at the surface is 90 times the sea-level pressure on Earth. The thick atmosphere of Venus is mostly carbon dioxide, which traps solar energy (the greenhouse effect). Therefore, the surface temperature of Venus (about 500°C) is actually hotter than the daytime temperature on Mercury. In addition, the clouds of Venus are composed of droplets of highly corrosive sulfuric acid. This would be deadly to the kind of creatures that live on Earth.

Venus takes longer to rotate than any other planet. In fact, Venus takes longer to rotate on its axis than it takes to revolve around the sun. (See Figure 27-1 on page 665.) So much for our "twin planet."

PHASES OF MERCURY AND VENUS   As viewed from Earth, the portion of the lighted surface of Mercury and Venus changes in a predictable cycle. Like the moon, these two planets show a full range of phases. Galileo was the first astronomer to document the phases of Venus as he observed them with his telescopes.

When Mercury and Venus pass between Earth and the sun, they, like the new moon, appear dark as viewed from Earth. This is illustrated in Figure 27-3, on page 668. Then, the lighted portion of

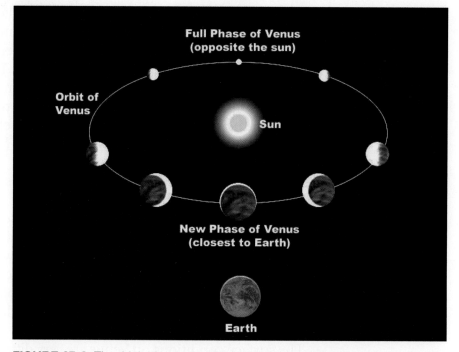

FIGURE 27-3. The thick clouds that cover Venus make it appear featureless in visible light. Depending upon the positions of the sun, Earth, and Venus, the apparent size and lighted part of Venus (the phase) change.

each planet increases until it is fully lighted as the planets pass behind the sun. This change in phase is accompanied by a change in apparent size. Mercury and Venus appear largest at their closest approach to Earth in the crescent phase. As they move farther from Earth and the lighted portion increases, these planets appear smaller in angular diameter. The orbits of the other planets are outside Earth's orbit. The night side of Earth faces the lighted side of these planets. Therefore, they never show the dark, or new, phase. They always appear mostly lighted as observed from Earth.

EARTH   Oceans cover about 70 percent of Earth's surface. They average about 4 km (2.5 mi) in depth. In fact, Earth is the only planet scientists know to have large amounts of water in all three states: ice, liquid water, and water vapor. Surface temperatures are moderate from a record low of −89°C (−128°F) in Antarctica to 58°C (136°F) in Africa's Sahara Desert.

Earth's unique surface conditions have supported the evolution of many forms of life. In addition, living things changed the planet.

Earth's original atmosphere probably was mostly carbon dioxide like that of Venus. The carbon dioxide content of the atmosphere is now less than 1 percent (0.03 percent). Most of the original carbon of the atmosphere has dissolved in the ocean or become part of plants, animals, or rocks. Weathering and erosion along with plate tectonics constantly change Earth's surface.

MARS Ancient astronomers named Mars after the Roman god of war because of its red color. Of the eight planets in our solar system, the conditions on Mars come closest to the favorable conditions for life found on Earth. Like Venus, the atmosphere of Mars is mostly carbon dioxide. However, the atmosphere of Mars is much thinner than that of Venus or Earth. Mars can warm up to a comfortable 20°C (68°F) near the Martian equator. However, at night, about 12 h later, it cools to −60°C (−76°F) because the thin atmosphere does not absorb and radiate heat. Due to the tilt of the rotational axis of Mars, it has seasons like Earth. However, the polar regions become so cold that solid carbon dioxide (dry ice) forms on the surface. Landing modules sent to Mars have transmitted photographs and other data from its red surface.

Some Martian surface features look as if they were formed by large rivers. However, there is no liquid water on the surface of Mars today. This has led some astronomers to theorize that most of the atmosphere of Mars has been lost and that the climate on Mars was warmer in the past. If the water is still there, it is probably in the form of ice under the planet's surface. All forms of life on Earth need water. So scientists theorize that primitive life forms may exist below the Martian surface.

STUDENT ACTIVITY 27-2 —DESIGN A LANDING MODULE

1: ENGINEERING
DESIGN 1

NASA has hired your group to design an instrument module that will send back information about Mars after it reaches the planet. Your module may include only easily obtainable instruments that you can understand and operate. Make a drawing of the module and write an explanation of what each instrument is as well as what information it will gather. Radio communication will be your only link. NASA is not sending people or animals to Mars on this mission.

The Outer Planets

The outer planets are Jupiter, Saturn, Uranus, and Neptune. These Jovian planets are larger and less dense than the terrestrial planets. They are composed mostly of hydrogen and helium rather than the heavier elements (iron, silicon, and oxygen) of the terrestrial planets. The Jovian planets have small, rocky cores surrounded by a liquid mantle and a thick gaseous shell. However, they are still the most massive planets because of their huge size. Each Jovian planet also has a system of rings around it. These rings are made of small particles that orbit independently in the plane of the planet's equator.

JUPITER   The first of the outer planets, Jupiter is the largest planet in the solar system. Jupiter alone makes up two-thirds of the total mass of the eight planets. Jupiter is more than 99 percent hydrogen and helium. This is more like the composition of a star than the composition of the terrestrial planets. The pressure at the center of Jupiter is about 10 times the pressure within Earth. However, it is still not enough pressure to support nuclear fusion, which powers the sun and other stars.

Jupiter's rapid rotation causes a noticeable bulge at its equator. One of Jupiter's other noticeable features is the Great Red Spot. Some astronomers think that this may be a permanent, raging storm. Others think it is an area of calm within the turbulent atmosphere. With binoculars or a small telescope, you can see four of Jupiter's many moons from Earth. They sometimes look like a line of stars that runs through the equator of Jupiter. These four satellites are known as the Galilean moons, named after their discoverer, Galileo.

SATURN   The rings of Saturn make it one of the most unusual objects in the night sky. Saturn's rings are made of ice particles and dust. They might have become a moon if this debris were not so close to Saturn and broken by Saturn's gravitational field. These rings are sometimes visible through binoculars from Earth, as in Figure 27-4. Rings of the other Jovian planets are less dense and therefore more difficult to see. Second in size only to Jupiter, Saturn is also composed primarily of hydrogen and helium. It is the least dense planet with an average density less than that of water.

FIGURE 27-4. Saturn's rings are composed of particles of rock and ice. There are seven gaps in the rings, created where moons have swept up debris with their weak gravitational attraction. In this view Saturn is eclipsing the sun.

URANUS AND NEPTUNE   Uranus and Neptune are almost alike. They are composed mostly of gas. Although less than half the diameters of Jupiter and Saturn, they are much larger than Earth. Uranus and Neptune are not as bright as the other planets. Therefore, they were not discovered until telescopes were turned to the heavens about the time of Galileo. Uranus, the brighter of the two, is barely visible without binoculars or a telescope. They are dim because they are so far from the sun. In addition, they are not as large as Jupiter and Saturn. The rotational axis of Uranus is tilted nearly 90°, so it seems to spin on its side. Neptune is very similar in size and composition to Uranus, but nearly twice as far from the sun.

As the most distant planets, Neptune and Uranus are the coldest planets. At the cloud tops, which we see as the outer part of these planets, both have temperatures below −220°C.

STUDENT ACTIVITY 27-3 —THE SOLAR SYSTEM TO SCALE

6: MAGNITUDE AND SCALE 3

Use the data in the *Earth Science Reference Tables* to construct a scale model of the solar system. If you use a single scale for the size of the planets and their distance from the sun, you will probably need to work outside. For smallest planets to be dots on a sheet of paper, the distance to Neptune may need to be greater than the size of your school building.

WHAT OTHER OBJECTS ORBIT THE SUN?

4: 1.2c, 1.2d

From its discovery in 1930 until 2006, Pluto was the ninth planet. However, in many ways it is different from the outer planets. For a long time, Pluto had puzzled astronomers because it is so different from the four outer planets. Unlike these gas giants, Pluto is small and relatively dense. Pluto has a more eccentric orbit than any of the planets. In fact, Pluto is sometimes closer to the sun than is Neptune.

Recently, astronomers have discovered many objects in the outer solar system that are similar to Pluto. In 2003, they discovered Eris, which is 27 percent more massive than Pluto and farther from the sun.

Dwarf Planets

In 2006, the IAU (International Astronomical Union) defined the term *planet* for the first time. This definition left out Pluto. The IAU created a new group of solar system objects: the dwarf planets. The IAU classified Ceres, Eris, and Pluto as dwarf planets. Ceres is in the asteroid belt between Mars and Jupiter. Its radius is about 10 percent of Earth's. Pluto and Eris have larger orbits than any of the planets. The radius of each is about 20 percent of Earth's. Makemake was added in 2008. All four are nearly as dense as rocks found near Earth's surface. Ceres, Pluto, Makemake, and Eris are shown in proper order in Figure 27-2 on page 666. With better telescopes and technology, astronomers are discovering more dwarf planets.

Asteroids

The asteroid belt is a ring of debris that separates the terrestrial and Jovian planets. **Asteroids** are irregularly shaped rocky objects that are smaller than planets. They orbit the sun in the same direction as the planets. Perhaps the asteroids are debris left over from the formation of the solar system. Earth's gravity has pulled some asteroids to its surface. Evidence from these asteroids indicates that they may be the remains of a planet that broke into many fragments when it was struck by another object. The spacing between Mars and Jupiter is larger than the pattern observed among other planets. This supports the idea of a missing planet. Visit the following Web site to take the Asteroid Challenge: Target Earth, a virtual lab where you will learn about the near-Earth asteroid, Apophis: *http://spaceclass.org/apophis/*. Visit the following Web site to answer the question: Did an asteroid cause a tsunami in New York 2300 years ago (with animation)? *http:// dsc.discovery.com/news/2008/11/20/asteroid-tsunami-print.html*

Comets

Occasionally, we can see **comets** in the night sky. They appear as a small spot of light with a long tail that points away from the sun. (See Figure 27-5.) Comets have highly eccentric orbits. This means that they spend the great majority of their time in the far outer reaches of the solar system. Many comets orbit the sun beyond the orbit of Neptune. Some comets seen in the inner solar system were probably pulled into highly elliptical orbits around the sun by the gravity of a passing star or other object in space. When comets enter the inner part of the solar system near Earth and the sun, they move quickly under the influence of the sun's strong gravitational field. The brightest comets are usually visible for several weeks. Then, they return to the outer portion of their orbits.

Comets are sometimes called dirty snowballs because they are made of ice and rock fragments. When comets come close to the sun, some of the ice and dust escapes. These particles form the comets' distinctive tails. Only about 30 comets per century are bright enough to be seen without a telescope. The best known is Halley's (HAL-ease) comet. It returns to the inner solar system about every 75 years, and it usually is visible without a telescope. Halley's Comet was visible in 1986, and will return in 2061.

FIGURE 27-5. Comets are icy objects with highly eccentric orbits. Comets form visible tails as they pass through the inner parts of the solar system. Most comets are very dim. They do not streak quickly through the sky like meteorites. They can remain visible in the same part of the sky for weeks.

Meteors

On a clear night, it is sometimes possible to see streaks of light that move across a portion of the sky, usually in a fraction of a second. These are solid bits of rock from outer space that enter the upper atmosphere at great speed. Friction with the atmosphere creates a streak of light known as a **meteor**. Table 27-1 lists the times of the year when meteors are especially numerous. Visit the following Web site to watch a large meteor as it streaks across the

Table 27-1 Meteor Showers

Name/Source	Date of Maximum	Duration Above 25% of Maximum	Approximate Limits	Number per Hour at Maximum
Quadrantids	Jan. 4	1 day	Jan. 1–6	120
Aquarids	July 27–28	7 days	July 15–Aug. 15	35
Perseids	Aug. 12	5 days	July 25–Aug. 18	90
Orionids	Oct. 21	2 days	Oct. 16–26	30
Geminids	Dec. 14	3 days	Dec. 7–15	120

FIGURE 27-6. Aerial and ground-level views of Meteor Crater, Arizona. Known as the world's most spectacular impact structure, Meteor Crater is 1 km (0.7 mi) wide and 170 m (570 ft) deep. The white arrow shows where the ground-level photograph on the right was taken. For scale, note the spectators looking down into the crater from the platform at the lower right.

FIGURE 27-7. The Willamette Meteorite was found in Oregon and acquired by the American Museum of Natural History in 1906. It is an iron meteorite weighing over 15.5 tons, the sixth largest meteorite in the world. The holes were caused by millions of years of weathering in the wet Oregon climate.

sky in Canada: *http://www.youtube.com/watch?v=Os9FrsVMZew &feature=related*

As they move through space, the bits of rock are called **meteoroids**. Most meteoroids burn up before they reach the ground. However, the larger ones often survive their fiery trip through the atmosphere. Those that strike Earth's surface are called **meteorites**. Meteorite impacts have left many craters. One is Arizona's famous Meteor Crater (Barringer Crater), seen in Figure 27-6. Scientists study meteorites to learn about the origin of the Earth and solar system. Nearly all meteorites are debris left over from the formation of the solar system.

Astronomers divide meteorites into two groups. The most common are the stony meteorites. They are composed of minerals common in Earth's mantle layer. Other meteorites are mostly iron. They are thought to be similar to the composition of Earth's core. Figure 27-7 shows an iron meteorite. Visit the following Web site to learn more about the Willamette Meteorite: *http://www.amnh. org/rose/meteorite.html*

CHAPTER REVIEW QUESTIONS

Part A

1. Which planet is approximately 10 times farther from the Sun than is Earth?

   (1) Mars   (3) Saturn
   (2) Jupiter   (4) Uranus

2. Which object has the greatest density?

   (1) Earth   (3) Jupiter
   (2) the moon   (4) the sun

3. What is the approximate inferred age of our solar system?

   (1) 1.3 billion years   (3) 14 billion years
   (2) 4.6 billion years   (4) 149 billion years

4. Which bar graph below correctly shows the orbital eccentricity of the eight planets of our solar system?

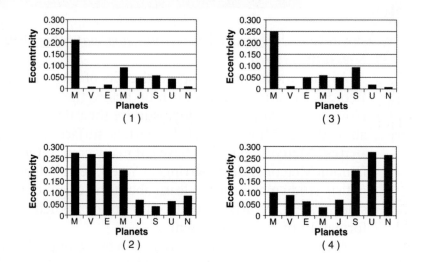

5. Which object is closest to Earth?

   (1) the sun   (3) Venus
   (2) the moon   (4) Mars

6. Large craters found on Earth support the hypothesis that impact events have caused

(1) a decrease in the number of earthquakes and an increase in sea level
(2) an increase in solar radiation and a decrease in Earth radiation
(3) the red shift of light from the most distant stars in the universe
(4) mass extinctions of life-forms and global climate changes

Part B

Base your answers to questions 7 and 8 on the diagram below, which shows the inferred sequence of changes by which our solar system formed from space debris.

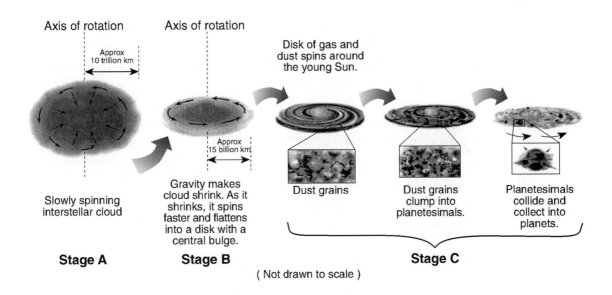

7. How did the young sun form between stages *B* and *C*?

(1) Gravity caused the center of the cloud to contract.
(2) Gravity caused heavy dust particles to split apart.
(3) Outgassing occurred sending jets of material outward.
(4) Outgassing from Earth provided material to make the sun.

8. After the young sun formed, the disk of gas and dust

   (1) became a large sphere
   (2) split into two massive objects
   (3) became larger in diameter
   (4) formed into the planets

Base your answers to questions 9 through 11 on the passage below and your knowledge of Earth science.

A Newly Discovered Planet

Scientist studying a sun-like star named *Ogle-Tr-3* discovered a planet that is, on average, 3.5 million km from the surface of the star. The planet was discovered as a result of observing a cyclic decrease in the brightness of *Ogle-Tr-3* every 28.5 h. The changing brightness is a result of the planet blocking some of the starlight when it passes between the star and Earth. This observation enabled astronomers not only to find the planet, but also to determine its mass and density. The mass has been calculated to be approximately 159 times the mass of Earth. The new planet is only 20 percent as dense as Jupiter. Scientists think that the low density is a result of the planet being very close to the star *Ogle-Tr-3*.

9. Compared to the periods of revolution of Mercury and Venus, the newly discovered planet's period of revolution is

   (1) shorter than both Mercury's and Venus's
   (2) longer than both Mercury's and Venus's
   (3) shorter than Mercury's but longer than Venus's
   (4) longer than Mercury's, but shorter than Venus's

10. The density of this newly discovered planet is approximately

   (1) 0.1 g/cm³   (3) 1.3 g/cm³
   (2) 0.3 g/cm³   (4) 2.0 g/cm³

11. Which similar light-decreasing event in our solar system is the result of the moon being between Earth and the sun?

   (1) summer solstice
   (2) winter solstice
   (3) solar eclipse
   (4) lunar eclipse

Part C

Base your answers to questions 12 through 15 on the diagram below, which shows the orbits of the four inner planets of our solar system. The dot on each orbit shows where the planet is closest Sun.

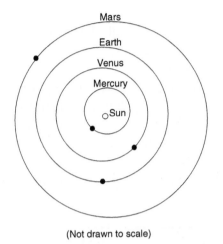

(Not drawn to scale)

12. Make a copy or sketch of the diagram above and write a *W* on the orbit of Mars where the gravitation attraction between the sun and Mars is *weakest*.

13. Circle the name of the largest of the four inner planets.

14. How long does it take Mercury to spin once on its axis?

15. What is the precise shape of these planetary orbits and the shape of all orbits of satellites around their primaries?

Base your answers to questions 16 through 18, on page 680, on the diagram below, which shows the inferred interiors of the four terrestrial planets.

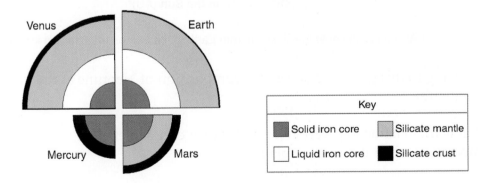

16. What are the two most common elements in the crusts of Mercury, Venus, Earth, and Mars?

17. Which two planets would allow S-wave earthquake waves to pass through the core to the opposite side of the planet?

18. How are the densities of the terrestrial planets different from the densities of the Jovian planets?

Base your answers to questions 19 and 20 on the graphic below, which shows the relationship between the distance of a planet from the sun and the temperature of formation. (1 AU is Earth's distance from the sun.) The shaded zones indicate planetary composition when it first formed.

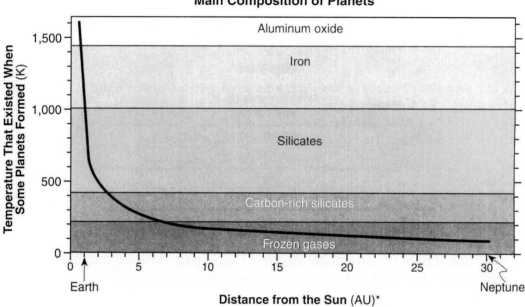

Main Composition of Planets

* 1 AU equals the average distance from Earth to the Sun or 149.6 million kilometers.

19. According to this graph, what was the composition of Neptune at the time of its formation?

20. What is the relationship between the distance of a planet from the sun and the temperature when it became a planet?

28 Stars and the Universe

WORDS TO KNOW

big bang	galaxy	nuclear fusion
cosmic background radiation	light-year	redshift
Doppler effect	luminosity	star
frequency	Milky Way galaxy	

This chapter will help you answer the following questions:

1. How do astronomers investigate stars?
2. Are stars different from other celestial objects?
3. How do stars generate energy?
4. What kinds of stars are there?
5. How do stars change?
6. What are the structure, history, and future of the universe?

HOW DO ASTRONOMERS INVESTIGATE STARS?

1: ENGINEERING
DESIGN 1
7: CONNECTIONS 1

Stars are extremely hot and have no solid surface. Scientists cannot send instruments to land on stars. Any devices would melt and probably vaporize long before reaching the visible surface of a star. Furthermore, other than the sun, stars are too far away to reach by spacecraft. With our present technology, it would take tens or even hundreds of thousands of years for a spacecraft to reach even the next nearest star. Therefore, most of the information astronomers have about stars comes from light and other electromagnetic energy they radiate into space.

The Electromagnetic Spectrum

To most of us, visible light is the most important kind of electromagnetic radiation. Light allows us to see. Visible light can be broken up into its component colors by a prism. This is shown in Figure 28-1.

Our eyes are sensitive to only a narrow band of wavelengths that we know as visible light. However, there are other electromagnetic radiations, which make up the electromagnetic spectrum. Look back at Figure 19-12 on page 452 from the *Earth Science Reference Tables*. It shows the full range of electromagnetic radiation, also known as the electromagnetic spectrum.

STUDENT ACTIVITY 28-1 —MAKING A SPECTRUM

1 SCIENTIFIC
INQUIRY: 1, 2, 3

You can separate sunlight into its spectrum of colors with a glass prism. This works best in a darkened room where windows face the sun. Close the shades so that a narrow slit of direct sunlight enters the room. Place the prism near the narrow opening that admits sunlight. The prism will bend the light beam and separate it into its colors. You may need to rotate the glass prism to project a visible spectrum. The spectrum can be projected onto a sheet of white paper. The stronger the light and the closer the paper is held to the prism, the brighter the spectrum will be. To increase the size of the spectrum, move the paper screen away from the prism. What two changes in the spectrum do you observe as the paper is moved away from the prism?

Elements can absorb and emit light. Each element has its own characteristic absorption lines, or absorption spectrum. Stars are composed primarily of hydrogen and helium. White light that passes through these gases shows dark lines in the orange, yellow, green, and blue areas of the spectrum that characterize hydrogen

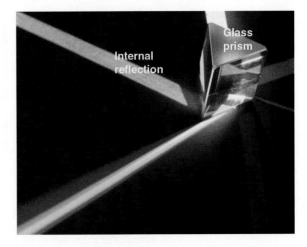

FIGURE 28-1. A beam of white light enters this glass prism from the top. Some of the light is reflected. However, the light that passes through the glass is split into a rainbow of colors. This shows that white light is actually a mixture of colors.

and helium. These spectral lines correspond to the energy that electrons absorb when they move to higher energy levels within the atoms. The atoms give off the same colors when the electrons fall to lower or inner energy levels. Each element has a unique set of energy levels. Therefore, these "spectral fingerprints" allow astronomers to identify the composition of distant stars. Visit the following Web site to decode cosmic spectra: *http://www.pbs.org/ wgbh/nova/origins/spectra.html*

Optical Telescopes

Astronomers use optical telescopes to study the stars. (See Figure 1-5 on page 10.) These telescopes concentrate the light of stars. Optical telescopes allow them to see objects that are too dim to be seen with unaided eyes. Some people think that the most important feature of a telescope is how much it magnifies. However, the stars are so distant that even the most powerful telescopes show nearly all stars as points of light. When an image is magnified too much, it becomes dim, unclear, or fuzzy.

Other factors are more important than magnification in telescope construction and use. First is the size, or diameter, of the front lens (or light-gathering mirror). This determines the dimmest object that can be observed. The larger the diameter of the front lens, the dimmer the objects you can see. The farther astronomers look into space, the dimmer the objects become.

FIGURE 28-2. (*l.*) The largest optical telescopes are placed in observatories. The observatories are built on remote mountains to lift them above as much of atmospheric pollution as possible. (*r.*) The Hubble Space Telescope orbits Earth far above the distorting effects of Earth's atmosphere. The Hubble telescope has taken spectacular images of distant stars and galaxies, including the Eagle Nebula pictured on page 604.

The second factor is the quality of optics of the telescope. If the lenses or mirrors are not made with great precision, magnified images will not be sharp. Earth's atmosphere is also a limiting factor. This is why major observatories are built on high mountains, where the atmosphere is thin and has less effect on the light. Figure 28-2*l* shows several buildings containing large telescopes on a mountaintop in Arizona. The Hubble Space Telescope (Figure 28-2*r*) orbits Earth above the distorting effects of Earth's atmosphere. This telescope was a major step forward in observational astronomy. Visit the following Web site to see the universe through the Hubble Space Telescope: *http://news.xinhuanet.com/english/ 2006-12/18/content_5501941.htm*

STUDENT ACTIVITY 28-2 —MAKING A TELESCOPE

1: ENGINEERING
DESIGN 1

You can build a simple telescope using two convex lenses. A tube to hold the lenses at the proper distance and alignment is helpful but not necessary. By using lenses with more or less curvature, you can change magnification. Moving the lens that is closest to your eye adjusts the focus.

FIGURE 28-3. This group of radio telescopes in western New Mexico uses 27 giant receiving surfaces to amplify weak radio signals from nearly any direction in space.

Radio Telescopes

Radio telescopes gather long-wavelength radio energy rather than light. Radio telescopes, like those in Figure 28-3, are not affected by our atmosphere or by clouds of dust and gas in space that block light. They also can detect objects that do not produce light. Radio telescopes do not form sharp images. This makes it difficult to tell the exact position of a radio source. However, using radio telescopes astronomers make observations that would not be possible with optical telescopes.

Other Telescopes

Other kinds of telescopes use electromagnetic wavelengths shorter than visible light, such as x-rays and gamma rays. These telescopes must be located in orbit above Earth's atmosphere, which filters out these forms of radiation.

Technology has changed the ways astronomers use telescopes. The first telescopes were used for direct observations. If astronomers wanted a permanent record of what they saw, they had to draw it by hand. Film cameras allowed astronomers to take pictures through their telescopes. Today, the more advanced telescopes use electronic sensors like those in digital cameras along with computers to create better quality images than ever before possible.

Spectroscope

The spectroscope is one of the most important tools that astronomers use. A spectroscope separates light into its component colors (wavelengths). A spectroscope works something like a prism. When starlight passes through a spectroscope, dark lines appear in certain parts of the spectrum. These dark lines are produced when certain wavelengths of light are absorbed by gaseous elements within the outer parts of the star.

THE SEARCH FOR EXTRATERRESTRIAL LIFE

1: SCIENTIFIC INQUIRY 1, 2, 3

In 1996, scientists announced that a meteorite recovered from the ice of Antarctica might contain evidence of life outside Earth. Scientists studied the meteorite with a microscope. They saw what might be fossilized bacteria. They also identified organic compounds thought to be the result of biological processes near this fossil.

Most scientists agree that the meteorite came from Mars. However, no evidence of life on Mars has been found by remote observations or by spacecraft. If life exists there, it is probably primitive, microscopic organisms. These organisms most likely live under the planet's surface. Some scientists suggest that the fossil-like shape and the compounds thought to be of biological origin could be the result of inorganic processes. Scientists are still discussing whether the features in this meteorite are our first direct evidence of life outside Earth.

ET, Phone Earth

Another attempt to discover extraterrestrial life is project SETI (Search for Extraterrestrial Intelligence). SETI seeks to find radio transmissions coming from outside our solar system. Considering the billions of stars in the universe, it seems possible that planets with conditions similar to those on Earth could be orbiting some stars. It also seems possible that on some of these planets there may be technological civilizations like our own.

The huge distances in space may keep people from traveling among the stars. However, radio waves can travel long distances at the speed of light. Radio waves are long-wavelength electromagnetic radiation that we cannot see. Astronomers use large radio telescopes to explore the universe using radio waves. Radio waves can penetrate dust and clouds of gas that prevent visual observations.

How would scientists know if they received signals from another civilization? The signals probably would not be in a familiar language. Astronomers look for patterns in radio transmissions that have no known natural source. These patterns could be signals from another civilization. For several decades, some of the largest radio telescopes have been listening for intelligent transmissions. Figure 28-3 on page 685 is a group of radio telescopes designed to receive long-wave electromagnetic radiation.

At first, analyzing these signals was a problem. How could scientists separate intelligent communications from the large amount of radio noise generated by stars? This is where computers come in. Computers can quickly find patterns in radio signals. Faster and more powerful computers have let astronomers scan far more observations than humans could ever analyze. However, since SETI began in about 1985, no signals identified as likely forms of intelligent communication have been found.

What Could This Mean for Humans?

If scientists did detect intelligent communications, how could it affect Earth? Perhaps the information would provide us new insights into mathematics, science, or technology. Perhaps people could learn more about the promise and the dangers of a developing civilization. On the other hand, there is always the danger of conquest by an unknown military power. The future is always unknown. Fortunately, we have learned to apply new technologies and discoveries to improve our lives. Experience has taught that, in the long run, the developments of science, for the most part, benefit humans.

WHAT IS A STAR?

4: 1.2b

A **star** is a massive object in space that creates energy and radiates it as electromagnetic radiation. The sun is a star. If you compare the sun with the thousands of stars known to astronomers, the sun appears to be a typical star. Actually, most of the stars we observe in the night sky are larger and brighter than the sun. At the same distances as the visible stars, there are far more stars smaller than the sun that are too dim to see from Earth. Observations of the sun give astronomers insights into most other stars and their observations of other stars help them understand the sun. **Never look directly at the sun**.

STUDENT ACTIVITY 28-3 —LIGHT INTENSITY AND DISTANCE

1: SCIENTIFIC
INQUIRY 1, 2, 3

You can use a light meter to measure the change in light intensity with distance. Place a lightbulb in a dark room. Measure the intensity of the light at different distances from the lightbulb. Make a data table of the light intensity at various distances from the lightbulb. Graph your data. What other factor shows a similar change of intensity with distance?

Starlight

For centuries, astronomers have wondered how the sun could produce the large amounts of energy it radiates into space. They knew that the brightest light sources on Earth produced light by chemical changes such as burning. Fuels such as wood and coal rapidly combine with oxygen in the atmosphere to produce heat and light. If the sun burned coal or wood to produce energy, it would run out of fuel very quickly. As scientists became aware of the age the solar system, it became clear that a very different process was taking place in the sun.

Advances in science in the early twentieth century showed that matter could be changed into energy. You may have heard of Albert Einstein's famous equation $E = mc^2$, in which E is energy, m is mass, and c is the speed of light. The speed of light is a very large number. Therefore, the square of that number is enormous. The point of this formula is that great amounts of energy are created by the conversion of a small amount of mass. In the nearly 5 billion years since the solar system originated, scientists estimate that the sun has only lost about one-third of 1 percent of its total mass.

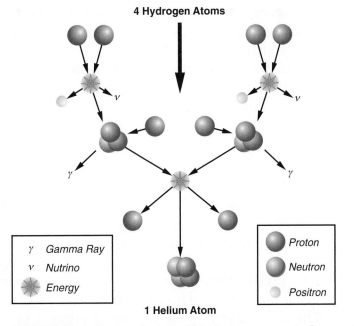

4 Hydrogen Atoms

1 Helium Atom

γ Gamma Ray
ν Nutrino
✳ Energy

● Proton
● Neutron
○ Positron

FIGURE 28-4. Nuclear fusion within the sun converts four protons (hydrogen nuclei) into one atom of helium. This process occurs in several steps. The sun's mass is decreasing by about 4 million tons/s. This loss of mass is expected to continue for another 5 billion years with plenty of mass left over.

Nuclear Fusion in Stars

Most of the mass of the sun or a similar star is hydrogen, the lightest element. When four hydrogen nuclei join to make a helium nucleus, they lose about 1 percent of their mass. The process by which light elements join to make heavier elements is called **nuclear fusion**. (See Figure 28-4.) While 1 percent may seem like a small loss of mass, it is enough to create a great amount of energy. However, nuclear fusion can occur only under extreme conditions of heat and pressure. Although the gas giant Jupiter is the largest planet in our solar system, it is not quite large enough to support fusion and become a star.

Energy Escapes from Stars

Energy created deep in the sun moves to the sun's visible surface by radiation and convection. From the solar surface, the energy escapes as electromagnetic radiation. The surface temperature of the star determines the kind of electromagnetic energy it radiates into space. The sun is a yellow star because its roughly 6000°C surface radiates most intensely as yellow light in the visible part of the spectrum.

Based on observations of other stars, astronomers predict that the sun will continue to radiate energy as it now does for about another 5 billion years.

HOW ARE STARS CLASSIFIED?

1: SCIENTIFIC
INQUIRY: 1
4: 1.2b

In the early twentieth century, astronomers in Denmark and the United States discovered a way to classify stars. They based their classification on the amount of electromagnetic energy the stars generate and their temperature. The total energy output of a star is its **luminosity**, or absolute brightness. Apparent brightness, or stellar magnitude, is how bright the star looks as seen from Earth. The closer a star is, the brighter it appears to us. The brightness of a star depends on its absolute magnitude, or luminosity, and its distance from the observer.

You may have noticed that when you turn off an incandescent lightbulb the color of the hot wire briefly changes to red before it goes dark. Red is the coolest color of light visible to our eyes. If a material is heated beyond red-hot, it becomes white and then blue. Continued heating would push the radiation into the ultraviolet part of the spectrum and beyond.

Hertzsprung-Russell Diagram

Astronomers use the Hertzsprung-Russell, or H-R, diagram to classify stars. This graph was named in honor of the two men who developed it. Figure 28-5 is from the *Earth Science Reference Tables*. Unlike other graphs, this graph is usually plotted with the temperatures *decreasing* to the right along the bottom axis. (Usually, values increase to the right as well as upward on the vertical axis.) Visit the following Web site to compare different magnitude levels with lightbulbs: *http://www.ioncmaste.ca/homepage/resources/web_resources/CSA_Astro9/files/multimedia/unit2/magnitudes/magnitudes.html*

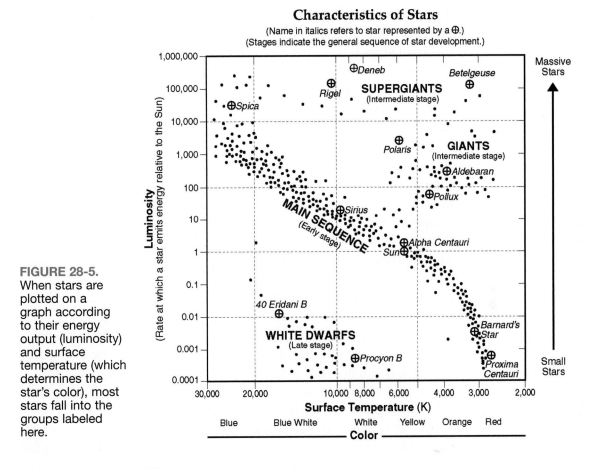

Characteristics of Stars
(Name in italics refers to star represented by a ⊕.)
(Stages indicate the general sequence of star development.)

FIGURE 28-5.
When stars are plotted on a graph according to their energy output (luminosity) and surface temperature (which determines the star's color), most stars fall into the groups labeled here.

Main Sequence Stars

When plotted on this graph, most stars fall into distinct groups. The greatest number of stars fall into an elongated group that runs across the luminosity and temperature diagram from the upper left to the lower right. This region of the graph is the main sequence. The position of a star along the main sequence is primarily a function of the mass of the star.

RED DWARF STARS The smallest stars, such as Barnard's Star, are red dwarf stars, which are barely large enough to support nuclear fusion. They are red because they are relatively cool. These stars are so dim that even the red dwarfs closest to us are difficult to see without a telescope. In fact, about 80 percent of the night stars closest to Earth are too dim to be visible to the unaided eye.

This leads astronomers to infer that red dwarf stars are more numerous than all other groups of stars.

Small stars last longer than larger stars. The lower temperature and pressure in these stars allow them to conserve hydrogen fuel and continue nuclear fusion much longer than larger stars. The combination of small size and slow production of energy makes them very dim.

BLUE SUPERGIANT STARS At the other end of the main sequence are the blue supergiants. These massive stars last a small fraction of the lives of the smaller stars. The extreme conditions of temperature and pressure at the center of these stars cause rapid decrease in their large quantities of hydrogen. Some of them are a million times brighter than the sun. They are also much hotter than the sun, giving them a blue color. These largest stars are not nearly as common as the smaller stars, in part because they burn out quickly. The most massive stars last less than one-thousandth of the life of the sun. Rigel, a bright star in the winter constellation Orion, is 10,000 times as luminous as the sun. The blue color of Rigel is apparent if you compare it with Betelgeuse, another bright star in Orion. Betelgeuse is a red giant star on the opposite side of the same constellation.

White dwarfs, red giants, and the supergiants are the most common star groups outside the main sequence. Most other stars fall into one of theses three groups on the temperature-luminosity graph.

HOW DO STARS EVOLVE?

4: 1.2b

Different sizes of stars have different life cycles. However, the evolution of stars can be illustrated by considering a star about the size of the sun.

Birth of a Star

Star formation begins when a cloud of gas (mostly hydrogen) and dust begins to come together (condense) under the influence of gravity. There are two sources for the material in the cloud. Some of it is hydrogen and helium left over from the formation of the universe about 14 billion years ago. The rest is the debris from the explosions of massive stars that formed earlier in the history of the universe. This initial phase takes about 50 million years. (The process is faster for larger stars and slower for smaller stars.)

As the material comes together, heat is generated by the collapse of the matter and by friction. This heat causes the temperature to rise until it is hot enough and the pressure is high enough to support nuclear fusion. When fusion starts, the star becomes visible. The star produces and radiates large amounts of energy. With binoculars or a small telescope, you can see several young stars below the belt of Orion shining through a giant cloud of gas that surrounds them. Visit the following Web site to watch the life cycle of a star: *http://aspire.cosmic-ray.org/labs/star_life/support/HR_animated_real.html*

Middle Age

The star becomes less luminous after it fully condenses. An average star spends most of its life on the main sequence region of the luminosity and temperature graph. Gravitational pressure balanced by heat from nuclear fusion keeps the star from shrinking further. Middle age is the longest and most stable phase of stellar evolution.

Death of an Average Star

After about 10 billion years, a star the size of the sun runs low on hydrogen. Fusion slows. The core of helium collapses, which causes the outer part of the star to expand quickly. The star becomes a red giant. Fusion of helium and other heavier elements replaces hydrogen fusion. The outer shell of gases expands and cools in the red giant stage. Eventually, all that is left is a dense, hot core, which is a white dwarf star.

Death of a Massive Star

Stars with more than about 10 times the mass of the sun end their period in the main sequence more violently. These stars create a variety of heavier elements before they collapse. The collapse process of larger stars generates so much energy that these stars end their life in an explosion called a supernova. They briefly generate more energy than the billions of stars that make up the whole galaxy. Most of the mass of the star is blown into space. The core of the star may form an extremely dense object called a neutron star. Some stars are so massive that they form an object with gravity so strong that not even light can escape: a black hole. Black holes cannot radiate energy, but they can be detected because energy is given off by matter that falls into the black hole. They can also be located by their gravitational effects on other objects. Astronomers have recently detected a black hole at the center of the Milky Way and many other galaxies.

WHAT IS THE STRUCTURE OF THE UNIVERSE?

4: 1.2a

Early astronomers noticed fuzzy objects in the night sky. They called these objects nebulae (singular, nebula). The word *nebula* comes from the Latin word for cloud. Unlike the stars, these objects looked like dim fuzzy patches of light.

Nebulae and Galaxies

Telescopes revealed that some nebulae are regions of gas and dust where stars are forming. In addition, some nebulae were farther away than any known stars. Astronomers eventually realized that some nebulae are huge groups of stars held together by gravity. These objects are called **galaxies**. The whole Andromeda galaxy is visible as a small, faint patch of light high in the autumn sky. Powerful telescopes revealed that the Andromeda galaxy, like billions of other galaxies, is a gigantic group of billions of stars. Galaxies are separated by vast distances that contain relatively few stars. Figure 28-6 is a typical spiral galaxy. Visit the following Web site to learn what a billion (or a trillion) pennies would look like: *http://www.kokogiak.com/megapenny/default.asp*

Astronomers have discovered that collisions of galaxies are relatively common. However, the matter within the galaxies is so spread out that the two galaxies generally pass right through each other without individual stars colliding.

FIGURE 28-6. Galaxy NGC 4414 is a typical spiral galaxy composed of billions of stars. The Milky Way galaxy and its relatively nearby twin, the Andromeda galaxy, are similar spiral galaxies. The location of Earth and the solar system in the Milky Way galaxy is in one of the spiral arms far from the galactic center. The white arrow in this image shows such a position.

The Milky Way

Astronomers realized that all the stars visible in the night sky are a part of the group of stars called the **Milky Way galaxy**. The sun and solar system are part of the Milky Way galaxy. This name came from the faint, white band of light (Figure 28-7 on page 696) that can be seen stretching across the sky on very dark, moonless nights. (You cannot see the Milky Way in cities. In these areas light pollution prevents the night sky from being dark enough to let it be seen.) This broad band is actually made of thousands of stars.

Radio telescopes helped astronomers map the Milky Way galaxy. Astronomers estimate that it is composed of roughly 100 billion stars. Clouds of dust and gas obscure most of them. These clouds are also a part of our galaxy. The center of the galaxy is located in the direction of the summer constellation Sagittarius. The shape of the Milky Way galaxy, like NGC 4414 and the Andromeda galaxy, is a flattened spiral. The sun and solar system are located about two-thirds of the way from the center to the outer edge, as shown in Figure 28-6. As stars orbit the center of the galaxy, inertia keeps gravity from drawing them together.

Orbiting the center of the galaxy is another one of Earth's cyclic motion in space. Our planet rotates on its axis in a 24-h cycle. It

FIGURE 28-7. On dark, moonless nights, ancient people saw a band of faint light crossing the sky. They called it the Milky Way because it reminded them of spilled milk. They did not realize that they were actually seeing millions of distant stars, many of them much brighter than the sun.

also revolves around the sun each year. The solar system revolves around the center of the Milky Way galaxy in about 220 million years. Although this is a long time, the Milky Way galaxy is so large that this motion is actually about 10 times faster than Earth's revolution in its orbit around the sun.

Clusters and Superclusters

Galaxies are not the largest structures in the universe. The Milky Way and Andromeda galaxies are part of a cluster of about 30 galaxies. This cluster is called the Local Group. Astronomers are now mapping superclusters of galaxies and even larger structures of matter. Why the matter of the universe is so unevenly distributed is one of the most important questions that astronomers are trying to answer. Visit the following Web site to model galaxy collisions on your computer: *http://burro.cwru.edu/JavaLab/GalCrashWeb/ main.html*

WHAT IS THE HISTORY OF THE UNIVERSE?

4: 1.2a

When you look at very distant objects in the universe, you are looking back in time. This is because light has a limited speed. You learned in an earlier chapter that you could estimate the distance to a lightning strike by counting the seconds between seeing the flash and hearing the thunder. Due to light's speed, you see the flash at essentially the same time it occurred. Light travels so fast that it could circle Earth about 7 times in 1 s. Visit the following Web site to watch the PowerPoint Presentation—Can We Rewind the Clock to Catch a Glimpse Near the Beginning of Time? *http:// www.bnl.gov/physics/colloquium/PHENIX_Final.ppt*

Using Light as a Yardstick

Distances in space are so great that light cannot reach Earth instantly. Light and other electromagnetic energy take about 3 s to travel from the moon to Earth. It takes about 8 minutes for this energy to travel from the sun to Earth. Light takes more than 4 years to arrive from the nearest star, Proxima Centauri. The most distant object visible to the unaided eye is the Andromeda Galaxy. Light from the Andromeda galaxy takes about 2 million years to reach us.

Astronomers use light to measure the universe. A **light-year** is the distance that any form of electromagnetic energy can travel in 1 year. That distance is about 10 trillion km (6 trillion mi). Although the light-year may sound like a measure of time, it is a measure of distance.

When astronomers look at distant objects in space, they see them as they were when the light started its journey toward Earth. The deeper astronomers look into space, the farther back in time they see. At present, astronomers estimate that the most distant objects they can see are about 13 billion light-years away. Astronomers can see what the universe looked like about a billion years after its origin. They estimate the origin occurred about 14 billion years ago.

Redshift

Astronomer Edwin Hubble examined the spectra of distant galaxies in the early 1900s. He compared the dark absorption lines, or absorption spectra, of light from these faraway galaxies to the absorption spectra of nearby stars. Nearby stars had absorption spectra similar to those produced in the laboratory. However, the dark lines in the spectra of distant galaxies were shifted toward the red end of the spectrum. Hubble reasoned that the motion of distant galaxies away from Earth causes the **redshift** of spectral lines. Figure 28-8, on page 698, illustrates the redshift of spectral lines.

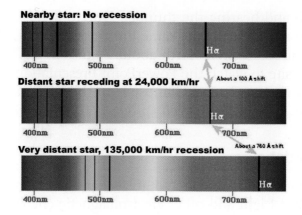

FIGURE 28-8. When we look at the signature elements of stars, the dark lines that mark hydrogen and other elements are clearly visible. The more distant stars and galaxies show the dark lines shifted toward the red end of the spectrum.

If the galaxies were moving toward Earth, the spectral lines would shift toward the blue end of the spectrum. The shift toward the red end of the spectrum indicates that the galaxies are moving away from Earth. Visit the following Web site to see the absorption lines of elements in the Periodic Table: *http://jersey.uoregon.edu/vlab/elements/Elements.html*

You can experience a similar change with sound. If you stand next to an automobile racetrack, the high-pitched sound of the approaching car changes to a lower pitch as the car speeds past you. Pitch is related to **frequency**, which is a measure of how many waves pass a given point in a given period of time. This apparent change in frequency and wavelength of energy that occurs when the source of a wave is moving relative to an observer is called the **Doppler effect**. It was named for Christian Johann Doppler, the scientist who explained it in 1842. The change in the frequency and wavelength of sound waves is similar to the changes that Hubble observed with light. The motion of a star away from us at a significant fraction of the speed of light extends the electromagnetic waves. The extended (stretched) waves make the star look redder than if it were not receding. Figure 28-9 shows how the motion of a source toward or away from an observer affects the wavelength. The greater the redshift, the faster the object is moving away. Astronomers have found that the most distant galaxies are moving away the fastest. Visit the following Web site to watch animated simulations of the Doppler effect on sound: *http://www.colorado.edu/physics/2000/applets_New.html*

Two other factors supported Hubble's hypothesis of an expanding universe. The redshift of light of distant galaxies is observed in all directions. In addition, the dimmer galaxies, which are

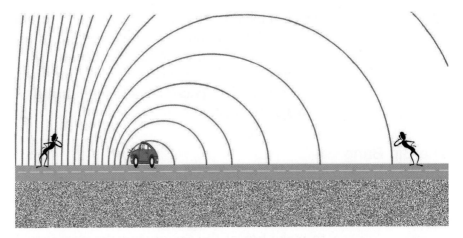

FIGURE 28-9. The Doppler effect is a difference in wave frequency depending upon whether the source is moving toward or away from the observer. This occurs with both sound and light waves. Distant galaxies appear more red to us because the light waves are spread out by the motion of these galaxies away from us.

thought to be dim because they are farther from Earth, showed greater redshift. He also reasoned that this kind of motion is a characteristic of an explosion. You might think that the motion of distant galaxies away from Earth in all directions means that we are at the center of the expansion. However, from every position within the expanding matter of an explosion, matter is moving away in all directions.

STUDENT ACTIVITY 28-4 —A MODEL OF THE BIG BANG

6: MODELS 2

> Inflate a round balloon to the size of a tennis ball. Draw several small dots on the balloon's surface. Notice that as the balloon is inflated more, the dots always move apart. If observers were located anywhere on the surface of the balloon or even inside the balloon, as the balloon is inflated, they would see the dots moving away in all directions. No matter what location is chosen, it would appear that the observer is at the center of the expansion. Therefore, there is no way that astronomers can find the center of the universe.

In the 1960s, Arno Penzias and Robert Wilson were working on long-distance radio communications for the Bell Telephone Company. They built a special outdoor receiver to detect weak radio signals. However, the device picked up annoying radio signals that they could not eliminate. They investigated the source of these radio signals and found that the energy they were picking up was billions of years old. They were listening to the origin of the universe. These radio signals, the **cosmic background radiation**, are weak electromagnetic radiation left over from the formation of the universe.

The Big Bang

The outward motions of distant galaxies and the cosmic background radiation are evidence that the universe began as an event now called the **big bang**. The name was first proposed as a joke to make fun of the theory, but the name stuck. This theory proposes that at its origin, the universe was a concentration of matter so dense that the laws of nature as we know them today did not apply. This matter expanded explosively, forming the universe. Even the most extreme conditions that exist within the largest stars do not compare with the beginning of the universe.

Experiments and the theories of physics confirm that the speed of light is the greatest velocity possible for matter or energy. The speed of light is about 300 million m/s. Like the temperature of absolute zero (0 K), the speed of light is one of the absolute limits known to science. Astronomers reason that the universe is expanding at this rate.

Using the speed of light, astronomers work back to the time of the big bang. They estimate that the universe began with the big bang 14 billion years ago. Expanding outward in all directions, the universe today could be as much as 28 billion light-years across. This is so gigantic that it is nearly impossible for humans to comprehend how large this is. Earth, the solar system, and even the huge Milky Way Galaxy are incredibly small compared with the size of the universe. Visit the following Web site and take A Cosmic Journey: A History of Scientific Cosmology and click on The Expanding Universe: *http://www.aip.org/history/cosmology/*

WHAT IS THE FUTURE OF THE UNIVERSE?

4: 1.2a
6: PATTERNS OF
CHANGE 5

From our point of view, Earth, as a planet, is very stable. Earth's orbit around the sun is unchanging. Scientists expect the sun to continue its energy production on the main sequence for billions of years. Collisions with large objects from space, such as those thought to mark the ends of past geologic eras, are possible. Fortunately, these events are becoming less likely as the solar system ages. However, the very-long-term future of the universe is not clear.

The Future of the Universe

Scientists once thought the universe had three possible futures. Some scientists proposed that the expansion of the universe might be slowing due to gravity. However, it is possible that there is not enough gravity to stop the expansion. In this case, the universe would continue to expand without limit. Other scientists proposed that there may be enough gravity to just stop the expansion, leading to a steady state. If the universe has enough gravity to reverse its expanding phase, it could fall back together in a very distant event that some astronomers call the "big crunch."

A good way to understand this is to consider a baseball thrown straight up. Gravity brings the ball back to the ground. However, if you could propel the ball fast enough, it would continue upward into space and never return to Earth. Figure 28-10 illustrates the big bang followed by the big crunch. Visit the following Web site to

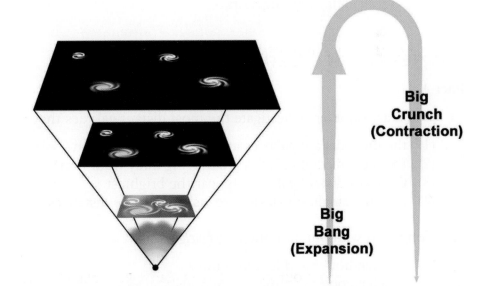

Big Crunch (Contraction)

Big Bang (Expansion)

FIGURE 28-10. Scientists know that the universe has expanded over the past 14 billion years. The expansion began with the "big bang." Will expansion stop and gravity pull the universe back into a black hole? Recent evidence seems to favor a model with continued growth and even accelerating expansion.

watch the video of Neil deGrasse Tyson discussing Death by Black Hole: *http://fora.tv/2008/02/19/Neil_DeGrasse_Tyson_Death_by_Black_Hole*

In recent years, astronomers have found evidence that the universe is not only expanding, but that it is expanding at an increasing rate. What force could work against the force of gravity to cause this? It is as if you threw a ball up into the air and it did not fall back to Earth. In fact, it is as if the ball flew upward faster and faster with time. This would be surprising, indeed. Astronomers find these observations just as surprising.

Astronomers have named the mysterious cause of this accelerating expansion "dark energy." However, they cannot explain it. Nor can they explain the source of gravitational force that holds the rapidly spinning galaxies from breaking apart. This force is attributed to the gravitational attraction of "dark matter," which astronomers think makes up about 90 percent of the matter in the universe. Dark matter and dark energy—the mysteries of science just keep coming.

The ultimate future of the universe depends upon the balance between the expansion of the big bang, gravity, and dark energy. To date, astronomers have not been able to determine which one will dominate. This remains one of many questions that guide scientific investigation.

CHAPTER REVIEW QUESTIONS

Part A

1. When astronomers investigate the most distant galaxies they see evidence that

   (1) the most distant galaxies are moving toward Earth
   (2) the most distant galaxies are moving away from Earth
   (3) the most distant galaxies appear the brightest
   (4) only the smallest galaxies are visible at these distances

2. Compared with our sun, the star *Betelgeuse* is

   (1) smaller, hotter and less luminous
   (2) smaller, cooler and more luminous
   (3) larger, hotter and less luminous
   (4) larger, cooler and more luminous

3. The symbols below are used to represent different parts or objects in space.

Universe = ☐   Earth = ○   Galaxy = ⟋⟋   Solar system = ○

Which diagram below best shows the relationship between our solar system, the universe, and our galaxy?

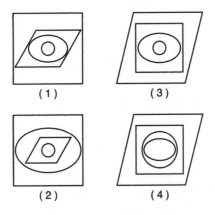

(1)   (3)

(2)   (4)

4. Astronomers have observed that light from a faint galaxy designated TN J0924-2201 has its spectral lines shifted toward the red end of the spectrum. This shift is evidence that the

   (1) galaxy contains only very hot stars
   (2) galaxy contains only relatively cool stars
   (3) galaxy is moving toward Earth
   (4) galaxy is moving away from Earth

5. Which list shows stars in order of increasing surface temperature?

   (1) Barnard's Star, Polaris, Sirius, Rigel
   (2) Aldebaran, sun, Rigel, Procyon B
   (3) Rigel, Polaris, Aldebaran, Barnard's Star
   (4) Procyon B, Alpha Centauri, Polaris, Betelgeuse

6. Most scientists believe that the Milky Way galaxy is

   (1) spherical in shape
   (2) 4.6 million years old
   (3) composed of stars the revolve around Earth
   (4) one of billions of galaxies in the universe

7. Which star gives off the *least* light?

(1) Deneb
(2) Proxima Centauri
(3) the sun
(4) Sirius

8. Approximately how much brighter than the star Aldebaran is the star Betelgeuse?

(1) Betelgeuse is approximately twice as bright as Aldebaran.
(2) Betelgeuse is approximately 10 times as bright as Aldebaran.
(3) Betelgeuse is approximately 100 times as bright as Aldebaran.
(4) Betelgeuse is approximately 10,000 times as bright as Aldebaran.

Part B

Base your answers to questions 9 through 12 on the table below, which contains inferred information about the evolution of the universe.

Data Table

Stage	Description of the Universe	Average Temperature of the Universe (°C)	Time From the Beginning of the Universe
1	The size of an atom	?	0 s
2	The size of a grapefruit	?	10^{-43} s
3	"Hot soup" of electrons	10^{27}	10^{-32} s
4	Cooling allows protons and neutrons to form	10^{13}	10^{-6} s
5	Still too hot to allow atoms to form	10^{8}	3 minutes
6	Electrons combine with protons and neutrons, forming hydrogen and helium atoms; light emission begins	10,000	300,000 years
7	Hydrogen and helium form giant clouds (nebulae) that will become galaxies; first stars form	−200	1 billion years
8	Galaxy clusters form and first stars die; heavy elements are thrown into space, forming new stars and planets.	−270	13.7 billion years

9. How soon did protons and neutrons form after the beginning of the universe?

   (1) 0.000000006 s
   (2) 0.000006 s
   (3) 0.6 s
   (4) −6 s

10. What is the most appropriate title for this data table?

    (1) Evidence of the Law of Superposition
    (2) Biological Evolution of Life-forms
    (3) History According to the Big Bang Theory
    (4) Progressive Shrinking of the Universe

11. According to the table, which statement best describes the changes in the temperature of the universe?

    (1) The temperature of the universe is known to have decreased continuously.
    (2) It seems likely that the temperature of the universe has decreased since an age of 10^{-32} s.
    (3) The temperature of the universe is known to have increased continuously.
    (4) It seems likely that the temperature of the universe has increased since an age of 10^{-32} s.

12. Approximately how old was the universe when planets formed?

    (1) 10^{-32} s
    (2) 13,700,000 s
    (3) 13,700,000 years
    (4) 13,700,000,000 years

13. The diagram below represents the spectral lines of hydrogen observed in a laboratory experiment and hydrogen lines observed from three stars.

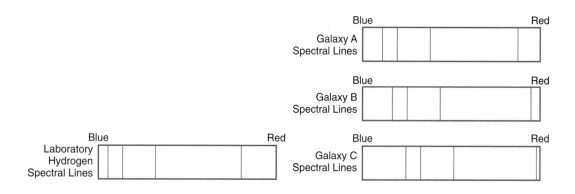

What is the best inference that can be made concerning the movement of galaxies *A, B,* and *C*?

(1) Galaxy *A* is moving away from Earth, but galaxies *B* and *C* are moving toward Earth.
(2) Galaxy *B* is moving away from Earth, but galaxies *A* and *C* are moving toward Earth.
(3) Galaxies *A, B,* and *C* are all moving toward Earth.
(4) Galaxies *A, B,* and *C* are all moving away from Earth.

Part C

Base your answers to questions 14 and 15 on the passage below and on your knowledge of Earth science.

The Future of the Sun

Hydrogen gas is the main source of fuel that powers the nuclear reactions that occur within the sun. But like any fuel, the sun's hydrogen is in limited supply. As the hydrogen gas is used up, scientists predict that the helium created as an end product of earlier nuclear reactions will begin to fuel different nuclear reactions. When this happens, the sun is expected to become a red giant star with a radius that would extend out past the orbit of Venus, and possibly as far as Earth. Conditions on Earth would certainly become impossible for humans and other life-forms. No need to worry at this time. This event is not expected to happen for at least a few billion years.

14. What specific nuclear reaction converts hydrogen into helium within the sun?

15. According to this paragraph, what is the range of estimates for the solar radius when it becomes a red giant? (Use kilometers as your units of measure.)

16. Explain how a red star such as Aldebaran can give off more light than the sun, which is much hotter than Aldebaran.

Base your answer to question 17 on the incomplete table below.

Star Classification

Star	Color	Classification
Sun	Yellow	Main sequence
Procyon B		

17. What is the color of the star Procyon B, and to what group of stars does it belong?

18. What do x-rays, infrared radiation, and radio waves have in common?

19. What force could slow the expansion of the universe?

20. What is the surface temperature of the North Star?

Appendices

APPENDIX A: LABORATORY SAFETY

Laboratory work is an important part of science. However, many materials, equipment, and procedures that scientists use are potentially hazardous. Working in the science lab requires more attention to safety than do most other activities in the school setting. Work and conduct yourself with care in the laboratory. You should be familiar with the rules below, and be alert to any hazards.

1. Follow all safety precautions as directed by the teacher and stated in printed directions. If you do not understand the directions, ask your teacher to clearly explain them to you.

2. Notify the teacher at once if there is an accident or dangerous situation such as an electrical problem; an injury; or a spill of water, chemicals, or other substances. The teacher will supervise the cleanup of spilled chemicals, broken glass, or broken equipment.

3. Never perform a laboratory procedure unless the teacher approves it.

4. Touch equipment and materials only as directed. If you do not understand how to perform a procedure, ask for help.

5. Wear laboratory safety goggles whenever you mix chemicals, heat liquids, perform procedures that could result in flying particles, or perform any tasks that could injure your eyes. When performing these procedures, everyone at the lab station must wear safety goggles.

6. When using a flame or a hot plate, keep combustible materials including paper, clothing, jewelry, and hair safely away from the heat source. If you are heating something in a container, be sure the container has an opening and point the open end away from people.

7. Hot plates, burners, glassware, and other materials remain hot for a long time. Use special caution during the cooling of these objects. Do not hesitate to ask for help, or for tongs or other special materials to handle these objects.

8. The use of acids, bases, and many other laboratory chemicals requires special precautions. You should handle these chemicals only as instructed, and their use must be supervised by your teacher.

9. Use sharp or pointed objects such as scissors, knives, and pins in such a way as to avoid puncture injuries.

10. Dispose of materials as directed. Do not dispose of solids in sinks or put hazardous materials into wastebaskets.

11. Keep the work area neat and clean. At the end of the procedure, return equipment and materials as you are instructed. Unplug electrical devices and turn off gas and water unless instructed otherwise by your teacher.

12. Know the location of and how to use safety equipment and materials such as running water and fire-retarding devices. Also know how to notify another adult if the teacher is unable to respond to events.

APPENDIX B: A FORMAT FOR LABORATORY REPORTS

This lab report format is presented as a general guide to help students organize lab activities and written reports. Reports for laboratory activities are required for admission to the Regents examination. For some lab activities, your teacher may present labs in a written guided format. At other times, you may not be given special written instructions about how to conduct and report on a particular lab. Or, you may be asked to completely plan and conduct your own procedure. For those cases, this format is offered to help you organize, conduct, and report on your laboratory activity. Be sure you understand whether each student must submit a report or if one report for the whole lab group is sufficient. Different lab groups should not work together or produce virtually identical lab reports. In general, it is best to keep lab reports neat, and brief, but complete.

Open-Ended Laboratory Report Format

Title: The name of the lab. This may be a brief indication of what is being investigated.

Date: The date the lab procedure was performed.

Names: List all members of the lab group.

Objective(s): A clear statement of what your lab procedure is intended to explore.

Materials: Make a list of all the special equipment and supplies needed to conduct the laboratory procedure.

Procedure: In a sequence of numbered steps, tell how this lab was performed. These steps should be clear enough that a person who is not familiar with the activity can follow them and obtain good results. Please include any safety precautions or other warnings that may be necessary.

Observations/Data/Graphs: List the outcome of your procedure in a neat and organized format. This will include the data you collect, any mathematical steps taken with the data, and graph(s) to clarify your findings.

Conclusions or Discussion: The conclusion should relate your observations and results to the objectives at the beginning of the lab report.

APPENDIX C: THE INTERNATIONAL SYSTEM OF UNITS (S.I.)

S.I. units were based on European measures commonly known as the metric system. They are used by the people of every major nation of the world except the United States. Scientific publications usually use S.I. or metric units. Most S.I. units are related to larger and smaller units by a factor of 10. The system of measures most commonly used in the United States is sometimes known as United States customary units.

Basic Units

Basic units (meter, liter, gram) are not composed of other units. Derived units are a combination of basic units. Examples of derived units include volume, which can be measured in meters cubed (distance³); speed, which is measured in meters per second (distance/ time); and density, which is measured in grams per cubic centimeter (mass/volume).

Quantity	S.I. Unit	Symbol	Approximate U.S. Equivalent
length	meter	m	39.37 inches (about 3 inches longer than 1 yard)
mass	kilogram	kg	2.204 pounds
time	second	s	(Identical units)
temperature	kelvin*	K	(Note the conversion chart to degrees Fahrenheit or Celsius in the *Earth Science Reference Tables*)

*°Celsius is often used in metric measures. Kelvins should be expressed without a degree (°) sign.

S.I. and Metric Prefixes with Examples

Prefix	Symbol	Multiple	Example (length)
kilo	k	1000	kilometer (1 km = 1000 m)
centi	c	1/100	centimeter (1 cm = 0.01 m)
milli	m	1/1000	millimeter (1 mm = 0.001 m)

Prefixes for other multiples of ten are sometimes used, but the three shown above are the most common prefixes.

Conversions Between S.I. (Metric) and U.S. Customary Units

These conversion factors are limited to a convenient number of decimal places.

Length

1 centimeter = 0.394 inch   (1 inch = 2.54 centimeters)
1 meter = 39.37 inches   (1 foot = 0.341 meter)
1 kilometer = 0.62 mile   (1 mile = 1.61 kilometers)

Volume and Capacity

1 liter = 1.06 quarts   (1 quart = 0.95 liter)
1 milliliter = 0.035 ounce   (1 ounce = 28.4 milliliters)

The cubic centimeter is sometimes used as the equivalent of the milliliter.

Mass
1 gram = 0.035 ounce   (1 ounce = 28.4 grams)
1 kilogram = 2.204 pounds   (1 pound = 0.4536 kg = 453.6 grams)

APPENDIX D: PHYSICAL CONSTANTS

Speed of light: 300,000,000 m/s (186,000 mi/s)
[1 light-year is about 6 trillion
(6×10^{12}) mi or 10 trillion km]
Lowest possible temperature: 0 K ($-273°$C or $-460°$F)
Density of water: 1 g/mL (at 3.98°C)
Mass of hydrogen atom: 1.67×10^{-27} kg

APPENDIX E: GRAPHS IN SCIENCE

It is often easier to understand data when it is presented in the form of a simple graph. Making and reading graphs are important skills to scientists and others. There are many kinds of graphs, but the two-dimensional line graph, bar graph, and pie graph are the most commonly used forms. A line graph is used to show continuous change. A bar graphs and pie graphs shows unconnected data values.
    Chapter 1 (pages 2 to 26) has a section on making graphs.
    If you need to make a graph, follow these rules to get the best results.

1. Remember that the purpose of a graph is to make data more meaningful. Keep the graph simple and neat, but include all necessary information.

2. Label the two axes with the quantity (such as mass, temperature, or whatever you are graphing) and the appropriate units (such as grams or °C).

3. Label the axes with numbers that are appropriate for the range of data. The data (line or bars) should extend at least half way along each of the two axes. Graphs that simply represent general relationships may be shown without including numbers and units on the axes. It should be understood that values on graphs usually increase to the right along the *x* axis and upward along the *y* axis.

4. Usually, the independent variable is graphed along the bottom, or *x* axis. For example, if your graph shows the amount of rock weathering that occurs through time, time should increase along the bottom axis. Because time influences the amount of weathering, rather than weathering affecting the passage of time, time is considered the independent variable. Values of the independent variable are usually selected to show how the dependent variable changes.

5. When the data has been plotted, a clear pattern may become visible. Sometimes the data points form a line that can be drawn with a straight edge. In other graphs, a smooth curve better fits the data. Draw the graph line accordingly.

6. Logically, some graph lines pass through the origin (0, 0), and others do not. Consider what you are graphing and draw the graph line accordingly.

7. All measurements include some error. Error can be made very small, but error cannot be eliminated. For this reason, it is possible that the most appropriate line will not pass through the center of each data point. It may be a straight line or a gentle curve that clearly shows the trend and estimated values of the data. This is sometimes called a "best fit" line.

8. A title, usually shown near the top of the graph, helps to identify the purpose of the graph. For example, a graph titled "Average Monthly Temperatures at Buffalo, New York" makes the meaning of the graph clear.

9. The guidelines above are not always followed. Graphs may violate one or more of these rules. As long as the graph makes the data more meaningful, it has fulfilled its principal function.

APPENDIX F: EARTH SCIENCE REFERENCE TABLES

In recent years between a third and a half of the questions on the Regents exam have involved the use of the Earth Science Reference Tables. *Furthermore, that document has many words you will need to understand. You can download and print the 2010 edition of the* Earth Science Reference Tables *at http://www.emsc.nysed. gov/osa/reftable/earthscience-rt/esrt2010-engw.pdf*

If you have your own copy of the Reference Tables, *you can use a marker to highlight the terms and tables you do not understand. Ask your teacher to explain them to you.*

Glossary

This glossary can be used as a study aid. The terms that follow are helpful in learning Regents Earth science. Please be aware that the primary importance of these terms is their applications to understanding Earth systems. If you are not familiar with some of these terms, you may wish to use the index of this book to find them within their scientific context. Definitions alone are of limited use.

The entries for the terms are intended to help you understand how the words are applied in this specific course. Broader and more precise definitions can be found in textbooks that are more advanced and other reference materials.

The abbreviation ESRT is used to highlight terms used and sometimes clarified in the *Earth Science Reference Tables*. Many important terms that are not listed here, such as specific rock types and names of geologic ages, may be understood using charts in the *Reference Tables*.

abrasion: The grinding away of rock by friction with other rocks

absolute age: An age expressed as a specific amount of time, absolute age always includes a unit of time; numerical age (ESRT)

absolute humidity: The mass of water vapor in each cubic unit of air

acid precipitation: Precipitation (snow or rain) with corrosive (low pH) chemical properties, generally the result of pollution from the burning of fossil fuels

agents of erosion: Moving water, wind, or ice that causes the transport of weathered materials

air mass: A large body of air that is relatively uniform in temperature and humidity (ESRT)

altitude: The angular elevation of an object above the horizon

angle of insolation: The angle between Earth's surface and incoming rays of sunlight; angle of the sun above the horizon

Antarctic Circle: The latitude (66.5°S) south of which the sun does not rise on the Southern Hemisphere's winter solstice; the latitude (66.5°S) south of which the sun is in the sky for 24 h on the Southern Hemisphere's summer solstice (ESRT)

anticyclone: A region of relatively high atmospheric pressure

aquifer: An underground zone of porous material that contains useful quantities of groundwater

arctic air mass: A large body of very cold air that originated in the Arctic (ESRT)

Arctic Circle: The latitude (66.5°N) north of which the sun does not rise on the Northern Hemisphere's winter solstice; the latitude (66.5°N) north of which the sun is in the sky for 24 h on the Northern Hemisphere's summer solstice (ESRT)

arid climate: A climate that has little rain and low humidity

asteroid: An irregularly shaped rocky mass that is smaller than a planet and occupies an orbit around the sun; most are found between the orbits of Mars and Jupiter

asthenosphere: The upper part of the mantle, capable of slow deformation and flow under heat and pressure (ESRT)

astronomy: The study of Earth's motions and the objects beyond Earth, such as planets and stars

atmosphere: The layer of gases that surrounds a celestial body (ESRT)

avalanche: a large amount of snow and rock that moves rapidly downhill over a steep slope

axis: An imaginary line that passes through Earth's North and South poles

azimuth: The compass direction specified as an angle. Azimuth starts at 0° at due north and progresses through east (90°), south (180°), west (270°), and back to north (360°, or 0°).

banding: The light- and dark-colored layers of mineral that form parallel to foliation in metamorphic rocks (ESRT)

barometer: An instrument used to measure air pressure

barrier islands: Offshore features, similar to sandbars, that rise above sea level

bed load: The sediments that roll or bounce along the bottom of a stream

bedrock: The solid, or continuous, rock that extends into Earth's interior

big bang: The theory that the universe formed as a concentration of matter expanded explosively

bioclastic sedimentary rocks: Rocks composed of materials made from or by living organisms (ESRT)

biological activity: The actions of plants and animals that cause weathering

blizzard: A winter snowstorm that produces heavy snow and winds of 35 mph (56 km/h) or greater

boiling: The change in state from liquid to gas (vapor) at the boiling temperature

caldera: A large, bowl-shaped depression formed when the top of a volcano collapses into the emptied magma chamber

capillarity: The tendency of a substance to pull water into tiny spaces, or pores, by adhesion

celestial objects: Objects in the sky that are beyond Earth's atmosphere

chemical change: A change, such as rusting, that results in the formation of a new substance

chemical weathering: A natural process that occurs under conditions at Earth's surface, forming new compounds

classification: The organization of objects, ideas, or information according to their properties

clastic: Sedimentary rocks that are composed of the weathered remains of other rocks; fragmental (ESRT)

cleavage: The tendency of some minerals to break along smooth, flat surfaces (ESRT)

climate: The average weather conditions over a long time, including the range of conditions

cloud: A large body of tiny water droplets or ice crystals suspended in the atmosphere

cloud-base altitude: The height at which rising air begins to form clouds

comet: An object made of ice and rock fragments that revolves around the sun usually in a highly eccentric orbit; it may be visible periodically in the night sky as a small spot of light with a long tail

compounds: Substances made up of more than one kind of atom (element) combined into larger units called molecules

condensation: The process by which a substance changes from a gas to a liquid (ESRT)

condensation nuclei: Tiny particles of solids suspended in the air on which water condenses to form clouds

conduction: The movement of heat that occurs as heated molecules pass their vibrational energy to nearby molecules

conservation: The careful use, protection, and restoration of our natural resources

contact metamorphism: The process in which an intrusion of hot, molten magma causes changes in the rock close to it (ESRT)

continental air mass: A large body of air that has relatively low humidity because it originated over land (ESRT)

continental climate: A climate characterized by large seasonal changes in temperature

continental glacier: A glacier that flows outward from a zone of accumulation to cover a large part of a continent

contour line: A line on a map that connects places having the same elevation (height above or below sea level)

convection: A form of heat flow that moves matter and energy as density currents under the influence of gravity (ESRT)

convection cell: The pattern of circulation that involves vertical and horizontal flow

convergence: The act of moving together (ESRT)

convergent plate boundary: A place where lithospheric plates collide (ESRT)

coordinate system: A grid in which each location has a unique designation defined by the intersection of two lines (ESRT)

Coriolis effect: The apparent curvature of the path of winds and ocean currents as they travel long distances over Earth's surface; caused by Earth's rotation

correlation: Matching bedrock layers by rock type or by age

cosmic background radiation: Weak electromagnetic radiation (radio waves) left over from the formation of the universe (big bang)

crater: A bowl-shaped depression at the top of a volcano caused by an explosive eruption or the impact of an object from space

crystalline sedimentary rocks: Sedimentary rocks that form by precipitation (ESRT)

cyclone: (1) A region of relatively low atmospheric pressure; (2) term applied to hurricanes in the Indian Ocean; (3) synonym for tornado

decay product: The stable, ending material of radioactive decay (ESRT)

decay-product ratio: A comparison of the amount of the original radioisotope with the amount of its decay product (ESRT)

deforestation: Cutting forests to clear the land for other uses

delta: A deposit of sediment built into a large body of water by deposition from a stream

density: The concentration of matter, or the mass per unit volume (ESRT)

deposition: The settling, or release, of sediments that have been carried by an agent of erosion (ESRT)

dew: Liquid water that forms by condensation on cold surfaces

dew point: The temperature at which air is saturated with water vapor (ESRT)

dew-point temperature: The temperature to which air must be cooled to become saturated with moisture (ESRT)

discharge: The amount of water flowing past a particular place in a specified time

divergence: The act of moving apart

divergent plate boundary: A place where lithospheric plates separate (ESRT)

Doppler effect: The apparent change in frequency and wavelength of energy radiated by a source as a result of the motion of the source or the observer

Doppler radar: A device that uses reflected radio waves to measure at a distance wind speed and direction

drainage divides: The high ridges, from which water drains in opposite directions, that separate one watershed from another

drainage pattern: The path of a stream, which is influenced by topography and geologic structures

drought: A long period of dry weather

drumlins: Streamlined hills of glacial origin aligned north-to-south that have steep sides, a blunt north slope, and a gentle slope to the south; made of till

dune: A hill or ridge of wind-blown sand

duration of insolation: The amount of time the sun is visible in the sky, or the number of hours between sunrise and sunset

dynamic equilibrium: The state of balance in which opposing processes take place at the same rate; a state of balance of events

Earth science: The science that applies the tools of the physical sciences to study Earth; including the solid Earth, its oceans, atmosphere, and core, and surroundings in space

earthquake: A sudden movement of Earth's crust that releases energy (ESRT)

eccentricity: A measure of the elongation of an ellipse (ESRT)

eclipse: (See *lunar eclipse* and *solar eclipse*.)

ecology: The branch of science that is concerned with the relationships between organisms and their environment

El Niño: The periodic replacement of upwelling cold water by warm water along the western coast of South America

elements: The basic substances that are the building blocks of matter (ESRT)

ellipse: A closed curve formed around two fixed points such that the total distance from any point on the curve to both fixed points is constant

epicenter: The place on Earth's surface directly above an earthquake's focus (ESRT)

equator: An imaginary line that circles Earth halfway between the North and South Poles (ESRT)

equilibrium: A state of balance

equinox: One of the two days each year on which the sun rises due east and sets due west, on which the length of day and night are equal, on which the sun's vertical rays are at the equator; the first day of spring or fall

erosion: The transportation of sediments by water, air, glaciers, or by gravity acting alone. (See also *agents of erosion*.) (ESRT)

erratics: Large rocks transported from one area to another by glaciers

escarpment: A steep slope or a cliff of resistant rock that marks the edge of a relatively flat area

evaporation: The change in state from liquid to gas when the temperature is below the boiling point

evolution: The gradual change in living organisms from generation to generation, over a long period of time

exfoliation: A type of physical weathering caused by expansion that breaks rock into large curved slabs

exponential notation: A method of expressing very large and very small numbers using powers of 10

extinction: The death and disappearance of every individual of a particular species (ESRT)

extrusion: The movement of magma onto Earth's surface (ESRT)

faults: Cracks in Earth's crust along which movement occurs

felsic: Describes light-colored minerals rich in aluminum or rocks made of these minerals; felsic rocks are rich in feldspar and quartz (ESRT)

field: A region in which a force, temperature, land elevation, or another quantity can be measured at any location (ESRT)

floodplain: A flat region next to a stream or river that can be covered by water in times of flood

flotation: The method by which particles that are too large to be carried in solution or by suspension float on water

fluid: Any substance that can flow, usually a liquid or a gas

focus: (1) The place where rock begins to separate during an earthquake, usually located underground. (2) Either of the two fixed points that determine the shape of an ellipse (ESRT)

fog: Very low clouds that reach the ground (ESRT)

foliation: The alignment of mineral crystals, caused by metamorphism (ESRT)

fossils: A record of prehistoric life preserved in rock (ESRT)

fracture: The way minerals break along curved surfaces (ESRT)

fragmental: Describes sedimentary rocks that are composed of the weathered remains of other rocks; clastic (ESRT)

freezing: The change in state from liquid to solid

freezing rain: Rain that freezes on contact with Earth's surface

frequency: A measure of how many waves pass a given point in a given period of time

front: A boundary, or interface, between air masses (ESRT)

frost: Ice crystals that form when water vapor comes in contact with surfaces whose temperature is below 0°C

frost wedging: A form of physical weathering caused by repeated freezing and thawing of water within cracks in rocks

galaxy: A huge group of stars held together by gravity

geocentric: A model of the universe that assumed that Earth is stationary and at the center of the universe

geologists: Scientists who study the materials, origin, history, and structure of Earth and how it changes

geology: The study of the rock portion of Earth, its interior, and surface processes

geosphere: The mass of solid and molten rock that extends more then 6000 km from Earth's solid surface to its center

glacier: A large mass ice that flows over land due to gravity

global warming: A long-term increase in the average temperature of Earth's atmosphere, it is the result of the increased concentration of carbon dioxide and other greenhouse gases in the atmosphere

graded bedding: Within a layer of sediment, the gradual change in sediment size from bottom (large) to top (small) showing the order in which particles settled; vertical sorting

gradient: The change in any field value per unit distance (ESRT)

gravity: The force of attraction between all objects

greenhouse effect: The process by which carbon dioxide and water vapor absorb heat radiation, increasing the temperature of Earth's atmosphere

Greenwich Mean Time: The basis of standard time throughout the world; based on measurements of the position of the sun in Greenwich, England

grooves: Furrows of glacial origin in bedrock that are deeper and wider than striations

groundwater: Water that enters the ground and occupies free space in soil and sediment as well as openings in bedrock, including cracks and spaces between grains

hail: Pellets of ice, which grow larger as they repeatedly become coated with water, and are then blown higher into cold air where the coating of water freezes; eventually the ice pellets become heavy enough to fall to the ground. (Hail is most common during thunderstorms.) (ESRT)

half-life period: The time it takes for half of the atoms in a sample of radioactive element to decay (ESRT)

hardness: The resistance of a mineral to being scratched (ESRT)

hazard: Unsafe conditions that pose a threat of property damage, injury, or loss of life

heliocentric: A model of the universe that places the sun at the center of the solar system

high-pressure system: A place where cool, dry air sinks lower into the atmosphere

horizontal sorting: A decrease in the size of sediment particles with distance from the shore, produced as a stream enters calm water

hot spot: A long-lived source of magma within the asthenosphere and below the moving lithospheric plates (ESRT)

humidity: The water-vapor content of air (ESRT)

hurricane: A large storm of tropical origin that has sustained winds in excess of 74 mph (120 km/h) (ESRT)

hydrologic cycle: A model that represents water movement and storage within Earth, on the surface, and within the atmosphere

hydrosphere: Earth's liquid water, including oceans, surface water, and groundwater

hygrometer: An instrument used to measure atmospheric humidity

igneous rocks: Rocks that form by the solidification of melted rock (ESRT)

inclusion: A fragment of one type of rock that is enclosed in another rock

index fossils: Fossils used to establish the age of rocks; they must be easy to recognize, found over a large geographic area, and have existed for a brief period of geologic time (ESRT)

inertia: The tendency of an object at rest to remain at rest or an object in motion to move at a constant speed in a straight line unless acted on by an unbalanced force

inference: A conclusion based on observations

infiltration: The process in which water soaks into the ground

insolation: Solar energy that reaches Earth (*in*coming *solar* radi*ation*)

intrusion: The movement of magma to a new position within Earth's crust. A body of rock that was injected into surrounding rock as magma

island arc: A curved line of volcanic islands that are the result of partial melting of a tectonic plate where it descends beneath another oceanic plate

isobars: Isolines (see also) that connect locations with the same atmospheric pressure on a weather map

isoline: A line on a field map that connects places having the same field quantity value

isotherm: A line on a field map that connects places having the same temperature

isotopes: Atoms of the same element that contain different numbers of neutrons in their nucleus (ESRT)

jet streams: Wandering currents of air far above Earth's surface that influence the path of weather systems (ESRT)

joule (J): Unit used to measure energy (ESRT)

Jovian planet: A planet whose composition is similar to Jupiter's; also know as a gas giant (ESRT)

kettle: A small closed basin formed in a moraine

lake-effect storms: Precipitation events that occur downwind from large lakes as the result of moisture that enters the air over the lake; especially common as early winter snow events

land breezes: Light winds that blow from the land to the water; they usually develop at night as the air over the land becomes cooler than the air over the water

landform: A feature of a landscape

landscape: A region that has landforms that are related by similarities in shape, climate, and/or geologic setting; the general shape of a large area of the land surface, such as plains, plateau, or mountain (ESRT)

landslide: The rapid, downslope movement of rock and soil

latent heat: Energy absorbed or released when matter changes state (ESRT)

latitude: The angular distance north or south of the equator (ESRT)

lava: Melted rock coming from a volcano, or such rock that has cooled and hardened

levees: Banks along a river of natural or human origin

lightning: Sudden electrical discharges within clouds, between clouds, and between clouds and the ground that are seen as flashes of light

light-year: The distance electromagnetic energy can travel in 1 year, approximately 6 trillion mi (10 trillion km)

liquefaction: The process in which strong shaking allows water to surround the particles of sediment, changing the sediments into a material with the properties of a thick fluid

lithosphere: The outermost, relatively brittle layer of solid Earth, which includes the crust and upper mantle (ESRT)

lithospheric plate: A major section of Earth's outer solid shell that generally moves as a rigid unit

logarithmic: A scale in which an increase of one unit translates to a 10-fold increase in the quantity measured

longitude: The angular distance east or west of the prime meridian (ESRT)

longshore transport: The motion of sediment parallel to the shore caused by waves

low-pressure system: A region of atmospheric convergence where low-density air rises, a cyclone

luminosity: The total energy output of a star; absolute brightness (ESRT)

lunar eclipse: A short-term darkening of the moon caused by the moon orbiting through Earth's shadow [Occurs only during the full-moon phase.]

luster: The way light is reflected and/or absorbed by the surface of a mineral (ESRT)

mafic: Describes dark-colored minerals rich in magnesium (ESRT)

magma: Hot, liquid rock within Earth (ESRT)

major axis: The distance across an ellipse measured at it widest point

maritime air mass: A large body of air that has relatively high humidity because it originated over the ocean or other large body of water (ESRT)

maritime climate: A humid climate that occurs over the oceans and in coastal locations

mass movement: The motion of soil or rock down a slope without the influence of running water, wind, or glaciers

meander: A curve that develops in the path of a river when the river flows over relatively flat land

mechanical weathering: The breaking up of rock into smaller particles without a change in composition; physical weathering

melting: The change in state from solid to liquid (ESRT)

Mercalli scale: A scale for measuring earthquake intensity based on the reports of people who felt the quake and observed the damage it caused

mesosphere: The layer of Earth's atmosphere directly above the stratosphere, in which temperature decreases with increasing altitude (ESRT)

metamorphic rocks: Rocks that form as a result of heat and/or pressure on other rocks causing chemical (mineral) or physical changes (ESRT)

meteor: A streak of light produced as a meteoroid burns due to friction with Earth's atmosphere

meteorite: a meteoroid (see also) that strikes Earth's surface

meteoroid: A piece of rock that moves through space

meteorologist: A scientist who studies the weather

meteorology: The study of Earth's atmosphere and how it changes

mid-latitude cyclone: An area of low pressure or a storm system, such as those that usually move eastward across the United States

mid-ocean ridges: A system of underwater mountain ranges that circles Earth like the seams on a baseball (ESRT)

Milky Way galaxy: The group of billions of stars that includes the sun and our solar system, it is visible as a faint band of light across the night sky

mineral: A natural inorganic, crystalline solid that has a specific range of composition and consistent physical properties (ESRT)

model: Anything that is used to represent something else

Moho: The boundary between Earth's crust and mantle (ESRT)

Mohs scale: A special scale of hardness used to identify minerals (ESRT)

monsoons: Seasonal changes in the direction of the prevailing winds, causing changes in temperature and rainfall

moraine: A deposit of unsorted glacial sediment (till) pushed into place by an advancing glacier

mountain landscape: A rugged landscape that has great relief from the top of the highest peaks to deep valleys, commonly underlain by resistant rock types and distorted structures including folds and faults

natural resources: Any material from the environment that is used by people

neap tides: The smallest tidal range, which occurs when the sun and moon are at right angles as observed from Earth

nonrenewable resources: Resources that exist in a fixed amount or for which the rate of regeneration is so slow that use of these resources will decrease their availability

nuclear fusion: The process by which the nuclei of light elements, such as hydrogen, under intense heat and pressure form the nuclei of heavier elements, such as helium

oblate: Slightly flattened at the poles

observations: Information gathered through the use of sight, touch, taste, smell, and hearing

ocean current: Flow of ocean water, usually horizontally, that transports energy and biological nutrients, and can influence the climate of nearby land areas

ocean trench: A deep-ocean location where old lithosphere moves back into Earth's interior; also called a subduction zone or a convergent plate boundary (see also) (ESRT)

oceanography: The study of the oceans that cover most of Earth

ores: Rocks that are mined to obtain a substance they contain of economic value

organic matter: The remains of living organisms in which plants can grow

origin: How something was formed

origin time: The time at which a fault shifted to produce an earthquake (ESRT)

original horizontality: The principle that no matter the present angle or orientation of sedimentary rock layers, the layers were originally horizontal and were tilted after deposition

outcrop: A place where bedrock is exposed at Earth's surface

outgassing: The process in which bubbles of hot gas escape from magma exposed to reduced pressure near Earth's surface

outwash: Sorted sediments deposited by water from a melting glacier

overland flow: The water from precipitation that flows downhill under the influence of gravity until it reaches a stream or seeps into the ground; runoff

paleontology: The study of fossils

paradigm: A coherent set of principles and understandings

percent error: A comparison of the size of an error with the size of the value being measured (ESRT)

permeability: The ability of soil or sediment to allow water to flow through it

phase: The observed shape of the lighted portion of a celestial object, for example, the moon or Venus

phases of matter: The states of matter: solid, liquid, and gas

physical weathering: The breaking up of rock into smaller particles without a change in composition; mechanical weathering (see also)

plains: Relatively flat landscapes, commonly at low elevation and usually underlain by flat-lying sedimentary rocks; the range of elevation is small (ESRT)

plastic: A material that is solid under short-term stress, but flows like a liquid when stress is applied over a long period of time

plate tectonics: A theory of crustal movements that combines sea-floor spreading with continental drift (ESRT)

plateau: A rolling landscape or elevated, comparatively flat region with modest topographic relief (ESRT)

plutonic: Describes igneous rocks that form deep underground (ESRT)

polar air mass: A large body of cold air that originated near one of Earth's poles (ESRT)

polar front: The boundary between two great convection cells; the most common path of the upper atmosphere polar jet stream. (ESRT)

polarity: The direction of a magnetic field determined with an instrument such as a magnetic compass

pollution: A sufficient quantity of any material or form of energy in the environment that harms humans or the plants and animals on which they depend

porosity: The ability of a material to hold water in open spaces, or pores

precipitation: (1) The settling of solids from solution, often the result of the evaporation of seawater (ESRT) (2) Water that falls to Earth as rain, show, sleet, or hail (ESRT)

prevailing winds: The most common wind direction and speed at a particular location and time of year (ESRT)

primary waves (P-waves): Longitudinal earthquake waves that cause the ground to vibrate forward and back along the direction of travel; the earthquake waves that travel the fastest; P-waves can travel through solids, liquids, and gases (ESRT)

prime meridian: The north-south line through Greenwich, England, from which longitude is measured (ESRT)

profile: A cross section, of an object

psychrometer: An instrument, made up of two thermometers mounted side-by-side on a narrow frame, that is used to determine the dew-point temperature and relative humidity; also known as a wet- and dry-bulb thermometer (ESRT)

radar: A method or device that uses reflected radio waves to locate or map distant objects or weather events; an acronym from *ra*dio *d*etection *a*nd *r*anging

radiation: The transfer of energy in the form of electromagnetic waves

radioactive: Describes atoms that break down spontaneously, releasing energy and/or subatomic particles to become different elements (ESRT)

radioisotope: An unstable isotope that breaks down spontaneously at a predictable rate

radiometric dating: determining absolute age with radioactive isotopes

rain: Liquid precipitation that falls quickly; precipitation droplets larger than drizzle (ESRT)

rain showers: Short periods of rain (ESRT)

redshift: A displacement of the spectral lines of very distant stars and galaxies, an increase in the wavelength of starlight caused by rapid relative motion of the star away from the observer (see *Doppler effect*)

reflection: The process by which light bounces off a surface or material

refraction: The bending of light and other energy waves as they enter a substance of different density

regional metamorphism: The process in which a large mass of rock is changed by heat and pressure due to large-scale movement of Earth's crust (ESRT)

relative age: The age of one thing compared to the age of another

relative humidity: A comparison of the actual water-vapor content of the air with the maximum amount of water vapor the air can hold at a given temperature (ESRT)

relief: The difference in elevation from the highest point to the lowest point on the land surface in a specific region

renewable resources: Resources that can be replaced by natural processes at a rate that will not decrease their availability

residual soil: Soil that formed in place and remains there

Richter scale: A scale for measuring earthquake magnitude based on measurements from seismographs

rock: A substance that is or was a natural part of the solid Earth, or lithosphere (ESRT)

runoff: The water from precipitation that flows downhill under the influence of gravity until it reaches a stream, or seeps into the ground; runoff may also include stream flow

sandbar: A low ridge of sand deposited along the shore by currents

satellite: An object in space that revolves around another object as a result of gravity

saturated air: The condition in which air is holding as much moisture as it can at a particular temperature

scattering: The reflection of light in many different directions

science: A universal method of gathering, organizing, analyzing, and using information about the environment

scientific notation: (See *exponential notation*.)

sea breezes: Light winds that blow from the water to the land that usually develop in the late morning or afternoon when the land warms; they continue into the evening until the land cools

sea-floor spreading: The process in which new lithosphere is made at the mid-ocean ridges, and adds on to older material that moves away from the ridges on both sides

secondary waves (S-waves): Transverse earthquake waves that cause the ground to vibrate side-to-side, perpendicular to the direction of travel; S-waves travel through solids, but not liquids or gases (ESRT)

sediment: The loose material created by the weathering of rock (ESRT)

sedimentary rocks: Rocks that form as a result of the compression and cementing of weathered rock fragments or shells of once-living animals (ESRT)

seismic moment: A scale for measuring the magnitude of an earthquake based on the total energy released by the earthquake

seismograph: An instrument that measures the magnitude of earthquakes

seismologists: Scientists who study earthquakes

seismology: A science that deals with earthquakes

silicate: A mineral that contains silicon and oxygen

sleet: A form of precipitation that consists of rain drops that freeze before they reach the ground; also known as ice pellets. Unlike hail, sleet does not require violent winds aloft (ESRT)

smog: A mixture of fog and air pollution particles, especially smoke from the burning of fossil fuels

snow showers: Short periods of snowfall (ESRT)

soil: A mixture of weathered rock and the remains of living organisms in which plants can grow

soil horizons: The layers of a mature soil

solar eclipse: A short-term darkening of the sun caused by the moon passing in its orbit between Earth and the sun [Occurs only during the new (dark) moon phase]

solar noon: The time at which the sun reaches its highest point in the sky

solar time: Time based on observations of when the sun reach its highest point and crosses a north-south line through the sky

solution: The method by which dissolved solids are carried in water

sorting: The separation of particles of sediment as a result of differences in their shape, density, or size

source region: The location in which an air mass originated

species: A group of organisms so similar that they can breed to produce fertile offspring

specific heat: The energy needed to raise the temperature of 1 gram of a substance 1 Celsius degree (ESRT)

spring: a place where groundwater flows onto the surface of the ground

spring tides: The largest tidal range, which occurs when Earth, the sun, and the moon are in a line with one another (not related to Earth's seasons)

station model: A standard format used to display abbreviated weather data

strain: The elastic bending of rocks in response to stress

stratosphere: The layer of Earth's atmosphere directly above the troposphere, in which the temperature increases with increasing altitude (ESRT)

streak: The color of the powdered form of a mineral (ESRT)

stream: Flowing water, such as a brook, river, or even an ocean current

stream system: All the streams that drain a particular geographic area

stress: Force that tends to distort rock, resulting in slow bending or fracture

striations: Parallel scratches in bedrock that were made by rocks transported by glaciers

subduction zone: A region in which Earth's crust is destroyed as it is pulled down into the mantle (ESRT)

summer solstice: The name generally applied to the day of the year with the longest period of sunlight (For observers in the Northern Hemisphere, this occurs near June 21. The Northern Hemisphere summer solstice occurs when the vertical rays of the sun are at the Tropic of Cancer. In the Southern Hemisphere, the summer solstice occurs in December when the vertical rays of the sun are at the Tropic of Capricorn.)

superposition: The concept that, unless rock layers have been moved, each layer is older than the layer above it and younger than the layer below it

surf zone: An area on the shore that extends from where the waves' base touches the ocean bottom to the upper limit the waves reach on the beach

suspension: How small particles that settle very slowly are carried by water or how cloud particles stay in the sky

tectonics: Large-scale motions of Earth's crust that are responsible for uplift and mountain building (ESRT)

temperate climate: A moderate climate that has large seasonal changes in temperature

temperature: A measure of the average kinetic energy of the molecules in a substance (ESRT)

terminal moraine: Irregular, hilly deposits of till formed where a glacier stopped advancing and began to melt back

terrestrial coordinates: Coordinates based on Earth's system of latitude and longitude

terrestrial planet: A rocky planet whose composition is similar to Earth's (ESRT)

texture: The surface characteristics of a rock that are the result of size, shape, and arrangement of mineral grains (ESRT)

thermometer: An instrument used to measure temperature

thermosphere: The highest layer of Earth's atmosphere, located directly above the mesosphere, in which temperature rises with increasing altitude (ESRT)

thunderstorm: A rainstorm that produces thunder, lightning, strong winds, and sometimes hail (ESRT)

tidal range: The difference between the lowest water level and the highest water level

tides: The twice- (or once-) daily cycle of change in sea level caused by the gravitational influence of the moon and sun on Earth's oceans

till: Unsorted sediments deposited by a glacier

topographic map: An isoline map on which the isolines, called contour lines, connect places having the same elevation

topography: The shape of the land

tornado: A small, usually short-lived storm that has extremely high winds (ESRT)

transform boundary: A place where two lithospheric plates move past each other without creating new lithosphere or destroying old lithosphere (ESRT)

transpiration: The process by which plants release water vapor to the atmosphere, largely through pores in their leaves

transported soil: Soil that formed in one location and was moved to another location

travel time: The time between the breaking of the rocks that causes an earthquake and when the event is detected at a distant location (ESRT)

tributary: A stream that flows into a larger stream

Tropic of Cancer: The greatest latitude north of the equator reached by the sun's vertical ray; 23.5°N (ESRT)

Tropic of Capricorn: The greatest latitude south of the equator reached by the sun's vertical ray; 23.5°S (ESRT)

tropical air mass: A large body of warm air that originated close to the equator (ESRT)

troposphere: The lowest layer of Earth's atmosphere, in which temperature decreases with increasing altitude (ESRT)

tsunami: A series of waves caused by an earthquake or underwater landslide that can cause damage and loss of lives in coastal locations

unconformity: A buried erosion surface that represents a gap in the record of Earth's history

uniformitarianism: The concept that the geological processes that took place in the past are similar to those that occur now

urbanization: The development of heavily populated areas

valley glaciers: Glaciers that begin in high mountain areas and flow through valleys to lower elevations

vaporization: The change in state from liquid to gas (vapor) at any temperature (ESRT)

velocity: Speed; change in distance divided by change in time; sometimes velocity is used to include both speed and direction (ESRT)

vent: A place where lava comes to the surface

vertical rays: Sunlight that strikes Earth's surface at an angle of 90°

vertical sorting: Within a layer of sediment, the gradual change in sediment size from bottom (large) to top (small) showing the order in which particles settled; graded bedding

vesicular: Rocks that contain gas pockets, or vesicles (ESRT)

volcanic: Fine-grained, extrusive igneous rocks (ESRT)

volcano: An opening in Earth's surface through which molten magma (lava) erupts

water table: The upper limit of the underground zone of saturation or the top surface on an aquifer

watershed: The geographic area drained by a particular river or stream; drainage basin

weather: The short-term conditions of Earth's atmosphere at a given time and place (ESRT)

weather hazard: Weather events that generate strong winds, excessive precipitation, and other hazards

weathering: The physical and chemical changes in rocks that occur when they are exposed to conditions at Earth's surface

winter solstice: The name generally applied to the day of the year with the shortest period of sunlight (For observers in the Northern Hemisphere, this occurs near December 22. The Northern Hemisphere winter solstice occurs when the vertical rays of the sun are at the Tropic of Capricorn. In the Southern Hemisphere, the winter solstice occurs in June when the vertical rays of the sun are at the Tropic of Cancer.)

zenith: The point in the sky directly over an observer's head

zone of aeration: The part of the rock and soil in which air fills most of the available spaces

zone of saturation: The part of the rock and soil where all available spaces are filled with water

Index

A

Abrasion, 90, 166, 167
  resistance to, 169–170
  weathering by, 191
Absolute age, 382, 400, 401
Absolute humidity, 476, 482, 483
Absolute time, establishing, 395–401
Absolute zero, 439, 442
Absorption spectrum, 682
Acid pollution, 172
Acid precipitation, 172, 570
Adhesion, 238
Adirondack Mountains, 299
  acid damage in, 571
  climate in, 588, 593
  as dome structure, 294
  faulting in, 295
  garnet mines in, 133
  radial drainage in, 219
  rocks in, 276, 425, 641
  scenic and recreational value in, 289–290
  snow in, 271
Advisory (weather), 566
Agate, 91–92, 102
Age
  absolute, 382, 400, 401
  relative, 381–382
Agents of erosion, 188–192
Agriculture, ground pollution in, 156
Air, 87
  cooling of, 486–487
Air masses, 522–524
  arctic, 524
  codes for, 524
  continental, 524, 554–555
  maritime, 524
  polar, 524, 560
  sources regions for, 523–524
  symbols for, 527
  tropical, 524, 552, 560
Air pollution, 149–150, 158–159, 570–571
Air pressure, 444–446
  altitude and, 496
  humidity, winds and, 499–500

measuring, 444–446
  temperature, winds and, 498–499
  on weather map, 537
Air temperature on weather map, 536
Alaska current, 591
Alaska oil pipeline, preparation against earthquake threats, 364
Aleutian Islands, 350
Allegheny Plateau, 276, 296, 299
Almandine garnet, 91, 133
Alps, 293
Altitude, 607
  air pressure and, 496
Aluminum, 89, 120, 148
Amazon River, 216
American bison, 416
American Plate, 354
Amethyst, 102
Amphiboles, 91, 102, 115, 123, 132, 136
Amplitude, 451
Anchorage, Alaska, 307
Andes Mountains, 350
Andromeda, 696, 697
Anemometer, 446, 447, 495
Angle(s)
  divisions of, 41–43
  of insolation, 454–457
Animals. *See also specific by name*
  air-breathing, 421
  role of, in soil formation, 176
Annular drainage, 219
Anorthosite, 641
Antarctica, 274, 282
Antarctic Circle, 623
Anthropocene Era, 423–424
Anticyclones, 525
Apollo program, 639–640
Appalachian Plateau, 294
Aquifers, 239
Archaeopteryx, 423
Arches National Park, Moab, Utah, 39
Arctic air masses, 524
Arctic Circle, 623
Argon, 36

Arid climate, 580
Armstrong, Neil, 639
Asteroids, 672–673
Asthenosphere, 344, 345, 348
Astrology, 606–607
Astronauts, 34, 661, 663
Astronomy, 675
  celestial objects in, 607, 625–628
  defined, 5–6, 606
  Earth's motions in, 620–629
  geocentric model in, 618–619
  heliocentric model in, 619–620
  internal clock and, 608–609
  stars in, 628–631
  sun's path across sky in, 609–618
  telescopes in, 685
Atacama Desert, 591
Atlantic Ocean, 216, 588
Atmosphere, 35, 36–38, 327, 437
  carbon dioxide in, 37, 566
  changes in, 566–571
  greenhouse effect and, 461
  measuring water in, 476–479
  science of, 439
  storage of energy in, 470–474
  visibility of sun and, 458
  water vapor in, 474–476
Atmospheric pressure, 444, 446, 496
Atoms, 87, 439
  structure of, 395–396
Autumn, 613–614, 623, 625
Autumnal equinox, 613
Avalanche, 374–375
Avalanche shed, 374
Axis, 41
Azimuth, 60, 607

B

Balance scales, 9, 14
Banding, 132
Bar graphs, 22
Barnard's Star, 691
Barometer, 444–446
Barometric pressure, 444, 446

Barometric trends on weather map, 537–538
Barrier islands, 262
Basalt, 121, 122, 124, 125, 252, 325, 502
Bay of Fundy, 256
Beach, 259, 262
  sand on, 167
  source of sediment for, 259–260
Bed load, 189
Bedrock, 147, 166, 570
  inclusions in, 366–367
  New York, 425–426
  permeability of, 234
Big bang, 700, 701
Big Dipper (constellation), 44, 625, 627, 631
Bigsby Pond, topographic map of, 69
Big Thompson River, flood of, 561
Bioclastic rocks, 129–130
Biological activity in physical weathering, 168–169
Biological evolution, 7, 414
Biological productivity, 152
Biology
  environmental, 7
  evolutionary, 7, 414
Biosphere, 85
Biotite, 91, 103, 115, 123, 132
  crystal shape of, 98
Black holes, 694
Blizzards, 554–555
Blue planet, Earth as, 228, 247–248
Blue supergiant stars, 692
Body fossils, 407, 408–409
Boiling, 472
Brass, 147
Breccia, 127, 641
Breezes
  land, 504
  sea, 502–503
Bronx, 298
Bronze, 147
Brooks, 208
Buffalo (animal), 416
Buffalo, New York, 468
Building codes, weather hazards and, 564

C

Calcite, 91, 101, 104, 115, 570
  as cementing material, 126, 127, 139
  crystal shape of, 97, 98
  hardness of, 96
  special properties of, 100
  weathering of, 169, 171
Calcium, 89
Calcium carbonate, 104, 171–172
Caldera, 371
Calendar time, 609
California
  climate in, 361–362
  geologic hazards in, 361–362
California Current, 590
Cape Cod, Massachusetts, 176
Capillarity, 237–238
Carbon-12, 395–396, 400
Carbon-14, 397, 400
Carbon dioxide, 461
  in the atmosphere, 37, 566
  greenhouse effects and, 568
  human-caused emissions of, 8
Carbonic acid, 171–172
Cascade Mountains, 346
Cassiopeia (constellation), 625, 631
Castor, 625
Catastrophic events, 429, 431
Catskill Mountains, 274, 296, 299
  climate of, 593
Catskills-Allegheny Plateau, 299
Celestial objects. *See also specific*
  Earth's motions and, 625–628
  position of, 607
Cells
  convection, 309, 500–501, 505, 512
  first, 413
Celsius scale, 441–442
Cenozoic Era, 423–424
Centimeters, 14
Ceres, 672
Chalk, 411
Chemical changes, 171
Chemical precipitates, 129
Chemical reactions, 90, 91
Chemical weathering, 171–175, 249
Chemistry, 5, 6
Chile, 2010 earthquake in, 364, 366

Chlorofluorocarbons (CFCs), 158, 453
Chlorophyll, 130
Chromosomes, 415
Cinder cone volcanoes, 370
Circadian rhythm, 608
Classification, 118–119
Clastic (fragmental) rocks, 126–129
Clay, 104–105, 129, 131, 191
  as cementing material, 126, 127, 139
  chemical weathering of, 173
  crystal shape of, 99
  landslides and, 373–374
  luster of, 94
Cleavage, 97–99, 116
Climate, 438–439
  arid, 580
  changes in, 578–580
  continental, 589
  defined, 580–581
  effect of latitude on, 582–585
  geographic factors affecting, 586–587, 593
  influence of, on landscape development, 295
  maritime, 589
  monsoon, 590
  ocean currents and, 254–255
  polar, 580
  rain shadow, 587
  showing of, on graphs, 594–596
  temperate, 580
Climate shifts, 7
Clocks, 9
Clock time, 46
  establishment of, 608–609
Closed system, 137
Cloud cover, 448
  on weather map, 537
Cloud droplets, 448–449
Clouds
  defined, 486
  formation of, 486–488
Clusters, 696
Coal, 87, 115, 129–130, 138, 149–150, 155
Coastlines, changes in, 259–264
Coefficient, 10
Cold fronts, 527–528
Colorado Plateau, 185, 292
Colorado River, 153, 185, 204, 214

Color of minerals, 91–92, 116
Columbia Plateau, 292
Comets, 567, 673
Compass directions, 59–60, 607
Composite volcanoes, 370
Compounds, 89
Concrete, 150–151
Condensation, 230–231, 472–473, 486, 487
Condensation nuclei, 487–488
Conduction, 308, 342, 458–459, 460, 502
Conglomerate, 127–128, 139
Conservation, 154–155
Constellations, 44, 625–626, 627, 631
Construction materials, 150–151
Contact metamorphism, 135, 389
Continental air masses, 524, 554–555
Continental climate, 589
Continental Divide, 207
Continental drift, 4, 335–336
Continental glaciers, 177, 185, 274–275, 282, 423
Continents, movement of, 334–343
Contour interval, 64, 67
Contour lines, 64, 66–67
Convection, 230, 308, 342–343, 459–460
Convection cells, 309, 500–501, 505, 512, 585
Convergence, zones of, 505
Convergent boundaries, 349–350
Coordinate system, 21
Copernicus, 619
Copper, 94, 146–147, 148, 502
   as element, 88–89
Coprolites, 410
Coral, 115
Coriolis, Gaspard-Gustave, 506
Coriolis effect, 253–254, 506–509, 560, 621
   mid-latitude cyclones, 525
   wind and, 514
Correlation, 427
Corundum, 93
Cosmic background radiation, 700
Crater Lake, 371
Craters, 369
Creeks, 208
Crescent phase, 649

Cross-cutting relationships, 387–389
Crushed stone, 150
Crystalline ice, 90
Crystalline sedimentary rocks, 129
Crystals, 87–88
   shape of, 96–97
Currents
   density, 253
   ocean, 254–255, 590–591, 621
   turbidity, 253
Customary system of measurement, 13
Cyclones, 525
   mid-latitude, 525, 533–534

D

Daily apparent motions, 628–629
Dams, failure of, from earthquakes, 365
Darwin, Charles, 414–415
Data on weather maps, 534–541
Decay product, 396
Decay-product ratio, 399
Deep-focus earthquakes, 349–350
Deforestation, 592
Delaware River, flood of, 562
Delta, 193, 210
Dendritic drainage, 218–219
Density
   defined, 15
   determination of, 15–18
   erosion and, 190
   formula for, 16
   of minerals, 100
Density currents, 253
Dependent variable, 21, 23
Deposition, 185, 193–198, 473
   defined, 193
   effect of particle density on, 195
   effect of particle shape on, 195
   effect of particle size on, 194
   by glaciers, 277–280
   by ice, 197–198
   by meltwater, 280–281
   running water in sorting sediments, 195–196
   by wind, 196–197
Depositional features of coastlines, 262–264
Desert pavement, 191, 192

Deserts
   rocky versus sandy, 191
   streams in, 228
Devonian rocks, 417
Dew, 473
Dew point, 230, 476–479
   cooling of air and, 486–487
   relative humidity and, 483
   on weather map, 536
Dew-point temperature, 476–477
Diamond, 89, 93, 326
   hardness of, 96
Diamond anvil, 327
Diatomaceous earth, 411
Diatomite, 411
Dinosaur National Monument, 407, 408
Dinosaurs, 417, 423
Directions, compass, 59–60
Discharge, 216
Divergence, 505
Divergent boundaries, 348–349
Divisions of angles, 41–43
DNA, evolution and, 415–416
Dolomite, 129
Dolostone, 129, 294
Doppler, Christian Johann, 698
Doppler effect, 698
Doppler radar, 495, 565
Double Arch (Arches National Park), 39
Double stars, 627
Doyle, Sir Arthur Conan, 380
Drainage divides, 207
Drainage patterns, 218–219
Drizzle, 449
Droughts, 563
Drumlins, 279–280
Dry-bulb temperature, 478–479, 483
Dunes, 196–197
Duration of insolation, 457–458
Dust storms, 154, 487
Dwarf planets, 672
Dynamic equilibrium, 199, 273

E

Eagle Nebula, 604
Early warning systems, 565–566
Earth
   beginning of life on, 412–413
   as Blue Planet, 228, 247–248
   changes in geography of, 353–355

Earth (Cont'd.)
chemical composition of, 35
circulation of solar energy
over, 458–462
core of, 327, 346
crust of, 35, 148, 154, 325, 339
density of, 346
diameter of, 326
insolation and, 453–454
interior of, 117, 324–329,
344–347
pressure in, 346–347
temperature of, 345–346
internal energy of, 308–309
magnetic field of, 60–62,
338–339
magnetic poles of, 338–339
mantle of, 325–327
moon and, 650–652
parts of, 35–40
revolution of, 620
rotation of, 61, 606, 620–621
seasons and motion of,
622–625
shape of, 1, 31–34
size of, 34–35
surface conditions of, 668–669
temperature of, 116–117
view of motion of, from space,
621–622
warming of, by sun, 451–458
water on, 228–231, 248
Earthquakes, 7
causes of, 306–310
deep-focus, 349–350
defined, 307
finding origin times, 316–318,
323–324
as geologic hazard, 362–368
locating, 318–323
measurement of, 311–314
plate boundaries and, 347–348
preparedness for, 367–368
radiation of energy from,
314–318
selected historic, 365
shadow zone of, 328–329
subduction zone, 349
as trigger for landslides, 364
Earthquake scales, 311–314
Earth science
branches of, 5–6
defined, 5–6
reasons for studying, 7–9

relationship to other sciences,
6–7
Eccentricity, 644
calculating, 644–646
Eclipse, 652
lunar, 32, 653, 654
predicting, 653–654
solar, 652–653, 654
Ecology, 6
Einstein, Albert, 688
Electrical energy, 151
Electromagnetic energy, 460
Electromagnetic radiation, 606,
689–690
Electromagnetic spectrum,
682–683
Electromagnetic wavelength, 685
Electromagnetic waves, 451–452
Electron microscopes, 87
Electrons, 395
Elements, 36, 88, 682. See also
specific elements
Elevation, climate and, 586
Ellipse, 642
El Niño, 255
Emergency planning, 564–565
Emperor Chain of islands, 352
Energy
electrical, 151
electromagnetic, 460
kinetic, 206, 439, 475, 648
orbital, 648
potential, 648
solar, 151, 451, 454, 458–462
storage of, in atmosphere,
470–474
wind, 151
Enfield Glen, 185
Engineering decisions, geologic
hazards and, 375
Enhanced Fujita scale of tornado
intensity, 560
Environmental awareness, 8–9
Environmental biology, 7
Environmental factors, effect of,
on weathering, 174–175
Epicenter, 310
Equator, 33–34, 41
Equilibrium, 460–461
between deposition and
erosion, 198–199
dynamic, 199, 273
Equinox, 458, 611
autumnal, 613
vernal, 612, 622

Eratosthenes, 34–35
Erie Canal, 206
Erie-Ontario Lowlands, 296, 299
Eris, 672
Erosion, 118, 412, 427
agents of, 188–192
defined, 185–186
equilibrium between
deposition and, 198–199
glacier, 192, 275–276
water, 188–191
wind, 154, 191
Erratics, 277
Errors in measurement, 19–20
Estimation, 14–15
of density, 17
Eurypterid, 421, 422
Evaporation, 230, 474–475, 482
defined, 475
energy supporting, 475
kinetic energy and, 475
rate of, 475–476
Evolution
biological, 7, 414
dinosaur, 423
DNA and, 415–416
organic, 7, 413–417
principles of, 414–415
Evolutionary biology, 7
Exfoliation, 167–168
Exponential notation, 10–11
Extinction, 416
mass, 422
Extraterrestrial life, search for,
686–687
Extrusion, 121, 389
Eye of the storm, 556
Eye wall, 557
Eyjafjallajokull, eruption of, 373

F

Fahrenheit scale, 441
Faults, 293, 294–295, 310
Feldspar, 89–90, 94, 102, 104,
132, 136, 173
crystal shape of, 98
hardness of, 96
Felsic lavas, 370
Felsic magma, 350
Felsic minerals, 120, 350
Felsic rocks, 337, 341, 369
Fields, 62–64
Finger Lakes, 276, 278, 289, 299,
588

Finger Lakes region, 185, 276
Fire Island, 262
Flash floods, 561–562
Floodplains, 209–210, 291
Floods, 7, 207, 553, 561–563
Flotation, 188–189
Flowchart, identifying minerals with, 105–106
Fluids, 87, 459
    convection in, 342–343
    defined, 342
Fluorite, hardness of, 96
Focus, 310, 642
Fog, 448, 487
Folds, 293
Foliation, 132, 133
Food chain, 152
Formula for density, 16, 57
Formula for eccentricity, 644
Formula for gradient, 71
Fossil fuels, 149–150, 568
    burning of, 172
    costs of, 158–159
Fossil limestone, 115
Fossil mold, 409–410
Fossils, 7, 126, 407–417
    body, 407, 408–409
    comparing types of, 427–429
    defined, 407
    index, 417, 420, 427
    trace, 407, 410
Fossil wood, 411
Foucault, Jean, 620
Foucault pendulum, 620–621
Fracture, 97–99
Freezing, 473–474
Freezing rain, 449, 554
Frequency, 698
Freshwater, 39, 153, 154
Fronts, 526
    cold, 527–528
    occluded, 531
    polar, 532–533
    stationary, 531–532
    symbols for, 527
    warm, 528–531
Frost, 90, 473–474
Frost wedging, 166–167
Full moon, 649
Fusion, nuclear, 451, 689

G

Gabbro, 125, 252
Galaxies, 694

Andromeda, 696, 697
    Milky Way, 695–696, 700
Galena, 93, 94
    crystal shape of, 98
    density of, 100
Galileo, 619, 667, 671
Galileo thermometer, 17
Galveston, Texas, hurricane damage in, 550–551
Gamma rays, 452, 685
Garnet, 94, 133, 136
    crystal shape of, 99
Gases, 87
    chlorofluorocarbon, 158
    greenhouse, 461
    natural, 149
Geiger counter, 101
Gemini (constellation), 625
Genesee River, 185, 290, 415
Geocentric model, 618–619
Geographic features, effect of, on climate, 593
Geographic Information System (GIS), 25–26
Geography, changes in Earth's, 353–355
Geological event sequencing, 381–401
    contact metamorphism in, 389
    cross-cutting relationships in, 387–389
    determining, 381–382
    establishing absolute time in, 395–401
    inclusions in, 386–387
    interpreting geologic profiles, 392–395
    original horizontality in, 384–386
    superposition in, 383–384
    unconformities in, 389–392
    uniformitarianism in, 383
Geologic factors, influence of, on landscape development, 294–295
Geologic hazards, 361–375
    earthquakes as, 362–368
    engineering decisions and, 375
    mass movements as, 373–375
    volcanoes as, 368–373
Geologic profiles, interpreting, 392–395
Geologic record, 416
Geologic time
    Anthropocene Era in, 423–424

Cenozoic Era in, 423–424
    division of, 417–424
Geologic time (Cont'd.)
    Mesozoic Era in, 423, 431
    of New York State, 418–419, 424–426
    Paleozoic Era in, 421–422, 425
    Permian Period in, 425
    Precambrian time in, 420–421, 425
Geologists, 327, 381
    correlation of rock layers by, 426–431
Geology, 5
Geosphere, 40, 437
Gibbous phase, 649
Glacial ice, shrinkage of, 579
Glacial moraines, 297–298
Glacial striations, 274, 276
Glacial till, 299
Glacier National Park, 578
Glaciers, 153, 197–198
    continental, 177, 185, 274–275, 282, 423
    coverage of, New York State by, 270–271
    defined, 192, 271–272
    deposition by, 277–280
    erosion by, 192, 275–276
    polar, 579–580
    valley, 272–273
Glaze, 92
Glen Canyon, 214
Global climate change, measuring, 579
Globalization, 144
Global marketplace, 144–146
Global Positioning System (GPS), 26, 49–50
Global warming, 158–159, 568–569, 579–580
Globe, 57
Glomar Challenger (oceanographic research ship), 340
Gneiss, 132, 139
Gold, 146, 147, 148
    density of, 100
    as element, 88
Google Earth, 42
Gorge, 185
Graded bedding, 194
Gradient, 71–73, 209
    stream, 213
Grand Canyon, 170, 185, 380

Granite, 85, 115, 120, 122, 123, 124–125, 139, 502
  chemical weathering of, 173
  common minerals in, 86
  formation of, 167–168
Graphite
  as element, 89
  special properties of, 100
Graphs
  bar, 22
  guidelines for making, 23–24
  line, 21, 22
  making and using, 20–24
  pie, 22, 23
  showing climate on, 594–596
Gravel, 150
Gravity, 62, 444, 647–648
  tides and, 256–257
Great Bear (constellation), 626
Great Bear Lake, topographic map of, 65
Great Lakes, 206, 299, 468, 501, 588
Great Plains, 153, 291
Great Red Spot, 670
Great Salt Lake, 153, 216
Greenhouse effect, 461–462, 568
Greenhouse gases, 461
Greenland, 274, 282
Greenwich, England, Royal Observatory in, 42
Greenwich Mean Time (GMT), 46, 48, 49
Greenwich meridian, 42
Grooves, 274, 276
Ground cover, 231
Ground pollution, 156–157
Groundwater, 228–229
  availability of, 238–239
  movement of, 234–238
  occurrence of, in specific zones, 232–233
  problems with, 239–241
Gulf of California, 310
Gulf Stream, 254, 590
Gypsum, 90, 129
  hardness of, 96

H

Hail, 449–450, 552, 553
Haiti, 2010 earthquake in, 363–364
Half-life periods, 396

working with, 398–400
Halite, 90, 115, 188
  crystal shape of, 98
  special properties of, 100
Halley's comet, 673
Harbor Hill Moraine, 280
Hardness, 95–96, 116
Hawaii, Big Island of, 351, 370
Hawaiian Islands, 351–352
  volcanoes on, 369–370
Hazards, 362
  geologic, 361–375
  weather, 563–566
Heat
  latent, 470
  specific, 470
Heliocentric model, 619–620
Helium, 682, 683
Hematite, 93
Hertzsprung-Russell diagram, 690
Hess, Harry, 338
High-pressure systems, 505, 508
Hilo, Hawaii, tsunami damage in, 366
Himalaya Mountains, 293, 294, 350
Hoover Dam, 214
Horizontal sorting, 196
Hornblende, 102, 115
Hornfels, 135
Hot spots, 351–352
Howe Caverns, 172
Hubbert, M. King, 84
Hubble, Edwin, 697, 698–699
Hubble Space Telescope, 604, 684
Hudson Highlands, 295, 296, 298, 425
Hudson-Mohawk lowlands, 205
Hudson River, 214, 216, 290, 295, 296, 298, 299
  meaning to New York State, 205–206
Hudson Valley, 298
  climate of, 593
Humidity, 442–444
  absolute, 476, 482
  air pressure, winds and, 499–500
  describing, 524
  relative, 443, 472–484, 514
Hurricane Ike, 551, 558
Hurricane Katrina, 212, 550, 561, 563

Hurricanes, 7, 212, 556–559, 564
  damage caused by, 550–551
  wind speed during, 495
Hydrogen, 439, 682
Hydrologic cycle, 229–231, 249
Hydrosphere, 35, 36, 38, 39, 327, 437
  chemical composition of, 35
Hygrometer, 478

I

Ice, 90, 271–272
  deposition by, 197–198
Ice ages, 281–283
Icebergs, 274–275
Ice caps, polar, 229
Iceland, 2010 volcanic eruption in, 373
Ice sheets, 274
Igneous rocks, 116–117, 118, 119, 137–138
  age of, 339
  classification of, 120–124
  common, 124–125
Impressions, 409–410
Inclusions, 366–367
Independent variables, 21, 23
Index fossils, 417, 420, 427
Indian-Austrian Plate, 350, 367
Indian Ocean, earthquake in, 366–367
Indonesia, eruption of Tambora in, 429
Inertia, 646
Inference, 10, 345
Infiltration, 176, 231, 232
Information, scientists in gathering and analyzing, 9–15
Insolation, 453–454, 613
  angle of, 454–457
  duration of, 457–458
Intergovernmental Panel on Climate Change, 569
Internal clock, 608–609
International system of measurement, 12–14, 609
Intrusions, 120, 389
  igneous, 387
Ionizing radiation, 101
Iron, 36, 120, 125, 147–148, 171, 327, 502
Iron oxide, 171, 173
Island arc, 350

Isobaric maps, 514–515
Isobars, 496, 508
Isoline, 62–64, 496
Isoline maps, 64, 75, 76
Isotherms, 62–64
Isotopes
    defined, 396
    radioactive, 395–397

J

Japan, islands of, 350
Jet streams, 512–514
Joints, 219
Jones Beach, 262
Joules (J), 470
Jovian planets, 666, 670–671
Jupiter, 664, 665, 670

K

Kauai, 351, 352
Kelvin scale, 442
Kepler, Johannes, 619
Kettle lakes, 279, 298
Kettles, 278–279
Kilauea, eruptions of, 351, 369
Kilograms, 14
Kimberlites, 326
Kimberly, South Africa, 326
Kinetic energy, 206, 439, 648
    evaporation and, 475

L

Lagoons, 262
Lake-effect storms, 468–469,
    554, 593
Lake Erie, 299, 425, 468, 588,
    593
Lake Meade, 214
Lake Ontario, 294, 299, 425, 468,
    588, 593
Lake Ronkonkoma, 298
Land breezes, 504
Landforms, 290
Landmarks, changes in natural,
    164
Landscapes, 290–293
    factors influencing
        development of,
        294–295
    mountain, 292–293
    in New York State, 296

Landslides, 186, 373–374
    earthquakes as trigger for, 364
Latent heat, 470
Latitude, 41
    effect of, on climate, 582–585
    salinity and, 250
    sun's path and, 617–618
Lava, 116, 123
    felsic, 370
Lead, 502
Legend, 67
Lepus the rabbit (constellation),
    625
Letchworth Gorge, 185, 290, 299
Levees, 212–213, 563
Lichens, 169
Life
    atmosphere and, 567–568
    beginning of, on Earth,
        412–413
Life science, 5, 7
Lightning, 552–553
Light-year, 697
Limestone, 115, 133, 138,
    171–172, 570
    weathering of, 169
Limestone caverns, 172
Line graphs, 21, 22
Liquefaction, 364
Liquids, 87
Lithosphere, 35, 38, 39–40, 340,
    344, 348
Lithospheric plates, 341–343
Local Group, 696
Location, determination of,
    40–50
Lockport dolostone, 294, 383
Lodestone, 60
Logarithmic scale, 314
Long Island, 240–241, 262,
    297–298, 425, 501
    beaches on, 280–281, 289
    climate of, 588, 593
    weather history in, 564
Long Island Sound, 588
Longitude, 42
    finding, 46–50
    sun's path and, 617
Longshore transport, 261–262
Love Canal, 156
Low-pressure systems, 504–505,
    508–509
Luminosity, 690
Lunar eclipse, 32, 653, 654
Lunar surface, 641–642

Luster, 93–95, 116
    metallic, 93–94
    nonmetallic, 94
Lyra the harp (constellation), 625

M

Mafic minerals, 120, 350
Mafic rocks, 337
Magma, 116, 326
    felsic, 350
Magnesium, 36, 89, 120,
    125
Magnet, Earth as a, 60–62
Magnetism, 60–61
Magnetite, 91
    density of, 100
    special properties of, 100
Main sequence stars, 691–692
Major axis, 644
Makemake, 672
Mammals, 423–424
Mammoth Caves, 172
Manhattan, 298
Manhattan Prong, 298
Mantle, 327, 328, 344
Maps, 56–59
    isobaric, 514–515
    isoline, 64, 75, 76
    political, 57
    road, 57
    star, 627–628
    topographic, 64–77
    weather, 62, 75
Map scale, 67–70
Marble, 133–134, 172, 570
Maria, 642
Mariner's sextant, 44
Maritime air masses, 524
Maritime climate, 589
Mars, 604, 665, 669
    atmosphere of, 567
Mass movements, 186–187,
    373–375
Mastodon, 408–409
Materials, specific heats of
    common, 502
Mathematical models, 56
Matter, 439
    density of, 15–18
    phrases of, 470
    states of, 470–471
Maui, 351
Mauna Kea, 370
Meanders, 210–212

Measurement
  customary system of, 13
  errors in, 19–20
  metric system of, 12–14, 609
Melting, 471–472
Meltwater, deposition by, 280–281
Mercalli scale, 311–312, 314
Mercury, 444
Mercury (planet), 640, 664, 665, 666–667
  phases of, 667–668
  surface of, 462
Meridians, 42
Mesosphere, 38
Mesozoic Era, 423, 431
Metallic luster, 93–94
Metal ores, 146–148
Metamorphic rocks, 117–118, 119, 138, 299
  formation of, 130–137
Metamorphism, 138, 412
  contact, 135
  regional, 134–135
Meteor Crater, 674, 675
Meteorites, 327, 412, 675
  composition of, 346
Meteoroids, 675
Meteorologists
  in describing air masses, 522
  in predicting weather, 485–486, 522
  in tracking weather fronts, 526–527
Meteorology, 5, 439
Meteors, 674–675
  impact of, 417
Metersticks, 9
Methane, 461
Metric rulers, 14
Metric system, 12–14
Mexico, 1985 earthquake in, 363
Miami, Florida, hurricane damage in, 550
Mica, 103, 104
Microscopes, 9
  electron, 87
Mid-Atlantic Ridge, 346, 348, 354
Mid-latitude cyclones, 525
  evolution of, 533–534
Mid-latitude temperatures, 583–584
Mid-ocean ridges, 337–338, 340, 342, 348–349
Milky Way galaxy, 695–696, 700

Mineral identification tests, types of, 91
Minerals
  characteristics of, 86–88
  cleavage of, 97–99, 116
  color of, 91–92, 116
  common, 101–107
  crystal shape of, 96–97
  defined, 85–86, 88–90
  density of, 100
  felsic, 350
  fracture of, 97–99
  hardness of, 95–96, 116
  identifying, by observable properties, 90–101
  identifying with a flowchart, 105–106
  luster of, 93–95, 116
  mafic, 350
  properties of common, 107
  special properties of, 100–101
  streak in, 92–93
  using, 148–149
Mississippi River, 188, 213, 216
  delta in, 193
  flooding of, 562
Mississippi Valley, 291
Moab, Utah, Arches National Park in, 39
Models
  defined, 56–57
  mathematical, 56
  scale, 57
Mohawk River, 299
Moho, 325
Mohs, Friedrich, 95
Mohs scale of hardness, 95–96
Molecules, 89, 439, 442
Molokai, 351
Monsoon climate, 590
Monsoons, 511–512
Moon
  apparent size of, 654–655
  Earth and, 651–652
  eclipse of, 32, 653, 654
  exploration of, 639–640
  history of, 640–642
  lunar surface of, 641–642
  orbit of, around Earth, 650–651
  origin of, 641
  phases of, 648–652
  tides and, 257
Moraines, 277–279
  glacial, 297–298

Moss, 169
Mountains, 118, 292–293
  climate and, 586–587
Mount Chimborazo, 586
Mount Kilimanjaro, 586
Mount Marcy, 299
Mount McKinley, 293
Mount St. Helens, eruption of, 350, 370–371, 372, 429
Mount Vesuvius, eruption of, 177
Mount Washington, 494, 495
Mount Whitney, 305
Mudflow, 187
Muscovite, 103
  crystal shape of, 98
Mutations, 415

N

Na-Ca feldspar, 102
National Weather Service, 565–566
Native copper, 89
Natural disasters, 7
  cost of, 550–551
Natural gas, 149
Natural resources, 7–8, 84, 145–146
Nazca Plate, 349, 354
Neap tides, 258
Nebula, 628, 663, 694
Nebular contraction, 662–663
Neptune, 664, 665, 671, 673
Neutrons, 395
Neutron star, 694
Newark Basin, 298
New Orleans, 212
  flooding in, 563
New York bedrock, 425–426
New York City
  Central Park in, 274, 276
  climate of, 590
  determining location in, 40
New York State
  climate in, 295
  coverage of, by glaciers, 270–271
  effect of geographic features on climate in, 593
  fossils in, 409–410
  garnet as official mineral of, 133
  geologic history of, 418–419, 424–426
  landscapes of, 296

meaning of Hudson River to, 205–206
natural wonders in, 289–290
seasonal winds in, 512
stromatolites in, 412–413
unconformities in, 390
wind farms in, 151
New York State Thruway, 206
Niagara Escarpment, 294
Niagara Falls, 156, 166, 289
profile of rock layers at, 383–384
Niagara River, 294
Nile River delta, 193
Nitric acid, 172, 570
Nitrogen, 499
in atmosphere, 36
Nitrogen oxides, 172
Nitrous oxide, 461
Nonmetallic luster, 94
Nonrenewable resources
construction materials as, 150–151
defined, 146–151
fossil fuels as, 149–150
metal ores as, 146–148
using mineral resources, 148
North American Plate, 351, 362
North Atlantic Current, 254
North Pole, 33, 43, 61, 609
North Star, 33, 43–45
Notation
exponential, 10–11
scientific, 10
Nuclear fusion, 451, 689
Nuclear physics, 7

O

Oahu, 351
Oblate shape, 34
Observations, 2–3, 9
qualitative, 9
quantitative, 9
Obsidian, 123, 125
Occluded fronts, 531
Ocean currents, 254, 590–591, 621
climate and, 254–255
Oceanographers, 250
Oceanography, 5
Oceans
circulation of water in, 253–255
investigating, 250–252

Ocean trenches, 341
Ocean water
composition of, 248–249
origin of, 248
source of salt in, 249–250
Ogallala aquifer, 154
Oil, 84, 149. *See also* Petroleum
Old Man of the Mountain, 166
Olivine, 89, 104, 120
On the Origin of Species by Means of Natural Selection (Darwin), 414
Onyx, 102
Open clusters, 627
Optical telescopes, 683–684
Orbital energy, 648
Orbits, describing, 642–646
Ores, 146
Organic evolution, 7, 413–417
Organic matter, 175
Organic rocks, 129–130
Origin, 116
Original horizontality, 384–386
Origin time, finding, for earthquakes, 316–318, 323–324
Orion the hunter (constellation), 625, 631
Oswego, 296
Outcrop, 385, 427
Outer planets, 670–671
Outgassing, 566
Outwash, 280
Overland flow, 208, 249
Oxygen, 89, 101, 102, 439, 499
for air-breathing animals, 421
in Earth, 36
Ozone, 158, 421, 452–453, 567

P

Pacific Ocean, 153
Pacific Plate, 351, 354, 362
Paleontology, 407, 420
Paleozoic Era, 421–422, 425
Palisades, 298
Pangaea, 335, 422
Paradigm, 334
shift in, 334–335
Parallels, 41
Parkfield, California, 305
Particles
effect of density, on deposition, 195

effect of shape, on deposition, 195
effect of size, on deposition, 194
Passenger pigeons, 416
Pegmatite, 123
Penzias, Arno, 699–700
Percent deviation, 19
Percent error, 19
Permeability, 231, 234–235
Permian Period, 425
Peru-Chile Trench, 349–350
Petrified wood, 409
Petroleum, 87, 149, 155
Phases
of matter, 470
of the moon, 648–652
Photosynthesis, 421, 453, 567, 592
Phyllite, 132
Physical weathering, 166–169, 174
Physics, 5, 6
Pie graphs, 22, 23
Pitch, 698
Plagioclase feldspar, 102, 115, 123, 124
special properties of, 100
Plains, 291, 299
Planets
dwarf, 672
grouping of, 665–671
outer (Jovian), 666, 670–671
properties shared by, 663–665
terrestrial, 665, 666–669
Plastic, 345
Plateaus, 291–292
Plate boundaries, 347–353
types of, 348–351
Plate tectonics, 4, 334–355, 663
changes in Earth's geography, 353–355
internal structure of Earth, 344–347
movement of continents in, 334–343
plate boundaries in, 347–353
Pleistocene Epoch, 281
Pluto, 671–672
Plutonic rocks, 120
Polar air masses, 524, 560
Polar climate, 580
Polar fronts, 532–533
Polar glaciers, 579–580
Polar ice caps, 153, 229

Polaris, 33, 43–45, 622, 627, 629, 631
Polarity, 339, 340
Polar regions, temperatures in, 584
Political maps, 57
Pollution
    acid, 172
    air, 149–150, 158–159, 570–571
    defined, 156
    ground, 156–157
    water, 157–158, 207, 239
Pollux, 625
Polychlorinated biphenyls (PCBs), 157
Ponds, 279
Population growth, 145
Porcelain, 94
Porosity, 235–237
Portage Glacier, 273
Potassium, 89
Potassium-40, 397
Potassium feldspar, 102, 115, 123
Potential energy, 648
Precambrian time, 420–421, 425
Precipitation, 129, 231, 448–450, 584–585
    acid, 172, 570
    forms of, 449–450, 467–469
    measuring, 449
    predicting, 485–486
    on weather map, 537
Pressure, Earth's internal, 346–347
Prevailing winds, 510, 589–590
Primary waves, 314–315, 318, 320, 323, 328
Prime meridian, 42
Producers, 152
Profile, 73–75
Protons, 395
Proxima Centauri, 697
Psychrometer, 478
Pumice, 123, 125
Pyrite, 91, 93, 94
Pyroxene, 91, 102, 120, 124, 132

Q

Qualitative observations, 9
Quantitative observations, 9
Quarter moon, 649
Quartz, 89, 91, 94, 97, 102, 115, 123, 132, 136
    chemical weathering of, 173

    hardness of, 96
    rose, 91
    smoky, 91
    weathering of, 169
Queenston Shale, 383

R

Radar, 495
    Doppler, 565
Radial drainage, 219
Radiation, 308, 451, 502
    cosmic background, 700
    electromagnetic, 606, 689–690
    of energy from earthquakes, 314–318
    ultraviolet, 453
Radioactive dating, 400
Radioactive decay, 396–397
Radioactive isotopes, 395–397
    selecting best, 400–401
Radioactive wastes, 156–157
Radioactivity, 396
Radiometric dating, 641
Radio telescopes, 685, 687, 695
Radio waves, 452, 686–687
Rain, 449, 552. See also Precipitation
    freezing, 449, 554
Rain shadow climate, 587
Rain shower, 552
Ratio, 57
Real-time satellite coverage, 565
Recycling, 148, 155
Red dwarf stars, 691–692
Redshift, 697–699
Reduction, 154–155
Reference tables, 106
Reflection, 454
Refraction, 328, 453–454, 458
Regional metamorphism, 134–135
Relative age, 381–382
Relative humidity, 443, 472–484, 514
Remote sensing, 328
Renewable resources, identifying important, 151–154
Replacement, 155, 409
Residual soil, 177
Resistance to abrasion, 169–170
Resources. See also Natural resources; Nonrenewable resources
    conservation of, 154–155

Revolution, 620
Rhyolite, 122, 125
Richter scale, 312–313, 314
Rigel, 692
Ring of Fire, 371
River rocks, 166, 167
River system, 206–213
    deltas in, 210
    floodplains in, 209–210
    levees in, 212–213
    meanders in, 210–212
    watersheds in, 206–208
Road maps, 57
Rochester, 290, 296
Rock cycle, 137–139, 165–166, 663
Rock gypsum, 129
Rocks
    age of, 340–341
    categories of, 116–137
    correlation of layers, by geologists, 426–431
    defined, 115
    felsic, 337, 341, 369
    igneous, 116–117, 118, 119, 137–138
    mafic, 337
    metamorphic, 117–118, 119, 138, 299
    river, 166, 167
    sedimentary, 117, 119, 138, 291
Rock salt, 115, 188
Rocky Mountains, 153, 204, 293
Rose quartz, 91
Rotation, 61, 606, 620
Royal Observatory in Greenwich, England, 42
Rubidium-87, 397
Running water in sorting sediments, 195–196
Runoff, 208
Rusting, 171

S

Saffir-Simpson scale of hurricane intensity, 557
Sagan, Carl, 36
Salinity, latitude and, 250
Salt, source of, in oceans, 249–250
Salt water, 39, 153
Saltwater invasion, 240–241

San Andreas Fault, 305, 310, 351, 354, 362
Sand, 150
  beach, 262
  shatter, 167, 168
Sandbars, 262
Sandstone, 115, 128, 139
  weathering of, 169
San Francisco, 310
  1906 earthquake in, 364
Satellite, 640
  determining orbit of, 646–648
Saturated air, 443
Saturn, 664, 665, 670
Scale models, 57
Scatter, 454
Schist, 132
Schoharie Creek, washout of bridge over, 375
Science(s)
  branches of, 5
  defined, 2–4
  facts of, 2
  inquiry in, 26
  methods of, 2–3
  relationship of Earth science to, 6–7
Scientific notation, 10
Scientists
  effect of technology on work of, 25–26
  in gathering and analyzing information, 9–15
  making and using graphs, 20–24
  use of metric system by, 13
Scoria, 123, 125
Scorpio, 631
Sea breezes, 502–503
Seafloor spreading, 337–341
Sea scorpion, 421
Seasons
  Earth's motions and, 622–625
  path of sun and, 610–615
Secondary waves, 314–315, 318, 319–320, 323, 328
Secret Caverns, 172
Sediment, 117, 166, 188–189
  glacial, 277
  source of, for beaches, 259–260
Sedimentary rocks, 117, 119, 138, 291
  deposition of, in layers, 384–386
  formation of, 126–130

Sediments, sorting of, by running water, 195–196
Seismic moment, 313
Seismic waves, 40, 314–315, 325, 327, 344
Seismographs, 312–313
Seismology, 311
Septic system, 239
SETI (Search for Extraterrestrial Intelligence), 686–687
Sewage, 239–240
Shale, 128, 131–132, 291
Shatter sand, 167, 168
Shield volcanoes, 370
Shorelines, management of active, 264–265
Sierra Nevada Mountains, 586
Silica, as cementing material, 126, 127, 139
Silicate, 101
Silicon, 89, 101, 102
Silicon dioxide, 89
Siltstone, 128
Silver, 94, 146
Sleet, 449, 553–554
Smith, Tilly, 367
Smog, 570
Smoky quartz, 91
Snow, 90, 231, 271, 449, 467–469
  lake-effect storms and, 468–469
Snow shower, 554
Sodium, 89
Sodium chloride, 188
Soil, 152–153
  defined, 175
  formation of, 175–177
  residual, 177
  transported, 177
Soil horizons, 176
Solar eclipse, 652–653, 654
  annular, 655
Solar energy, 151, 451, 454
  circulation of, over Earth, 458–462
Solar noon, 46, 47, 609
Solar system
  data on, 665
  origin of, 662–663
  planets in, 663–665
Solar time, 46, 47
Solids, 87
Solution, 188
Sorting, 195–196

horizontal, 196
vertical, 194
Source regions, 523–524
Southern Cross (constellation), 45
  stars of, 34
South Pole, 33, 45, 61, 609
Space
  Apollo program and, 639–640
  colonizing, 661–662
  view of Earth's motion from, 621–622
Species, 414
Specific heat, 470, 502
Spectroscopes, 686
Spirit Lake, 370
Spit, 262
Spring, 612, 622
Springs, 228
Spring sides, 257–258
Starlight, 688
Star maps, 627–628
Stars, 687–690
  birth of, 692–693
  blue supergiant, 692
  classification of, 690–692
  death of a massive, 693–694
  death of an average, 693
  defined, 687
  effect of Earth's motions on, 628–629
  energy escape from, 689–690
  investigation of, 681–686
  main sequence, 691–692
  middle age, 693
  nuclear fusion in, 689
  red dwarf, 691–692
Staten Island, 298
States of matter, 470–471
  changes in state, 471–474
Stationary fronts, 531–532
Station model, 535
Steel, 147–148
  hardness of, 96
Storm surges, 558, 564
Storm warning, 566
Storm watch, 566
Strain, 309
Stratosphere, 38
Streak in minerals, 92–93
Streak plate, 92, 96
Streak test, 92
Streams, 206
  features of, 209–213
  gradient of, 213

Streams, (*Cont'd.*)
  size of, 216–217
  source of water in, 227–228
  velocity of, 213–215
Stream system, 206
Stress, 309
Striations, 192
Stromatolites, 412–413
Subduction zones, 340–341, 342, 371
  earthquakes in, 349
  volcanoes in, 371
Submersibles, 252
Subsidence, 241
Sulfur, 88–89, 91
Sulfuric acid, 570
Sumatra, damage from tsunami in, 366–367
Summer, 613, 622–623
Summer monsoon thunderstorms, 562
Summer solstice, 613
Sun
  apparent path of, across the sky, 609–610
  apparent size of, 654–655
  changes of path with the seasons, 610–615
  objects orbiting, 671–675
  path of, and observer's location, 617–618
  role of, in warming Earth, 451–458
  tides and, 257–258
Superclusters, 696
Superposition, 383–384, 427
Surface area, chemical weathering and, 172–174
Surface ocean currents, 58
Surf zones, 261
Suspension, 188
Susquehanna River, flood of, 562

T

Tableland, 291
Taconic Mountains, 298
Talc, 94
  hardness of, 96
  special properties of, 100
Tambora, eruption of, 429
Tasmanian tiger, 416
Technology, effect of, on work of scientists, 25–26
Tectonic plates, 59

Tectonics, 341
Telescopes, 9
  optical, 683–684
  radio, 685, 687, 695
Temperate climate, 580
Temperature, 62, 439–442
  air pressure, wind and, 498–499
  defined, 439
  describing, 524
  dew-point, 476–477
  Earth's internal, 345–346
  latitude and, 582–584
  measuring, 439–442
  wet-bulb, 478–479, 483
Temperature scales, 441–442
Terminal moraine, 278
Terrestrial coordinates, 40–43
Terrestrial planets, 665, 666–669
Terrestrial winds, 584–585
Texture, 120
Thermometers, 9, 439–442
  Galileo, 17
Thermosphere, 38
Thunder, 552
Thunderstorms, 552–554
Tidal range, 255–256
Tides
  causes of, 255–259
  gravity and, 256–257
  moon and, 257
  sun and, 257–258
Till, 277, 280
  glacial, 299
Time zones, 46–47, 48
Tin, 147
Topaz, 93
Topographic maps, 64–77
  common features of, 66–71
Topographic profiles, 73–75
Topography, 66
Tornadoes, 7, 559–561, 566
  wind speed in, 495
Toxic chemicals, 156
Trace fossils, 407, 410
Transform boundaries, 350–351
Transpiration, 230, 475, 592
Transplanted soil, 177
Travel time, 318
Trenches, 252
Tributary, 208
Trilobites, 417, 422
Tropical air masses, 524, 552, 560
Tropic of Cancer, 583, 609, 623

Tropic of Capricorn, 583, 609, 623
Tropics, temperature in, 582–584
Troposphere, 37–38, 552
  chemical composition of, 35
Tsunami, 366–367
Tug Hill Plateau, 299, 468
Tully Valley, landslide in, 373–374
Turbidity currents, 253
Twain, Mark, 521

U

Ultraviolet radiation, 453
Unconformities, 389–392
Uniformitarianism, 383, 427
United Nations, 569
United States Geological Survey (USGS), 66
  topographic maps, 64
Universe
  future of, 700–702
  history of, 696–700
  structure of, 694–696
Uplift, 118, 134, 168
Uranium, 396
Uranium-235, 396
Uranium-238, 396, 397
Uranus, 665, 671
Urban heat islands, 592–593
Urbanization, 592
Ursa Major (constellation), 626, 631
Ursa Minor (constellation), 631
U-shaped valleys, 276, 290
Utah Slot Canyons, 185

V

Valdez, Alaska, 307
Valley glaciers, 272–273
Valley Heads Moraine, 278
Valleys, 275–276
  U-shaped, 276, 290
  V-shaped, 209, 210, 275
Vaporization, 472
Vapor trails, 488
Variables
  dependent, 21, 23
  independent, 21, 23
Vegetation, 592
Velocity, stream, 213–215
Ventifacts, 191, 369
Venus, 640, 664, 665, 667

atmosphere of, 567
phases of, 667–668
surface of, 462
Vernal equinox, 612, 622
Vertical ray, 622
Vertical sorting, 194
Vesicles, 122–123
Vesicular texture, 122
Visibility on weather map, 536
Volcanic glass, 123, 125
Volcanic rocks, 121
weathering of, 177
Volcanoes, 118, 177, 368–373.
    See also specific by name
defined, 368
eruptions of, 7, 177, 350, 351,
    369, 370–371, 372, 373,
    417, 429, 487, 566
in Hawaii, 351
as hazards, 371–373
mafic, 350
types of, 369–371
Vollmer, Frederick, 6
V-shaped valleys, 209, 210, 275

W

Wallace, Alfred Russel, 414
Warm fronts, 528–531
Water, 153–154
    in carrying excess heat energy,
        470–471
    climate and large bodies of,
        588–589
    density of, 17
    on Earth, 228–231, 248
    erosion by, 188–191
    in hydrosphere, 36
    measuring, in atmosphere,
        476–479
    properties of, 473
    sources of, 475–476
    in streams, 227–228
Water cycle, 229–231
Water pollution, 157–158, 207,
    239
Watersheds, 206–208
Water table, 232–233

Watertown, 296
Water vapor, 37, 123, 248,
    442–444, 461, 500
    entrance into atmosphere,
        474–476
Watkins Glen, 185, 186
Wavelength, 451–452
Waves, 260
Weather
    defined, 438
    elements of, 438–450
    on weather map, 536
Weather events, hazards posed
    by, 551–563
Weather forecasting, 521–522
Weather fronts, 526–527
Weather hazards, protection
    from, 563–566
Weather history, identifying
    area's, 564
Weathering, 118, 152–153, 166
    by abrasion, 191
    chemical, 171–175, 249
    effect of environment on,
        174–175
    mechanical, 166
    physical, 166–169, 174
    reasons for, 165–166
    of volcanic rocks, 177
Weather instruments, placement
    of, 450
Weather maps, 62, 75
    data shown on, 534–539
    drawing, 541–543
    gathering data for, 539–541
    in making predictions,
        543–544
    symbols on, 535
Wegener, Alfred, 4, 335–336, 337,
    338
Weight, 647
Wet-bulb temperature, 478–479,
    483
Whirlpool Sandstone, 384
Willamette Meteorite, 675
Wilson, J. Tuzo, 341
Wilson, Robert, 699–700
Wind, 446–447

causes of, 496–500
deposition by, 196–197
direction of, on weather map,
    536
humidity, air pressure and,
    499–500
local, 500–504
measuring, 446–447
prevailing, 510, 589–590
rate of evaporation and, 476
regional, 504–515
speed of, 494–495
speed of, on weather map,
    536–537
temperature, air pressure and,
    498–499
terrestrial, 584–585
Wind erosion, 154, 191
Wind vane, 446
Winter, 614–615, 623, 625
Winter solstice, 614
Wood, 152, 155
    fossil, 411
    petrified, 409

X

X-rays, 452, 685

Y

Yearly apparent motions, 631
Yellowstone National Park, 308

Z

Zenith, 607
Zero, absolute, 439, 442
Zinc, 147
Zone
    of aeration, 232
    of convergence, 505
    of saturation, 232
    of subduction, 252
Zoning, weather hazards and, 564

Photo Credits

All Photographs, except those listed below were provided courtesy of the author, Thomas McGuire.

Unit 1, Earth from space, courtesy of NASA; **Figure 1-3**, Geologist, courtesy or Dr. Frederick Vollmer, State University of New York, New Paltz; **Figure 2-6,** Man using sextant, courtesy of Dominique Prinet, by Des Ramsay; **Figures 3-8, 3-11, 3-Q1,** courtesy of USGS; **Figure 5-1**, Geologist on lava, courtesy of Monica Buckner; **Figure 6-4**, Average daily solar insolation, courtesy of *project sol.aps.com/solar/data insolation.asp*; **Figure 6-Q8**, Image of New York City, courtesy of NASA/GSFC/ METI/ERSDAC/JAROS, and U.S./Japan ASTER Science Team; **Figure 8-9**, Adapted from NASA images; **Figure 11-Q19**, Beach damage, courtesy of USGS; **Figure 15-1**, Old map, Public Domain; **Figure 16-1**, US Seismic Hazard Regions, based on USGS on-line publication *http://pubs.usgs.gov/fs/ 2003/fs017-03*; **Figure 16-2**, Destruction in Sumatra caused by a tsunami, U.S. Navy Photo, Photographer's Mate 2nd Class Phillip A. McDaniel; **Figure 16-4**, Mount St. Helens, Washington, 1980 eruption, courtesy of USGS, Joseph Rosenbaum; **Figure 18-5 (r.)**, Stromatolites in Shark Bay Australia, Dr. J. Richard Kyle, University of Texas at Austin; **Figure 19-7**, Anemometer, courtesy of U.S. Department of Energy's Atmospheric Radiation Measurement Program; **Figure 19-10**, Weather station, courtesy of NOAA; **Figure 19-14**, Earth's energy budget, image courtesy of NASA's ERBE (Earth Radiation Budget Experiment) program; **Figure 20-1**, Average seasonal snowfall in Western NYS, courtesy NOAA; **Figure 21-1**, Isobaric map, modified from NOAA data map; **Figure 21-11**, Northern and Southern Hemisphere hurricanes, based on NASA hurricane images; **Figure 22-1**, Air mass sources adapted from NASA image; **Figure 22-Q4**, Weather map adapted from NOAA image; **Figure 22-Q10**, Weather map of the Northeast, courtesy of NOAA; **Figure 23-1**, 1900 Galveston hurricane damage, courtesy NOAA; **Figure 23-7**, Tornado, courtesy of NOAA; **Figure 23-11**, Graph of Mauna Loa Monthly Mean Carbon Dioxide, courtesy of NOAA; **Figure 23-Q6**, Saffir-Simpson Hurricane Scale, courtesy of NOAA; graphic Design by R.L. Sheperd; **Figure 23-Q9**, Map showing Number of Billion Dollar Natural Disasters by state, 1980–2009, courtesy of NOAA; **Figure 24-1**, Graph of mean temperature 1000 to 2010, Public Domain; **Figure 24-4**, Climate regions adapted from NOAA map; **Unit 8**, Eagle Nebula, courtesy of NASA; **Figure 25-Q15**, Graph of the altitude of the sun at local noon, Cornel University; **Figure 26-1**, Astronaut John W. Young on the moon, courtesy of NASA; **Figure 26-5**, Phases of the moon, courtesy of U.S, Naval Observatory; **Figure 27-4**, Saturn, courtesy of NASA; **Figure 27-5**, Comet of 1882, Public Domain; **Figure 28-1**, White light passing through a prism © Adam Hart-Davis; **Figure 28-2 (r.)**, Hubble Space Telescope, courtesy of NASA; **Figure 28-6**, Spiral galaxy NCG 4414, courtesy of NASA; **Figure 28-7**, Summer Milky Way, courtesy Jens Hackman, *http://www.kopfgeist.com*